N&N Science Series

EARTH SCIENCE
MODIFIED PROGRAM

Architecture & Content Construction
Wayne Garnsey & **Virginia Page**

Graphic Design & Engineering
Eugene Fairbanks & **Wayne Garnsey**

Art Creation & Craftsmanship
Eugene B. Fairbanks

Ownership & Production
N&N Publishing Company, Inc.
18 Montgomery Street
Middletown, New York 10940-5116

For Ordering & Information

1-800-NN 4 TEXT
Internet: www.nandnpublishing.com
email: nn4text@warwick.net

To the Teacher

N&N Science Series: Earth Science – Modified Program is structured according to the Earth science program modification syllabus. It is a comprehensive review and supplement for a one-year general introduction to Earth Science course on the secondary school level.

The cover design is allegorical (forces at work on the Earth) and a somewhat parodical representation of a modified Earth. Although this book is centered on Earth's "natural" processes, environmental awareness is an integral part of the presentation. Therefore, the effects on Earth systems and landscape produced by humankind is discussed in relation to the Earth's own dynamic processes.

If you are familiar with the other books in the *N&N Science Series: Biology, Chemistry, Physics,* and *Earth Science* (traditional program), you are aware of our narrative writing philosophy (that is, tutorial – one-on-one). Our hundreds of illustrations are basic and easy to understand, just as a teacher illustrates a point – avoiding the superfluous. Topical questions are asked within the text to stimulate and direct the student's attention and encourage thinking. Hundreds of multiple-choice, 130 free-response questions, and practice exams challenge the student's learning and provide extended practice for the final exam.

Vocabulary is always a part of our books. After all, how can the student read for comprehension when she/he does not understand "the language?" To help, our glossary and index are combined. The glossary is comprehensive and provides concise definitions, explanations, and examples. And, if the student needs more understanding of the term, the text page references are provided right there – with the term, not in a separate index.

Beginning on page 4 is *Your Construction Guide to modified "Tri-Sense(able) Learning"* – **AN INFORMAL CONVERSATION WITH THE STUDENT**. Encourage your students to read and make the *"Tri-Sense(able) Learning"* a part of their study program. It's not magical – just common sense, and it works.

All of your colleagues at N&N wish for your students, as well as ours, good learning, good grades, and good application of the knowledge and processes studied here. If you have criticisms, comments, and/or suggestions, we want to hear from you. Send us a letter, email us at *nn4text@warwick.net*, or call us at *1-800-nn4text*.

Dedicated to Our Students

WITH THE SINCERE HOPE THAT OUR BOOK WILL FURTHER ENHANCE THEIR EDUCATION
AND BETTER PREPARE THEM WITH AN APPRECIATION AND UNDERSTANDING
OF THE SCIENTIFIC PRINCIPLES THAT SHAPE OUR WORLD.

Our sincere thanks to the many teachers and their students who contributed their knowledge, skills, experience, and assistance in the preparation of this manuscript.

Special thanks to Paul Stich for his editorial prowess.

No part of this book may be reproduced by any mechanical, photographic, or electronic process, nor may it be stored in a retrieval system, transmitted, or otherwise copied for public or private use, without the prior written permission of the publisher.

© Copyright 1999
N&N Publishing Company, Inc.

SAN # - 216-4221   ISBN # - 0935487-63-8
2 3 4 5 6 7 8 9 0 BookMart Press 2000 1999

Table of Contents

Table of Contents .... page 03

Construction Guide: Tri-Sense(able) Learning 04

Prologue 07

Unit One	***Earth Dimensions***	19
Unit Two	***Minerals & Rocks***	31
	Optional Topic A – ***Rocks, Minerals, & Resources***	43
Unit Three	***Dynamic Crust***	51
	Optional Topic B – ***Earthquakes & Earth's Interior***	61
Unit Four	***Surface Processes & Landscapes***	71
	Optional Topic C – ***Oceanography***	91
	Optional Topic D – ***Glacial Geology***	99
Unit Five	***Earth's History***	107
Unit Six	***Meteorology***	121
	Optional Topic E – ***Atmospheric Energy***	137
Unit Seven	***Water Cycle & Climate***	145
Unit Eight	***Earth in Space***	161
	Optional Topic F – ***Astronomy***	175
Unit Nine	***Environmental Awareness***	183
	Reference Tables	195
	Glossary & Index	211
	Practice Modified Earth Science Exam 1	223
	Practice Modified Earth Science Exam 2	
	Practice Modified Earth Science Exam 3	

DEAR TEACHER, THE FOLLOWING IS NOT FOR YOU. IT IS AN INFORMAL CONVERSATION WITH THE STUDENT.

YOUR CONSTRUCTION GUIDE TO MODIFIED "TRI-SENSE(able) LEARNING"

So, you don't like to read directions. You like to "cut to the chase," "skip the boring stuff," and "can't be bothered with details."

But, can you agree that if something can save you hours of tedious effort, it is worth a few minutes of your valuable time?

Well, if you have "wisdom beyond your years," you just agreed. Great, you are ready to begin *Tri-Sense(able) Learning*.

Agree or Disagree: "A mind is a terrible thing to waste." Whether you want to or not, common sense tells you that you have to agree. So, let's utilize your best asset in life – your brain. What follows will help you succeed not only in this course, but in all of life's situations.

Try it. There's "no cost or obligation." Nothing additional to buy. (You already have the book.) Nothing to lose – everything to gain!

If you "can't talk the talk, you can't walk the walk." It's true, you must be able to communicate, and to do that, you must know the words. So, start your learning with the **Vocabulary Words** at the beginning of each Unit. Look up words that you don't know or you can't use in a sentence that others will understand. Use the glossary at the end of this book.

Once you are satisfied that you understand the key words of the Unit, begin reading the material. Read to learn (comprehension), but understand that this kind of reading is different than reading for pure pleasure.

You may be one who enjoys learning about the Earth, or you may have very little interest in Earth Science. In either case, becoming proficient in your knowledge of Earth Science or any other school subject is work. Learning requires you to study. Studying requires **concentration**, **planning**, **time**, and **hard work**. So, if you want to be "the most that you can be," you must make the commitment.

At sometime in your life, you've been deep into something of real interest. Just as you're getting to "the good part," you are interrupted. You get angry. Why? Your concentration was broken. When the irritation was gone, what did you have to do? That's right! You started all over again – from the beginning! So, you ask, "What's that got to do with anything?"

You won't like this, but studying with the television on, boom-box blasting, or another distraction just doesn't work. You need a quiet place where you can "talk it through" to yourself or even out loud. Commit yourself to a study time without interruptions such as the phone. Concentration needs quiet. Even a computer's CPU, at 300$^+$MHz, becomes confused and corrupted and freezes when conflicts *interrupt its thinking!*

Did you know that you have an Intranet (similar to the Internet) within yourself? Sounds radical, perhaps silly, but remember that it is your senses that input into your brain –

Page 4   N&N© SCIENCE SERIES – EARTH SCIENCE – MODIFIED PROGRAM

your massive, unlimited hard drive. It's a fact that Intel's Pentium© "megahertz brains" can't come close to the computing abilities of your brain – your memory center. Think of using your brain much like using your home or school computer. Inputting the information is easy. Remembering where you put it, and getting it back out, is difficult. On the Internet, you use the *search engines* into which you put "key words" or descriptions. Yahoo©, Excite©, and WebCrawler© do the rest.

Your brain is already equipped with an excellent *search engine*, called *recollection*. Your brain's search engine works much the same way as the computer's does with "key words." If you organize how you store information, it becomes easy to retrieve it when needed, such as on a test.

For example, suppose you want a certain CD. You head for the mall, then to a particular store in that mall. If you don't know where the store is located, you check the mall's directory. Once inside the music store, you search for a sign with your category of music on it. Then you search for another sign with music type or artist. You have the idea. It's not at all unlike using the computer with hard drives, partitions, folders, and files.

With your study environment (no distractions) and determination to learn ("just do it") in good shape, let's get down to it. Here's how ***Tri-Sense(able) Learning*** works.

First, use your eyes. Start by telling your brain what it is that you are going to read about. Don't skip headings – skim over them so you know what you will be reading about. Then, read the material, one paragraph at a time. STOP! Tell your brain what you just read – the topic and brief description of the information in the paragraph.

Give this a try: read the paragraph in the following box using the procedure suggested above. See if your results are similar to mine.

A. WEATHERING

The physical and chemical processes that change the characteristics of rocks on the Earth's surface are called **weathering**. In order for weathering to occur, the physical environment of the rocks must change and the rocks must be exposed to the air, water in some form (ice, snow, or liquid), or the acts of humans or other living things. Therefore, weathering is the response of rocks to the change in their environment.

Now tell yourself what you read without looking back at the paragraph.

Topic of paragraph – "Weathering"
Paragraph talked about – "change causes weathering"
What causes change? – "physical and chemical processes"
What are physical processes? – "exposure to air, water, humans, and others"
KEY WORDS – "weathering, physical process, chemical process, environment, rocks, exposure, change"

Couldn't do it? Don't quit, try another ***Tri-Sense(able) Learning*** tool – using your fingers to write! That's right, take notes as you read. DO NOT WRITE DOWN SENTENCES – just the "key words."

Don't worry, your brain will put the words together right, IF you have input-ed the words to the brain.

That's the secret. *Tri-Sense(able) Learning* follows the old "KISS" principle, "**Keep It *Simple* S**tudent."

"Just the facts, M'am," said *Dragnet's* Sergeant Friday, "Just the facts." But, how do you get the facts? Ask! You can easily learn what a paragraph is telling you. When you finish reading the paragraph, ask, How? What? Where? When? and Why? questions of the paragraph's topic – in this case, "Weathering."

Examples:

A HOW question? – "How does weathering occur?"
  Answer – "By exposing rocks to air, water, and acts of man or living things."
A WHAT question? – "What causes weathering?"
  Answer – "physical and chemical processes."
A WHERE question? – "Where does weathering happen?"
  Answer – "On rocks on the Earth's surface."
A WHEN question? – "When does weathering take place?"
  Answer – "When rocks respond to a change in their environment."
A WHY question? – "Why do processes cause weathering?"
  Answer – "Physical and chemical processes change the characteristics of rocks on the Earth's surface."

Still not sure? Try yet another addition to the *Tri-Sense(able) Learning* input routine. Now, this really sounds foolish, perhaps even embarrassing, but try using your ears!

If you study the paragraph as above and still cannot tell yourself what you just read, then adjust your input by adding sound. In other words, read out loud! Add to sight and touch, go with the "surround sound!" By doing this, your brain will consider the incoming information very important.

Why? (Get ready, this is a six part explanation.)

When you read out loud, you must (1) **input** the words through your eyes, (2) **process** the data within your brain, (3) **convert** this processed data from vision to audio, (4) **coordinate** nerves and muscles for speech, (5) use your ears to **input** the sound, and finally, (6) **re-process** the data within your brain.

This more than triples the input of data in the brain – as three senses are utilized, sight, touch, and hearing. The resulting complex brain function makes the information "more important" to your memory center, allowing you to more easily learn and, most importantly, *recall* the information studied.

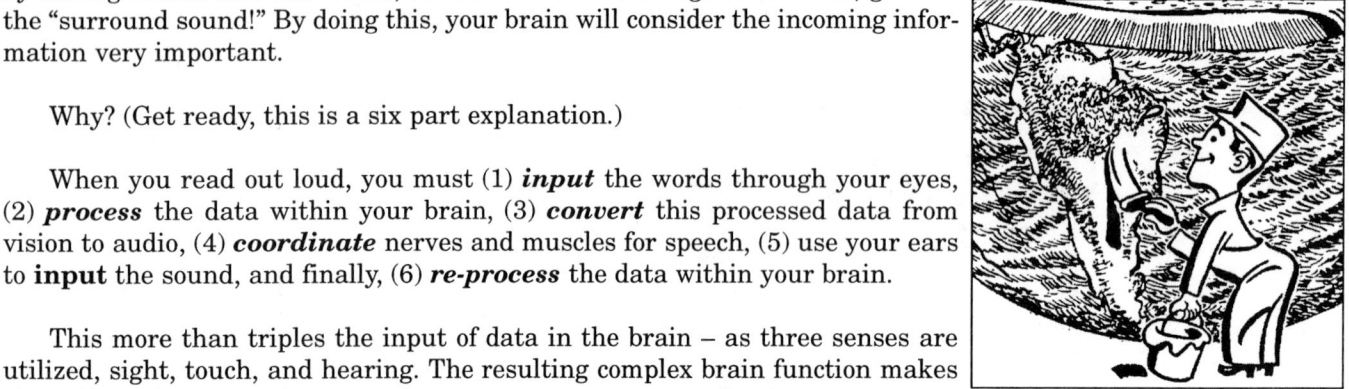

Put it all together, and what do you get? *Tri-Sense(able) Learning* with fantastic results – success! Sight (vision input/processing), touch (processing/writing output), and speaking/hearing (sound output/re-input) work together in a marvelous way – by the Rule of Squares (2x – 4x – 16x). That is, **vision** input and processing gives you twice the data input/retrieval. Add **touch** processing and writing and you square the result to four times the learning power. But, if you add the third, **sound** output and re-input, you square the 4 times to 16 times. Now that's human MEGAHERTZ power. Like the odds? 1 gets you 2, 2 gets you 4, but 3 gets you 16!

Now, put this constructive learning into practice.

Turn to The Prologue. Begin reading. First, use the Tri-sense approach to read and study each Unit, then answer the questions. You'll be absolutely amazed at the benefits of practicing this simple but guaranteed to succeed, *Tri-sense(able) Learning* technique.

PROLOGUE
MODIFIED EARTH SCIENCE PROGRAM

VOCABULARY TO BE UNDERSTOOD

Classification	Inferences (Interpretations)	Percent Deviation (Error)
Cyclic Change	Instruments	Phases of Matter
Density (Mass/Volume)	Interface	Sensory Perception
Dynamic Equilibrium	Mass	Time
Energy Flow	Measurements	Volume
Events	Observation	Weight

A. EVIDENCE & FACTS

Observations involve the interaction of a person's senses, such as sight, hearing, taste, touch, and smell, with the environment (surroundings). Observations directly involve **sensory perception**. Since our powers of observation are limited by our senses, it is often necessary for an observer to use **instruments** to extend his or her ability to observe and collect data. Any piece of information determined directly through the senses is called an observation.

How can the local environment be observed?

Instruments are used to improve upon one's powers of observation. For example, it is possible to observe the moon with the human eye, but a telescope increases the amount of detail that can be observed.

Instruments are used to extend one's ability to collect data, which cannot be detected by human senses. For example, the human body has no sense receptors that can determine the presence of potentially dangerous radiation, such as x-rays and atomic radiation. However, using a detection device like the Geiger counter allows a person to make observations beyond the capabilities of human senses.

Inferences are interpretations based on observable properties. In other words, an inference is an **interpretation** or a **conclusion** (**hypothesis** or "educated guess") based on observations. For example, students observe a smooth stone along a stream. They infer that the smoothness is due to the eroding action of water on the stone. This inference *may* or *may not* be a fact.

In the study of science, inferences may become "facts" due to the discovery of additional collaborating evidence. The same scientist, or another scientist, may make observations which prove that the water did cause the stone to become smooth.

It is important to understand the difference between an observation (what is discovered with the senses) and an inference (what is interpreted by the mind). Without sufficient observations, inferences may be incorrect.

A system of **classification** is based on the properties of an observed event or object. Events or objects that are similar in their properties are generally grouped (organized) together, allowing for more meaningful study. For example, all naturally formed solid objects composed of one or more minerals may be classified as rocks. This helps to organize the study of Earth materials. However, it does not accurately describe all rock types. Therefore, rocks are "sub-classified" into many other groups, such as sedimentary, igneous, and metamorphic rocks. In turn, these rock types are again "sub-classified" into further, more distinct groups. For example, sedimentary rocks may be classified as clastics, evaporites, or organic rocks.

How can observations within the environment be classified?

B. PROPERTIES OF OBJECTS

UNITS OF MEASUREMENTS

Measurements contain at least one basic dimensional quantity and describe the properties of objects numerically. These include quantities such as mass, length, or time.

How can properties of the environment be measured?

Mass is the amount (quantity) of matter which an object contains and is usually measured in **grams** ($1/1000$ kilogram). Mass is distinguished from weight and should not be confused with it.

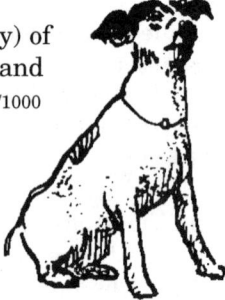

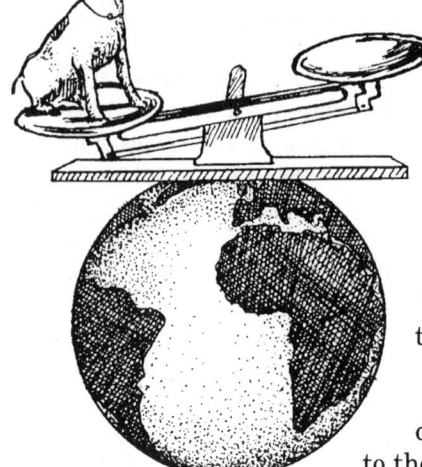

Weight of an object is determined by the pull of gravity on the mass of an object. Therefore, the object's (e.g., dog's) mass remains constant regardless of the amount of gravitation acting upon it; whereas, the weight of the object (e.g., dog) varies according to the gravitational force.

Length is the distance between the ends or sections of an object, or the total distance between two determined points of that object and is usually measured in **meters**, **centimeters** ($1/100$ meter), or **millimeters** ($1/1000$ meter).

Time is the measurable period during which an action, process, or condition exists, continues, or occurred. It may be a measurement such as a second, or a season, a schedule, an age, or a generation. Time may be a less definite measurement since it is often a relative event.

DIMENSIONAL QUANTITIES & COMPARISONS

Some properties of matter cannot be measured by basic, single units of measurement. Instead, mathematical combinations of the basic dimensional quantities must be used. Examples include:

> **Density** is a measurement involving the mass (quantity) of a substance or material per unit of volume.
>
> **Pressure** is the amount of force compared to the surface area.
>
> **Volume** is space occupied, expressed in cubic units. It is a combination of three dimensions, including length, width, and height.

ERRORS

Since all measurements are made by senses or by extensions of senses (instruments) and are actually approximations, measurements can not be expected to be "exact." Therefore, a small margin of error is expected. For example, a centimeter scale may be used to measure the length of an object. The actual length of the object may be slightly more or less than the centimeter increments on the scale. Therefore, a slight error may occur when the measurement is taken.

Percent Deviation (Percent Error)

$$\text{percent error} = \frac{\text{difference from accepted value}}{\text{accepted value}} \times 100\%$$

Percent Deviation (**percent error**) is obtained when mathematical calculations are used to solve problems involving measured and accepted values (see bottom of previous page).

For example, if a student using a balance determines the mass of an object to be 95 grams, but the mass of that object is actually 100 grams, there is an error. To determine the percent error, the student compares the observed mass and the actual mass, then finds the difference between them (substitution).

100 grams − 95 grams = 5 grams
(actual mass)   (measured mass)   (difference)

The percentage error is determined as follows:

$$\text{percent error} = \frac{5 \text{ grams}}{100 \text{ grams}} \times 100 = 5\%$$

SAMPLE PROBLEMS

1. A student's measurement of the mass of a rock is 30 grams. If the accepted value for the mass of the rock is 33 grams, what is the percent deviation (percent of error) of the student's measurement?

   (1) 9%   (3) 30%
   (2) 11%   (4) 91%

2. A student determines that the density of an aluminum sample is 2.9 grams per cubic centimeter. If the accepted value for the density of aluminum is 2.7 grams per cubic centimeter, what is the student's approximate percent deviation?

   (1) 0.70%   (3) 7.4%
   (2) 0.20%   (4) 20%

What are some characteristics of the properties of the environment?

Density is the concentration of the matter found within an object and is independent (does not depend) on the size and the shape of the material. Density is the mass of the object divided by the volume of the object, which is usually given in **g/cm³**:

$$\text{density} = \frac{\text{mass}}{\text{volume}}$$

PHASES OF MATTER

Earth materials may exist in three main forms: solid, liquid, or gas. These three states or phases of matter are dependent on the pressure or temperature conditions in which the material is placed. Generally, the material will change from a gas to a liquid to a solid as the temperature is lowered or as pressure is increased. As the temperature is increased or the pressure is lowered, the material will change state from a solid to a liquid to a gas.

There are several factors which affect the density, therefore the state, of a substance. Factors include:

- *The density of a gas varies with pressure and temperature.* For example, in a weather system, warmer air rises through cooler air because the warmer air molecules are farther apart (less dense). The cooler air is more compressed (air molecules are closer together, having greater density, having higher pressure). Compared to the cool air, warm air has lower pressure. Therefore, the less dense warm air is pushed up by the more dense cool air which is sinking. Increasing the pressure on a gas causes the gas to contract (molecules move closer together). Decreasing the pressure produces the opposite effect. This principle also applies to liquids and solids, but to a lesser extent.

- *The maximum density of most materials occurs in the solid phase.* For example, a granite rock will sink in a melt of granitic magma (liquid rock).

- Water is a noted exception to the above. Water has its maximum density at a temperature of approximately 4°C, when it is a liquid. Therefore, the solid form of water, ice, will float on liquid water. When water freezes to form ice, it expands. The water molecules move farther apart, making ice less dense than liquid water.

C. NATURE OF CHANGE

CHARACTERISTICS OF CHANGE

The Earth environment is in a constant state of change. Any change in an Earth system or object can be described as the occurrence of an **event**. This event (change) may occur suddenly (such as lightning during a thunderstorm). Change may also be observed to take very long periods of time (for example, the formation of a mountain).

GRAPHING – A WAY TO SHOW DATA

Factors involved in change are called **variables**. For example, if the temperature was measured and recorded every hour for twelve hours, time and temperature would be the variables.

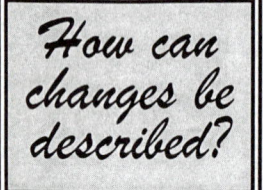

How can changes be described?

Graphing of data (the variables) is often used to show the kind and rate of change that occurs between the variables. The variable you know before you begin (in this case, time in months) is graphed on the horizontal axis of the graph.

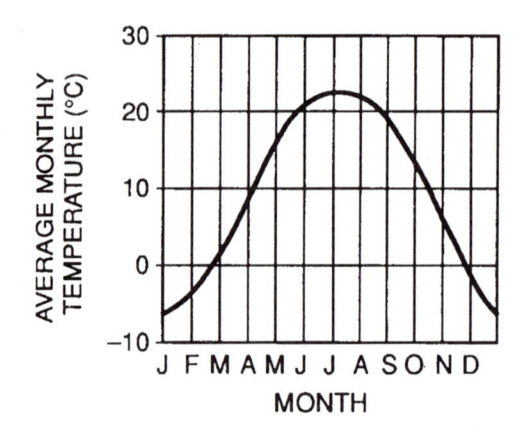

The shape of the graph line shows what kind of change occurred. The steepness of the graph line indicates the rate of change. The steeper the line, the faster the rate of change. In this case, the graph illustrates a rapid increase of temperature during the months of March, April, and May, then a slowing in June and basically no change in July. In August there is a slow decrease in temperature, followed by a rapid decrease from September to November.

If both variables change in the same direction, the graph will show a **direct relationship**. The graph at the right, comparing stream velocity and particle suspension, illustrates this. *The direct relationship, according to the graph, is as stream velocity increases, the amount of suspended particles it can carry increases, as well.*

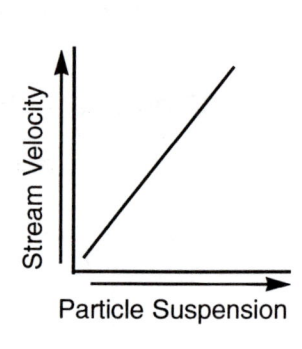

If one variable increases as the other variable decreases, the graph will show an **inverse relationship**. The graph at the right, comparing stream velocity and particle discharge, illustrates this. The inverse relationship, according to the graph, is *as stream velocity increases, the amount of particles settling decreases*.

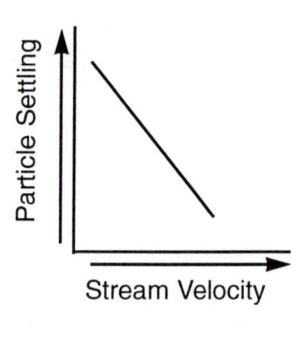

Graphs showing repeating patterns indicate **cyclic changes** in the variables. The graph below compares time of day with height of the tide. The relationship between the variables represents a cyclic change. As time increases, the height of the tide increases, then decreases, then increases, and so on. Graphing data in this way helps make accurate predictions.

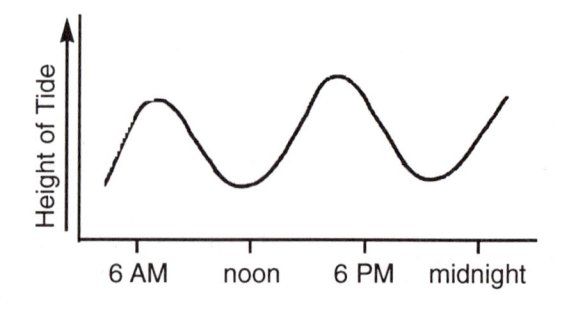

In addition to line graphs, pie graphs and bar graphs may also be used to show data.

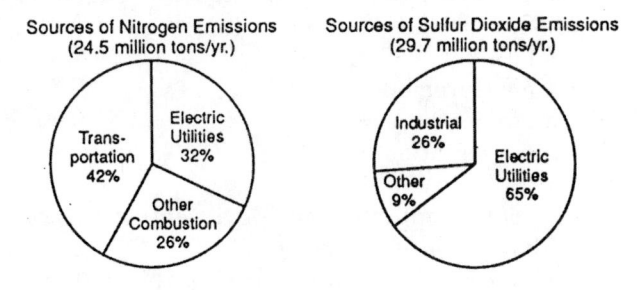

Sources of Nitrogen Emissions (24.5 million tons/yr.)
- Transportation 42%
- Electric Utilities 32%
- Other Combustion 26%

Sources of Sulfur Dioxide Emissions (29.7 million tons/yr.)
- Industrial 26%
- Other 9%
- Electric Utilities 65%

The pie graphs above compare the sources (in percent) of Nitrogen and Sulfur Dioxide emissions.

The bar graph at the right shows the percentage distribution of the Earth's surface elevation above and depth below sea level.

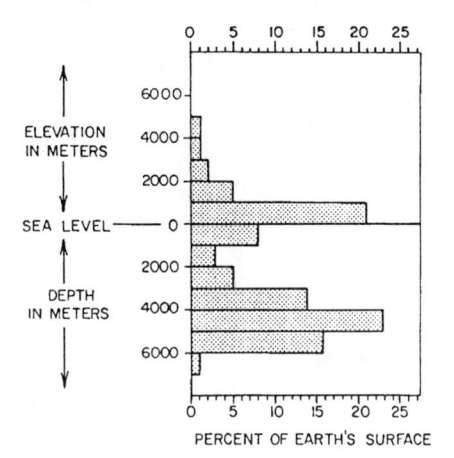

Graphs can also be used to compare more than one item using the same variables. The graph below from the *Earth Science Reference Tables* compares the travel time for earthquake P– and S–waves. Note that the variables, Travel Time (vertical axis) and Epicenter Distance (horizontal axis), are used for both waves.

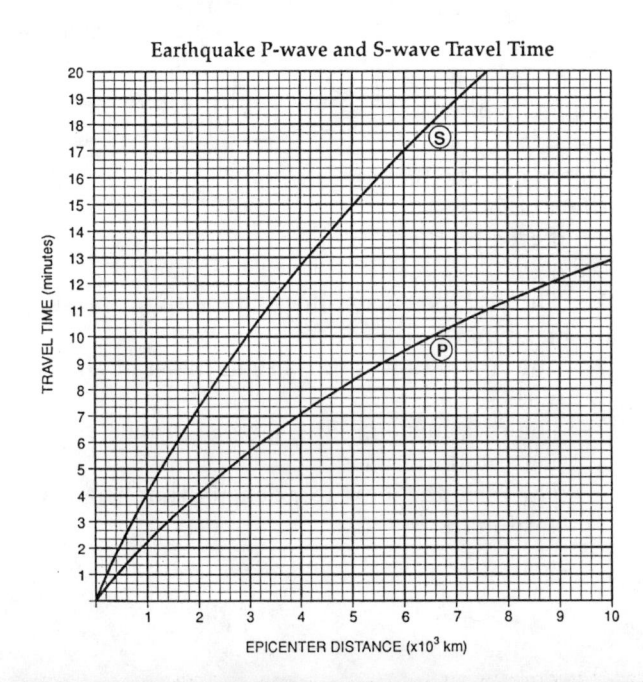

Earthquake P-wave and S-wave Travel Time

CYCLIC & NON-CYCLIC CHANGES

Most changes in the environment are cyclic. **Cyclic changes** involve events that repeat in time and space, usually in an orderly manner. For example:

- Through alternating changes in state (phase) (e.g., evaporation and condensation), water is cycled and purified between the ground, water reservoirs, and the atmosphere.

- Seasonal changes (also cycles), such as freezing and thawing, produce predictable changes such as the primary type of weathering that is dominant.

- The Sun, Moon, stars, planets, and other celestial objects have definite cyclic motions.

General weather patterns repeat in an orderly manner. However, not all changes are cyclic. There are "one-direction" events, such as the radioactive decay of certain elements, the extinction of a species, or the impact of a meteorite on the Moon.

D. ENERGY & CHANGE

ENERGY FLOW & EXCHANGE

When environmental change occurs, energy is lost by one part (source) of the environment and gained in another part (sink). This change occurs *simultaneously* (at the same time). For example, as the energy of the Sun (the source) strikes the Earth's crust (the sink) and is absorbed, the radiant energy produces heat to warm the surface of the Earth. The energy within the high speed winds (the source) of a hurricane or tornado is absorbed by trees, buildings, and other things (the sink), causing destruction and erosion.

What is the relationship of energy to change?

This exchange of energy (loss or gain) occurs at an **interface** (location or boundary) between the affected parts of the environment. Change in the environment occurs at an interface between materials, such as air to rock. Rocks weather on the outside where they come in contact with moisture and gases in the air.

PROLOGUE – MODIFIED EARTH SCIENCE PROGRAM – N&N© — Page 11

PREDICTABILITY OF CHANGE

If there is sufficient evidence and knowledge of the nature of the environmental change, it may be possible to predict the scope (type and amount) and direction that future changes will take.

Meteorologists are able to predict general storm tracks and astronomers are able to predict eclipses because of the many years of collecting data. It should be noted that general events are fairly predictable (such as the sunrise/sunset), whereas individual occurrences are much more difficult to precisely predict (such as the weather).

It is usually easier to make accurate predictions when there are many observations and few variables involved in the change.

ENVIRONMENTAL BALANCE

Our environment is in a **state of equilibrium** that tends to remain unchanged because, when equilibrium is disturbed nature works to establish a new one. Change is also constant, so what appears to be unchanged, may indeed be slowly changing. For example, if a stream is eroding and depositing sediment at the same rate in a stream bed, there appears to be no change at that location. The erosional-depositional rate of the stream is said to be in dynamic equilibrium or balance.

Should there be a sudden change in the climatic or weather conditions, such as severe rainfall causing flooding, there could be a vast change in the stream. Erosion could change the stream's banks, its course, and its contents.

QUESTIONS FOR PROLOGUE

1 In order to make observations, an observer must always use
　1　experiments
　2　the senses
　3　proportions
　4　mathematical calculations

2 While in the classroom during a visual inspection of a rock, a student recorded four statements about the rock. Which statement about the rock is an observation?
　1　The rock formed deep in the Earth's interior.
　2　The rock cooled very rapidly.
　3　The rock dates from the Precambrian Era.
　4　The rock is black and shiny.

3 A person observes a sediment consisting of clay, sand, and pebbles and then states that this material was transported and deposited by an agent of erosion. This statement is
　1　a fact
　2　a measurement
　3　an inference
　4　an observation

4 Which statement made during a weather report is most likely an inference?
　1　The record low temperature for this date was set in 1957.
　2　Hot and humid conditions will continue throughout the week.
　3　The high temperature for the day was recorded at 2 p.m.
　4　The current barometric pressure is 29.97 in.

5 A predication of next winter's weather is an example of
　1　a measurement
　2　a classification
　3　an observation
　4　an inference

6 Using a ruler to measure the length of a stick is an example of
　1　extending the sense of sight by using an instrument
　2　calculating the percent of error by using a proportion
　3　measuring the rate of change of the stick by making inferences
　4　predicting the length of the stick by guessing

7 The diagram at the right shows a rock and a standard mass on a double pan balance. Which statement about the rock is best supported by the diagram?
　1　The rock has a smaller volume than the standard mass.
　2　The rock has less mass than the standard mass.
　3　The rock has a greater force of gravity than the standard mass.
　4　The rock has a greater density than the standard mass.

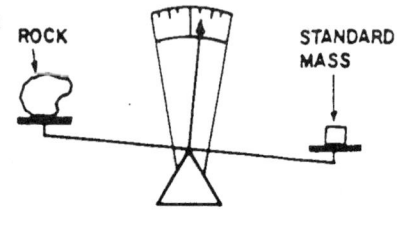

8 If each side of the cube shown at the right has the same length as the measured side, what is the approximate volume of the cube?
   (1) 2.20 cm³
   (2) 4.84 cm³
   (3) 6.60 cm³
   (4) 10.65 cm³

9 A classification system is based on the use of
   1 human senses to observe properties of objects
   2 instruments to observe properties of objects
   3 observed properties to group objects with similar characteristics
   4 inferences and conclusions

10 Which property was probably used to classify the substances in the chart below?

Group A	Group B	Group C
water	aluminum	water vapor
gasoline	ice	air
alcohol	iron	oxygen

   1 chemical composition
   2 state (phase) of matter
   3 specific heat
   4 abundance within the Earth

11 The table at the right identifies four density groups. According to this classification system, a sample of quartz with a mass of 27 grams and a volume of 10 cubic centimeters should be placed in group
   1 A
   2 B
   3 C
   4 D

Group	Density (g/cm³)
A	1.0-3.9
B	4.0-7.9
C	8.0-11.9
D	12.0-15.9

12 Student A finds the density of a piece of quartz to be 2.50 grams per cubic centimeter. Student B finds the density to be 2.80 grams per cubic centimeter. The actual density of quartz is 2.65 grams per cubic centimeter. Which is a true statement about student A's percent of error (percent deviation)?
   1 It is less than student B's percent of error.
   2 It is greater than student B's percent of error.
   3 It is the same as student B's percent of error.
   4 It cannot be determined.

13 A student determines the density of a rock to be 2.2 grams per cubic centimeter. If the accepted density of the rock is 2.5 grams per cubic centimeter, what is the percent deviation (percentage of error) from the accepted value?
   (1) 8.8 %
   (2) 12.0%
   (3) 13.6%
   (4) 30.0 %

14 A student calculates the period of Saturn's revolution to be 31.33 years. What is the student's approximate deviation from the accepted value? [refer to Earth Science Reference Tables]
   (1) 1.9%
   (2) 5.9%
   (3) 6.3%
   (4) 19%

15 Which factor can be predicted most accurately from day to day?
   1 chance of precipitation
   2 direction of the wind
   3 chance of an earthquake occurring
   4 altitude of the Sun at noon

16 Over a 30-day period, an observer would have the most difficulty measuring the
   1 rotation of the Earth
   2 discharge of a river
   3 changing phases of the Moon
   4 weathering of a mountain

17 The diameter through the equator of Jupiter is about 143,000 kilometers. What is this distance written in scientific notation [powers of 10]?
   (1) 143×10^2 km
   (2) 1.43×10^3 km
   (3) 1.43×10^5 km
   (4) 143×10^5 km

18 In which phase (state) do most Earth materials have their greatest density?
   1 solid
   2 liquid
   3 gas

19 At which temperature does water have its greatest density?
   (1) –7°C
   (2) 0°C
   (3) 96°C
   (4) 4°C

20 The diagrams below represent two solid objects, A and B, with different densities.

Object A (Density = 0.8 g/cm³)
Object B (Density = 1.2 g/cm³)

What will happen when the objects are placed in a container of water (water temperature = 4°C)?
1 Both objects will sink.
2 Both objects will float.
3 Object A will float, and object B will sink.
4 Object B will float, and object A will sink.

21 The diagrams below represent two differently shaped blocks of ice floating in water. Which diagram most accurately shows the blocks of ice as they would actually float in water?

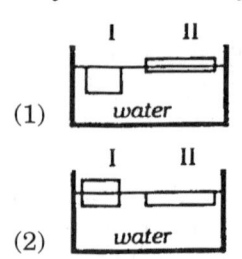

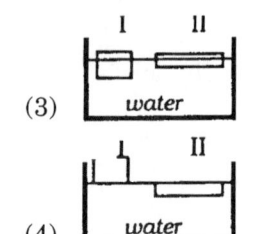

(1), (2), (3), (4)

22 Compared to the density of liquid water, the density of an ice cube is
1 always less
2 always greater
3 always the same
4 sometimes less and sometimes greater

23 If a wooden block were cut into eight identical pieces, the density of each piece compared to the density of the original block would be
1 less
2 greater
3 the same

24 During a laboratory activity, four students each determined the density of the same piece of granite. The results are shown in the table at the right.

Student	Density Determined
1	2.69 g/cm³
2	2.71 g/cm³
3	2.72 g/cm³
4	2.69 g/cm³

The accepted value for the density of granite is 2.70 grams per cubic centimeter. Therefore, the results of this activity indicate that
1 the accepted density of granite is incorrect
2 the balance used by student 3 was broken
3 each student determined the exact accepted value for the density of granite
4 the density determined by each student contains a small error

25 The graph at the right shows the relationship between the mass and volume of a mineral.

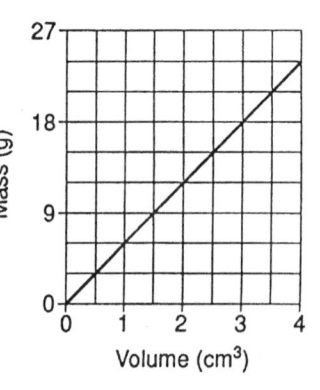

What is the density of this mineral?
(1) 6.0 g/cm³
(2) 9.0 g/cm³
(3) 3.0 g/cm³
(4) 4.5 g/cm³

26 As shown at the right, an empty 1,000.-milliliter container has a mass of 250.0 grams. When filled with a liquid, the container and the liquid have a combined mass of 1,300. grams.

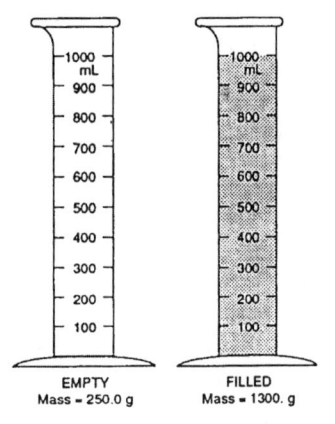

What is the density of the liquid? [refer to the Reference Tables]

(1) 1.00 g/mL
(2) 1.05 g/mL
(3) 1.30 g/mL
(4) 0.95 g/mL

27 A mineral expands when heated. Which graph best represents the relationship between change in density and change in temperature when that mineral is heated?

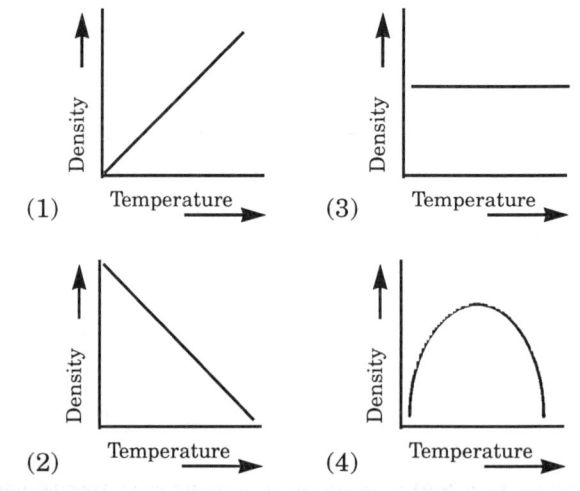

(1), (2), (3), (4)

28 A student calculates the densities of five different pieces of aluminum, each having a different volume. Which graph best represents this relationship?

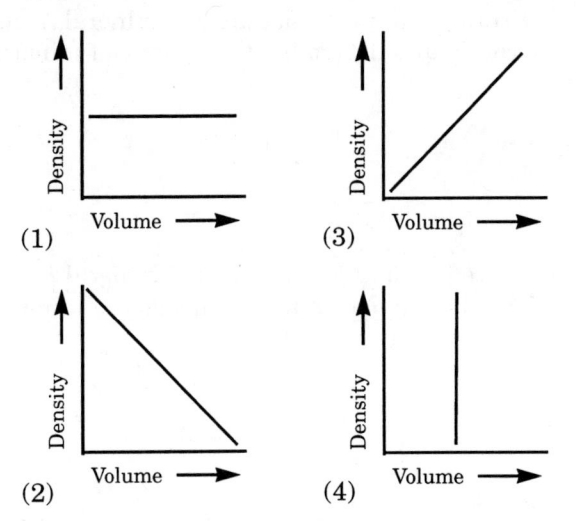

(1) Density vs Volume — horizontal line
(2) Density vs Volume — decreasing line
(3) Density vs Volume — increasing line
(4) Density vs Volume — vertical line

29 Generally, what is the correct relationship between pressure and temperature when a material changes from a gas to a liquid to a solid?
1 temperature increases, only
2 temperature lowers and/or pressure increases
3 both temperature and pressure decrease
4 both temperature and pressure increase

30 When the amounts of water entering and leaving a lake are balanced, the volume of the lake remains the same. This balance is called
1 saturation
2 transpiration
3 equilibrium
4 permeability

31 An interface can best be described as
1 a zone of contact between different substances across which energy is exchanged
2 a region in the environment with unchanging properties
3 a process that results in changes in the environment
4 a region beneath the surface of the Earth where change is not occurring

32 Which line best identifies the interface between the lithosphere and the troposphere?
1 line A
2 line B
3 line C
4 line D

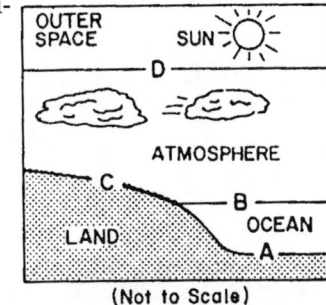

33 Future changes in the environment can best be predicted from data that are
1 highly variable and collected over short period of time
2 highly variable and collected over long periods of time
3 cyclic and collected over short periods of time
4 cyclic and collected over long periods of time

34 Which graph most likely illustrates a cyclic change?

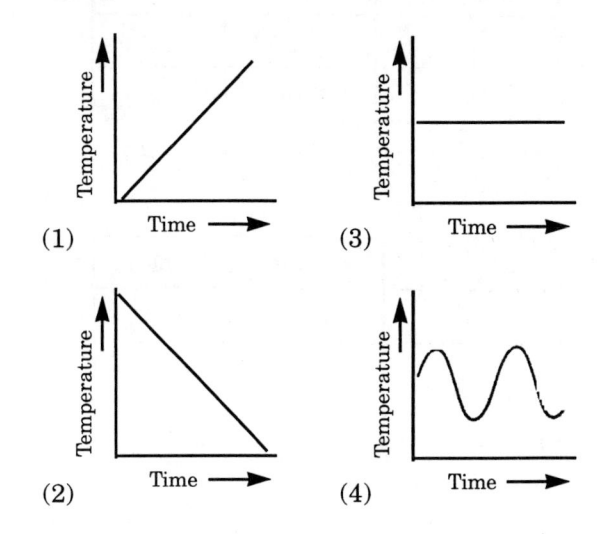

(1) Temperature vs Time — increasing line
(2) Temperature vs Time — decreasing line
(3) Temperature vs Time — horizontal line
(4) Temperature vs Time — wave

35 Which statement best explains why some cyclic Earth changes may not appear to be cyclic?
1 Most Earth changes are caused by human activities.
2 Most Earth changes are caused by the occurrence of a major catastrophe.
3 Most Earth changes occur over such a long period of time that they are difficult to measure.
4 No Earth changes can be observed because the Earth is always in equilibrium.

36 During a ten year period, which is a noncyclic change?
1 the Moon's phases as seen from the Earth
2 the Earth's orbital velocity around the Sun
3 the impact of a meteorite on the Earth
4 the apparent path of the Sun as seen from the Earth

SKILL ASSESSMENT

Base your answers to questions 1 through 10 on your knowledge of Earth science, the *Reference Tables* and the data in Tables I and II below. Tables I and II show the volume and mass of three samples of mineral *A* and three samples of mineral *B*.

Table I: Mineral *A*

Sample No.	Volume	Mass
1	2.0 cm³	5.0 g
2	5.0 cm³	12.5 g
3	10.0 cm³	25.0 g

Table II: Mineral *B*

Sample No.	Volume	Mass
1	3.0 cm³	12.0 g
2	5.0 cm³	20.0 g
3	7.0 cm³	28.0 g

Use the data to construct a graph on the grid provided below.

1. Mark an appropriate scale on the axis labeled "Mass (in grams)."

2. Plot a line graph for mineral *A* and label the line "Mineral *A*."

3. Plot a line graph for mineral *B* and label the line "Mineral *B*."

MASS v. VOLUME

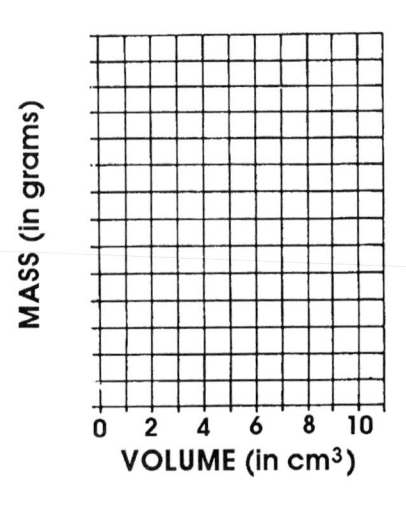

(grid: MASS (in grams) vs VOLUME (in cm³), 0 to 10)

4. Write the formula for density:

5. Substitute the data for sample 3 of mineral *A* into the formula and determine the density of mineral *A*.

   Density of *A*:

6. Substitute the data for sample 3 of mineral *B* into the formula and determine the density of mineral *B*.

   Density of *B*:

7. In one sentence tell what sample 2 of minerals *A* and *B* have in common.

8. In a sentence explain what happens to the masses of these minerals if their volume increases.

9. Predict the mass of a same amount of mineral *A* with a volume of 12.0 cm³.

10. Explain what would happen to the density of sample 1 of mineral *B* if it is heated until it melts.

Page 16 — N&N© SCIENCE SERIES – EARTH SCIENCE – MODIFIED PROGRAM

Base your answers to questions 11 through 15 on the diagram and information below and on your knowledge of Earth science.

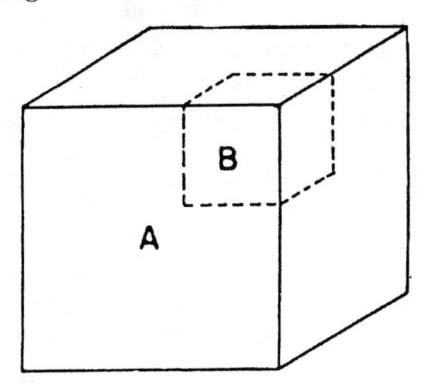

In the diagram, object A represents a solid cube of uniform material having a mass of 192.0 grams and a side that is 4.0 cm. long. Cube B is a part of cube A.

11 Substitute the appropriate data into the formula for density and calculate the density of Block A

12 If cube B is removed from cube A, what will be the value of the remaining part of cube A?

13 What is the density of cube B apart from cube A?

14 If pressure is applied to cube A until its volume is one-half of its original volume, what will its new density be?

15 A student measured the mass of cube B in order to calculate its density. The cube had water on it while its mass was measured. In one sentence predict how the calculated value for density would compare with the actual density.

16 A student finds the mass of an igneous rock sample to be 48.0 grams. Its actual mass is 52.0 grams. What is the student's percent deviation?

17 Devise a classification system for the following rocks: rock salt, dolostone, rhyolite, sandstone, basalt, rock gypsum, conglomerate, shale, and granite.

In one sentence tell what property you used.

18 In one sentence, tell what property is used to classify the land-derived sedimentary rocks listed in the *Earth Science Reference Tables*.

UNIT ONE
EARTH DIMENSIONS

VOCABULARY TO BE UNDERSTOOD

Altitude	Geographic Poles	Longitude
Atmosphere	Gradient	Meridians
Contour Interval	Gravitational Force	Model
Contour Line	Hydrosphere	Oblate Spheroid
Contour Map	Isoline & Iso-surface	Parallels
Coordinate System (Grid)	Latitude	Polaris (North Star)
Field	Lithosphere	Prime Meridian

Knowing the shape of the Earth and the properties and characteristics resulting from this shape is important to the overall understanding of the Earth and its systems.

A. EARTH'S SHAPE

The Earth appears to be the shape of a sphere (round in circumference), when observed from space or scaled down to a model, such as a classroom globe. However, by actual measurement, the Earth is not a perfect sphere. Instead, it has an oblate shape, having a larger circumference around the equator (0° latitude), than through the poles (0° longitude).

The circumference of the Earth at the equator is 40,076 kilometers, and the equatorial diameter is 12,757 kilometers. The circumference of the Earth through the poles is 40,008 kilometers, and the polar diameter is 12,714 kilometers. Therefore, the Earth is slightly "bulged" at the equator and slightly "flattened" at the poles. Thus, the true shape of the Earth is best defined as an **oblate spheroid**.

EVIDENCE FOR THE EARTH'S SHAPE

Observations of the North Star (Polaris). If a straight line passed from the South Pole, through the center of the Earth (polar axis) to the North Pole, and continued into space, it would very nearly pass through Polaris as well. Therefore, the star **Polaris** is called the **North Star**, as it lies in space practically over the geographic North Pole of the Earth. To understand why Polaris can be used to help determine the Earth's shape, it is necessary to understand the "geometry of a sphere" and the terms latitude and altitude.

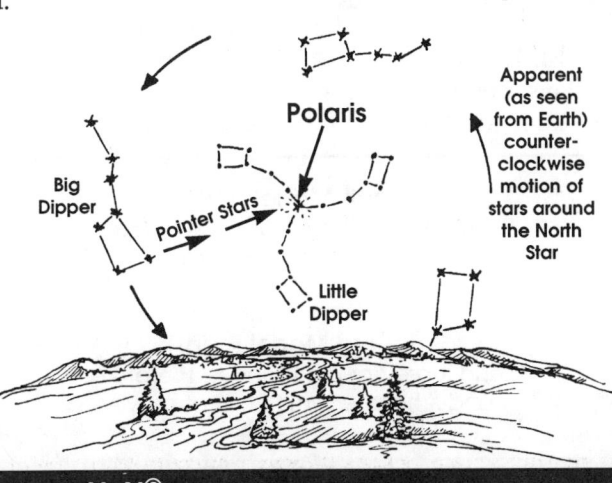

When observing an object in the atmosphere or space, the object's **altitude** refers to its angle (measured in degrees) above the horizon. When locating a point on the Earth's surface, the term **latitude** describes the point's position (measured as an angle in degrees) north or south of the Earth's equator.

According to a principle of the "geometry of a sphere," as a sphere is rotated, the angle of a fixed point outside of a sphere as compared to the surface of that sphere, is equal to the angle of the sphere's rotation. Therefore, if a sphere is rotated 45°, a fixed point outside of that sphere will be at an angle of 45° from its original point, before the sphere's rotation. (Note the illustration on the next page.)

How can Earth's shape be determined?

Considering Polaris to be a fixed point above the North Pole, its angle (altitude) to the Earth's polar axis should be the same as the angle of the Earth's tilt on that same axis. Indeed, the altitude of Polaris corresponds very closely with an

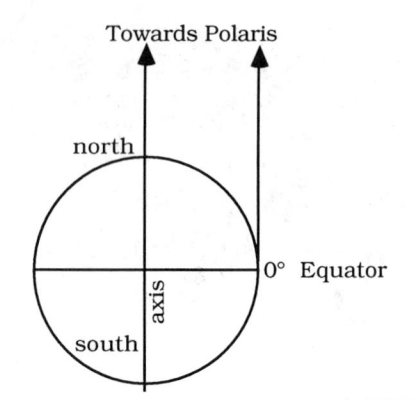

Disproving That Earth is a Perfect Sphere

As seen in these illustrations, the altitude of Polaris should be the same as the observer's latitude on the Earth, if the Earth is a perfect sphere.

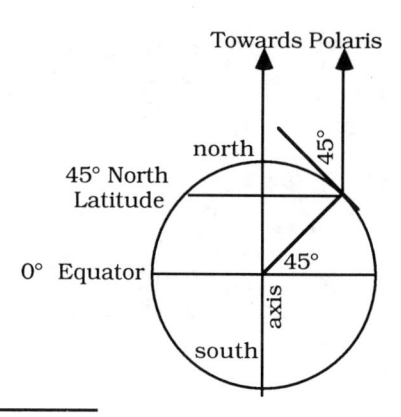

observer's latitude. But, when the angles are measured with precise instruments, the accurately measured altitude of Polaris is *not exactly* the same as the observer's latitude on Earth. Therefore, Polaris gives evidence that the Earth is not a *perfect* sphere; instead, it is *slightly* out of round, or **oblate**.

Photographs of the Earth from space. With the exploration of space, came the ability to take photographs of the Earth from great distances in space. When precisely measured, these photographs show the Earth to be larger at the equator and flatter at the geographic poles. However, the shape of the Earth, when drawn to scale on a sheet of paper, appears to be perfectly round.

Gravimetric (gravity) measurements. Gravity is the force of attraction between any two objects. Since the Earth has such a large mass, smaller objects with less mass are attracted (pulled) towards the Earth. If the Earth were a perfect sphere, it would be expected to exert an equal pull on objects anywhere on or above the Earth's surface at equal distances from the center of the Earth.

The **law of gravitation** states that a **gravitational force** is proportional to the square of the distance between the two centers of attracted objects. In other words, the weight of an object moved anywhere on the surface of the Earth (assuming Earth is a perfect sphere), should remain the same. However, precise measurements of objects indicate that the same object weighs more at the poles than at the equator of the Earth. Even accounting for the **centrifugal effect** produced by the Earth's rotation, there is still a greater gravitational pull at the poles than at the equator.

(Note, the Earth's oblateness in the left illustration below is greatly exaggerated.)

B. MEASURING EARTH'S SIZE

Earth's dimensions can be determined from observations of the Earth from space. The relative positions of the Earth and the Sun can be used to determine the exact size of the Earth.

How can Earth's size be determined?

Before the space age began, even as far back in history as 200 B.C., fairly good estimates of the Earth's dimensions could be made. For example, by measuring the altitude of the Sun at two different places on the Earth's surface at exactly the same time of day, the Earth's circumference can be mathematically determined.

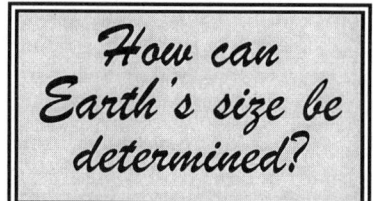

For example: Two observers are standing on the same meridian, a distance of 1000 kilometers apart. On the same day at the same time, each observer measures the altitude of the Sun. Finding the difference in the altitude of the Sun between two places gives the angle at the center of the Earth that separates the two observers. Assuming that the Earth is a perfect sphere with 360°, dividing the angle difference between the two observers into the 360° and multiplying that number by the 1000 kilometer distance between them, gives the total circumference of the Earth.

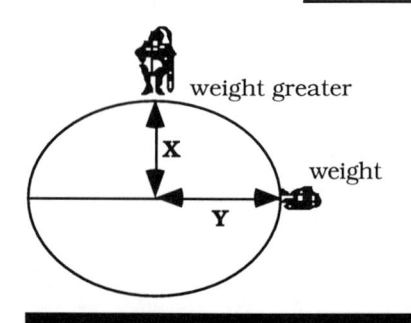

Earth Model (left)
Y greater distance than X; therefore, X has greater force of gravity.

Sphere Model (right)
Y and X are equal distances; therefore, Y and X have equal forces of gravity.

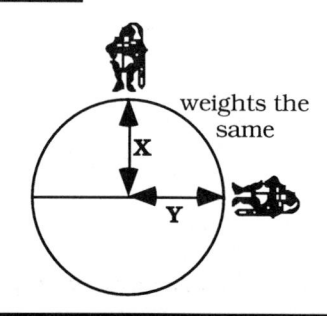

Page 20 — N&N© SCIENCE SERIES – EARTH SCIENCE – MODIFIED PROGRAM

Determining The Circumference Of The Earth

9° divided into 360° = 40° ($1/40$ of 360° circumference)
1000 km multiplied by 40 = 40,000 km (Earth's circumference)

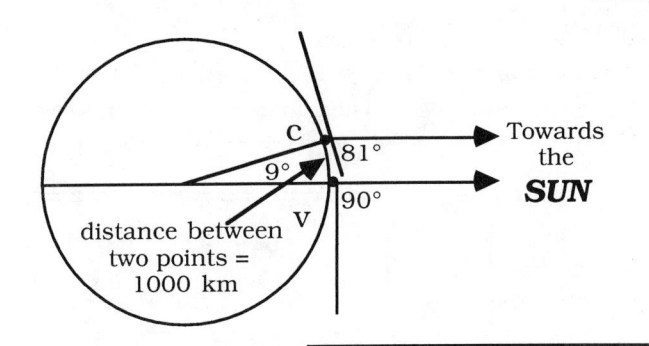

Determining by Proportion: The Circumference Of The Earth

$$\frac{9°}{1000 \text{ km}} = \frac{360°}{C}$$

$$C = \frac{1000 \text{ km} \times 360°}{9°}$$

$$C = 40,000 \text{ km}$$

When the circumference of the Earth is known, it is easy to calculate the surface area, radius, diameter, and volume of the Earth. These Earth dimensions are not exact because these calculations depend on the Earth's being a perfect sphere. However, we have learned that the Earth is slightly oblate.

The following formulas are used to calculate the dimensions of the Earth:

Where:

C is the circumference
r is the radius
D is the diameter
V is the volume
A is the surface area

$$r = \frac{C}{2\pi}$$

$$V = \frac{4}{3}\pi r^2$$

$$D = 2\pi$$

$$A = 4\pi r^2$$

(Note: π, "pi", approximately 3.14159, expresses the ratio of the circumference to the diameter of a circle.)

C. EARTH'S ATMOSPHERE

The atmosphere of the Earth is composed of the materials (solids, liquids, and gases) which form a thin envelope surrounding the Earth, held in place by gravitation and rotating with the Earth. The atmosphere extends several hundred kilometers above the Earth's surface into space and is the least dense of the Earth's three spheres: atmosphere, hydrosphere, and lithosphere.

What is the extent of the atmosphere, hydrosphere, and lithosphere?

The atmosphere is **stratified** (layered) into **zones**, each having distinct characteristics, including temperature and pressure ranges, composition, and effects produced on the Earth's surface. These zones include (from the Earth's surface towards space):

- **Troposphere.** Lowest region of the atmosphere between the Earth's surface and the tropopause, characterized by decreasing temperature with increasing altitude,
- **Stratosphere.** Region above the troposphere and below the mesosphere, where temperature increases with altitude due to the presence of ozone.
- **Mesosphere.** Portion of the atmosphere from about 30 to 80 kilometers above the Earth's surface, characterized by temperatures that decrease from 10°C to -90°C with increasing altitude, and
- **Thermosphere. Outermost** shell between the mesosphere and outer space, where temperatures increase steadily with altitude.

Earth's Hydrosphere (Oceans)

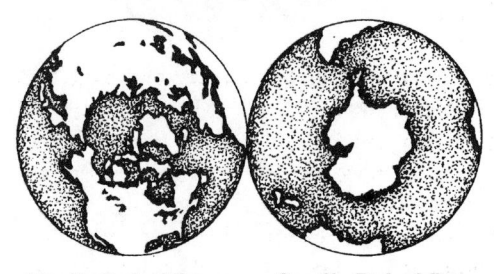

North Pole View   South Pole View

EARTH'S HYDROSPHERE

Almost three quarters (approximately 71%) of the Earth's surface is covered with a relatively thin film of water. Compared to the other zones of the Earth, it is very thin, only averaging 3.5 to 4.0 kilometers thick, much like the skin on an apple.

The **hydrosphere** includes all bodies of water, marine (salt-containing oceans) and fresh (inland lakes and rivers).

Earth's Atmospheric Zones
(diagram at right)

EARTH'S LITHOSPHERE

The most solid portion of the Earth is the rock near the Earth's surface and is a continuous solid shell, often under the hydrosphere. It accounts for the general underwater features of the Earth, including mountains, valleys, and the ocean floor.

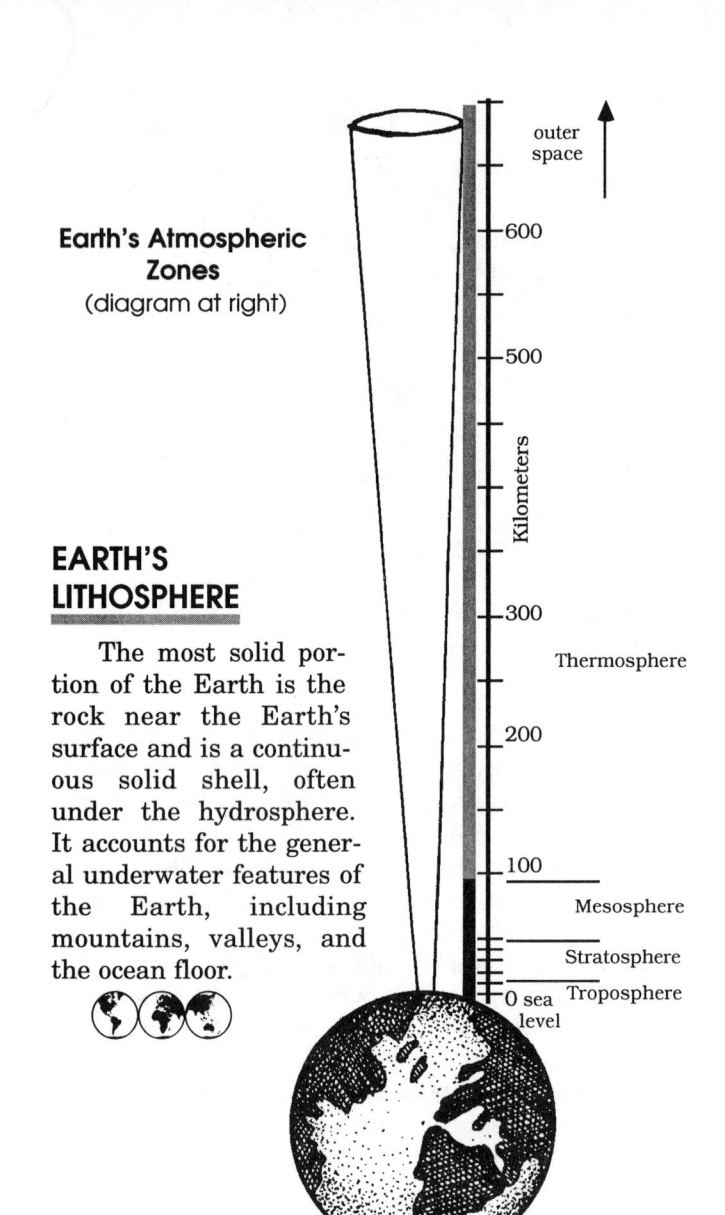

D. EARTH POSITIONS

COORDINATE SYSTEMS

There are many reference systems that can be used to determine positions on the Earth. A coordinate system uses a **grid** of imaginary lines and two points, called **coordinates**, to locate a particular position on the surface of the Earth (see illustration below). A fixed point can be located on a graph by identifying **axes** (the intersecting point of two lines).

How can a position on the Earth's surface be determined?

LATITUDE – LONGITUDE COORDINATE SYSTEM

The most commonly used coordinate system for the Earth is the **latitude – longitude system**, which is based on celestial observations or star angles. Since the Earth is an oblate sphere, it is not practical to use a flat graph as illustrated previously. Although the principle of locating a fixed point remains the same, the axes are imaginary lines called **parallels** and **meridians** circling the Earth.

PARALLELS OF LATITUDE

The **equator**, located halfway between the geographic poles, is a circle which divides the Earth into Northern and Southern Hemispheres. Lines, called parallels, are drawn on the Earth to measure latitude. These parallels decrease in size from the equator (0°) to the North Pole (90° N) and to the South Pole (90° S).

Locating a Fixed Point on a Graph (Coordinate System)

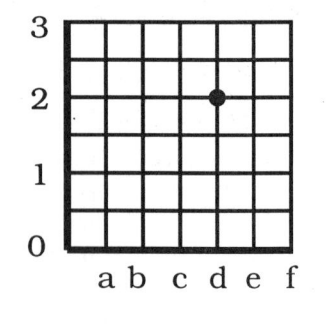

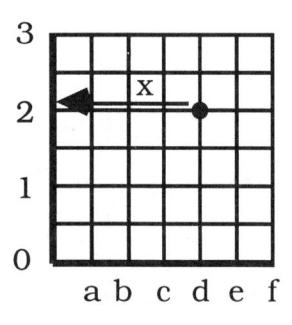

1st - Read the horizontal axis
x = d

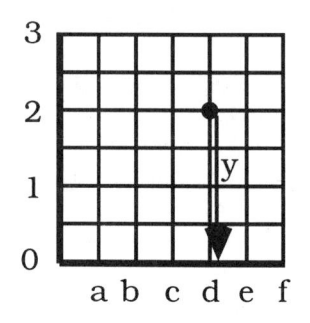

2nd - Read the vertical axis
y = 2

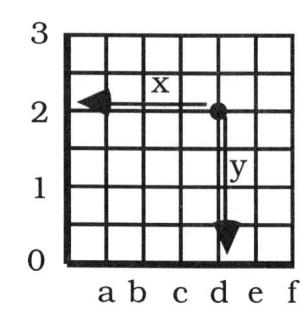

The location of the point is at
d,2.

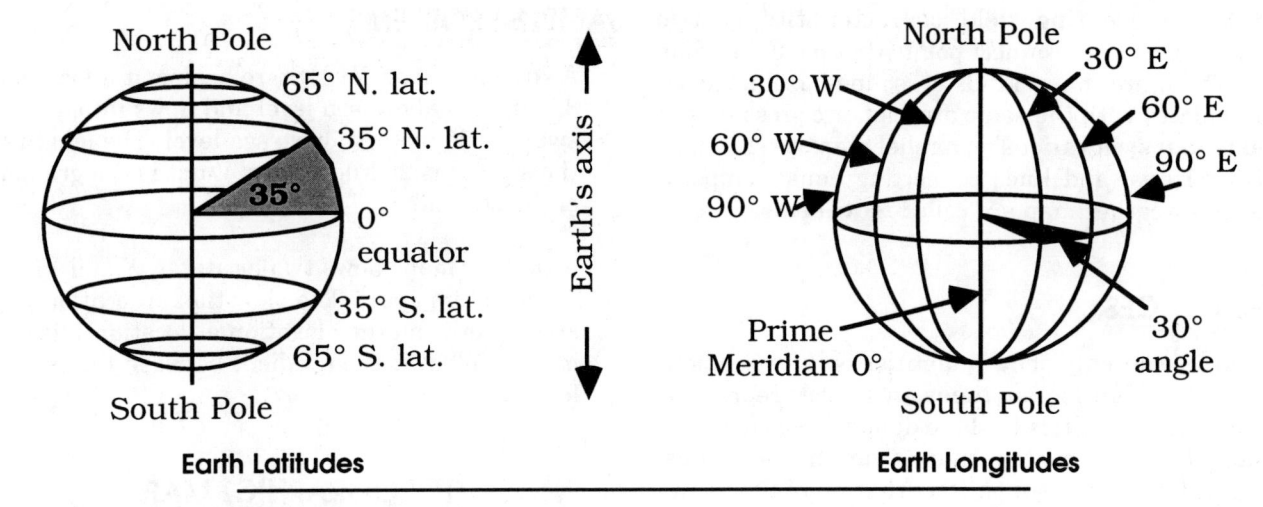

MERIDIANS OF LONGITUDE

The lines running between the North Pole and the South Pole are called **meridians**. The **Prime Meridian** runs through Greenwich, England and has a longitude of 0°. Longitude is measured using the meridians east or west of the Prime Meridian (0°) to a maximum longitude of 180° (**International Dateline**).

LATITUDE & LONGITUDE MEASUREMENTS

Both latitude and longitude are measured in degrees. Latitude is distance measured in degrees north or south of the equator (0° to 90° North, 0° to 90° South). Longitude is distance measured in degrees east or west of the Prime Meridian (0° to 180° meridian).

Latitude. The altitude of Polaris above the horizon is often used to determine a particular latitudinal position in the Northern Hemisphere. In the Southern Hemisphere other stars are used. Since travel in space, in air, and on the water require very precise navigation, astronomical tables (as found in *Reed's Almanac*) are also used. They account for variations in the positions of heavenly bodies based on time and date.

Longitude. The Earth rotates from west to east, 360° in a 24 hour period of time, or a distance of 15° of longitude per hour. Comparing your local time with the time at Greenwich, England (0°), enables you to determine your longitude. For every hour difference in time between an observer and **Greenwich Mean Time** (**GMT**), the observer is 15° away from the Prime Meridian.

If the Sun is on the Prime Meridian, it is 12 noon at Greenwich, England. Since the Earth rotates west to east, the Sun has not yet reached it highest point (noon time) to the west of Greenwich; therefore, time is earlier to the west. However, to the east of Greenwich, the Sun has already passed its highest point; therefore, time is later to the east of Greenwich.

For example, a ship sailing in the Atlantic Ocean maintains a clock (chronometer) with Greenwich Mean Time. If a time comparison is taken when the Sun is at its highest point over the ship (the ship's "local noon") and found to be 3 hours earlier than Greenwich Mean Time; then we know that the ship is at 45° West Longitude, since the ship's local time is 3 hours earlier than the same time in Greenwich (3 x 15° = 45°).

Modern navigational techniques such as **GPS** (**Global Positioning System**) used on land, air, and sea, employ the use of signals from both Earth and orbiting satellites.

E. POSITION CHARACTERISTICS

How can the characteristics of a position be measured and described?

FIELDS

A **field** is a region of space that contains a measurable quantity at every point. For example, every point in a region would have a temperature. Other examples of fields include: gravity, magnetism, relative humidity, and atmospheric pressures.

ISOLINES

In order to visualize field quantities, maps can be drawn by using **isolines**. Maps containing isolines are

models representing field characteristics in two dimensions. Isolines connect points that have the same value. There are many kinds of isolines. Isolines connecting places with the same atmospheric pressure are called **isobars**, elevations on a relief map are shown as **contour lines**, and lines connecting equal temperatures on a weather map are called **isotherms**.

ISO-SURFACES

Isolines can only show quantities on a two-dimensional surface, but **iso-surfaces** are models representing field characteristics in three dimensions, such as a magnetic field or a contour map of varying elevations, temperatures, or pressures. Note that all of the points of an iso-surface must have the same field value.

FIELD CHANGES

Fields are generally not *static* (unchanging). The characteristics of fields are usually *dynamic*. They change with the passage of time.

For example, the temperatures and air pressures over the surface of the Earth are in a constant state of change. Therefore, a weather map shows only those field conditions occurring during a specific time and date. Also, a contour map made several years ago of a particular landscape would have to be updated from time to time to remain accurate due to the changes produced by the forces of nature and humankind.

GRADIENT

A **gradient** or average slope within the field expresses the rate of change of the field quantity from one place to another place. The following formula is used to determine the rate of change:

$$\text{gradient} = \frac{\text{change in the field value}}{\text{change of distance}}$$

Example: If a weather map shows a change in atmospheric pressure from 996 mb to 1004 mb between towns that are 400 kilometers apart, then the rate of change (gradient) is 8 mb per 400 kilometers or 0.02 mb/km.

SAMPLE PROBLEMS

1 A stream in New York State begins at a location 350 meters above sea level and flows into a swamp 225 meters above sea level. The length of the stream is 25 kilometers. What is the gradient of the stream?

2 A contour map shows two locations, X and Y, 5 kilometers apart. The elevation at location X is 800 meters and the elevation at location Y is 600 meters. What is the gradient between the two locations?

CONTOUR (TOPOGRAPHIC) MAP

The topographic map uses isolines called **contour lines** to connect points of same elevation (usually based on sea level measurements). Generally, every 5th line is an **index contour line**. It is printed darker and is interrupted to give the elevation.

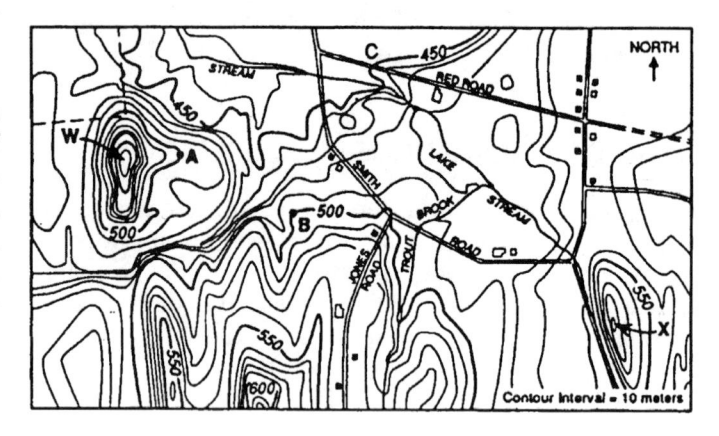

The difference in elevation between two consecutive contour lines is the **contour interval**. Common intervals are 5, 10, 20, 50 and 100 meters. Enclosed depressions are shown with hachured contour lines (the hachure marks ┬┬┬┬ pointing down-slope).

The **map legend** gives **distance scales** along with a **key** of symbols representing natural and cultural (man-made) features of the field, such as cities, roads, buildings, marshes, and forests.

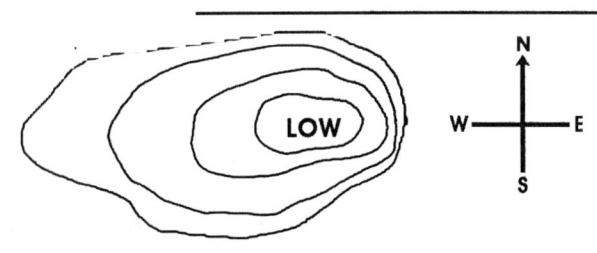

Isolines Indicating a Gradient
Note: Gradient on the east side of the LOW is steep; whereas, the gradient on the west side is gentle.

QUESTIONS FOR UNIT 1

1. An observer watching a sailing ship at sea notes that the ship appears to be "sinking" as it moves away. Which statement best explains this observation?
   1. The surface of the ocean has depressions.
   2. The Earth has a curved surface.
   3. The Earth is rotating.
   4. The Earth is revolving.

2. The true shape of the Earth is best described as a
   1. perfect sphere
   2. perfect ellipse
   3. slightly oblate sphere
   4. highly eccentric ellipse

3. According to the *Earth Science Reference Tables*, the radius of the Earth is approximately
   (1) 637 km
   (2) 6,370 km
   (3) 63,700 km
   (4) 637,000 km

4. At sea level, which location would be closest to the center of the Earth?
   (1) 45° South latitude
   (2) the Equator
   (3) 23° North latitude
   (4) the North Pole

5. At which location would an observer find the greatest force due to the Earth's gravity?
   1. North Pole
   2. New York State
   3. Tropic of Cancer ($23\frac{1}{2}$°N.)
   4. Equator

6. Which statement provides the best evidence that the Earth has a nearly spherical shape?
   1. The Sun has a spherical shape.
   2. The altitude of Polaris changes in a definite pattern as an observer's latitude changes.
   3. Star trails photographed over a period of time show a circular path.
   4. The length of noontime shadows change throughout the year.

7. Which diagram most accurately shows the cross-sectional shape of the Earth drawn to scale?

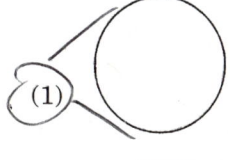

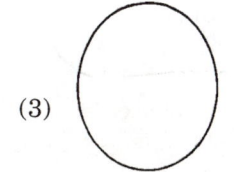

8. Which object best represents a true scale model of the shape of the Earth?
   1. a Ping-Pong ball
   2. a football
   3. an egg
   4. a pear

9. The best evidence that the Earth has a spherical shape would be provided by
   1. the prevailing wind direction at many locations on the Earth's surface
   2. the change in the time of sunrise and sunset at the single location during 1 year
   3. the time the Earth takes to rotate on its axis at different times of the year
   4. photographs of the Earth taken from space

10. The polar circumference of the Earth is 40,008 kilometers. What is the equatorial circumference?
    (1) 12,740 km
    (2) 25,000 km
    (3) 40,008 km
    (4) 40,076 km

11. The diagram at the right shows the altitude of the noon Sun as measured on March 21 by observers at locations A and B.

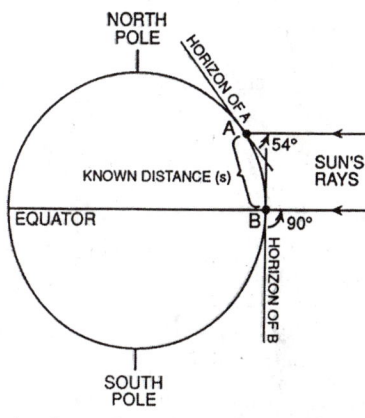

    According to the *Earth Science Reference Tables*, an observer can use the known distance, s, and the Sun's altitude at A and B to find the Earth's
    1. density
    2. eccentricity
    3. circumference
    4. oblateness

12. Based on the diagram below, what is the circumference of planet Y?

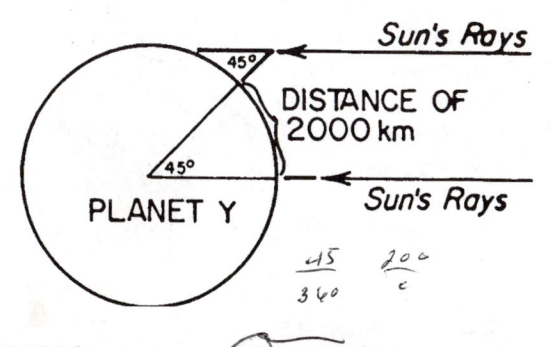

    (1) 9,000 km
    (2) 12,000 km
    (3) 16,000 km
    (4) 24,000 km

13 The water sphere of the Earth is known as the
    1   atmosphere   3   lithosphere
    2   hydrosphere   4   troposphere

14 Which statement most accurately describes the Earth's atmosphere?
    1   The atmosphere is layered, with each layer possessing distinct characteristics.
    2   The atmosphere is a shell of gases surrounding most of the Earth.
    3   The atmosphere's altitude is less than the depth of the ocean.
    4   The atmosphere is more dense than the hydrosphere but less dense than the lithosphere.

15 If the deepest parts of the ocean are about 10 kilometers and the radius of the Earth is about 6,400 kilometers, the depth of the ocean would represent what percent of the Earth's radius?
    1   less than 1%   3   about 25%
    2   about 5%   4   more than 75%

16 The graph at the right shows the percentage distribution of the Earth's surface elevation above and depth below sea level. Approximately what total percentage of the Earth's surface is below sea level? 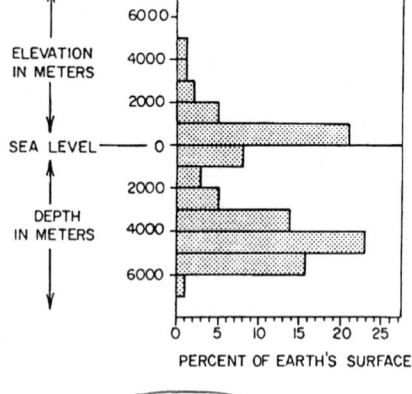
    (1) 30%
    (2) 50%
    (3) 70%
    (4) 90%

17 The diagram at the right shows an instrument made from a drinking straw, protractor, string, and rock. 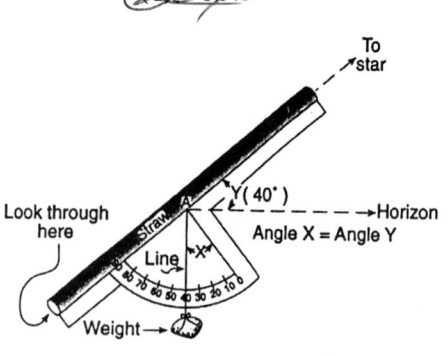
    This instrument was most likely used to measure the
    1   distance to a star
    2   altitude of a star
    3   mass of the Earth
    4   mass of the suspended weight

18 The Big Dipper, part of the star constellation Ursa Major, is shown below. 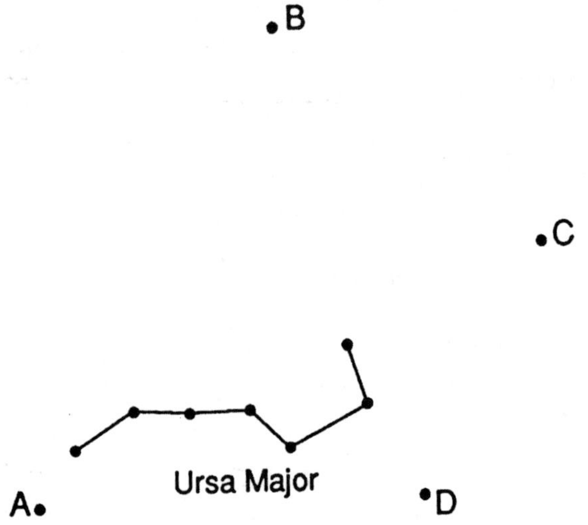
    Which letter represents Polaris (the North Star)?
    (1) A   (3) C
    (2) B   (4) D

19 An observer in New York State measures the altitude of Polaris to be 44°. According to the Reference Tables, the location of the observer is nearest to
    1   Watertown
    2   Elmira
    3   Buffalo
    4   Kingston

20 As an observer travels northward from New York State, the altitude of the North Star
    1   increases directly with the latitude
    2   decreases directly with the latitude
    3   increases directly with the longitude
    4   decreases directly with the longitude

21 To an observer on a ship at sea, at which latitude does the North Star appear closest to the horizon?
    (1) 5° N   (3) 50° N
    (2) 20° N   (4) 85° N

22 According to the Reference Tables, which city is located closest to 44°N latitude, 76° W longitude?
    1   Massena
    2   Binghamton
    3   Buffalo
    4   Watertown

23 As a ship crosses the Prime Meridian, the altitude of Polaris is 65°. What is the location of the ship?
    1   65° South latitude, 0° longitude
    2   65° North latitude, 0° longitude
    3   0° latitude, 65° West longitude
    4   0° latitude, 65° East longitude

24 The shaded area of the map below represents large areas of surface basaltic bedrock in the northwestern United States.

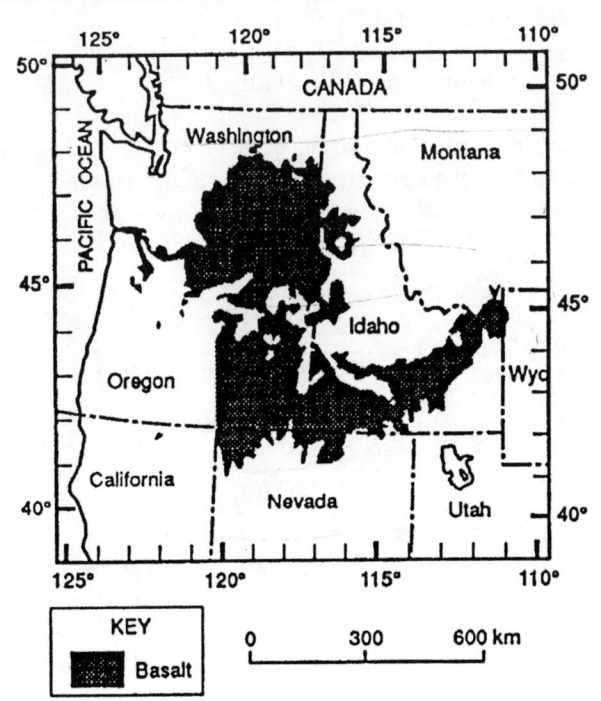

Which location is in the shaded area of surface basaltic bedrock?
(1) 40° N 120° W
(2) 44° N 122° W
(3) 46° N 120° W
(4) 48° N 116° W

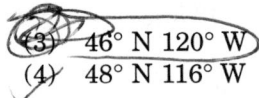

25 What is the elevation of the highest contour line shown on the map? 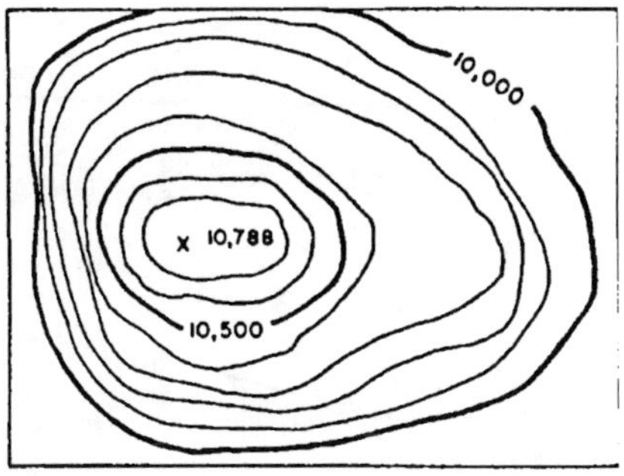
(1) 10,000 feet
(2) 10,688 feet
(3) 10,700 feet
(4) 10,788 feet

26 Isolines on the topographic map at the right show elevations above sea level, measured in meters. What could be the highest possible elevation represented on this map? 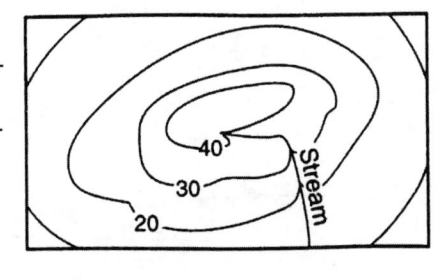
(1) 39 m
(2) 41 m
(3) 45 m
(4) 49 m

Base your answers to questions 27 and 28 on the diagram below which represents a contour map of a hill.

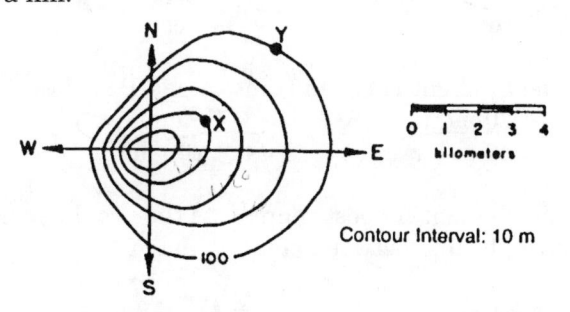

27 On which side of the hill does the land have the steepest slope?
1 north
2 south
3 east
(4) west

28 What is the approximate gradient of the hill between points X and Y?
(1) 1 m/km
(2) 10 m/km
(3) 3 m/km
(4) 30 m/km

Base your answers to questions 29 through 33 on the topographic map below and on your knowledge of Earth science. Points A, B, C, D, E, F, X, and Y are locations on the map. Elevation is measured in meters.

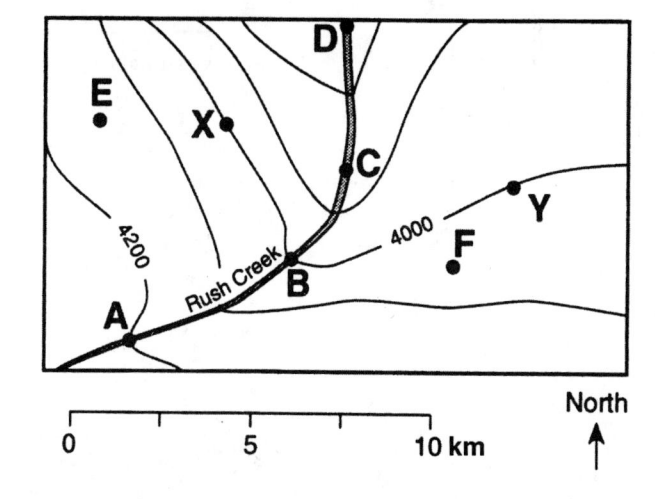

29 What is the contour interval used on this map?
(1) 20 m
(2) 50 m
(3) 100 m
(4) 200 m

30 Which locations have the greatest difference in elevation?
(1) A and D
(2) B and X
(3) C and F
(4) B and Y

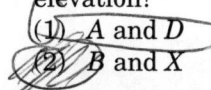

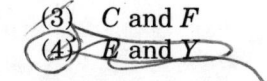

31. Between points C and D, Rush Creek flows toward the
   (1) north
   (2) south
   (3) east
   (4) west

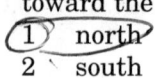

32. The gradient between points A and B is closest to
   (1) 20 m/km
   (2) 40 m/km
   (3) 80 m/km
   (4) 200 m/km

33. Which diagram best represents the profile along a straight line between points X and Y?

   (1) (2) (3) (4)

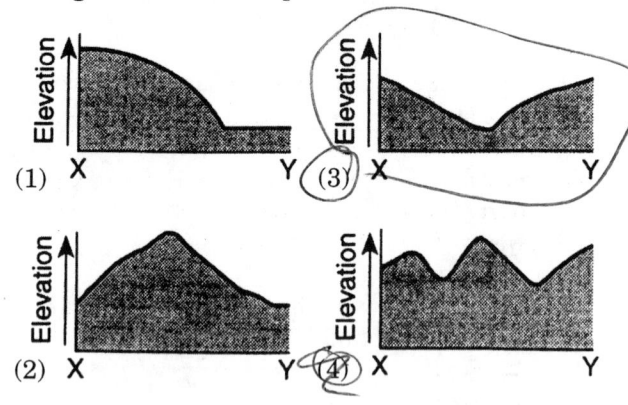

34. Which two towns received the same depth of volcanic ash?
   1. Pasco and Wenatchee
   2. Yakima and Ephrata
   3. Centralia and Ritzville
   4. Spokane and Ellensburg

35. Which equation should be used to determine the ash-fall gradient in millimeters per kilometer between Spokane and Ritzville?

   (1) Gradient = $\dfrac{55 \text{ mm} - 15 \text{ mm}}{75 \text{ km}}$

   (2) Gradient = $\dfrac{75 \text{ km}}{55 \text{ mm} - 15 \text{ mm}}$

   (3) Gradient = $\dfrac{55 \text{ mm} - 15 \text{ mm}}{100 \text{ km}}$

   (4) Gradient = $\dfrac{100 \text{ km}}{55 \text{ mm} - 15 \text{ mm}}$

36. If some of the ash was blown by the prevailing wind from the western edge of the ash-fall to the northeastern border of Washington, what was the approximate maximum distance that it traveled?
   (1) 150 km
   (2) 250 km
   (3) 500 km
   (4) 740 km

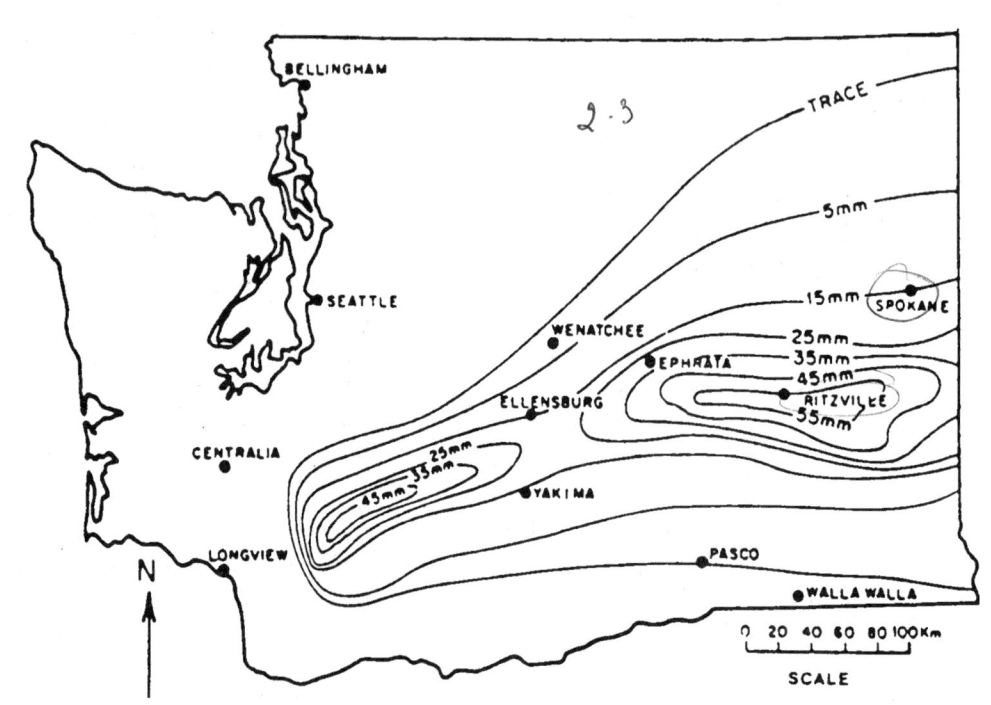

Base your answers to questions 34-36 on the outline map of the state of Washington below, the *Reference Tables*, and your knowledge of Earth science. The map shows isolines of ash-fall depths covering a portion of the State of Washington which resulted from a volcanic eruption of Mount St. Helens.

Page 28   N&N© Science Series – EARTH SCIENCE – Modified°Program

SKILL ASSESSMENTS

Base your answers to questions 1 through 5 on the topographic map of Cottonwood, Colorado, below.

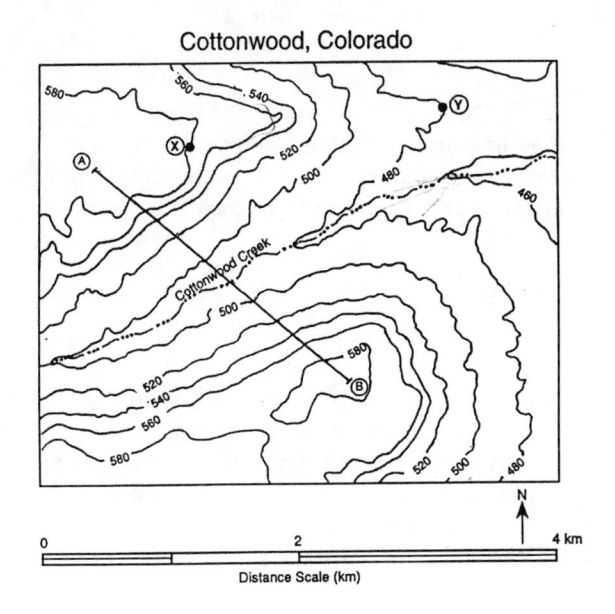

1. What is the contour interval of this map?

   Contour interval = 20

2. State the general direction in which Cottonwood Creek is flowing.

   North east

3. State the highest possible elevation, to the *nearest meter*, for point B on the topographic map.

   599

4. On the grid below, draw a profile of the topography along line AB shown on the map.

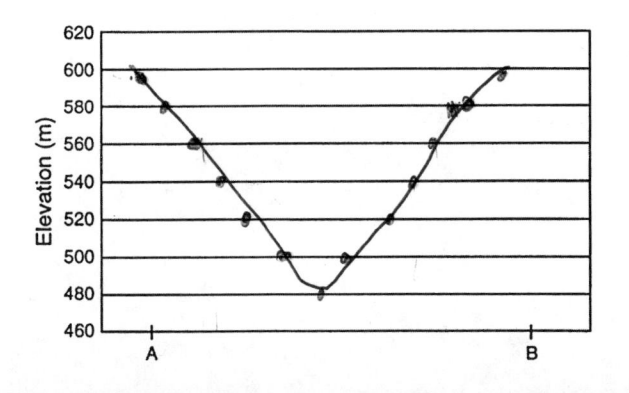

5. Calculate the gradient of the slope between points X and Y on the topographic map, following the directions below.

   a Write the equation for gradient.

   b Substitute data from the map into the equation.

   c Calculate the gradient and label it with the proper units.

Base your answers to questions 6 through 9 on the latitude-longitude system shown on the map. The map represents a part of the Earth's surface and its coordinate (grid) system. Point A through F represent locations in this area.

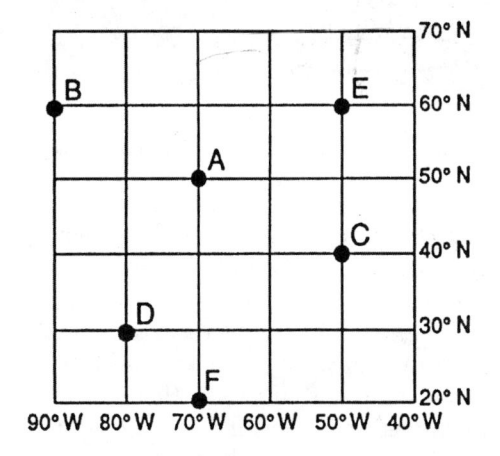

6. What is the latitude and longitude of point A?

7. What is the compass direction from point D toward point E?

8. In a sentence or two, explain where this map is located in relation to the Equator and Prime Meridian.

9. As a person travels from location E to location C, what will happen to the observed altitude of Polaris?

UNIT ONE – EARTH DIMENSIONS – N&N©   Page 29

Base your answers to questions 10 through 12 on the information and the field map below.

Pesticide Concentrations in Ground Water (ppb)

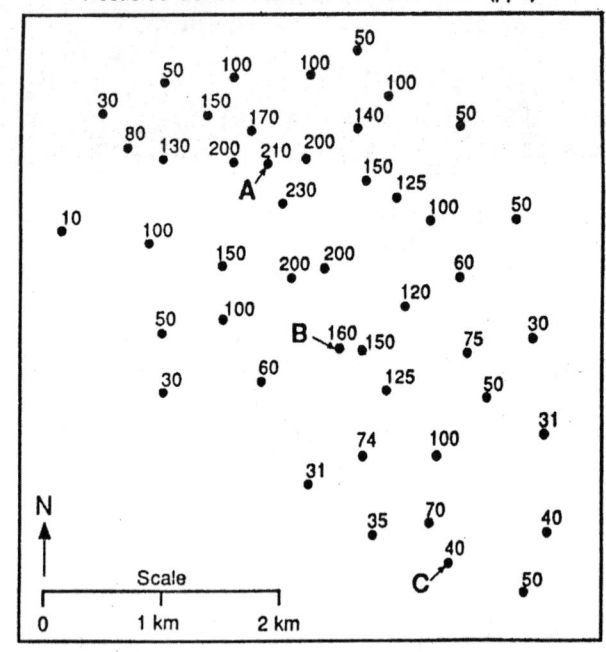

[Field map showing pesticide concentration data points with values ranging from 10 to 230 ppb. Location A is marked near the 230 value in the upper-left area; location B is marked near 160/150 values in the middle; location C is marked near 40/50 values in the lower portion. Scale shows 0 to 2 km. North arrow indicated.]

A water-soluble pesticide was applied to the ground surface at location A as shown on the map. Precipitation occurred and some of the pesticide that was carried into the ground dissolved in the water. As the ground water moved through the ground, it distributed the pesticide. A scientist took samples of the ground water and determined the concentration of the pesticide in parts per billion (ppb). The field map shows the concentrations of the pesticide.

10 Draw an accurate isoline map, following the directions below.
    a   Use an interval of 50 ppb.
    b   Begin your drawing with the 200-ppb isoline.

11 Using one or more complete sentences, briefly describe the general pattern of pesticide concentration as the distance from location A increases.

12 Use the isoline map you have made to determine the gradient of pesticide concentration in ppb between point B and point C, following the directions below.

    a   Write the formula for the gradient.

    b   Substitute data into the formula.

    c   Calculate the gradient.

    d   Label your answer with proper units.

UNIT TWO
MINERALS & ROCKS

VOCABULARY TO BE UNDERSTOOD

Cementation	Hardness	Precipitation
Cleavage	Igneous Rock	Recrystallization
Compression	Luster	Rock Cycle
Crystal	Mafic	Rock-forming Mineral
Crystalline Structure	Metamorphic Rock	Sedimentary Rock
Evaporite	Minerals	Solidification (Crystallization)
Felsic	Mohs' Scale	Specific Gravity
Foliation	Monomineralic	Streak
Fracture	Polymineralic	Texture

What is the composition of the Earth?

The solid portion of the Earth's crust, the lithosphere, is composed of naturally formed material made up of one or more minerals, called **rock**.

All rocks are composed of minerals. When only one mineral is found in a rock, the rock is said to be **monomineralic**. When a rock is composed of more than one mineral type, the rock is said to be **polymineralic**. Most rocks have a number of minerals in common.

A. MINERALS

Minerals are naturally occurring, inorganic, crystalline, solid materials with definite chemical composition, molecular structure, and specific physical properties. Of the more than 2,400 minerals on the Earth, about a dozen of them are so abundant that they comprise more than 90% of the lithosphere. These common minerals are called the "**rock-formers**."

CHARACTERISTICS OF MINERALS IN ROCKS

Minerals may be composed of single elements or compounds of two or more elements. Although there are more than 2,400 minerals, most of the Earth's crust is made of a relatively few common **elements** (composed of atoms which cannot be reduced to simpler substances). The most abundant element is oxygen (O_2 = 46.6% total Earth mass, 93.8% total Earth volume), and the second most abundant element is Silicon (Si = 27.7% total Earth mass, 0.9% total Earth volume). The other major elements are Aluminum (Al), Iron (Fe), Calcium (Ca), Sodium (Na), Potassium (K), and Magnesium (Mg).

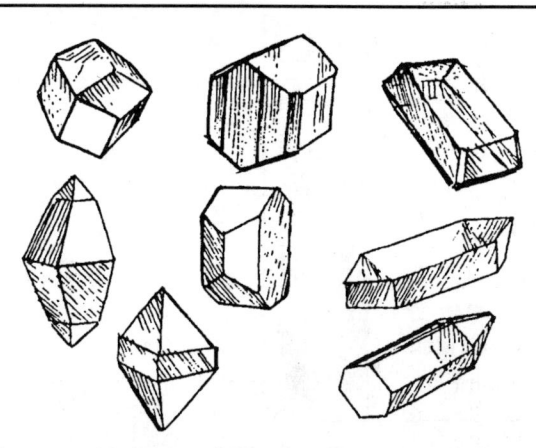

PHYSICAL AND CHEMICAL PROPERTIES OF MINERALS

What are some characteristics of minerals?

Minerals are identified on the basis of well-defined physical and chemical properties such as color, hardness, streak, luster, cleavage and fracture, crystal form and a variety of special properties such as odor, magnetism, and optical properties.

Color may be used for mineral identification. Whereas a few minerals have a distinctive color, many

UNIT TWO – MINERALS AND ROCKS N&N©   Page 31

minerals have the same color. Also, one mineral may have many colors due to traces of other elements. A good example is the mineral quartz, which may be white, pink, green, yellow, or purple.

Hardness is the resistance of a mineral to being scratched. Scratch two pieces of different minerals together. The harder mineral will make a scratch on the softer one. **Mohs' Scale of Hardness** lists ten minerals in order of hardness with #1 (talc) being the softest and #10 (diamond) being the hardest. By using one of these minerals of known hardness to scratch an unknown mineral, the relative hardness of the unknown mineral may be determined. Mineral hardness may also be compared to common objects like a fingernail, copper penny, or glass (see chart below).

MINERAL HARDNESS

Moh's Hardness Scale		Approximate Hardness of Common Objects
Talc	1	
Gypsum	2	Fingernail (2.5)
Calcite	3	Copper penny (3.5)
Fluorite	4	Iron nail (4.5)
Apatite	5	Glass (5.5)
Feldspar	6	Steel file (6.5)
Quartz	7	Streak plate (7.0)
Topaz	8	
Corundum	9	
Diamond	10	

Mohs' Scale of Hardness

(left) **"perfect basal cleavage"** as in mica which cleaves into thin sheets.

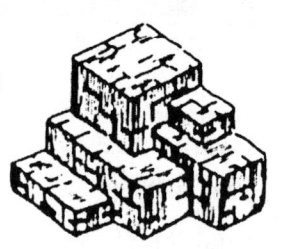

(right) **"cubic cleavage"** as in halite which breaks into small (rectangular) cubes.

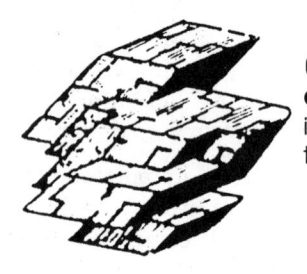

(left) **"rhombohedral cleavage"** as in calcite which splits in three directions parallel to their faces.

(right) **"concoidal fracture"** as in quartz which produces a wavy, curved shell-like surface.

Cleavage & Fracture in Minerals

Streak is the color of the powder of a mineral when crushed or scratched across a streak plate (a piece of unglazed porcelain). The color of a mineral may look very different than the color of its streak.

Luster is the appearance of light reflected from a mineral's surface. A mineral might shine like metal having a metallic luster. If a mineral does not look like a metal (nonmetallic) it might look glassy, waxy, greasy, pearly, earthy, or dull.

Some minerals tend to break along one or more smooth planes or surfaces. These minerals are said to have **cleavage**.

If a mineral has no well-defined cleavage it is said to have **fracture**. Minerals that have fracture usually break with uneven, splintery, or jagged surfaces.

The **crystal** form of a mineral is the external geometric form that results from the internal atomic structure of the mineral. Minerals similar in color may be distinguished by their difference in crystal form.

Specific gravity, the ratio of the weight of a mineral sample to the weight of an equal volume of water, is another property used to identify minerals. A mineral with a specific gravity of 3.0 would be 3.0 times as heavy as an equal volume of water. The specific gravity of a mineral is numerically equal to its **density**, because the density of pure water is 1 g/cm^3.

Some minerals have **special properties** such as calcite which will give off bubbles of carbon dioxide when a drop of cold, dilute hydrochloric acid is placed on it. The mineral halite tastes salty, magnetite is magnetic, and talc feels slippery.

B. IGNEOUS ROCKS

Rocks are classified as igneous, sedimentary, or metamorphic, depending on their composition and the environment in which they forms.

IGNEOUS ROCK FORMATION

Igneous rocks are formed from the solidification and crystallization of molten rock, making the rocks solid, compact, and/or hard.

How are igneous rocks formed?

Beneath the Earth's surface, molten rock material is called **magma**. When the magma reaches the Earth's surface, it is called **lava**. When magma or lava cools and solidifies, it forms **igneous rock**. As the liquid rock solidifies and becomes hard, mineral crystals may form, resulting in the igneous rock having a **crystalline texture**. Usually, there are many different minerals within this kind of rock.

The size of the crystals vary according to the conditions of time, temperature, and the pressure under which they form. In order for crystals to form (crystallization), the molten material must cool. The longer it takes for the molten material to cool, the larger the crystals that form and the more coarse the rock's texture (appearance and feel). By looking at the size of crystals in igneous rocks, the relative cooling rates of the rocks may be estimated.

The *Scheme for Igneous Rock Identification Chart* shows that texture, color, density, and mineral composition are used to identify igneous rocks. The light colored rocks are made primarily of the **felsic minerals** which are high in aluminum. Therefore, these rocks are lower in density than the dark colored **mafic minerals**, high in iron and magnesium.

Scheme for Igneous Rock Identification

ENVIRONMENT OF FORMATION							GRAIN SIZE	TEXTURE
INTRUSIVE (Plutonic)	Granite	Diorite	Gabbro	Peridotite	Dunite		1 mm or larger	Coarse
EXTRUSIVE (Volcanic)	Rhyolite	Andesite	Basalt Scoria	Rare	Rare		less than 1 mm	Fine
EXTRUSIVE (Volcanic)	Pumice *Obsidian	*Obsidian	Basalt Glass	Rare	Rare		Non-crystalline	Glassy

* Obsidian may appear black.

CHARACTERISTICS
- LIGHT ←— COLOR —→ DARK
- LOW ←— DENSITY —→ HIGH
- FELSIC (Al) ←— COMPOSITION —→ MAFIC (Fe, Mg)

MINERAL COMPOSITION (Relative by Volume)

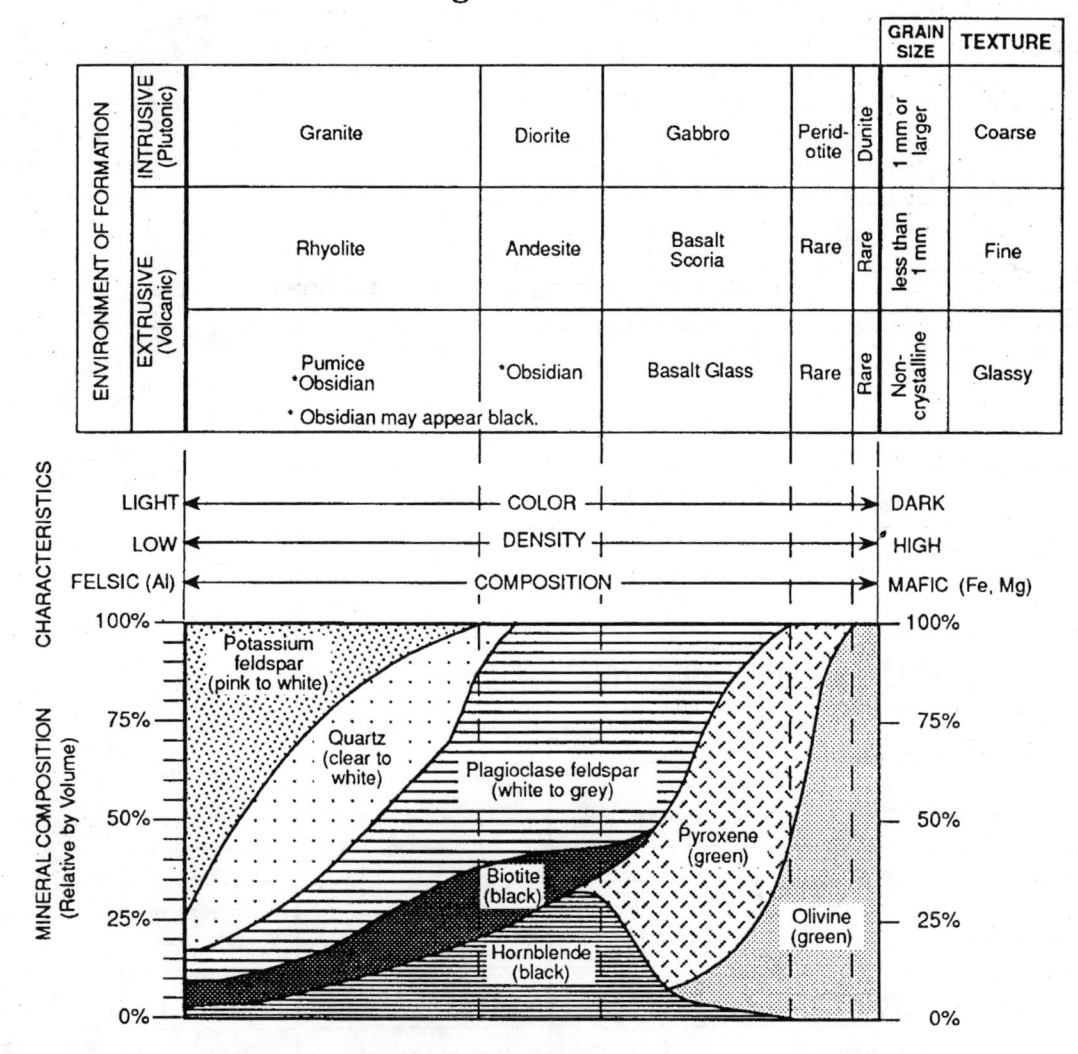

- Potassium feldspar (pink to white)
- Quartz (clear to white)
- Plagioclase feldspar (white to grey)
- Biotite (black)
- Hornblende (black)
- Pyroxene (green)
- Olivine (green)

Note: The intrusive rocks can also occur as exceptionally coarse-grained rock, Pegmatite.

An exception to the normal density identification is the glassy igneous rock, obsidian. A sample of obsidian may appear black, but is felsic in composition.

C. SEDIMENTARY ROCKS

The weathering processes break rock and produce sediments, which are transported by water, wind, and glaciers and deposited in different locations on land or under water. **Sedimentary rocks** usually contain rounded grains cemented in layers because running water is the major transporting agent of sediment. Many of the sedimentary rocks form under large bodies of water in the following ways.

How do sedimentary rocks form?

COMPRESSION, CEMENTATION

Some sedimentary rocks form when the pressure of water and other overlying sediments compress very small particles (clay and colloids) that have settled from the transporting agent (deposition). The pressure may itself be sufficient to form these fine sediments into rock. An example of a compressed (constricted, squeezed) sedimentary rock is shale, although cementing (binding particles and fragments of rocks) may occur also. In fact, some shales are layers made totally of natural cement.

Overlying Sediments

Some sediments are combined with mineral cements that **precipitate** out of (separated from solution) the ground water and result in cementation. Common mineral cements are iron, silica, and lime. Usually, cementation occurs with the larger sediments, such as sand, pebbles, and small rocks.

Sandstone is an example of a sedimentary rock in which the cemented particles are sorted in a uniform size. Most **conglomerates** are formed when the pebble sized particles are unsorted and cemented together. Frequently, the color of the rock is determined by the cementing agent.

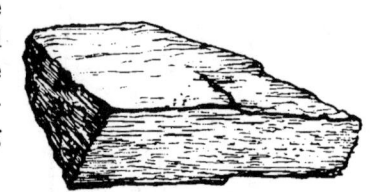

Uniform Particles

Scheme for Sedimentary Rock Identification

INORGANIC LAND-DERIVED SEDIMENTARY ROCKS					
TEXTURE	GRAIN SIZE	COMPOSITION	COMMENTS	ROCK NAME	MAP SYMBOL
Clastic (fragmental)	Mixed, silt to boulders (larger than 0.001 cm)	Mostly quartz, feldspar, and clay minerals; May contain fragments of other rocks and minerals	Rounded fragments	Conglomerate	
			Angular fragments	Breccia	
	Sand (0.006 to 0.2 cm)		Fine to coarse	Sandstone	
	Silt (0.0004 to 0.006 cm)		Very fine grain	Siltstone	
	Clay (less than 0.0006 cm)		Compact; may split easily	Shale	

CHEMICALLY AND/OR ORGANICALLY FORMED SEDIMENTARY ROCKS					
TEXTURE	GRAIN SIZE	COMPOSITION	COMMENTS	ROCK NAME	MAP SYMBOL
Nonclastic	Coarse to fine	Calcite	Crystals from chemical precipitates and evaporites	Chemical Limestone	
	Varied	Halite		Rock Salt	
	Varied	Gypsum		Rock Gypsum	
	Varied	Dolomite		Dolostone	
	Microscopic to coarse	Calcite	Cemented shells, shell fragments, and skeletal remains	Fossil Limestone	
	Varied	Carbon	Black and nonporous	Bituminous Coal	

CHEMICAL PROCESSES

Some sedimentary rocks form as a result of chemical processes, such as the **evaporation** and **precipitation** of a dissolved mineral out of evaporating water.

The precipitation of one mineral from solution forms monomineralic rocks, called **evaporites** (such as limestone, dolostone, gypsum, and rock salt).

BIOLOGICAL PROCESSES

Some sedimentary rocks form as a result of biologic processes. Biological materials include the remains of any living thing, including plants and animals. Coal forms from plant remains which are deposited in water, decayed, and then compressed.

Some water animals, such as coral, use minerals to form their shells. When they die, the minerals (often calcium and sea salts) are left behind, compress, and form sedimentary rocks (limestone). Fossil evidence is found nearly exclusively in sedimentary rocks.

Sedimentary Rock with Fossils

D. METAMORPHIC ROCKS

Metamorphic rocks are igneous, sedimentary, and other metamorphic rocks that have been changed in form usually deep within the Earth. The high pressures, temperatures, and chemical solutions deep in the Earth cause changes in the existing rocks, forming them into metamorphic rocks.

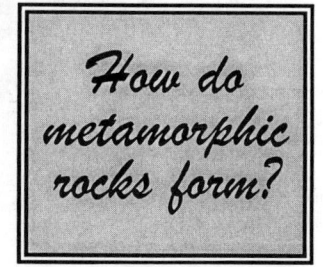

How do metamorphic rocks form?

Metamorphism is the result of the **recrystallization** of *unmelted* material under high temperature and pressure. These extreme conditions cause the mineral crystals to grow and new minerals to form without melting (a solid state reaction). The weathering and erosion of the Earth's surface eventually exposes the metamorphic rocks.

Metamorphic rocks are usually harder and more dense than the rocks from which they were formed. Original layers frequently become bent or distorted. The metamorphic rock may also show **foliation**, an alignment of minerals or the separation of minerals into platy (flaky) layers or light and dark bands.

Metamorphic Rock Distorted (bent) Foliation

Scheme for Metamorphic Rock Identification

TEXTURE	GRAIN SIZE	COMPOSITION	TYPE OF METAMORPHISM	COMMENTS	ROCK NAME	MAP SYMBOL
FOLIATED — Slaty	Fine	CHLORITE / MICA / QUARTZ / FELDSPAR / AMPHIBOLE / GARNET / PYROXENE	Regional	Low-grade metamorphism of shale	Slate	
FOLIATED — Schistose	Medium to coarse			Medium-grade metamorphism; Mica crystals visible from metamorphism of feldspars and clay minerals	Schist	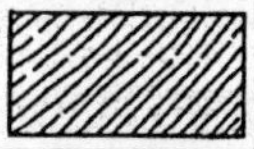
FOLIATED — Gneissic	Coarse		(Heat and pressure increase with depth, folding, and faulting)	High-grade metamorphism; Mica has changed to feldspar	Gneiss	
NONFOLIATED	Fine	Carbonaceous		Metamorphism of plant remains and bituminous coal	Anthracite Coal	
NONFOLIATED	Coarse	Depends on conglomerate composition		Pebbles may be distorted or stretched; Often breaks through pebbles	Meta-conglomerate	
NONFOLIATED	Fine to coarse	Quartz	Thermal (including contact) or Regional	Metamorphism of sandstone	Quartzite	
NONFOLIATED	Fine to coarse	Calcite, Dolomite		Metamorphism of limestone or dolostone	Marble	
NONFOLIATED	Fine	Quartz, Plagioclase	Contact	Metamorphism of various rocks by contact with magma or lava	Hornfels	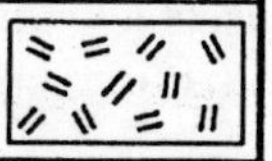

The presence or absence of foliation provides the basis for classifying metamorphic rocks. The nonfoliated metamorphic rocks are classified by composition.

Metamorphic Rock
Foliation with Banding

ENVIRONMENT OF FORMATION

The composition, structure, and texture of a rock depends on the environment (formation location) in which the rock forms.

DISTRIBUTION

Sedimentary rocks are found usually as a thin layer, or veneer, over large areas of continents. Igneous rocks, at or near the surface, are found in regions of volcanoes or mountains where **intrusion** (forced while molten into cracks or between other layers of rock) and **extrusion** (ejected at the Earth's surface) have occurred. Where intrusive igneous and metamorphic rocks are found exposed on the surface of the Earth, it is usually due to removing of over-lying rock through millions of years of weathering and erosion by water, wind, and glaciers.

E. THE ROCK CYCLE

The **rock cycle** is a model to show how closely the rock types are related. With the exception of meteorite material from space, the amount of rock material on Earth remains constant. The rocks themselves, however, gradually change responding to changes in their environment

How do rocks change?

Generally, igneous rock is considered to be the primary or parent rock of the Earth's crust. As rocks are attacked by the forces of weathering and erosion, the sediments formed are deposited, cemented, and formed into sedimentary rocks.

If these rocks are exposed to pressure and temperature extremes, they may be changed into metamorphic rocks. These rocks may then be subjected to additional Earth forces, melting them into magma and eventually solidifying into igneous rock. The rock cycle is then completed.

Any type rock, igneous, sedimentary, or metamorphic, may be changed into any other type depending upon the environment to which it is subjected. There is no preferred or predictable path that a rock will take within the environment.

EVIDENCE FOR THE ROCK CYCLE

The composition of some sedimentary rocks suggests that the components (sediments or rock particles and fragments) had varied origins. As previously discussed, sedimentary rocks may develop from a combination of sediments and organic remains that may have been transported great distances from several sources and deposited in one place. Then, the forces of the environment change these varied sediments into compacted and cemented rock.

The composition of some rocks suggests that the materials have undergone multiple transformations as part of the rock cycle. The Model of the Rock Cycle (from the *Earth Science Reference Tables*) shows how one form of rock can be processed into any number of other forms of rock. For example, starting with igneous rocks and following the outside pathway counterclockwise, igneous rock may be eroded – deposited – buried – cemented into sedimentary rock. That rock can be transformed by heat and pressure into metamorphic rock, and then be melted and solidify into igneous rock again.

Rock Cycle in Earth's Crust

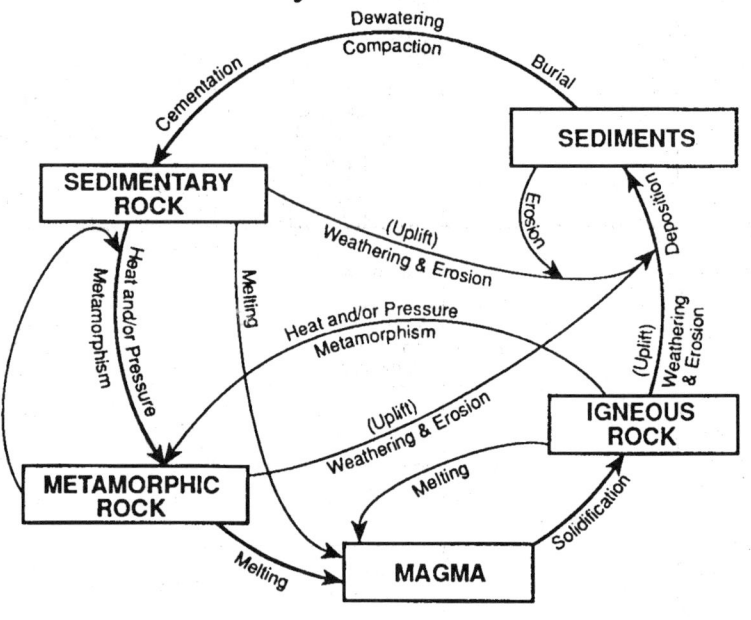

However, there are many other possible pathways in the rock cycle model.

The igneous rock may be recrystallized, forming metamorphic rock, which could then be eroded – deposited – buried – cemented into sedimentary rock. It is possible that sedimentary rock originally formed on the surface, could take a pathway through the rock cycle model, and eventually be part of a molten rock extrusion and finally become an extrusive igneous rock.

QUESTIONS FOR UNIT 2

1. The physical properties of a mineral are largely due to its
   1. volume
   2. melting point
   3. organic composition
   4. internal arrangement of atoms

2. A student crushes a small sample of mineral to see the color of its powder. The student is trying to determine the mineral's
   1. density
   2. luster
   3. chemical composition
   4. streak

3. To identify a mineral sample embedded in a block of plastic, which of the following physical properties could not be used?
   1. crystal shape
   2. hardness
   3. color
   4. luster

4. What information would you use to determine the density of a mineral?
   1. volume and shape
   2. volume and mass
   3. volume and age
   4. shape and mass

5. According to the *Reference Tables*, the most abundant element in the Earth's crust is
   1. oxygen
   2. silicon
   3. aluminum
   4. magnesium

Base your answers to questions 6 through 9 on your knowledge of Earth science, the *Reference Tables*, and the table of minerals (bottom of page) which shows the physical properties of nine minerals.

6. Which mineral has a different color in its powdered form than in its original form?
   1. pyrite
   2. graphite
   3. kaolinite
   4. magnetite

7. Which mineral contains iron, has a metallic luster, is hard, and has the same color and streak?
   1. biotite bica
   2. galena
   3. kaolinite
   4. magnetite

Mineral	Color	Luster	Streak	Hardness	Density (g/mL)	Chemical Composition
biotite mica	black	glassy	white	soft	2.8	$K(Mg,Fe)_3(AlSi_3O_{10})(OH_2)$
diamond	varies	glassy	colorless	hard	3.5	C
galena	gray	metallic	gray-black	soft	7.5	PbS
graphite	black	dull	black	soft	2.3	C
kaolinite	white	earthy	white	soft	2.6	$Al_4(Si_4O_{10})(OH)_8$
magnetite	black	metallic	black	hard	5.2	Fe_3O_4
olivine	green	glassy	white	hard	3.4	$(Fe,Mg)_2SiO_4$
pyrite	brass yellow	metallic	greenish-black	hard	5.0	FeS_2
quartz	varies	glassy	colorless	hard	2.7	SiO_2

Definitions
Luster: the way a mineral's surface reflects light
Streak: color of a powdered form of the mineral
Hardness: resistance of a mineral to being scratched
(soft – easily scratched; hard – not easily scratched)

Chemical Symbols
Al – Aluminum   Pb – Lead
C – Carbon   Si – Silicon
Fe – Iron   K – Potassium
H – Hydrogen   S – Sulfur
Mg – Magnesium   O – Oxygen

8 What do the minerals quartz, olivine, and biotite mica have in common with diamond?
   1 hardness   3 color
   2 luster   4 composition

9 Which mineral would be best to use in the manufacture of sandpaper?
   1 quartz   3 kaolinite
   2 graphite   4 biotite mica

10 The diagrams below represent samples of five different minerals found in the rocks of the Earth's crust.

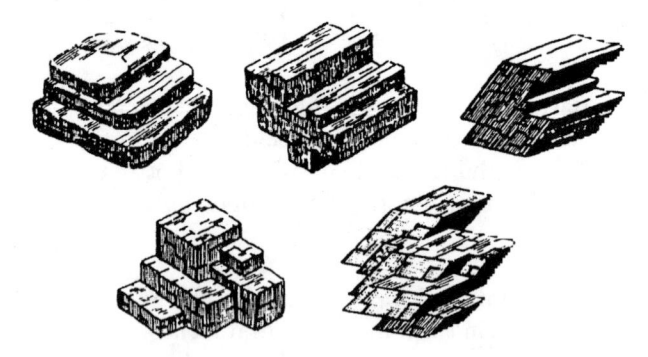

Which physical property of minerals is represented by the flat surfaces in the diagrams?
   1 magnetism   3 cleavage
   2 hardness   4 crystal size

11 The relative hardness of a mineral can best be tested by
   1 scratching the mineral across a glass plate
   2 squeezing the mineral with calibrated pliers
   3 determining the density of the mineral
   4 breaking the mineral with a hammer

12 Quartz mineral samples are best identified by their
   1 hardness   3 size
   2 color   4 mass

13 What causes the characteristic crystal shape and cleavage of the mineral halite as shown in the diagram at the right? 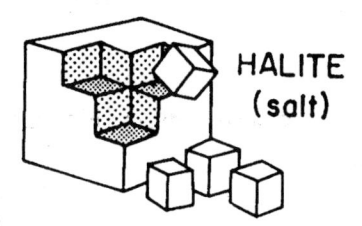
   1 metamorphism of the halite
   2 the internal arrangement of the atoms in halite
   3 the amount of erosion the halite has undergone
   4 the shape of other minerals located where the halite formed

14 When dilute hydrochloric acid is placed on the sedimentary rock limestone and the nonsedimentary rock marble, a bubbling reaction occurs with both. What would this indicate?
   1 The minerals of these two rocks have similar chemical composition.
   2 The molecular structures of these two rocks have been changed by heat and pressure.
   3 The physical properties of these two rocks are identical.
   4 The two rocks originated at the same location.

15 Which statement best describes a general property of rocks?
   1 Most rocks have a number of minerals in common.
   2 Most rocks are composed of a single mineral.
   3 All rocks contain fossils.
   4 All rocks contain minerals formed by compression and cementation.

16 Nine rock samples were classified into three groups as shown in the table below.

Group A	Group B	Group C
Granite	Shale	Marble
Rhyolite	Sandstone	Schist
Gabbro	Conglomerate	Gneiss

This classification system was most likely based on the
   1 age of the minerals in the rock
   2 size of the crystals in the rock
   3 way in which the rock formed
   4 color of the rock

17 Which two processes result in the formation of igneous rocks?
   1 solidification and evaporation
   2 melting and solidification
   3 crystallization and cementation
   4 compression and precipitation

18 According to the Reference Tables, which property would be most useful for identifying igneous rocks?
   1 kind of cement
   2 mineral composition
   3 number of mineral present
   4 types of fossils present

19 The diagram at the right represents the percentage by volume of each mineral found in a sample of basalt.

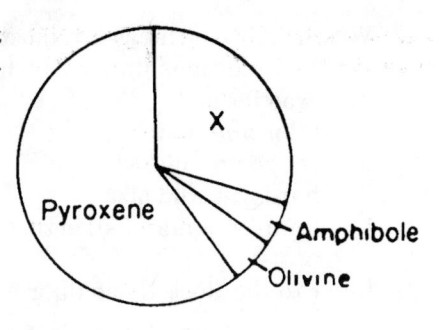

Which mineral is represented by the letter X in the diagram?
1 orthoclase feldspar
2 plagioclase feldspar
3 quartz
4 mica

20 According to the Reference Tables, rhyolite and granite are alike in that they both are
1 fine-grained
2 dark-colored
3 mafic
4 felsic

21 According to the *Reference Tables*, which graph best represents the comparison of the average grain sizes in basalt, granite, and rhyolite?

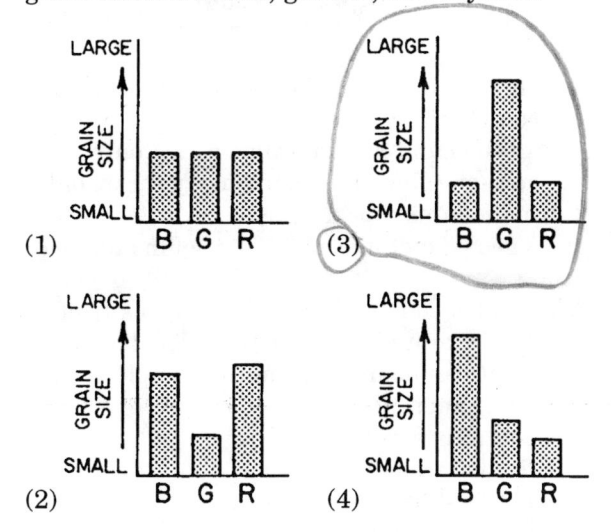

22 Which characteristic of an igneous rock would provide the most information about the environment in which the rock solidified?
1 color
2 texture
3 hardness
4 streak

23 Which property is common to most dark-colored igneous rocks?
1 high density
2 intrusive formation
3 abundant felsic minerals
4 coarse-grained texture

24 Which is the best description of the properties of basalt?
1 fine-grained and mafic
2 fine-grained and felsic
3 coarse-grained and mafic
4 coarse-grained and felsic

25 Large crystal grains in an igneous rock indicate that the rock was formed
1 near the surface
2 under low pressure
3 at a low temperature
4 over a long period of time

26 Which rock has cooled most rapidly from a molten state?
1 gabbro
2 granite
3 quartzite
4 obsidian

27 Where are the Earth's sedimentary rocks generally found?
1 in regions of recent volcanic activity
2 deep within the Earth's crust
3 along the mid-ocean ridges
4 as a thin layer covering much of the continents

28 Limestone can form as a result of
1 cooling of molten rock under the oceans
2 metamorphosis of conglomerate rock
3 precipitation from evaporating water
4 radioactive decay of dolostone

29 A chemically formed sedimentary rock composed of halite should be identified as
1 gypsum rock
2 rock salt
3 limestone
4 coal

30 The diagram below represents a conglomerate rock. Some of the rock fragments are labeled.

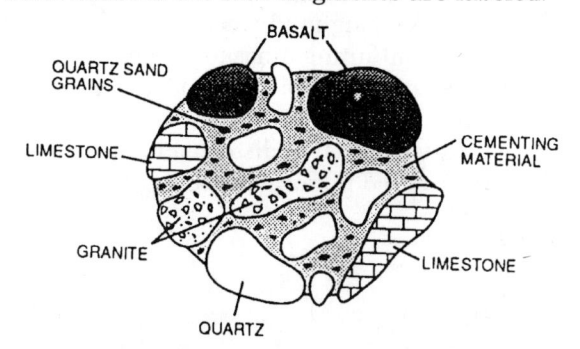

Which conclusion is best made about the rock particles?
1 They are the same age.
2 They originated from a larger mass of igneous rock.
3 They all contain the same minerals.
4 They have different origins.

31 According to the *Reference Tables*, which characteristic determines whether a rock is classified as a shale, a siltstone, a sandstone, or a conglomerate?
1   the absolute age of the sediments within the rock
2   the mineral composition of the sediments within the rock
3   the particle size of the sediments within the rock
4   the density of the sediments within the rock

32 Metamorphic rocks result from the
1   erosion of rocks
2   recrystallization of rocks
3   cooling and solidification of molten magma
4   compression and cementation of soil particles

33 While a geology student was walking along several outcrops, she found a rock specimen that showed the following characteristics:

Grain Size –   Coarse
Texture –   Foliated
Composition –   Quartz, feldspar, amphibole, garnet, and pyroxene

This specimen should be identified as
1   hornfels   3   gneiss
2   slate   4   anthracite

34 The metamorphism of a sandstone rock will cause the rock
1   to be melted
2   to contain more fossils
3   to become more dense
4   to occupy a greater volume

35 Which rock is most likely a metamorphic rock?
1   a rock showing mud cracks
2   a rock containing dinosaur bones
3   a rock consisting of layers of rounded sand grains
4   a rock composed of distorted light-colored and dark-colored mineral bands

36 Which characteristics are most useful for identifying the conditions under which a metamorphic rock was formed?
1   color and luster
2   shape and mass
3   hardness and size
4   composition and structure

37 Refer to the Rock Cycle diagram below.

Rock Cycle in Earth's Crust

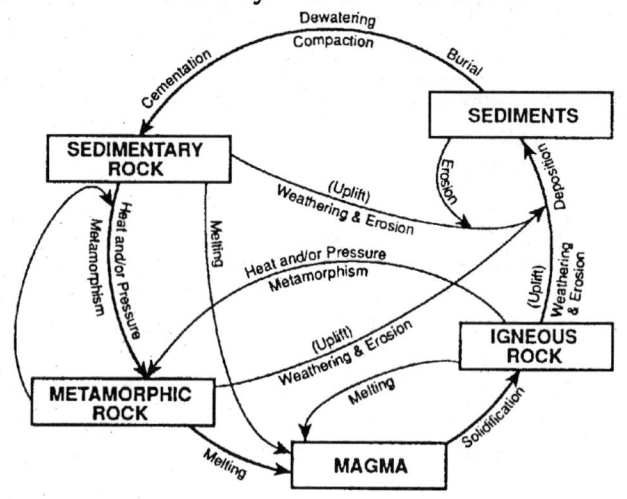

Which type(s) of rock can be the source of deposited sediments?
1   igneous and metamorphic rocks, only
2   metamorphic and sedimentary rocks, only
3   sedimentary rocks, only
4   igneous, metamorphic, and sedimentary rocks

38 A sample of conglomerate consists mostly of fragments of granite and sandstone. The best inference that can be made from the sample is that this conglomerate
1   contains fossils
2   formed from other rocks
3   resulted from solidification
4   formed during the Cambrian Period

SKILL ASSESSMENTS

Base your answers to questions 1-4 on your knowledge of Earth science, the *Reference Tables* and the data below for five different rock samples.

Data Table

ROCK SAMPLE	ORIGIN	CRYSTAL SIZE OR GRAIN SIZE	OTHER CHARACTERISTIC
1	igneous	no crystals	glassy
2	igneous	coarse	light color
3	igneous	fine	dark color
4	sedimentary	0.0003 cm in diameter	contains dinosaur footprints
5	metamorphic	coarse	shows banding

1. What sedimentary rock is represented by sample 4? During what geologic era was it likely to have been formed?

2. In a sentence, explain why rock sample one (1) has a glassy texture.

3. In a short paragraph, explain the probable cause of the banding characteristic seen in rock sample five (5).

4. Which rock sample is granite? Explain how you know.

Base your answers to questions 5 and 6 on the six illustrations below which represent six different rock types.

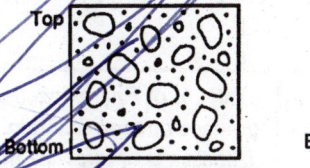

ROCK A: cemented sand and rounded pebbles

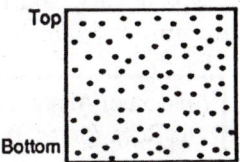

ROCK B: uniform smooth sand grains firmly cemented together

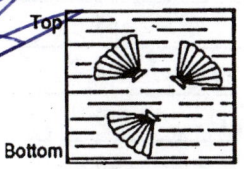

ROCK C: a matrix of fine colloidal-sized particles with shell fossils

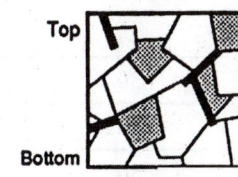

ROCK D: large intergrown mineral crystals

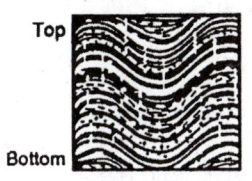

ROCK E: crumpled, distorted bands of different minerals

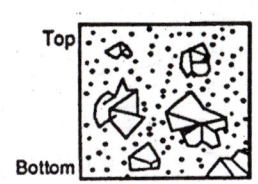

ROCK F: cemented sand and angular rock fragments

5. Classify each rock sample as igneous, sedimentary or metamorphic.

Rock A _____

Rock B _____

Rock C _____

Rock D _____

Rock E _____

Rock F _____

6. In one or two sentences, explain what characteristic(s) of the rock led to your classification.

UNIT TWO – MINERALS AND ROCKS N&N©

Base your answers to questions 7 through 9 on the data table below. The data table shows the relationship between the amount of aluminum in a type of rock and the energy needed to extract aluminum from that rock.

Amount of Aluminum in Rock (percent)	Amount of Energy needed to produce one ton of Aluminum (thousands of kilowatt-hours)
3	220
5	140
10	90
20	65
30	54
40	49
50	48

7 Mark an appropriate scale on the axis labeled "Amount of Aluminum in Rock." Plot a line graph using the data in the table above.

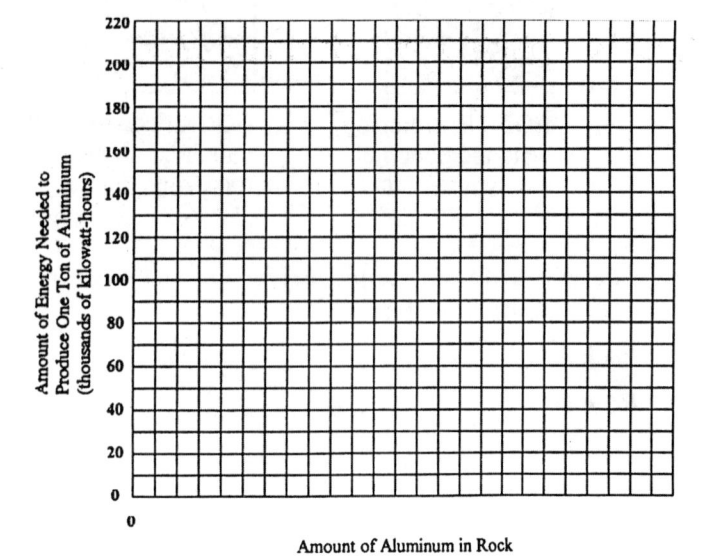

8 Based upon the graph you constructed, how much energy is needed to produce aluminum from rock that is 15 percent aluminum?

_____ thousands of kilowatt-hours

9 Using one or more complete sentences, describe how the amount of aluminum in a rock is related to the amount of energy needed to extract the aluminum from the rock.

TOPIC A
ROCKS, MINERALS, & RESOURCES

VOCABULARY TO BE UNDERSTOOD

Banding
Chemical Sedimentary Rock
Contact Metamorphism
Crystal
Crystalline Structure
Distorted Structure
Extrusive Igneous Rock
Fossil Fuel
Fragmental Sedimentary Rock
Intrusive Igneous Rock
Metamorphic Rock
Nonrenewable Resource
Organic Sedimentary Rock
Parent Rock
Precipitation
Recrystallization
Regional Metamorphism
Silicates
Silicon – Oxygen Tetrahedron
Texture
Transition Zone

A. MINERAL PROPERTIES

Minerals are said to be **crystalline** which means the atoms inside are bonded in a particular structure or pattern. Two minerals with the same chemical composition may have very different properties. A good example is diamond and graphite which are both made of carbon. Graphite is made of weakly bonded layers of carbon atoms and is very soft. Diamond is formed at great depths where extreme pressure causes the carbon atoms to have a compact structure so that diamond is the hardest mineral.

How do minerals differ from each other?

Silicon - Oxygen Tetrahedron

"Stick Figure"

KEY:
● SILICON ATOM
○ OXYGEN ATOM

Diagram illustrates relative atom size

Minerals may be grouped according to the elements of which they are made, or the compounds which they can form. For example, the **oxide** group is made of compounds of oxygen and one other element. The **carbonate** group is made of one or more metals combined with a carbon and three oxygen atoms (CO_3).

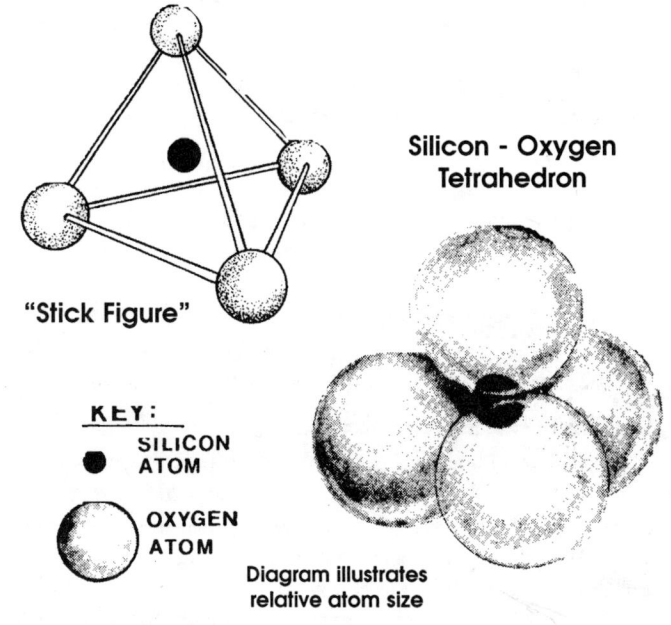

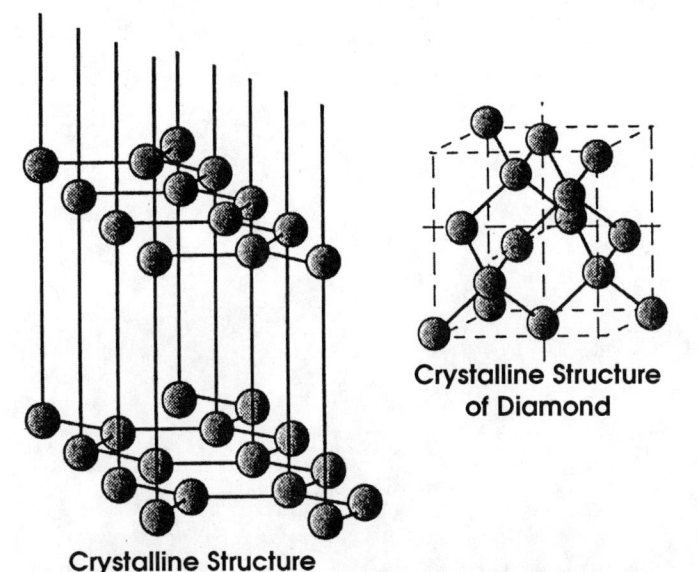

Since oxygen and silicon are the two most abundant elements in the Earth's crust, it is not surprising that the **silicates**, minerals containing both oxygen and silicon, are the largest group of minerals. Oxygen and silicon bond to form a **tetrahedral unit**, the basic building block of the silicates.

The **silicon-oxygen tetrahedra** have the ability to bond together in many different patterns and with many different elements forming silicate minerals with a wide variety of external properties.

Crystalline Structure of Graphite

Crystalline Structure of Diamond

Types of Sharing in Silicate Mineral Structures

Minerals	Olivine	Hornblende	Mica	Quartz
SiO$_4$ Patterns	Single Tetrahedron	Chains of Tetrahedra	Sheets of Tetrahedra	Networks of Tetrahedra
Chemical Composition	(Fe, Mg)$_2$SiO$_4$	Ca$_2$(Mg, Fe)$_5$(Si$_8$O$_{22}$)(OH)$_2$	K(Mg, Fe)$_3$(AlSi$_3$O$_{10}$)(OH)$_2$	SiO$_2$

← Increased Sharing of Oxygen →

← Fe + Mg Decreasing →

B. IGNEOUS ROCKS

What can be learned from igneous rocks?

The **texture** of the igneous rocks is dependent upon the rate of cooling. Slow cooling produces large crystals – rocks with a coarse texture (phaneritic). Rapid cooling produces small crystals – rocks with a fine texture (aphanitic).

Cooling is related to both temperature and pressure. Rapid cooling produces small crystals or a grainy texture. Very rapid cooling results in no crystal formation or a glassy texture. This is the result of a rapid drop in temperature or a pressure decrease. These conditions would exist near the surface of the Earth or when the molten material (lava) breaks through the surface, as during a volcanic eruption. When lava flows out and hardens on the Earth's surface, it is called an **extrusion**. The rock formed is called **extrusive igneous rock**.

However, deep within the Earth both temperature and pressure are much higher than at the surface. Therefore, cooling is slower and crystal size is larger. When magma hardens in the Earth, it is called an **intrusion**. The rock formed is called **instrusive igneous rock**.

As mentioned in the previous chapter, igneous rocks form by solidification and crystallization from magma or lava. To form magma, minerals must be melted. Some minerals, like quartz and muscovite mica, melt and solidify at relatively low temperatures. Others, like olivine and pyroxene, melt and solidify at much higher temperatures. In a body of cooling magma, the minerals that melt at high temperatures crystallize first and settle out of the melt. Those that melt at relatively low temperatures crystallize last, separating them from the minerals that crystallized first.

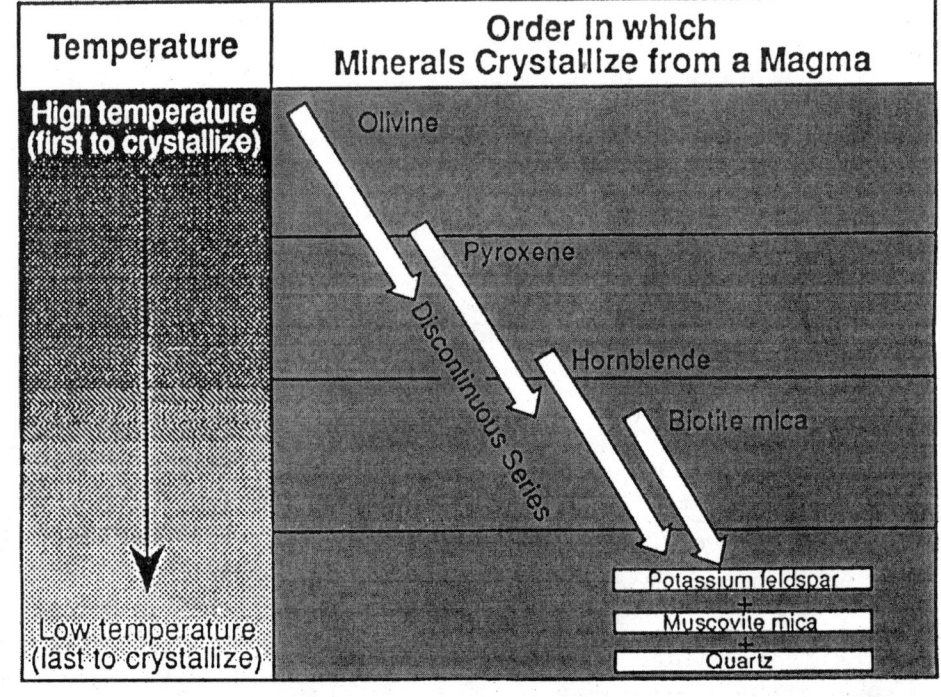

Temperature / Order in which Minerals Crystallize from a Magma

High temperature (first to crystallize) → Low temperature (last to crystallize)

Discontinuous Series:
- Olivine
- Pyroxene
- Hornblende
- Biotite mica
- Potassium feldspar + Muscovite mica + Quartz

This is why igneous rocks of different composition form from the same body of magma. This is the reason that igneous rocks containing olivine do not usually contain quartz.

Igneous rocks are common in both the continental and oceanic crust of the Earth. Low density, felsic, granitic igneous rock is common in the continental landmasses while the ocean basins are made primarily of high density, mafic, basaltic igneous rock.

C. SEDIMENTARY ROCKS

Sedimentary rocks are classified as **fragmental** (**clastic**), **chemical**, or **organic** depending upon how they were formed. The **fragmental sedimentary rocks** are composed of various sediments which have been compacted and cemented together. This kind of sedimentary rock is classified on the basis of grain size. As shown on the *Reference Tables*, if the particles range in size from 0.006 to 0.2 cm, the rock is sandstone.

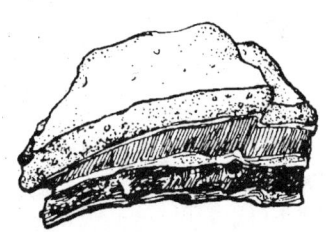

Discrete Layers Showing Different Sediments

The particle size in sedimentary rocks is related to the agent that transported and deposited the particles. The most important agent is running water which deposits sediment in horizontal layers. These horizontal layers are an important identifying characteristic of many sedimentary rocks.

A range of particle sizes, often resembling a stream bed or delta deposit, has larger, denser particles on the bottom and decreasing particle size and density towards the top of the rock. A predominance of one particle size in a layer indicates the deposition of one type, size, and density of sediment.

Chemically and **organically formed sedimentary rocks** are identified by composition and texture. The chemical sedimentary rocks tend to be made of one mineral such as rock salt made of the mineral halite. Although limestone is the most abundant chemically formed rock, nearly 90% of limestone is formed organically. Made of cemented shells, shell fragments, and skeletal remains.

Fossils found in sedimentary rocks provide evidence of the environment in which the rocks formed. Sedimentary rocks containing fossils of fish were formed in a marine environment whereas, fossilized impressions of land plants were not.

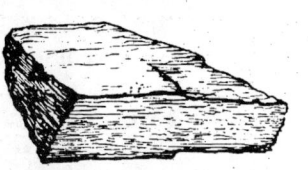

What information can be gained from sedimentary rocks?

Predominance Of One Particle Size

Organic (Fossil) Composition

D. METAMORPHIC ROCKS

Metamorphic rocks are those in which heat and pressure have caused recyrstallization often forming new minerals and larger crystals. The original rock from which a metamorphic rock forms is called the **parent rock**. Parent rock may be igneous, sedimentary, or metamorphic. Metamorphic rocks are often harder, denser, and less porous than the parent rock from which they formed.

What can we learn from metamorphic rocks?

Banding of different minerals give some metamorphic rocks a striped appearance, involving the segregation of light and dark minerals into layers. Generally, the more intense the temperature and pressure conditions, the thicker the mineral bands that are formed. Thick banding indicates a high degree of metamorphism.

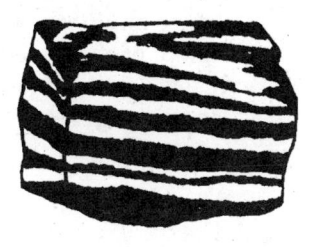

Banding In A Metamorphic Rock

Distorted structure is the folding and bending of rock layers, brought about by very strong Earth forces, causing the rock to have deformed structural appearance.

Distortion Of A Banded Rock Through Pressure

Contact metamorphism occurs where molten magma or lava comes in contact with other rocks and changes them. Since the change is caused by heat, this process is also called **thermal metamorphism**. There is no specific point that exists between the changed and unchanged rock; however, a transition zone from altered to unaltered rock may form.

Any rock may be altered by the action of molten material passing over or through it.

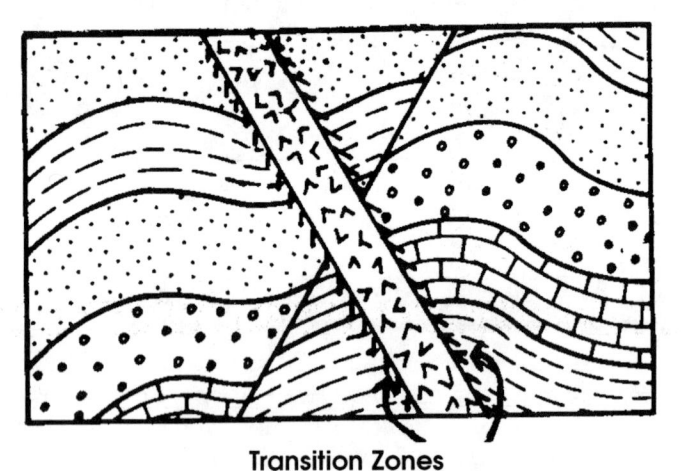

Transition Zones
(Boundary between sedimentary rock and intrusive igneous rock.)

The largest amount of metamorphic rock is produced by **regional metamorphism** associated with the extreme pressures that accompany **orogeny** or mountain building.

Different metamorphic rocks may be formed from the same parent rock depending upon the pressure/temperature environment in which it forms. For example, if a piece of shale (sedimentary) is metamorphosed, it progressively would form slate → phyllite → schist → gneiss as the **degree of metamorphism** increased.

E. RESOURCE CONSERVATION

With the Earth's population rapidly increasing, the demand for energy and products is rapidly increasing as well. At the present time, fossil fuels are the primary source of energy. **Fossil fuels**, coal, oil, and natural gas were formed from organic matter in the ancient past and are **nonrenewable resources**. They are nonrenewable, because they are being used much faster and in greater amounts than they can form. In addition to producing energy, petroleum derivatives are used in making plastics, medicines, cosmetics, fabrics and many other products. Much of the global economy depends upon the availability of these resources.

Uneven distribution of resources and increasing demand for dwindling resources results in higher prices, a decrease in the standard of living for many and in the past has even resulted in wars.

Why is the conservation of minerals and resources important?

Like the fossil fuels, most minerals are nonrenewable resources. In addition to using alternative energy sources where possible (solar, hydropower, wind, wave, geothermal, and nuclear energy), we must practice the *four "R's" of conservation* – **Reduce**, **Reuse**, **Recycle** and **Reclaim**. We must find ways to reduce our need, reuse what we can (less packaging and fewer disposables), use materials that can be recycled and reclaim usable material from things that have already been discarded.

QUESTIONS FOR TOPIC A

1 Which diagram best represents the model of a silicon-oxygen tetrahedron?

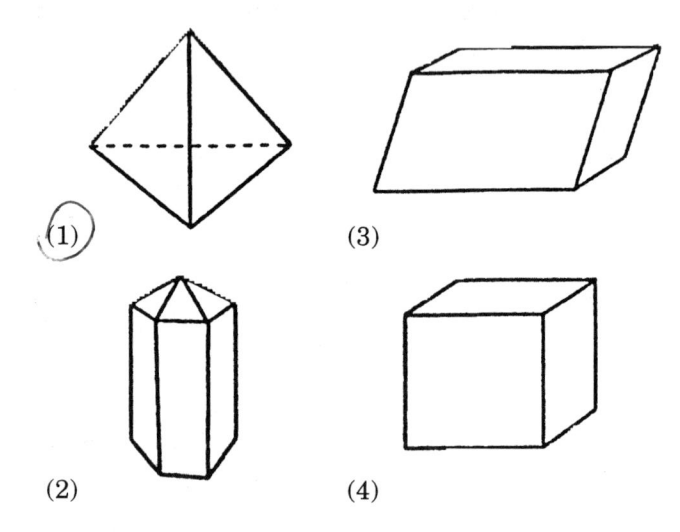

(1) (2) (3) (4)

2 The silicate tetrahedron model is most directly useful in explaining the properties of the materials in the
1 hydrosphere   3 atmosphere
2 Earth's crust   4 Earth's core

3   Which element combines with silicon to form the tetrahedral unit of structure of the silicate minerals?
    1   oxygen
    2   nitrogen
    3   potassium
    4   hydrogen

4   The data table at the right shows the composition of six common rock-forming minerals. The data table provides evidence that
    1   the same elements are found in all minerals
    2   a few elements are found in many minerals
    3   all elements are found in only a few minerals
    4   all elements are found in all minerals

Mineral	Composition
Mica	$KAl_3Si_3O_{10}$
Olivine	$(FeMg)_2SiO_4$
Orthoclase	$KAlSi_3O_8$
Plagioclase	$NaAlSi_3O_8$
Pyroxene	$CaMgSi_2O_6$
Quartz	SiO_2

5   Two mineral samples have different physical properties, but each contains silicate tetrahedrons as its basic structural unit. Which statement about the two mineral samples must be true?
    1   They have the same density.
    2   They are similar in appearance.
    3   They contain silicon and oxygen.
    4   They are the same mineral.

6   Why do diamond and graphite have different physical properties, even though they are both composed entirely of the element carbon?
    1   Only diamond contains radioactive carbon.
    2   Only graphite consists of organic material.
    3   The minerals have different arrangements of carbon atoms.
    4   The minerals have undergone different amounts of weathering.

7   The diagram at the right shows a sample of rock material that contains coarse-grained intergrown crystals of several minerals. [Mineral crystals are shown actual size.]
    This rock sample should be identified as
    1   rhyolite
    2   granite
    3   scoria
    4   basalt

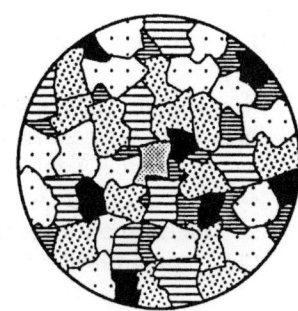

Base your answers to questions 8 through 10 on the diagrams below, on the *Earth Science Reference Tables*, and your knowledge of Earth science. Diagram I shows the order in which silicate minerals crystallize from magma. Diagram II shows the minerals that form as the number of linked tetrahedra increases.

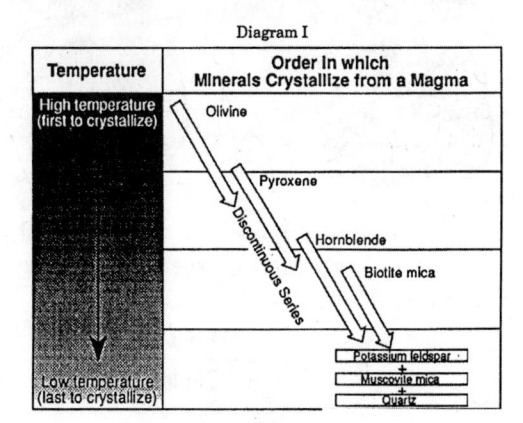

8   Which statement about the minerals quartz and olivine must always be true?
    1   They have the same form at the same temperature.
    2   They have the same density.
    3   They contain the elements silicon and oxygen.
    4   They contain the elements iron and magnesium.

9   The arrangement of the silicon-oxygen tetrahedra in the mineral mica is responsible for mica's
    1   black color
    2   colorless streak
    3   high crystallization temperature
    4   cleavage in one direction

10 Which statement about the process of crystallization of minerals from magma is true?
    1   Dark-colored rocks are the last to crystallize.
    2   Rocks crystallized at the highest temperatures are lower in density than rocks crystallized at low temperatures.
    3   The first rocks to crystallize will contain minerals composed of single tetrahedra.
    4   The rocks crystallized last contain the greatest amounts of iron (Fe) and magnesium (Mg).

11 Which granite sample most likely formed from magma that cooled and solidified at the slowest rate?

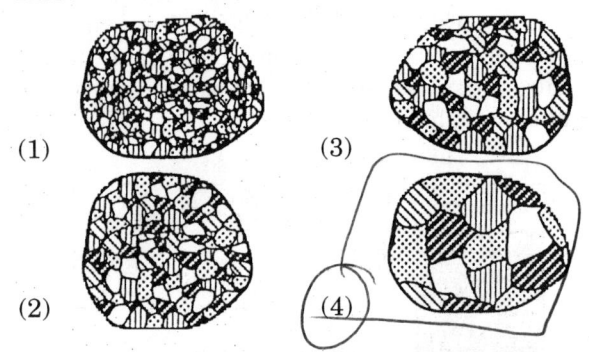

(1)   (2)   (3)   (4)

Base your answers to questions 12 through 15 on the diagram below which represents a geologic cross section.

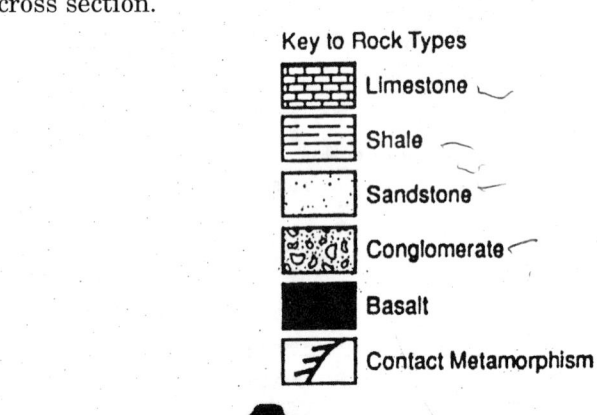

Key to Rock Types:
- Limestone
- Shale
- Sandstone
- Conglomerate
- Basalt
- Contact Metamorphism

12 In which location is a geologist most likely to find rock composed of intergrown crystals?
    (1) A   (3) C
    (2) B   (4) D

13 The rock at B most likely contains
    1 quartz, only
    2 quartz and potassium feldspar, only
    3 potassium feldspar, pyroxene, and olivine
    4 plagioclase feldspar, pyroxene, and olivine

14 Which rock is most likely organic in origin?
    1 limestone   3 basalt
    2 sandstone   4 conglomerate

15 At which location would quartzite most likely be found?
    (1) A   (3) E
    (2) B   (4) D

Base your answers to questions 16 and 17 on the diagram below which represents the formation of a sedimentary rock. [Sediments are drawn actual size.]

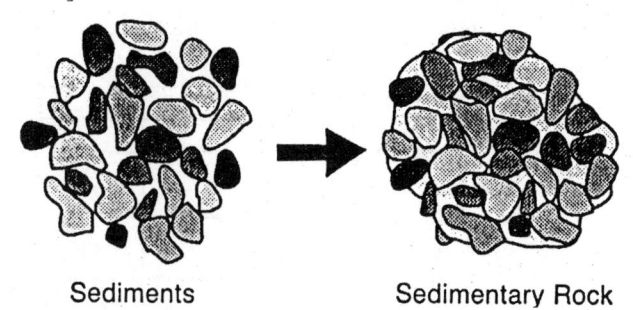

Sediments → Sedimentary Rock

16 The formation of which sedimentary rock is shown in the diagram?
    1 conglomerate   3 siltstone
    2 sandstone   4 shale

17 Which two processes formed this rock?
    1 folding and faulting
    2 melting and solidification
    3 compaction and cementation
    4 heating and application of pressure

18 A conglomerate contains pebbles of limestone, sandstone, and granite. Based on this information, which inference about the pebbles in the conglomerate is most accurate?
    1 They had various origins.
    2 They came from other conglomerates.
    3 They are all the same age.
    4 They were eroded quickly.

19 According to the *Reference Tables*, which sedimentary rock most likely formed as an evaporite?
    1 siltstone   3 gypsum
    2 conglomerate   4 shale

20 What rock is formed by the compression and cementation of sediments with particle sizes ranging from 0.08 to 0.1 centimeter?
    1 basalt   3 granite
    2 conglomerate   4 sandstone

21 Most of the surface bedrock of New York State formed as a direct result of
    1 volcanic activity
    2 spreading of the ocean floor
    3 melting and solidification
    4 compaction and cementation

22 Which characteristic of a sedimentary rock would furnish the greatest amount of information about the environment in which the rock was deposited?
    1 hardness   3 density
    2 color   4 texture

23 A Rock layer is composed of nonuniform particle sizes ranging in diameter from 0.9 to 23 centimeters. According to the *Reference Tables*, this rock layer should be represented by which symbol?

(1) 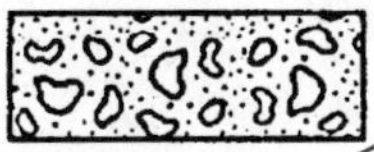   (3) 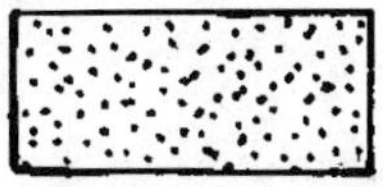
(2)   (4)

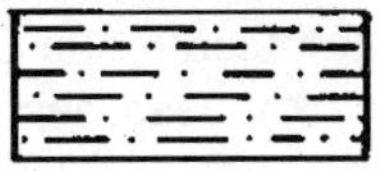

24 The geologic cross section below represents an igneous intrusion into layers of rock strata.

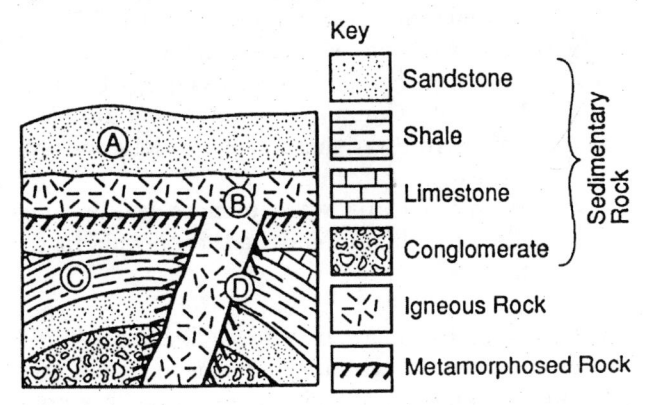

Key
- Sandstone
- Shale
- Limestone   } Sedimentary Rock
- Conglomerate
- Igneous Rock
- Metamorphosed Rock

Which letter indicates a point where unmelted rock changed as a result of increased temperature and pressure?
(1) A   (3) C
(2) B   (4) D

25 Which rocks would most likely be separated by a transition zone of altered rock (metamorphic rock)?
1 sandstone and limestone
2 granite and limestone
3 shale and sandstone
4 conglomerate and siltstone

26 A geologist was asked to identify two rocks composed of calcite. One rock appeared to be sedimentary, but the other showed evidence of having undergone metamorphism. Based on this information, the two rocks could be
1 quartz and limestone
2 gypsum and marble
3 limestone and marble
4 quartzite and gypsum

27 Which rock is usually composed of several different minerals?
1 rock gypsum   3 chemical limestone
2 quartzite   4 gneiss

Base your answers to questions 28 and 29 on the diagram below which shows the grade of metamorphism of felsic rocks and its relationship to temperature and depth. [Refer to the *Earth Science Reference Tables*.]

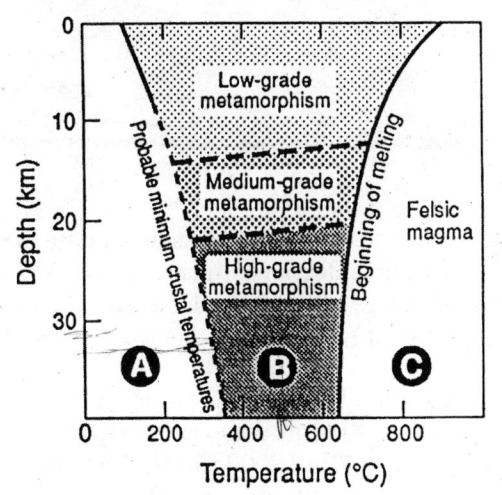

28 Which rock could be formed at a temperature of 500°C and a depth of 25 kilometers
1 gneiss   3 granite
2 schist   4 slate

29 Which igneous rock will most likely be formed if magma represented in region C is forced to the surface and cooled quickly?
1 gabbro
2 basalt
3 rhyolite
4 diorite

Base your answers to questions 30 and 31 on the diagram at the right which represents a cross section of a portion of the Earth's crust. Letters A through D indicate layers of different rock types.

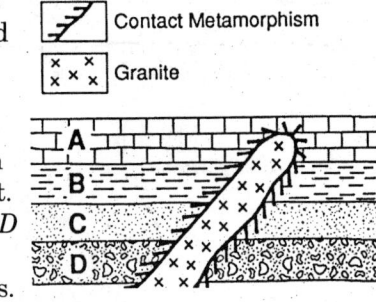

Key
- Contact Metamorphism
- Granite

30 According to the *Earth Science Reference Tables*, which rock layer is most likely to have the greatest range of particle sizes?
(1) A   (3) C
(2) B   (4) D

31 Which rock would most likely result from the metamorphism of the rock in layer C?
1 quartzite
2 marble
3 slate
4 metaconglomerate

TOPIC A — ROCKS, MINERALS, AND RESOURCES — N&N©

32 Which statement is best supported by the graph shown below?

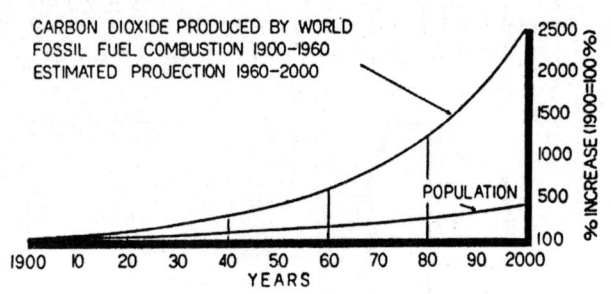

CARBON DIOXIDE PRODUCED BY WORLD FOSSIL FUEL COMBUSTION 1900-1960 ESTIMATED PROJECTION 1960-2000

1 From 1960 to 2000 it is anticipated that there will be a decrease in the use of fossil fuels.
2 From 1900 to 1960, the average person continuously used a greater quantity of fossil fuel.
3 By 1980 the world population was approximately 400 million.
4 From 1970 to 2000 the world population will remain relatively constant.

Base your answers to questions 33 and 34 on the world map and graph below. The graph shows the relationship between average annual income and average annual energy used per person for several countries.

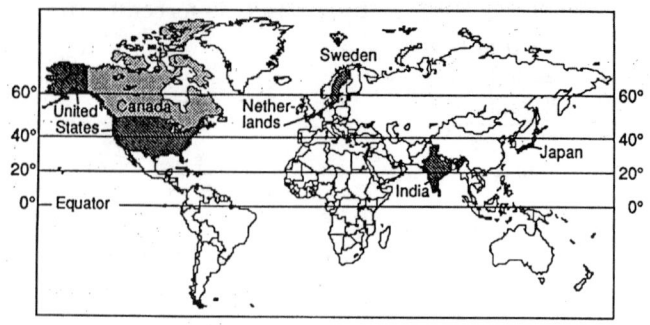

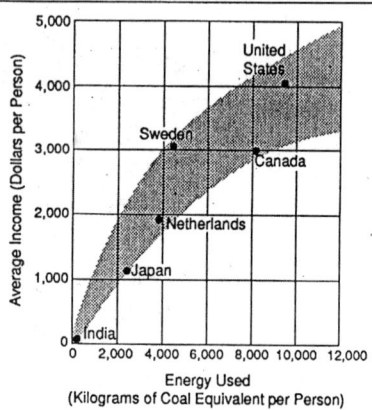

33 In which country is the average energy used per person the greatest?
1 India
2 Netherlands
3 United States
4 Japan

34 According to the graph, the average income of a person in Sweden is about the same as the average income of a person in Canada, but the average energy used per person in Sweden is about half the amount used per person in Canada. What is the most likely reason for this difference in energy use?
1 Sweden has more daylight per year than Canada and uses less energy for lighting.
2 Since Sweden's climate is much warmer than Canada's, Sweden uses less energy for heating.
3 Homes in Canada are better insulated than homes in Sweden and require less energy for heating.
4 People in Canada travel more and for longer distances by automobile than people in Sweden.

35 According to the Earth Science Reference Tables, in which New York State landscape region was most of the surface bedrock formed by regional metamorphism?
1 Atlantic Coastal Lowlands
2 Appalachian Uplands
3 Tug Hill Plateau
4 Adirondack Highlands

Unit Three
Dynamic Crust

Vocabulary To Be Understood

Bench Mark	Fault	Richter Scale
Continental Crust	Focus of an Earthquake	Seismic Waves, Seismograph
Continental Drift	Folded Strata	Strata
Convergent Plate Boundary	Fossils	Subsidence
Crust	Lithosphere	Tilted strata
Crustal Plates	Mid-Ocean Ridge	Transform Plate Boundary
Devergent Plate Boundary	Modified Mercalli Scale	Tsunami
Earthquake	Oceanic Crust	
Epicenter	Plate Tectonics	

A. CRUSTAL ACTIVITIES

The solid rock outer zone of the Earth is known as the **crust**. This crust is in a constant state of change, and there is much evidence to support the idea that the Earth's surface has always been changing.

What are some of the results of earthquakes and volcanoes?

Some of these changes can be directly observed, such as the results of earthquakes, crustal movements (both horizontal and vertical) along fault zones (and volcanoes on the surface of the Earth. Other evidence indicates that parts of the Earth's crust have been moving to different locations for billions of years.

AREAS OF CRUSTAL ACTIVITY

Crustal activities, such as earthquakes and volcanoes, occur for the most part in specific zones or regions of the Earth. Generally, they are along the borders of continents and oceans. The most active zones usually follow the continental borders of the Pacific Ocean, mid-ocean ridges, and across southern Europe and the Middle East into Asia. These zones mark the boundaries or edges of large pieces of the Earth's crust called **crustal plates**.

An **earthquake** is the sudden trembling or shaking of the ground, usually caused by movement along a

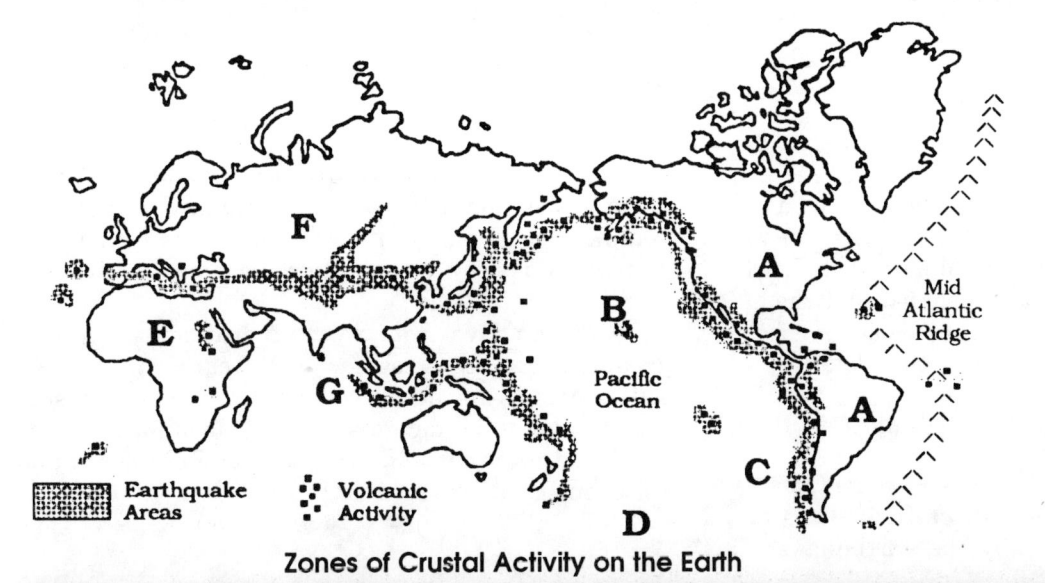

Key
Major Plate Areas:

A – American
B – Pacific
C – Nazca
D – Antarctic
E – African
F – Eurasian
G – Indian

Zones of Crustal Activity on the Earth

Unit Three – Dynamic Crust – N&N©   Page 51

break or fault in rock layers, releasing stress that has built up in the ground. When an earthquake occurs, **seismic waves** are generated and move out in all directions from the **focus** (point of origin). The point on the surface of the Earth directly above the focus is known as the **epicenter** of the earthquake. These seismic waves are detected and registered on delicate sensing instruments called **seismographs**.

The velocity of earthquake waves varies according to the density of the material through which they are traveling. The greater the density of the material, the greater their velocity. As they travel through materials of varying density, the variation in their velocity causes the waves to be bent or refracted. Since the density of the Earth gradually increases with depth, earthquake waves tend to increase in their velocity and are continually refracted as they travel down into the Earth.

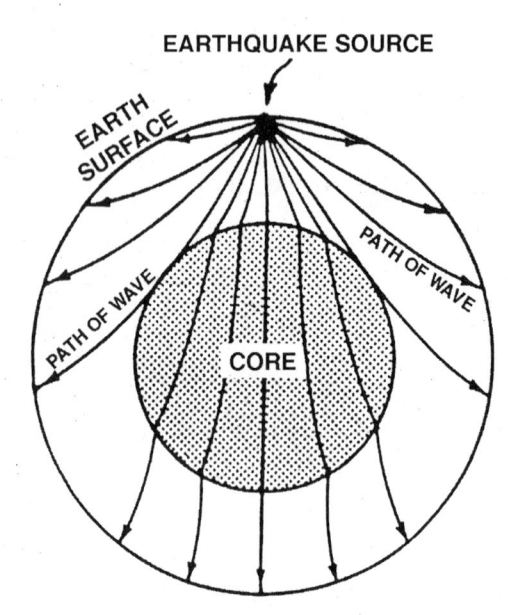

Refracted Waves in Earth's Interior
This diagram represents a cross section of the Earth showing the paths of earthquake waves from a single earthquake source.

Studying the transmission of these waves as they travel through the Earth allows scientists to make inferences about the composition and interior structure of the Earth. Scientists have concluded that the outer core of the Earth is liquid due to behavior of these waves (see *Extended Topic B* for further detail).

MEASURING EARTHQUAKE STRENGTH

The strength of earthquakes is estimated in a variety of ways. The **Modified Mercalli Scale** is based upon the damage inflicted by the earthquake. This **intensity scale** ranges from I to XII with I being felt by few people to XII resulting in total devastation to nearly all structures. Although this scale is still used to help locate the centers of earthquakes, it is not very precise, because building design and the nature of the surface materials may cause great variations in the resulting damage.

The **Richter Scale** is a **magnitude scale** used to describe the amount of energy released by an earthquake. Richter scale magnitudes range from 0 to 9. Those that are less than 2.5 are not usually felt by people. Each step up the magnitude scale indicates a release of 32 times more energy than the previous step. A magnitude 6 earthquake releases 32 times more energy than a magnitude 5 earthquake. A magnitude 6 earthquake may cause significant destruction if it occurs in a populated area. Approximately 20 major earthquakes in the magnitude 7.0-7.9 range occur each year and about once every 5-10 years an earthquake of 8.0 or more will devastate a region.

Both earthquakes and volcanoes present geologic hazards to people. Most people are killed or injured in an earthquake by falling debris or collapsing buildings. Unless they have been designed to withstand earthquakes, most buildings cannot withstand the side-to-side shaking resulting from earthquake waves. If the ground beneath the building is made of loose sediment, the shaking causes the sediment to break up and move, causing the building to fall over.

Fire is a big danger in the aftermath of an earthquake since fires often result from ruptured gas pipes. Firefighting equipment often cannot reach the fires because the streets are blocked with fallen debris and water lines supplying the hydrants have been broken by the shaking ground.

Why are earthquakes and volcanoes important?

Large submarine earthquakes or those that occur along a coastline may result in **tsunamis** or **seismic sea waves**. These giant waves have been known to completely destroy communities along low-lying coasts.

Although lava from volcanic eruptions may produce new land whose surface will weather into fertile soil, people living in actively volcanic areas are subjected to potential dangers. Flowing lava may destroy developed property, and violent explosions of ash and poisonous gases are an ever present danger.

B. DYNAMIC LITHOSPHERE

The minor crustal changes include the deformation and displacement of strata and fossils.

What evidence suggests that the lithosphere is dynamic?

DEFORMED ROCK STRATA

Sedimentary rocks appear to form in horizontal layers. However, observations of the Earth's surface indicate that the original formations of rock have changed through past Earth movements. These changes include **tilting**, **folding**, **faulting**, and the **displacement** of these layers of rock or **strata**.

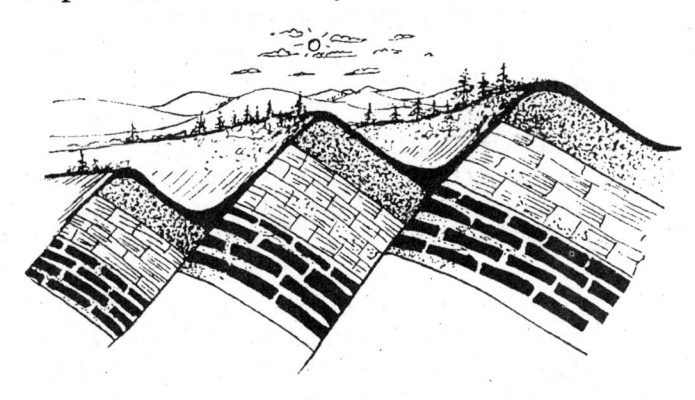

Displaced Fossils. Marine fossils, the remains or imprints of once living ocean organisms such as corals, fish, and other marine animals are found in layers of sedimentary rock in mountains, often thousands of feet above sea level. These marine fossils found at high elevations above sea level suggest the past **uplift of rock strata**.

Subsidence is the sinking or settling of rock strata. Observations at great depths in the oceans suggest that subsidence has occurred in the past, since fossils of shallow water organisms, marine animals that once lived near the ocean surface, have been found in the rocks of the ocean floor.

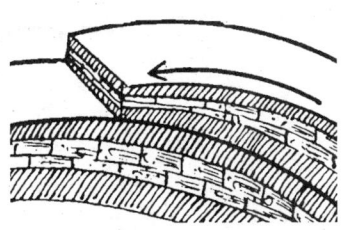

Thrust Faulting
The strata on the right is pushed over the strata on the left.

Displaced Strata. The displacement of strata provides direct evidence of crustal movement. These crustal movements may be either horizontal or vertical or a combination of both.

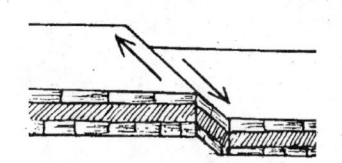

Horizontal Faulting
There is no uplifting of one section over another.

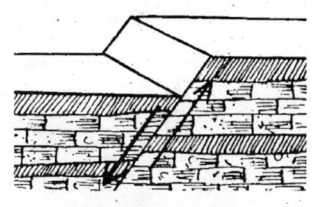

Vertical Faulting
There is no uplifting of one section over another.

Horizontal Displacement (faulting) occurs when the Earth's surface shifts sideways along a transform fault or crack in the crust. **Vertical displacement** (faulting) occurs when a portion of the Earth's surface is either uplifted or subsides, also along a fault. For example, in California, where earthquakes are fairly common result of crustal activity, both types of displacement can be observed along the San Andreas Fault Zone.

During the San Fernando earthquake of 1971, the horizontal displacement along the fault zones caused portions of roads to be displaced so that once road was separated into two. Because of the uplift and subsidence of adjacent land, cliffs and ravines were formed. The actual amount of movement can be calculated by comparing bench marks.

A **bench mark** is a permanent cement or brass marker in the ground indicating elevation above mean sea level and the latitude and longitude of that particular location. Following the San Fernando earthquake some bench marks were displaced several meters vertically and horizontally.

C. PLATE TECTONICS

Plate tectonics is the theory that the Earth's lithosphere is made of a number of solid pieces, called **plates**, and these plates move in relation to each other. The term "tectonics" refers to the forces that deform the Earth's crust. The **lithosphere** is made of the crust and the rigid upper part of the mantle. Although plate tectonics is a recent idea, it incorporates the earlier idea of **continental drift** put forth by Alfred Wegener in 1915. Wegener noted that the present continents appear to fit together as fragments of an originally larger landmass, much the same way the pieces of a jigsaw puzzle fit together. This is especially true if the edge of the continental shelves are used as the boundaries.

In the past, scientists discounted Wegener's idea, because they could not explain a mechanism that could move continents. However, over the years new evidence

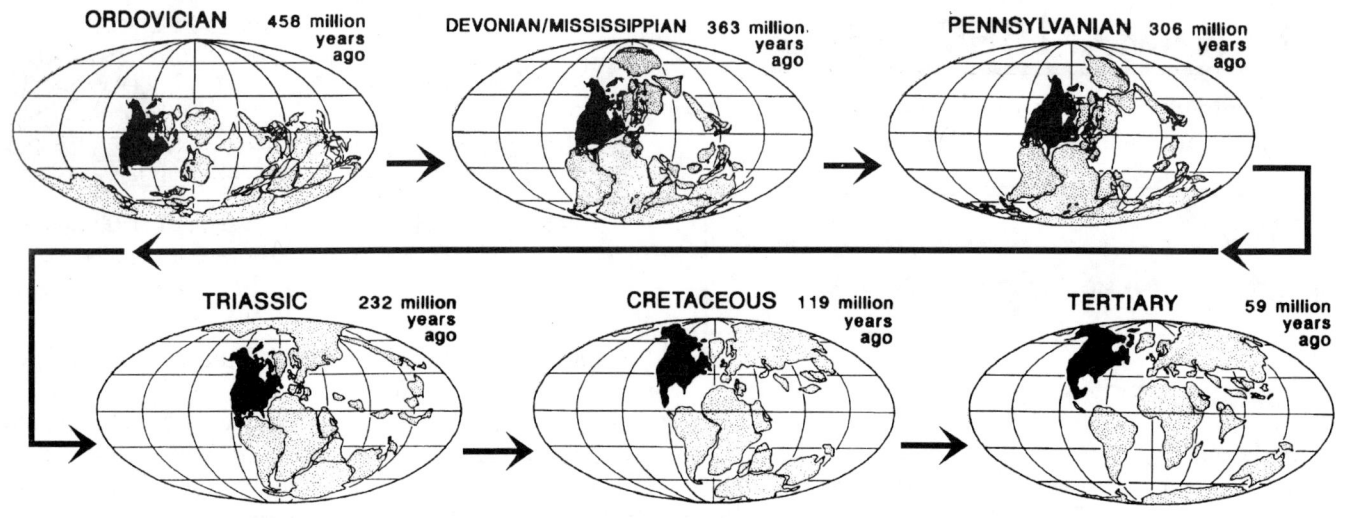

Inferred Position of Earth's Landmasses
Oldest inference is Ordovician (458 million years ago) to most recent is Tertiary (59 million years ago).

has been collected that indicates that approximately 200 million years ago, the major continents were connected, and that since that time they have been moving generally apart (see illustration above).

The lithospheric plates are various sizes, with some made only of ocean crust and others carrying continents on them. Three kinds of plate motion are associated with plate boundaries, **convergent**, **divergent**, and **transform**.

At **convergent plate boundaries**, plates collide with each other. If an ocean plate collides with a continental plate, the denser ocean plate made of basaltic material dives down (subducts) into the mantle forming a **subduction zone** with an ocean trench formed on the surface. Here, old crust is consumed by the mantle. The overriding continental plate forms mountains like the Andes of South America. If a continental plate collides with another continental plate, the edges of both plates are crumpled up forming folded mountains like the Himalayas of India.

At **divergent plate boundaries**, the plates have moved apart, which allowed heat and magma to flow up from below forming parallel ridges made of new crustal material. This can clearly be seen in the mid-Atlantic Ridge.

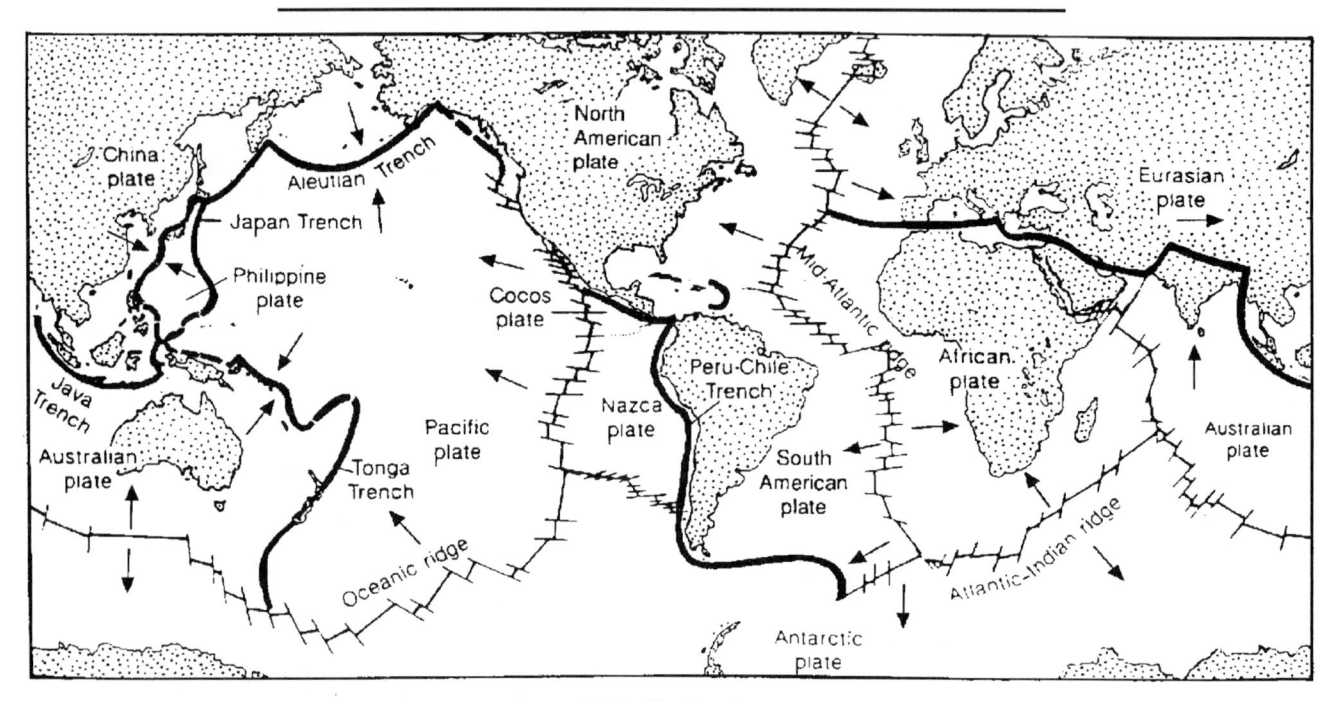

Plate Tectonics
This world map identifies the major plates, trenches, and ridges that are sites of major crustal activity.

Plate Boundaries

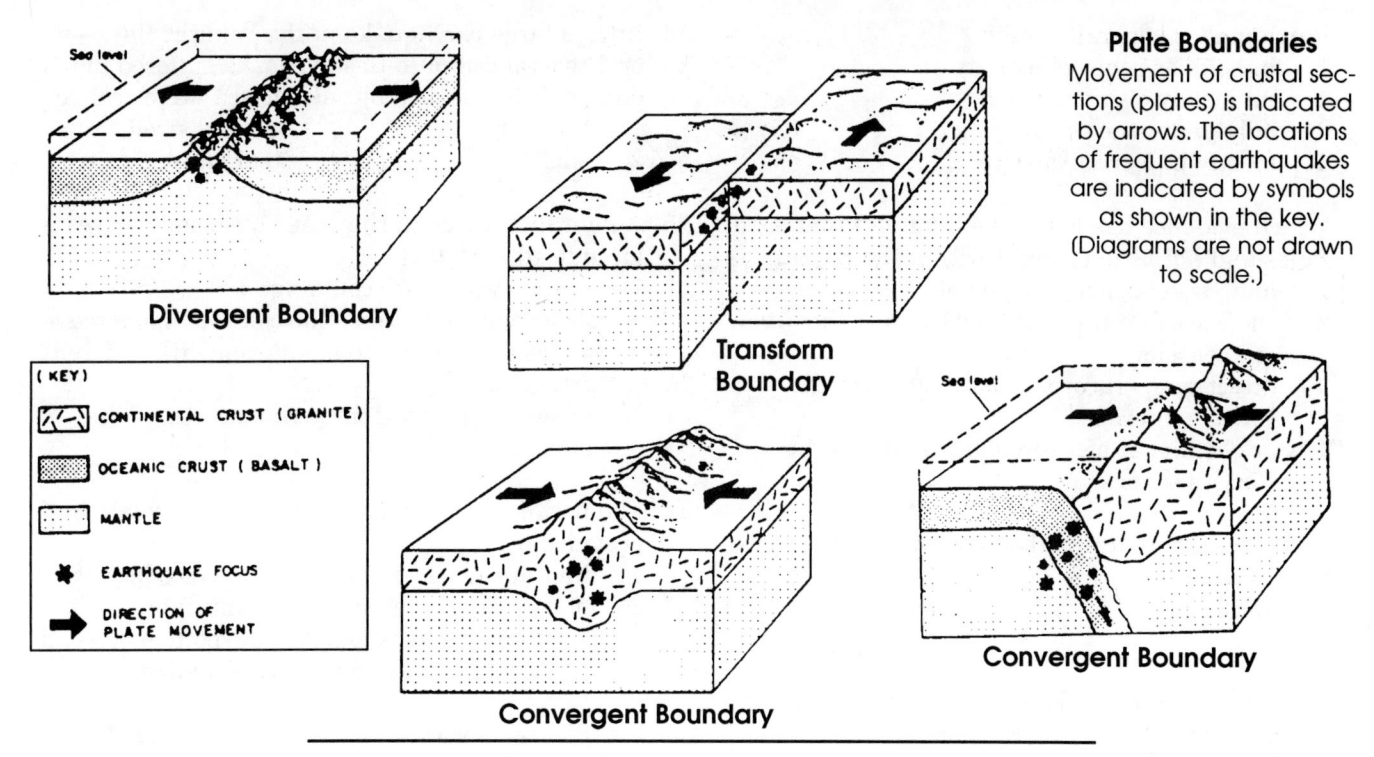

Movement of crustal sections (plates) is indicated by arrows. The locations of frequent earthquakes are indicated by symbols as shown in the key. (Diagrams are not drawn to scale.)

Key:
- Continental Crust (Granite)
- Oceanic Crust (Basalt)
- Mantle
- Earthquake Focus
- Direction of Plate Movement

Divergent Boundary
Transform Boundary
Convergent Boundary
Convergent Boundary

In places like the San Andreas Fault of California, plates grind slowly past each other. This is called a **transform plate boundary**. At this type of boundary crust is neither formed nor consumed.

Although plate motion is only a few centimeters a year, the interactions of the boundaries result in earthquakes, volcanism, and mountain building on a grand scale.

QUESTIONS FOR UNIT 3

1. A large belt of mountain ranges and volcanoes surrounds the Pacific Ocean. Which events are most closely associated with these mountains and volcanoes?
   1. hurricanes
   2. sandstorms
   3. tornadoes
   4. earthquakes

2. Which best describes a major characteristic of both volcanoes and earthquakes?
   1. They are centered at the poles.
   2. They are located in the same geographic areas.
   3. They are related to the formation of glaciers.
   4. They are restricted to the Southern Hemisphere.

3. Recent volcanic activity in different parts of the world support the inference that volcanoes are located mainly in
   1. the centers of landscape regions
   2. the central regions of continents
   3. zones of crustal activity
   4. zones in late stages of erosion

4. Which graph best represents the relationship between volcanic activity and earthquake activity in an area?

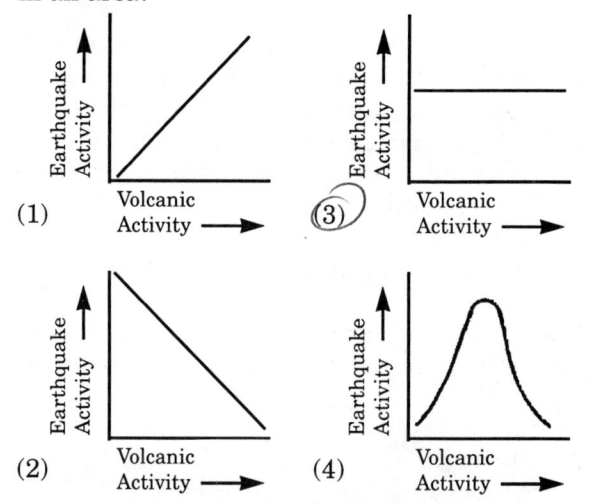

5. The immediate result of a sudden slippage of rocks within the Earth's crust will be
   1. a convergent plate boundary
   2. an earthquake
   3. erosion
   4. the formation of convection currents

6. Earthquakes generally occur
   1. in belts or zones of activity
   2. uniformly over the earth
   3. only under the oceans
   4. mostly in North America

7. Which evidence has led to the inference that solid zones and liquid zones exist within the Earth?
   1. analysis of seismic wave data
   2. the Earth's rotational speed during the different seasons
   3. direct temperature measurements of the Earth's interior
   4. gravitational measurements made at the Earth's surface

8. The place where an earthquake appears to originate is called the
   1. epicenter
   2. core
   3. focus
   4. zenith

9. Placing a seismograph on the Moon has enabled scientists to determine if the Moon has
   1. water
   2. an atmosphere
   3. radioactive rocks
   4. crustal movements

10. Seismic studies of the Moon have helped scientists to make inferences about
    1. water erosion on the Moon
    2. weathering on the Moon's surface
    3. radioactivity of the Moon's surface rocks
    4. the Moon's interior

11. Useful information regarding the composition of the interior of the Earth can be derived from earthquakes because earthquake waves
    1. release materials from within the Earth
    2. travel through the Earth at constant velocity
    3. travel at different rates through different materials
    4. change radioactive decay rates of rocks

12. The behavior of earthquake waves indicate that as depth in the Earth increases, the density generally
    1. decreases
    2. increases
    3. remains the same

13. The most destructive aspect of submarine earthquakes is the production of
    1. offshore trenches
    2. longshore currents
    3. giant sea waves along shores
    4. rip tides

14. After a large earthquake in the Yakutat Bay, several coastal communities in Alaska suffered much damage when struck by seismic sea waves called
    1. breakers
    2. surf
    3. ground swell
    4. tsunamis

15. An abrupt change in the speed of seismic waves is an indication that the
    1. seismic waves are colliding
    2. shear wave overtakes the compressional wave
    3. waves are going into a material with different properties
    4. waves are passing through material of the same density

16. The best evidence of crustal movement would be provided by
    1. dinosaur tracks found in the surface bedrock
    2. marine fossils found on a mountaintop
    3. weathered bedrock found at the bottom of a cliff
    4. ripple marks found in sandy sediment

17. The diagrams show cross sections of exposed bedrock. Which cross section shows the least evidence of crustal movement?

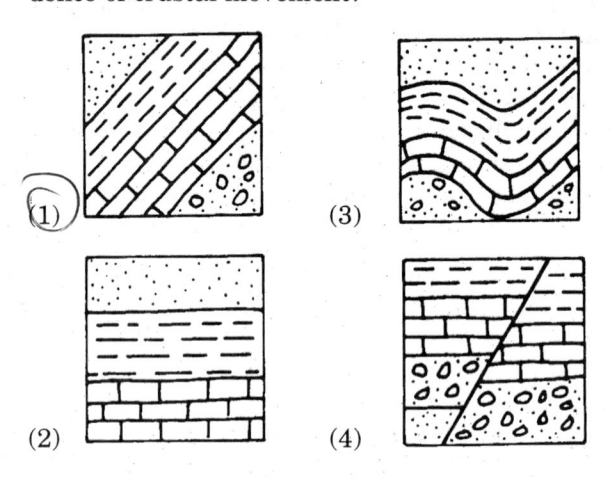

18. In an area of crustal activity, uplift is at the rate of 10 to 30 cm/100 yr. The erosion rate is 5 to 8 cm/100 yr. The elevation in this area is generally
    1. decreasing
    2. increasing
    3. remaining the same

19. What is the best evidence that the Earth's crust has been uplifted?
    1. younger fossils above older fossils in layers of rocks
    2. shallow-water fossils found at great ocean depths
    3. marine fossils found at high elevations above sea level
    4. marine fossils found in horizontal sedimentary layers

20 The landscape shown in the diagram is an area of frequent earthquakes.
This landscape provides evidence for
(1) converging convection cells within the rocks of the mantle
2 movement and displacement of the rocks of the crust
3 density differences in the rocks of the mantle
4 differential erosion of hard and soft rocks of the crust

21 Two geologic surveys of the same area, made 50-years apart, showed that the area had been uplifting 5 centimeters during the interval. If the rate of uplift remains constant, how many years will it take before there is a 50 centimeter change?
1   250 years   (3)   500 years
2   350 years   4   700 years

22 According to the *Earth Science Reference Tables*, during which geologic period were the continents all part of one landmass, with North America and South America joined to Africa?
1   Tertiary   3   Triassic
2   Cretaceous   (4)   Ordovician

23 Which of the following provides evidence for the theory of continental drift?
1   The shapes of the continents suggest that they could "fit together."
(2) All the continents contain igneous rocks.
3   There is a close correspondence between zones of earthquake and volcanic activity.
4   Rocks at the Earth's surface are constantly being eroded.

24 Contact zones between tectonic plates may produce trenches. According to the *Earth Science Reference Tables*, one of these trenches is located at the boundary between which plates?
(1) Australian and Pacific
2   South American and African
3   Australian and Antarctic
4   North American and Eurasian

25 The theory of plate tectonics suggest that
(1) the continents moved due to changes in the Earth's orbital velocity
2   the continent's movements were caused by the Earth's rotation
3   the present-day continents of South America and Africa once fit together like puzzle parts
4   the present-day continents of South America and Africa are moving toward each other

SKILL ASSESSMENTS

Base your answers to questions 1 through 4 on the information and map at the right and on your knowledge of Earth science.

An Earthquake occurred in the southwestern part of the United States. Mercalli-scale intensities were plotted for selected locations on a map, as shown at the right. (As the numerical value of Mercalli ratings increases, the damaging effects of the earthquake waves also increase.)

1   Using an interval of 2 Mercalli units and starting with an isoline representing 2 Mercalli units, draw an accurate isoline map of earthquake intensity.

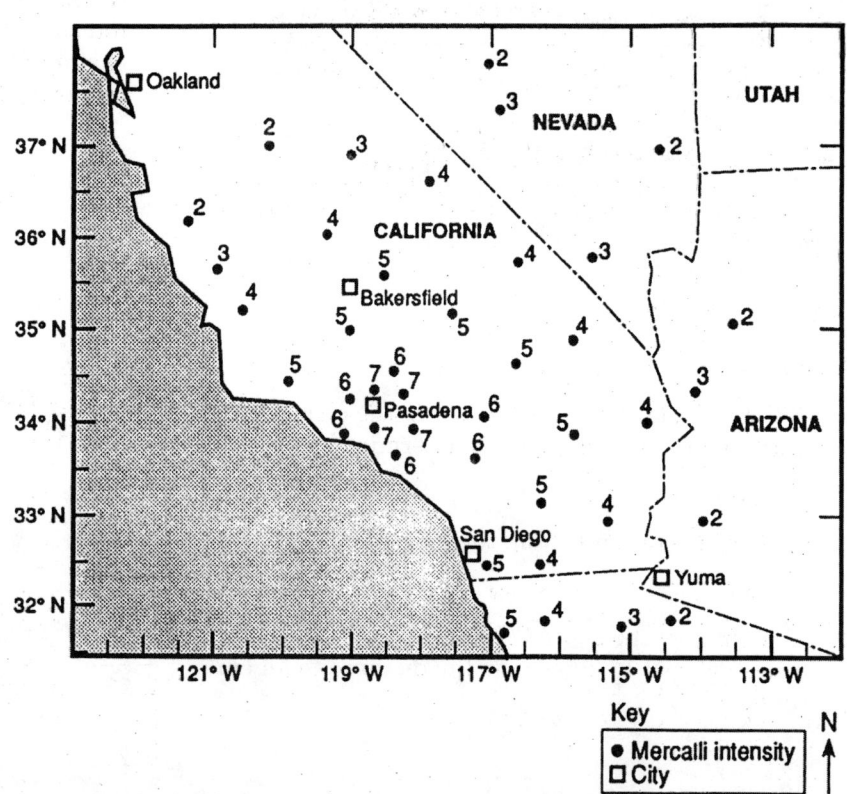

UNIT THREE – DYNAMIC CRUST – N&N©   Page 57

2   State the name of the city that is closest to the earthquake epicenter.

3   State the latitude and longitude of Bakersfield.

4   Using one or more complete sentences, identify the most likely cause of earthquakes that occur in the area shown on the map.

8   The radius of each circle on the map at the right represents the distance from each seismographic station to the epicenter. Label the points representing the three recording stations (A, B, C) to correctly illustrate the position of the seismographic stations relative to the earthquake epicenter.

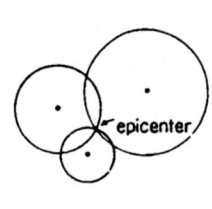

9   A fourth station recorded the same earthquake. The P-wave arrived, but the S-wave did not arrive. What is a possible explanation for the absence of the S-wave?

Base your answers to questions 5 through 10 on your knowledge of Earth science, the *Reference Tables*, and the three seismograms shown below. The seismograms were recorded at earthquake recording stations A, B, and C. The letters P and S on each seismogram indicate the arrival times of the compressional (primary) and shear (secondary) seismic waves.

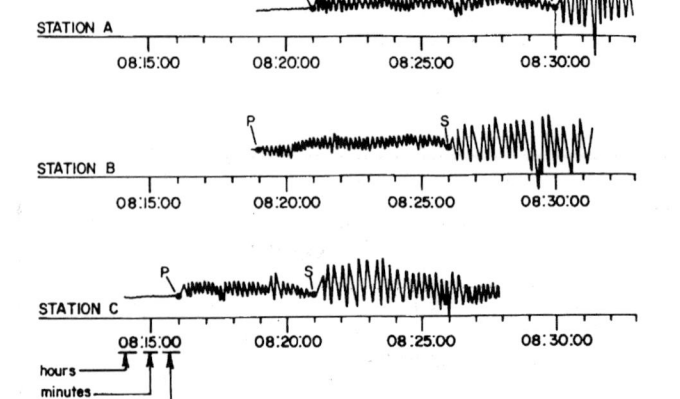

10 The epicenter distance from station A was calculated to be 7,600 kilometers. Approximately how long did the P-wave take to get to station A?

Base your answers to questions 11 through 14 on your knowledge of Earth science, the *Reference Tables*, and the diagrams which represent geologic cross sections of the upper mantle and crust at four different Earth locations. In each diagram, the movement of the crustal sections (plates) is indicated by arrows and the locations of frequent earthquakes are indicated by symbols as shown in the key. [Diagrams are not drawn to scale.]

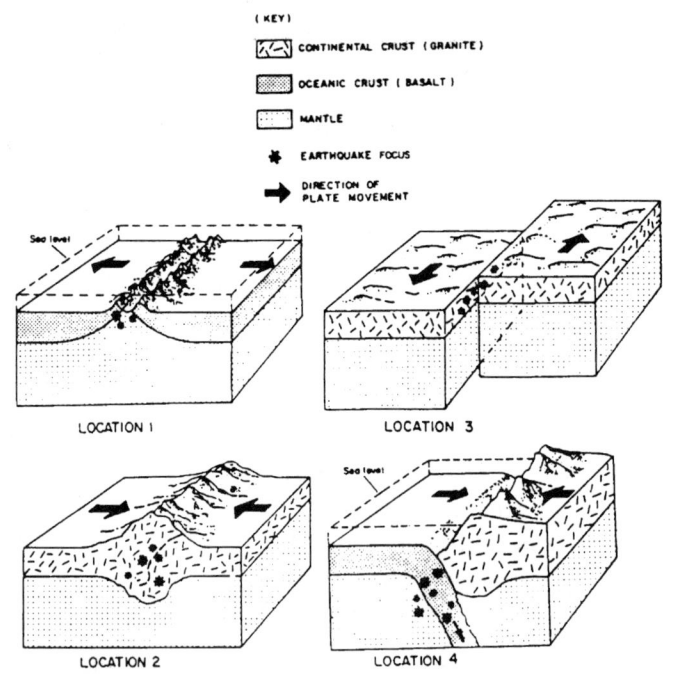

5   Tell which station is farthest from the epicenter and explain how that can be determined.

6   How far is station B from the epicenter?

7   Explain how the speeds of P- and S-waves enable a seismologist to determine the distance to an epicenter.

11 Explain why *Location 1* has primarily shallow focus earthquakes.

12 Which location would have some deep focus earthquakes? What is happening at this location that results in these deep focus earthquakes?

13 Which of the four locations most likely represents formation *X* in the profile shown below?

14 Which location might represent the formation of the Himalaya Mountains? Why?

15 On the diagram below, draw arrows illustrating the motion of the convection currents in the mantle at this location.

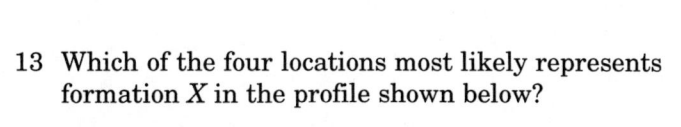

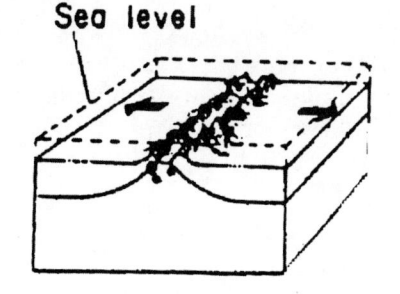

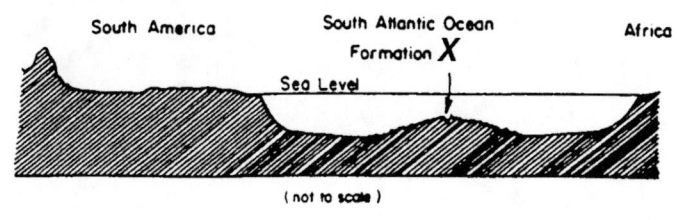

Topic B
Earthquakes and Earth's Interior

Vocabulary To Be Understood

Asthenosphere	Isostasy	Plate Tectonics
Compressional Wave (*P-wave*)	Mantle	Reversal of Magnetic Polarity
Continental Crust	Mantle Convection Cells	Sea (Ocean) – Floor Spreading
Continental Drift	Mid-Ocean Ridge	Secondary Wave (*S-wave*)
Crust	Moho	Seismic Waves
Earthquake	Oceanic Crust	Seismograph
Epicenter	Outer Core	Shear Wave (Secondary, *S-wave*)
Inner Core	Primary Wave (*P-wave*)	Shadow Zone

A. EARTHQUAKE WAVES

Earthquakes generate several kinds of seos,oc waves. Two types of earthquake waves that travel through the Earth are **compressional** and **shear** waves. **Compressional waves**, also called **Primary** or ***P-waves***, cause the material through which they pass to vibrate in the same direction as the wave is traveling. **Shear waves**, also called **Secondary** or ***S-waves***, cause the material through which they pass to vibrate at right angles to the direction in which the wave is traveling.

How can we learn about earthquakes?

Long waves or ***L-waves*** are waves that ripple the surface of the Earth causing the damage associated with earthquakes.

VELOCITIES OF WAVES

When traveling in the same material, primary waves travel at a greater velocity than secondary waves. Therefore, a seismograph records the primary waves before the secondary waves arrive. A single seismogram showing the arrival times of the primary and secondary waves may be used to determine the distance to the earthquake and its time of origin. By calculating the difference in arrival time between the primary and secondary waves on the travel-time graph in the *Earth Science Reference Tables*, the distance from the seismographic station to the epicenter may be determined.

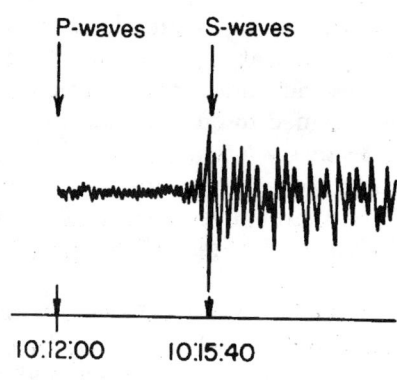

P-waves   S-waves

10:12:00   10:15:40

Primary Compressional Wave

Particle motion

Direction of wave travel

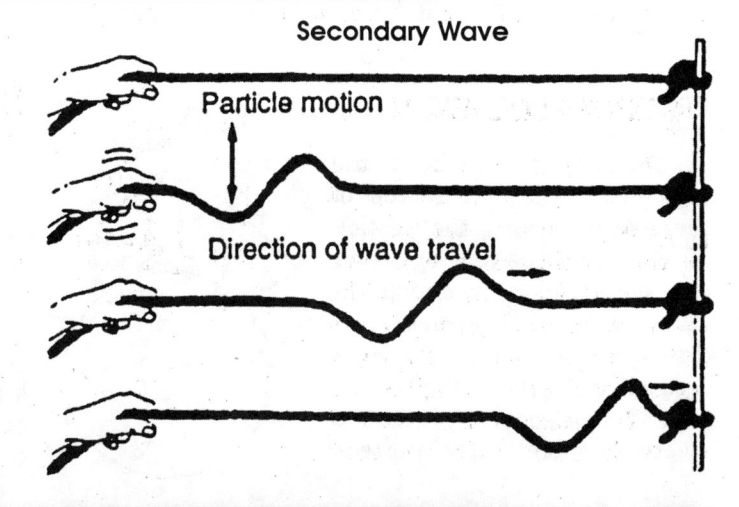

Secondary Wave

Particle motion

Direction of wave travel

Earthquake P-wave and S-wave Travel Time

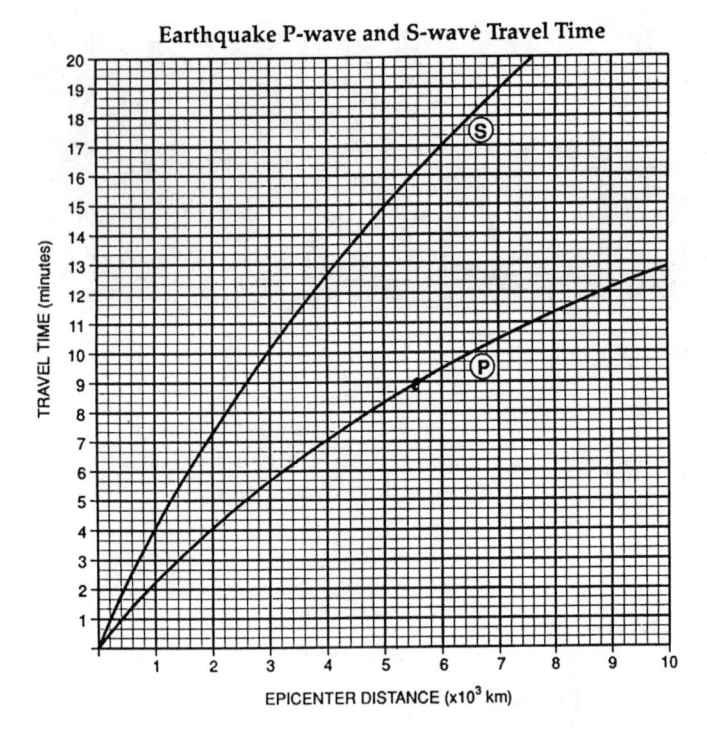

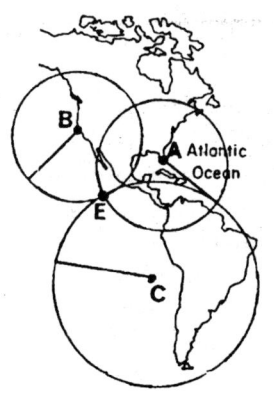

TRANSMISSION OF WAVES

Compressional or *P-waves* are transmitted through solids, liquids and gases, but *S-waves* are only transmitted through solids. This difference provides valuable information for scientists about the interior structure of the Earth. *S-waves* that penetrate the Earth to the depth of the outer core (2900 km) disappear. Since these waves are not transmitted by the outer core, the material of the outer core is assumed to be liquid.

Earthquakes generate *P-* and *S-waves* that move out from the earthquake focus in all directions. The seismographs that are located within 102° from the epicenter record both the *P-* and *S-waves*. Those seismographs that are farther away do not record the *S-waves* because they are not transmitted through the core. In a region known as the **shadow zone**, a band that runs from approximately 102° to 143° away from the earthquakes' epicenter (the opposite side of the Earth) neither the *P-* nor the *S-waves* are recorded. This is due to refraction of the *P-waves* out of that region.

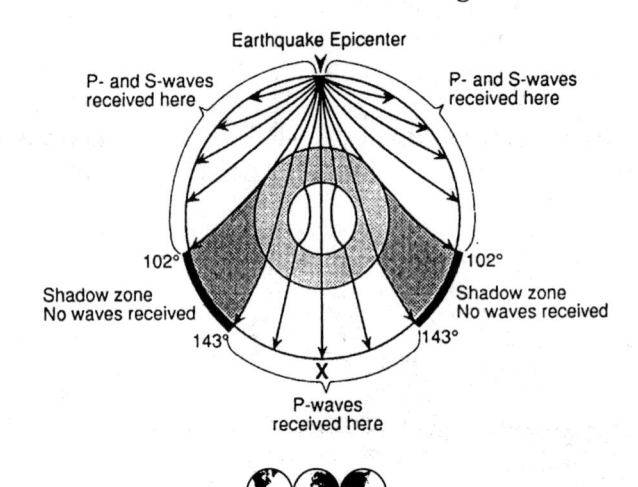

How can we investigate the structure of the Earth?

ORIGIN OF THE TIME OF AN EARTHQUAKE

The origin time of an earthquake can be inferred from the evidence of the epicenter distance and the travel time of the *P-waves* and *S-waves* (see the *Earthquake S-wave and P-wave Time Travel* graph above). The farther the recording station is from the epicenter, the longer it takes for the *P-waves* to reach the station. For example, a recording station receives the *P-wave* at 7:10 a.m. and the *S-wave* 5 minutes and 30 seconds later. If the epicenter distance has been determined to be 4,000 kilometers (determined by the difference – 5.5 minutes – in the arrival time of the *P-* and *S-waves*), then the travel time for the *P-wave* was 7 minutes. The earthquake's time of origin was 7:03 a.m. (7:10 less 07 minutes *P-wave* travel time).

The larger the difference in arrival times of the primary and secondary waves, the greater the distance to the earthquake epicenter. Additional information is needed to determine the direction to the epicenter.

EPICENTER LOCATION

By using the distance to the epicenter from a minimum of three seismograms, the location of the earthquake's epicenter can be calculated. On a map, the distance to each epicenter is drawn as the radius of a circle around each seismographic station. The epicenter (*E*) is located where the three circles intersect.

B. CRUST & INTERIOR PROPERTIES

There are four major Earth zones, three solid zones, and one liquid zone. The **crust**, the **mantle**, and the **inner core** are solid zones. The only liquid zone is the **outer core**.

Crustal Thickness. The crust of the Earth compared to the other zones is relatively thin, only a few kilometers in average depth. The average thickness of the **continental crust** is greater than the average thickness of the **oceanic crust** (under the ocean).

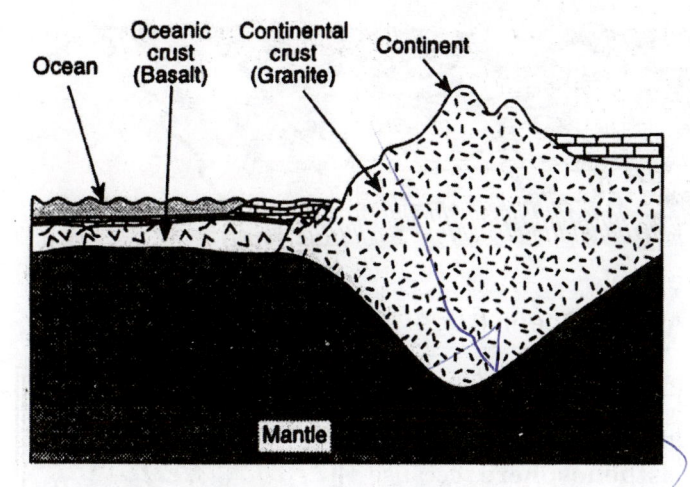

Crustal and Interior Composition. The oceanic and continental crusts have different compositions. The continental crust of the Earth is mainly composed of a low density felsic granitic material; whereas, the ocean crust is mafic or basaltic in nature. The mantle, on which the Earth's crust rests, accounts for the greatest part of the volume of the Earth. The crust-mantle boundary is called the **Mohorovicic Discontinuity** or the **Moho.** Below the mantle is the liquid outer core and the solid inner core. Evidence from the behavior of seismic waves and metallic meteorites suggests that the inner portion of the Earth is mainly a high density combination of the metallic elements iron (Fe) and Nickel (Ni).

What are the properties of the Earth's crust?

Interior Characteristics. The density, temperature, and pressure of the Earth's interior increase with depth (see figure on the Inferred Properties of the Earth's Interior, from the *Earth Science Reference Tables* at the right).

PLATE MOTION

Evidence to support the idea that the continents were together as one large landmass includes more than the shapes of the continents. For example:

1. Rock layers and fossils may be correlated across ocean basins. Many of the same rock types and their mineral composition, as well as the fossil types found in the rocks along the eastern coastline of South America match those along the western coastline of Africa.

2. Some mountain chains appear to be continuous from continent to continent (example, Appalachians and Caledonian).

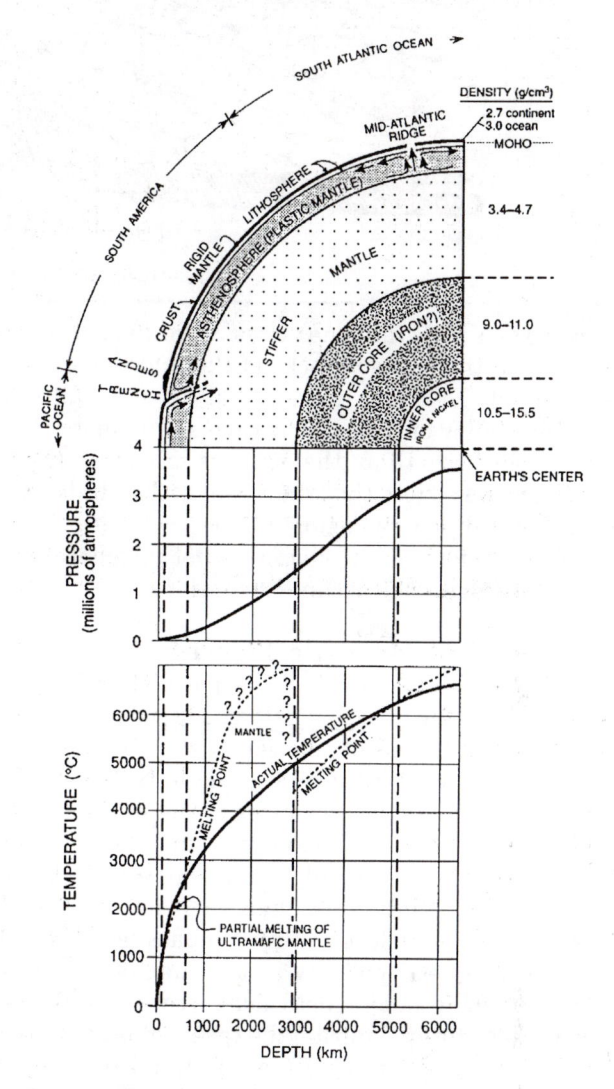

3. Rock and fossil evidence indicates ancient climates much different from those of today (example, Glacial deposits in tropical regions; coal deposits in the Arctic).

4. Rocks of the ocean basins are much younger than continental rocks.

The most conclusive evidence comes from the ocean basins.

SEA (OCEAN) FLOOR SPREADING

There is much evidence to indicate that the ocean floors are spreading out from the mountain regions of the oceans (mid-ocean ridges). The two major pieces of evidence are related to the **age of igneous ocean materials** and the **reversal of magnetic polarity**.

Igneous Ocean Rocks. The ocean crust is primarily composed of basaltic rocks that are formed

TOPIC B – EARTHQUAKES AND EARTH'S INTERIOR – N&N©

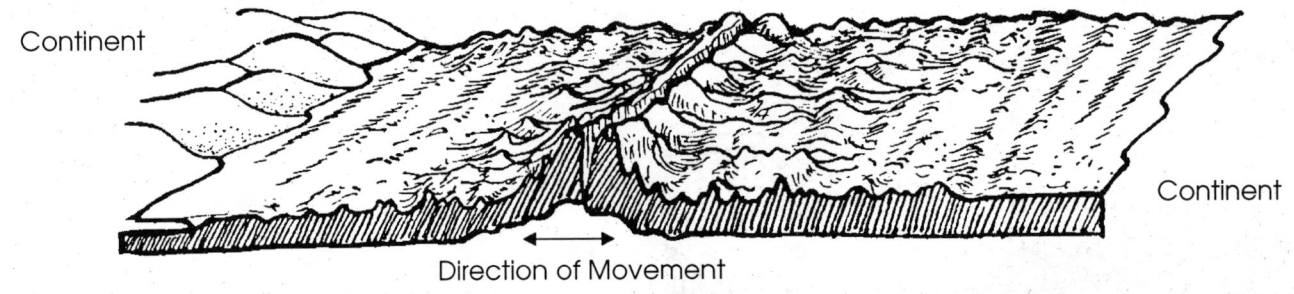

Movement of the Mid-Atlantic Ocean Ridge

when molten rock (magma) rises, solidifies, and crystallizes into the igneous rocks of the mid-ocean ridges (mountains). Evidence shows that igneous material along the center of the oceanic ridges is younger (more recently formed) than the igneous material farther from the ridges. Note that the age of rocks can be accurately determined by using radioactive dating techniques. Therefore, as new ocean crust is generated at mid-ocean ridges, the ocean floor widens.

Reversal of Magnetic Polarity. The strips of igneous (basaltic) rock which lie parallel to the mid-ocean ridges show matched patterns of magnetic reversals. Over the period of thousands of years, the magnetic poles of the Earth reverse (switch) their polarities (the north magnetic pole changes to the south magnetic pole and vice versa). When the basaltic magma flows up in the middle of the ridge and begins to cool, crystals of magnetic minerals align themselves with the Earth's magnetic field like tiny compass needles, thus recording magnetic polarity at the time of formation. Therefore, when the magnetic field is reversed, the new igneous rocks formed during the reversed polarity period have a reversed magnetic orientation from the previously formed rocks. These changes in magnetic orientation are found on both sides of the mid-ocean ridges, indicating that the development of the ocean floor is from the center of mid-ocean ridges outwards.

Rocks found in vastly different locations in the Earth's crust have recorded the position of the Earth's magnetic poles. This strongly supports the continental growth and mountain building concepts that are related to plate tectonics.

MANTLE CONVECTION CELLS

Although forces exist within the Earth that are powerful enough to move the lithospheric plates, the scientific community is not in total agreement on the specific mechanism involved.

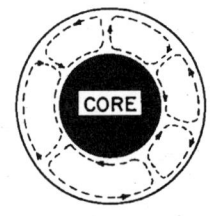

A **convection cell** is a stream of heated material that is moving due to density differences. Evidence suggests that convection cells exist within a part of the mantle called the **asthenosphere** because of the occurrence of heat flow highs in areas of mountain building and heat flow lows in areas of shallow subsiding basins. These convection cells may be part of the driving force which causes the continents to move.

What causes movements of the lithospheric plates?

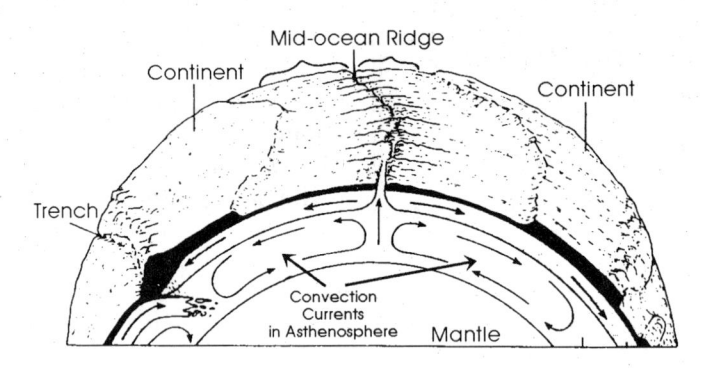

HOT SPOTS

Hot Spots are places on the Earth's surface with unusually high heat flow. Most of these occur along active plate margins, but some are found within the plates. The cause of these hot spots is thought to be plumes of magma rising up from the mantle producing sites of active volcanism. As a plate passes over a hot spot, a chain volcanic mountains forms, like the Hawaiian Islands. The only mountain that remains active is the mountain located directly over the hot spot.

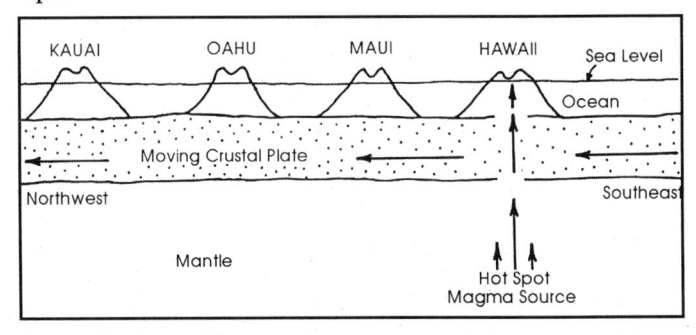

VERTICAL MOVEMENTS

Vertical movements can be related to things such as plate movement or isostasy. **Isostasy** is the condition of equilibrium or balance in the Earth's crust. Since the upper mantle acts like a very dense fluid, the crustal plates "float" on top of it. Any change in one part of the crust is offset by a corresponding change in another part of the crust.

If a piece of crust loses some of its material due to erosion, it becomes lighter and "floats" higher in the mantle. Where the erosional material is deposited the crust is weighted down causing that area to sink lower into the mantle.

QUESTIONS FOR TOPIC B

1. The epicenter of an earthquake is located near Massena, New York. According to the *Reference Tables*, the greatest difference in arrival times of the *P-* and *S-waves* for this earthquake would be recorded in
   1. Albany, NY
   2. Utica, NY
   3. Plattsburgh, NY
   4. Binghamton, NY

2. At a seismic station, the arrival of an earthquake's *P-wave* is recorded three minutes earlier than the arrival of its *S-wave*. Approximately how far from the station is the earthquake's epicenter?
   (1) 700 km
   (2) 1,400 km
   (3) 1,900 km
   (4) 5,600 km

3. The graph below shows the average velocities of *P-waves* traveling through various rock materials.

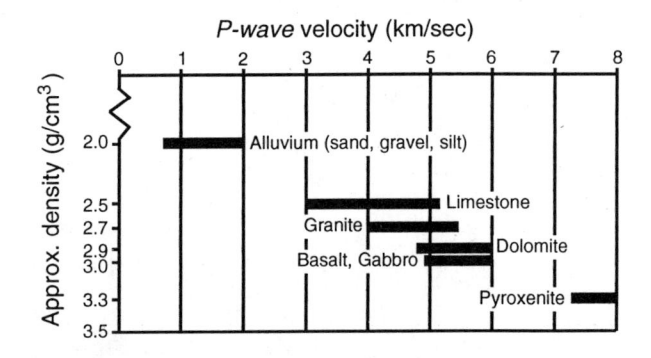

   The graph indicates that *P-waves* generally travel faster through rock materials that
   1. have greater density
   2. have undergone metamorphism
   3. are unconsolidated sediments
   4. are formed from marine-derived sediments

4. An earthquake occurred at 5:00:00 a.m. According to the *Reference Tables*, at what time would the *P-wave* reach a seismic station 3,000 km from the epicenter?
   (1) 5:01:40 a.m.
   (2) 5:04:30 a.m.
   (3) 5:05:40 a.m.
   (4) 5:10:15 a.m.

Base your answers to questions 5 through 9 on the *Earth Science Reference Tables*, the table below, and your knowledge of Earth science. The table shows some of the data collected at two seismic stations, A and B. Some data have been omitted.

Station	Arrival Time of P-Wave	Arrival Time of S-Wave	Difference in Arrival Times of P- and S-Waves	Distance to Epicenter
A	6:02:00 p.m.	6:07:30 p.m.	5 min 30 sec	— km
B	— p.m.	6:11:20 p.m.	7 min 20 sec	5,700 km

5. Which seismogram most accurately represents the arrival of the *P-* and *S-waves* at station *A*?

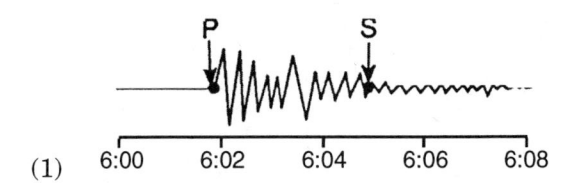

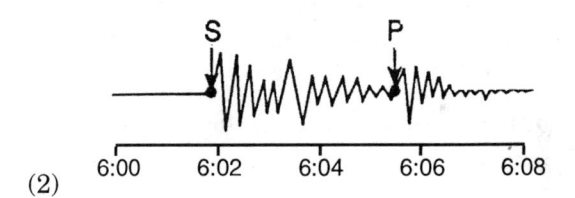

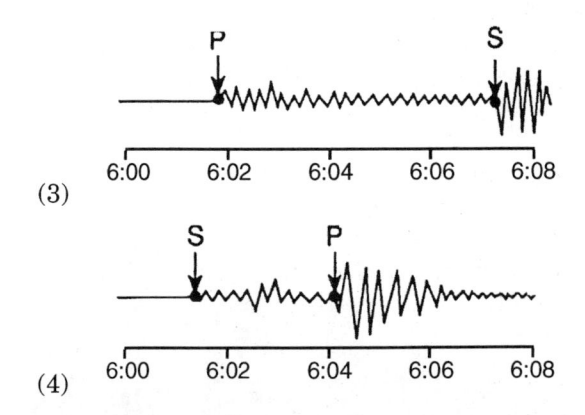

6. What is the approximate distance from the epicenter to station A?
   (1) 1,400 km
   (2) 1,900 km
   (3) 3,000 km
   (4) 4,000 km

7. What was the origin time of this earthquake?
   (1) 5:55:00 p.m.
   (2) 6:00:00 p.m.
   (3) 6:06:00 p.m.
   (4) 6:11:20 p.m.

Which statement best describes the seismic waves received at station B?
1. The P-wave arrived at 6:12 p.m.
2. The S-wave arrived before the P-wave.
3. The P-wave had the greater velocity.
4. The S-wave passed through a fluid before reaching station B.

9. What is the minimum number of *additional* stations from which scientists must collect data in order to locate the epicenter of this earthquake?
(1) 1
(2) 2
(3) 3
(4) 0

10. The circles on the map show the distances from three seismic stations, X, Y, and X, to the epicenter of an earthquake.

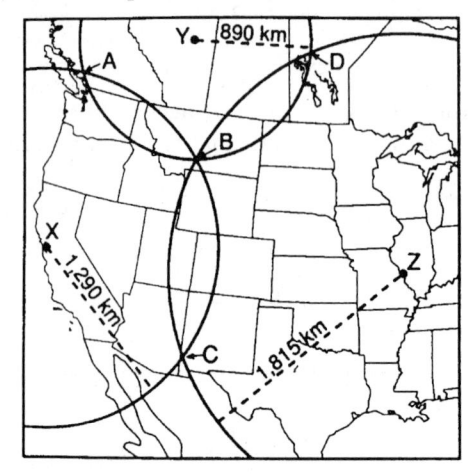

Which location is closest to the earthquake epicenter?
(1) A
(2) B
(3) C
(4) D

Base your answers to questions 11 and 12 on the diagram of the Earth showing the observed pattern of waves recorded after an earthquake.

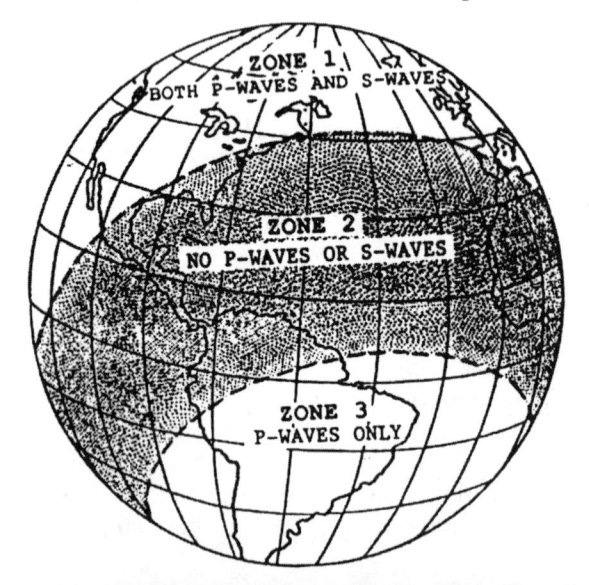

11. The lack of S-waves in Zone 3 can best be explained by the presence within the Earth of
1. density changes
2. mantle convection cells
3. a liquid outer core
4. a solid inner core

12. The location of the epicenter of the earthquake that produced the observed wave pattern most likely is in the
1. crust in Zone 1
2. mantle in Zone 2
3. crust in Zone 3
4. core of the Earth

Base your answers to questions 13 and 14 on your knowledge of Earth science, the *Reference Tables*, and the map which shows three circles used to locate an earthquake epicenter. Five lettered locations, A, B, C, D, and E, are shown as reference points. Epicenter distances from three locations are represented by r_1, r_2, and r_3.

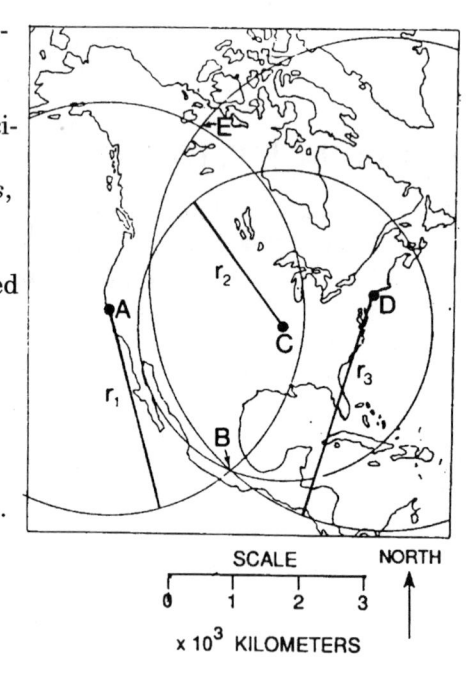

13. Location D is about 3,600 kilometers from the epicenter. What was the S-wave travel time to location D?
(1) 5 minutes 10 seconds
(2) 6 minutes 20 seconds
(3) 7 minutes 30 seconds
(4) 11 minutes 40 seconds

14. At which location could the seismogram at the right have been recorded?
(1) A
(2) B
(3) C
(4) D

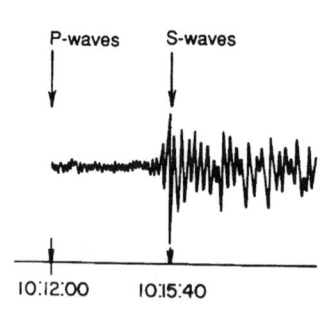

Modified Mercalli Scale of Earthquake Intensity

Intensity Value	Description of Effects
I	Usually detected only by instruments
II	Felt by a few persons at rest, especially on upper floors
III	Hanging objects swing; vibration like passing truck; noticeable indoors
IV	Felt indoors by many, outdoors by few; sensation like heavy truck striking building; parked automobiles rock
V	Felt by nearly everyone; sleepers awakened; liquids disturbed; unstable objects overturned; some dishes and windows broken
VI	Felt by all; many frightened and run outdoors; some heavy furniture moved; glassware broken; books off shelves; damage slight
VII	Difficult to stand; noticed in moving automobiles; damage to some masonry; weak chimneys broken at roofline
VIII	Partial collapse of masonry; chimneys, factory stacks, columns fall; heavy furniture overturned; frame houses moved on foundations

15 The map below represents intensity values of an earthquake according to the *Modified Mercalli Scale of Earthquake Activity*.

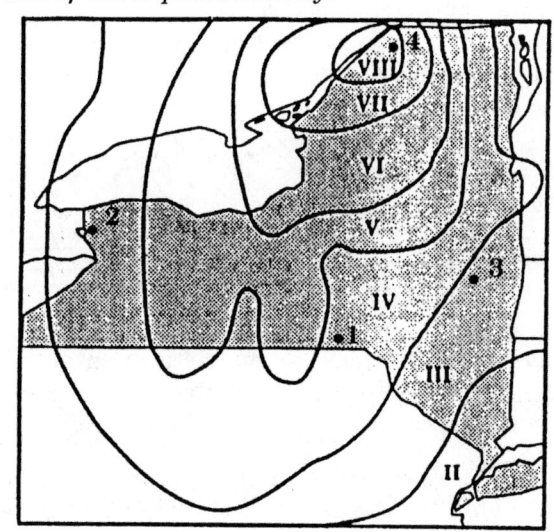

The epicenter of this earthquake is most likely located closest to
1 Binghamton (1)   3 Albany (3)
2 Niagara Falls (2)   4 Mt. Marcy (4)

16 The Mercalli-scale intensity of an earthquake that occurred before the invention of the seismograph can be inferred by
1 observing the local topography
2 reading historical observations of the event
3 measuring the rate at which heat escapes from the Earth
4 measuring the strength of the Earth's gravitational field

17 The cross-sectional diagram below of the Earth shows the paths of seismic waves from an earthquake. Letter X represents the location of a seismic station.

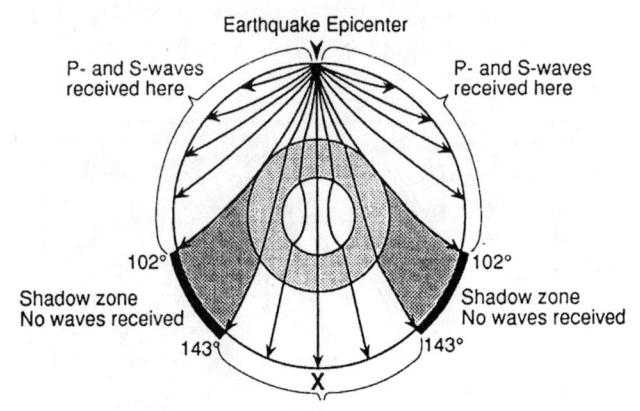

Which statement best explains why station X received only *P-waves*?
1 *S-wave* traveled too slowly for seismographs to detect them.
2 Station X is too far from the focus for *S-waves* to reach.
3 A liquid zone within the Earth stops *S-waves*.
4 *P-waves* and *S-waves* are refracted by the Earth's core.

18 Which information would be most useful for predicting the occurrence of an earthquake at a particular location?
1 elevation
2 climate
3 seismic history
4 number of nearby seismic station

19 Which characteristic of metallic meteorites would give the most useful information about the Earth's core?
1 mass and volume
2 reflectivity and color
3 composition and density
4 temperature and state of matter

20 According to the *Earth Science Reference Tables*, the rock located 1,000 kilometers below the Earth's surface is believed to be
1 liquid at approximately 1,800°C
2 solid at approximately 3200°C
3 liquid at approximately 4,000°C
4 solid at approximately 4,500°C

21 The primary cause of convection currents in the Earth's mantle is believed to be the
1 differences in densities of Earth materials
2 subsidence of the crust
3 occurrence of earthquakes
4 rotation of the Earth

Base your answers to questions 22 through 26 on the map below and your knowledge of Earth science. The map shows crustal plate boundaries located along the Pacific coastline of the United States. The arrows show the general directions in which some of the plates appear to be moving slowly.

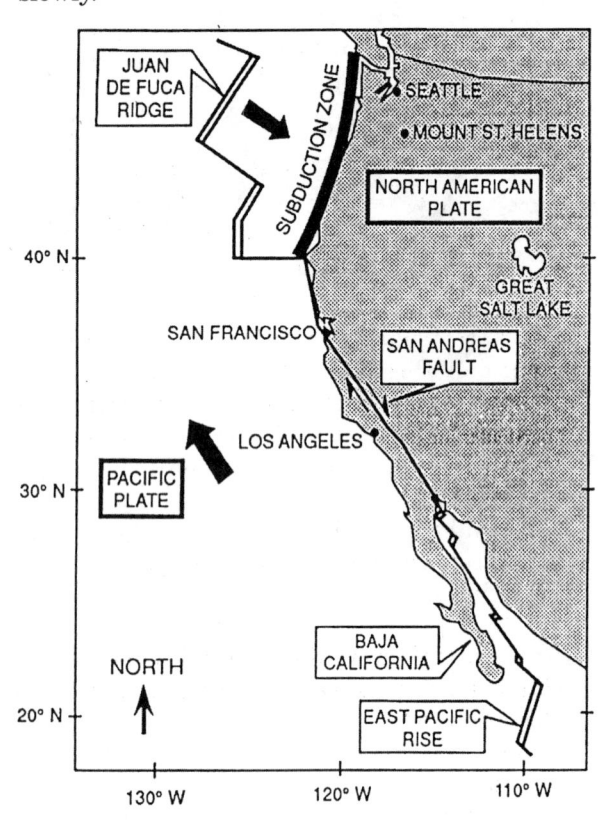

22 Which feature is located at 20° North latitude and 109° West longitude?
1 San Andreas fault
2 East Pacific rise
3 Baja, California
4 Juan de Fuca Ridge

23 Geologic studies of the San Andreas fault indicate that
1 many earthquakes occur along the San Andreas fault
2 the North American plate and the Pacific plate are locked in dynamic equilibrium
3 the subduction zone is the boundary at which the crustal plates are drifting apart
4 the age of the bedrock increases as distance from the fault increases

24 Which features are most often found at crustal plate boundaries like those shown on the map?
1 meandering rivers and warm-water lakes
2 plains and plateaus
3 geysers and glaciers
4 faulted bedrock and volcanoes

25 What would a study of the East Pacific rise (a mid-ocean ridge) indicate about the age of the basaltic bedrock in this area?
1 The bedrock is youngest at the ridge.
2 The bedrock is oldest at the ridge.
3 The bedrock at the ridge is the same age as the bedrock next to the continent.
4 The bedrock at the ridge is the same age as the bedrock at the San Andreas fault.

26 The best way to find the direction of crustal movement along the San Andreas fault is to
1 study the Earth's present magnetic field
2 observe erosion along the continental coastline
3 measure gravitational strength on opposite sides of the fault
4 match displaced rock types from opposite sides of the fault

Base your answers to questions 27 through 31 on the diagram below. The diagram shows a portion of the Earth's oceanic crust in the vicinity of the mid-Atlantic ridge. The stripes in the diagram represent magnetic bands of igneous rock formed in the oceanic crust. The orientation of the Earth's magnetic field at the time of rock formation is shown as arrows within each band. Letters A, B, C, D and E represent locations on the seafloor.

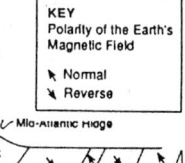

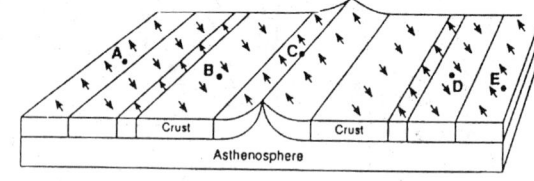

27 According to the diagram, which two locations have rock with the same magnetic orientation?
(1) A and B
(2) A and E
(3) B and E
(4) D and E

28 Heat flow measurements are made at locations A through E. Which graph best represents these measurements?

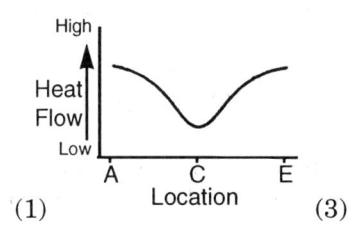

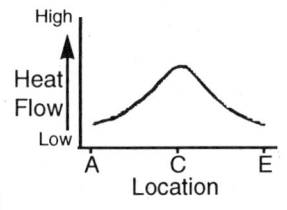

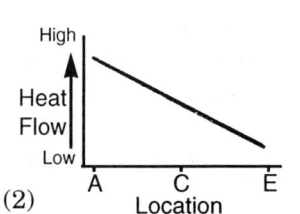

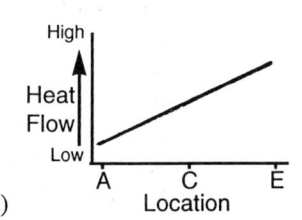

29 Rock samples taken from location *B* would most likely be composed of
1. granite
2. rhyolite
3. conglomerate
4. basalt

30 According to the *Theory of Plate Tectonics* the crust in this area is formed by convection currents in the asthenosphere below the mid-Atlantic ridge. Which cross-sectional diagram best represents the currents that formed the ridge?

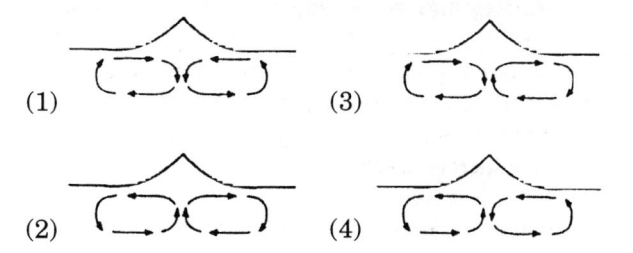

(1) (2) (3) (4)

31 Along a line drawn from location *A* to location *E*, the relative age of the oceanic crust would most likely
1. decrease from *A* to *E*
2. increase from *A* to *E*
3. decrease from *A* to *C* and increase from *C* to *E*
4. increase from *A* to *C* and decrease from *C* to *E*

Base your answers to questions 32 through 34 on the *Earth Science Reference Tables*, the map and information below, and your knowledge of Earth science.

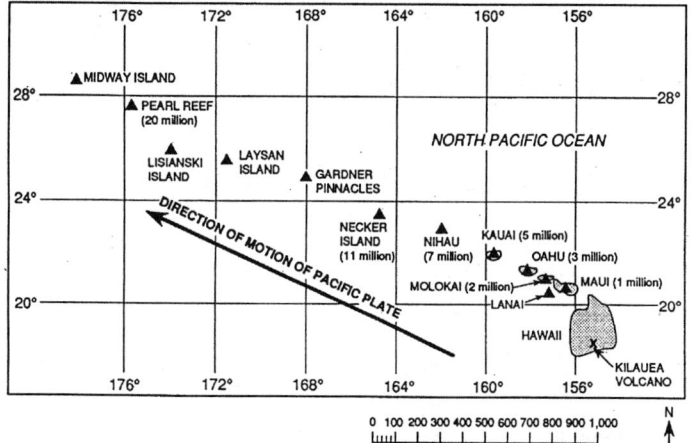

The map shows the location of major islands and coral reefs in the Hawaiian Island chain. Their ages are given in millions of years.

The islands of the Hawaiian chain formed from the same source of molten rock, called a hot plume. The movement of the Pacific Plate over the Hawaiian hot plume created a trail of extinct volcanoes that make up the Hawaiian Islands. The island of Hawaii (lower right) is the most recent island formed. Kilauea is an active volcano located over the plume on the island of Hawaii.

32 Approximately how far has the Pacific Plate moved since Necker Island was located over the hot plume at *X*?
(1) 300 km
(2) 1,100 km
(3) 1,900 km
(4) 2,600 km

33 Approximately how long will it take the noon Sun to appear to move from Kauai to Pearl Reef?
(1) 1 hour
(2) 2 hours
(3) 15 minutes
(4) 45 minutes

34 Which graph shows the general relationship between the age of individual islands in the Hawaiian chain and their distance from the hot plume?

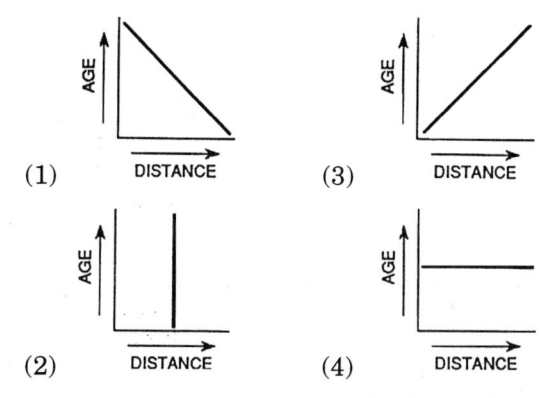

(1) (2) (3) (4)

35 Hot springs on the ocean floor near the mid ocean ridges provide evidence that
1. climate change has melted huge glaciers
2. marine fossils have been uplifted to high elevations
3. meteor craters are found beneath the oceans
4. convection currents exist in the asthenosphere

36 Evidence of subduction exists at the boundary between the
1. African and South American plates
2. Australian and Antarctic plates
3. Pacific and Antarctic plates
4. Nazca and South American plates

37 Mesozoic rocks and fossils found in Australia are most likely to match Mesozoic rocks and fossils found in
1. Europe
2. Antarctica
3. the Atlantic Ocean
4. North America

Base your answers to questions 38 through 40 on the *Earth Science Reference Tables*, the diagram below, and your knowledge of Earth science. The diagram represents three cross sections of the Earth at different locations to a depth of 50 kilometers bellow sea level. The measurements given with each cross section indicate the thickness and the density of the layers.

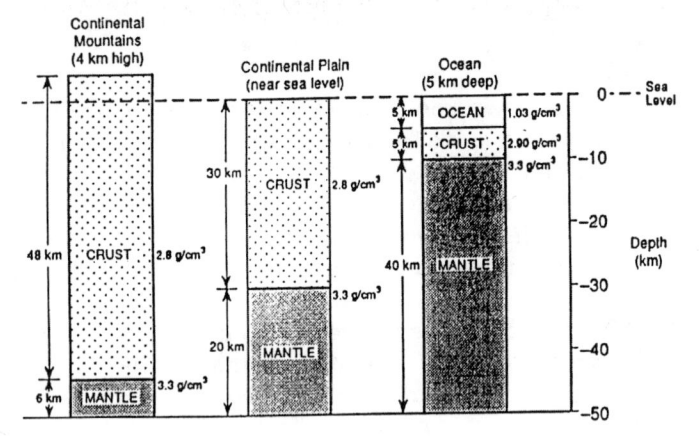

38 In which group are the layers of the Earth arranged in order of increasing average density?
1 mantle, crust, ocean water
2 crust, mantle, ocean water
3 ocean water, mantle, crust
4 ocean water, crust, mantle

39 Compared with the oceanic crust, the continental crust is
1 thinner and less dense
2 thinner and more dense
3 thicker and less dense
4 thicker and more dense

40 The division of the Earth's interior into crust and mantle, as shown in the diagram, is based primarily on the study of
1 radioactive dating
2 seismic waves
3 volcanic eruption
4 gravity measurements

UNIT FOUR — SURFACE PROCESSES AND LANDSCAPES

VOCABULARY TO BE UNDERSTOOD

Annular Drainage	Kettle hole	Settling Rate, Time
Carbonation	Landscape	Soil Conservation
Chemical and Physical Weathering	Leveling (Destructional) Forces	Soil Horizon
Colloid	Moraines	Solution
Dendritic Drainage	Mountain	Sorting of Sediments
Deposition	Outwash Plain	Suspension
Drumlin	Oxidation	Stream Bed
Dynamic Equilibrium	Plain	Stream Discharge
Erosion	Plateau	Transported Sediment of Soil
Escarpment	Precipitation	Transporting Agents
Graded Bedding	Radial Drainage	Trellis Drainage
Horizontal and Vertical Sorting	Residual Sediments or Soil	Uplifting (Constructional) Forces
Hydration	Rock Resistance	Weathering
Kame	Sediment	

A. WEATHERING

The physical and chemical processes that change the characteristics of rocks on the Earth's surface are called **weathering**. In order for weathering to occur, the physical environment of the rocks must change and the rocks must be exposed to the air, water in some form (ice, snow, or liquid), or the acts of humans or other living things. Therefore, weathering is the response of rocks to the change in their environment. Note: The longer the rocks are exposed to these weathering agents, the greater the degree of weathering.

Physical Weathering
Action of Freezing on Rocks

EVIDENCE OF WEATHERING

The Weathering Processes. When rocks are exposed to the hydrosphere and the atmosphere (water, air, and the substances within them) the physical and/or chemical composition and characteristics of the rocks can change. The end products of weathering are generally called **sediments** and are classified as boulders, cobbles, pebbles, sand, silt, clay, colloids, and dissolved particles (ionic minerals).

How is the Earth's crust affected by the environment?

Physical Weathering. In physical weathering, rocks are broken into smaller pieces without changing the chemical nature of the rock. For example, a section of the surface of a mountain may break off because water freezes in a crack in the rock surface, expanding and splitting the rock. The falling rock mass may shatter into boulders and rocks of various sizes. Even though the rock has changed in physical form, it has not changed chemically and still maintains its original composition. The action of wind or water-carried sediment (erosion) may further abrade (wear away by rubbing) the surface of the rocks forming fine particles.

Chemical Weathering. In chemical weathering, rocks are broken, and the rock material itself is *also* changed. Examples of chemical weathering include:

- **oxidation** occurs when oxygen from the air combines with the minerals of the rock to form oxides (for example, iron and oxygen form iron oxide, rust);

Weathering Agents and **Erosional Processes** tend to form caves in Limestone

- **carbonation** occurs when water containing carbonic acid dissolves the minerals of the rock (for example, the action of carbonic acid on limestone), and;

- **hydration** occurs when minerals, such as mica or feldspar, absorb water weaken and crumble to form clay.

CLIMATE AFFECTS WEATHERING

Climate affects both physical and chemical weathering. The chart below shows that climate tends to determine the amount, types, and/or rate of weathering. In cold and moist climates, physical weathering is dominant. In warm and humid climates, chemical weathering is dominant. In general, the more moisture available, the more weathering occurs.

WEATHERING RATES

The rate at which a material weathers varies inversely with the particle size. In equal amounts of the same kind of rock, the smaller the particle size, the greater the rate of weathering. This is due to a greater surface area in contact with the weathering agents.

Rock particles will weather at different rates depending on their mineral composition and its resistance to weathering agents. The harder the mineral, the slower the weathering. The softer the mineral, the faster the weathering.

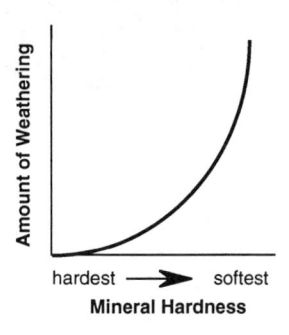

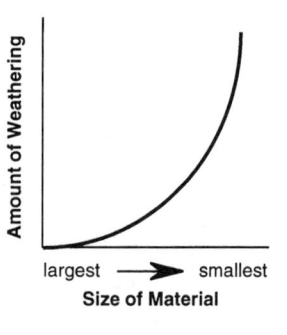

SOIL FORMATION

Soil is a combination of particles of rocks, minerals, and organic matter produced through weathering processes. Soil contains the materials necessary to support various plant and animal life.

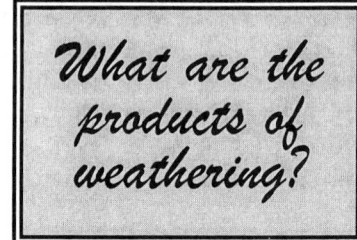

What are the products of weathering?

As a result of the weathering processes and biologic activity, soil horizons (layers) develop. These **soil horizons** vary in depth depending on the amount of weathering, the time over which the weathering occurs, and the climate. Normally, the longer the weathering occurs, the greater the depth of the soil formation.

In addition to weathering, the complex interrelationships of living organisms are significant factors in soil formation. For example, the breakdown and decay of leaves from plants and the life activities of earthworms and other small animals aid in the formation of soil. In addition, the effects caused by frost, rain, and air add to the process of weathering.

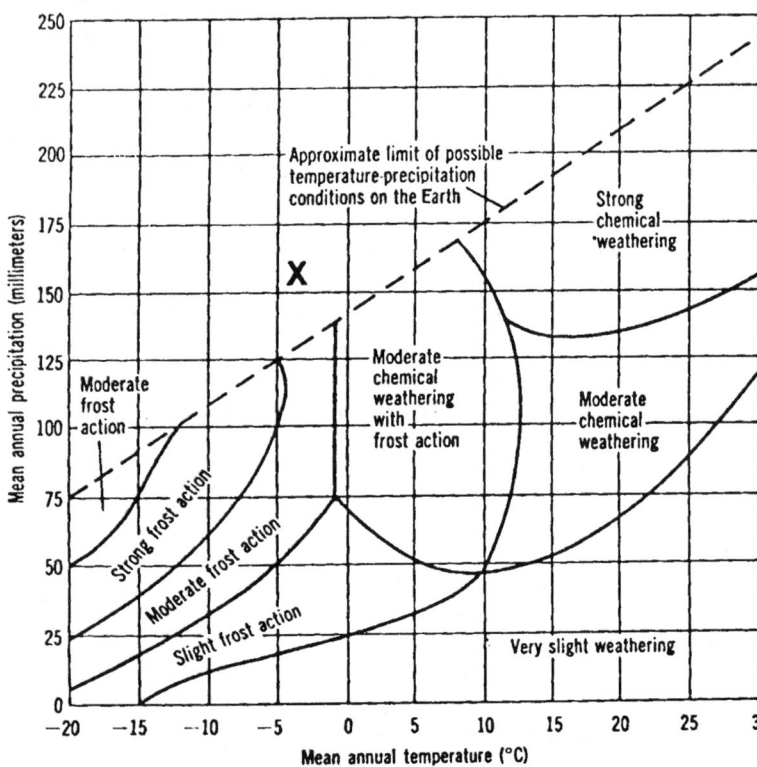

Weathering Determined by Climate

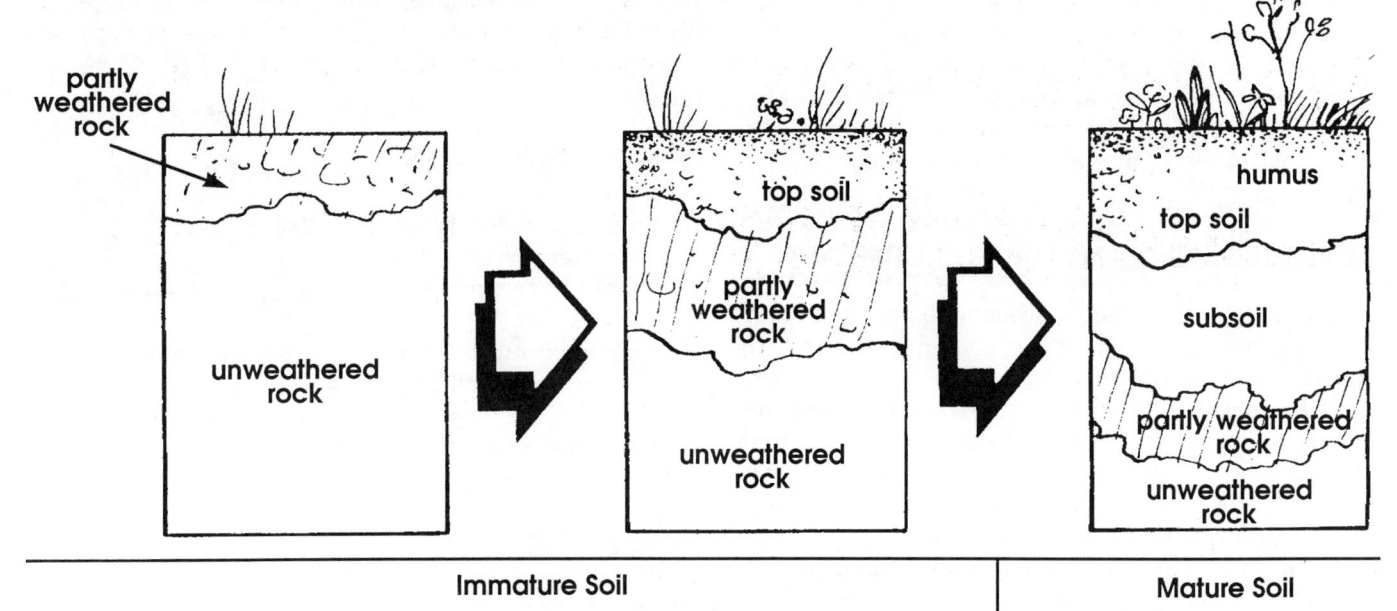

Soil Formation Diagram of Immature Soil to Mature Soil

Immature soil generally contains partially weathered and unweathered rock. Mature soil has various amounts of topsoil and subsoil containing organic matter, in addition to the weathering products, and partially weathered and unweathered rock. Until a soil has developed a subsoil, it is considered to be immature.

Soil associations differ in composition depending on the climate. The lack of large amounts of water in arid regions causes a general decrease in the vegetation and animal populations, thereby decreasing organic materials in the soil and producing only thin or poorly developed soils. Arid soils are often high in mineral salts since they are not dissolved and carried away (leached) by water. The soil of a humid region is usually high in organic materials, low in mineral salts, less sandy, and better capable of supporting large populations of plants and animals.

Different bedrock composition produces different soil composition. Although climate is the most important factor in the determination of soil type, soils formed from parent material of different minerals will have different chemical compositions.

Good soil is more valuable than gold, because fertile soil enables the production of food to support the Earth's growing population. Each year billions of tons of soil are eroded away or spoiled by human activities. Soils may be contaminated by salt from roads and the use of herbicides or other toxic chemicals. Construction, mining, and poor agricultural practices may cause rapid erosion. Since it frequently takes several hundred years to form one centimeter of soil, it is necessary for everyone to protect this essential resource by practicing proven methods of **soil conservation**.

B. EROSION

Erosion refers to the transportation of rock, soil, and mineral particles from one location to another by the action of water, wind, or ice. The motive power behind all of the agents of erosion is **gravity**. It is very important to note that a **transporting system** includes all of the agents involved in erosion and movement, including erosion, the transporting agent, energy, and the material moved.

What evidence suggests that rock materials are transported?

DISPLACED SEDIMENTS

The major evidence of erosion is the displacement of sediments from their source to another location. The beaches that border the coastlines of the United States are the result of transported sediment (sand) from inland mountains, by streams and rivers, to the oceans. The mineral composition of sediments and organic remains (fossils) found with the sediments are often indications of the sources of the erosional products.

Residual sediment is the material that remains at the location of the weathering, and **transported sediment** is the erosional product that has been moved from the source of weathering to another location. In New York State, there are far more transported sediments than there are residual sediments.

PROPERTIES OF TRANSPORTED MATERIALS

The actions that affect erosion and move sediments from one place to another are called **transporting agents**. Transported materials often possess properties distinctive of their transporting medium. Therefore, it is often possible to determine what eroding force and transporting agent moved the sediment.

For example, water generally tends to smooth and round the rock particles. Cobblestones, often used in Colonial times as road beds because of their round and smooth characteristics, were the results of stream erosion.

The longer the water or wind action, the smoother and rounder the sediment becomes. Sediments moved by winds are often pitted, or "frosted," as a result of abrasion during their travel.

Glacial products often have surface scratches due to being pushed and scraped by ice. When the action of gravity alone causes erosion, the sediments are sharp edged and angular. This evidence can be observed along highway road cuts and at the base of mountains.

Erosional and Transporting Agents

FACTORS AFFECTING TRANSPORTATION

Gravity, water, wind, ice, and human activities are the main factors affecting the transportation of the sediments of weathering.

Gravity. Most all transporting systems are caused by gravity. Gravity may also act alone, such as when loose pieces of rock on the side of a slope break away and fall down hill. Gravity also acts with other transporting agents, such as when water transports sediments in a stream.

How are the products of weathering transported?

Water. Running water is the predominant agent of erosion on the Earth. There are a number of factors which affect the way in which the running water of a stream transports sediment.

WATER VELOCITY AND SEDIMENTS

In a stream channel, the average velocity of the running water increases with an increase in the discharge. The **discharge** is the volume of water in the stream at any given location during a specific amount of time. Velocity and discharge of a stream are interdependent. In the spring, the streams usually move faster, due to the greater amount of water volume from melting snow and runoff.

The velocity of the water in a stream is directly proportional to the slope of a stream channel. As the slope of a stream increases, the velocity of the stream increases.

The fastest water flowing in a stream with a straight course is in the center just below the surface where friction is the least. However, in a meandering stream, the fastest water is on the outside of the curve or meander (such as A and C on the diagram on the next page), so that erosion is greatest on the outside of the meanders (again points A

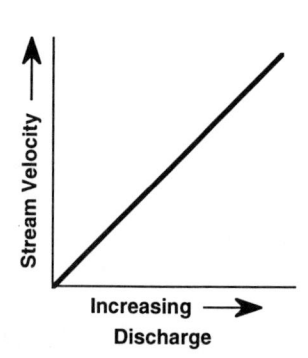

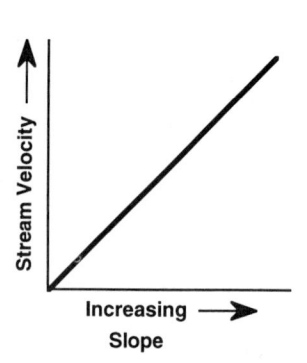

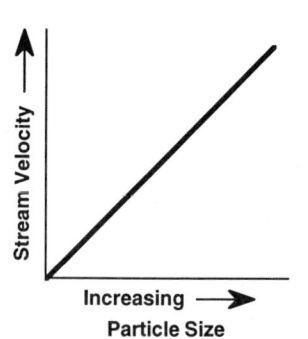

Stream Velocity
Relationship between Stream Velocity and
(1) Water Discharge,
(2) Slope of the Stream Bed, and
(3) the Size of Particles carried by the stream.

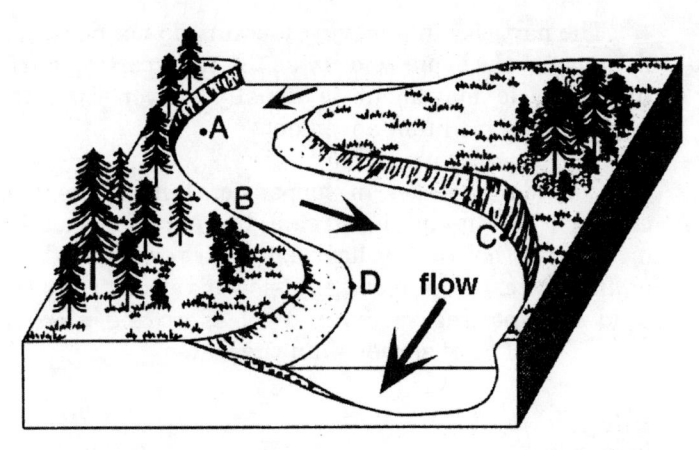

and *C*). Deposition takes place in the slower water on the inside of the meanders (such as at point *D*). Between the meanders is a "change-over" between erosion and deposition (such as point *B*).

Boundaries between erosion and deposition occur at various locations throughout the length of a stream. These interfaces are often found midstream in meanders and where changes in velocity occur near the mouths of streams. Where neither the process of erosion nor deposition is dominant, a state of dynamic equilibrium exists.

As the velocity of the stream water increases, the size of the particles that can be moved by the stream also increases. Streams carry material in **solution** (dissolved), in **suspension** (colloids, clay, silt, and sand), and by rolling, bouncing, or dragging along the stream bed (pebbles, cobbles, and boulders).

Relationship of Transported Particle Size to Water Velocity*

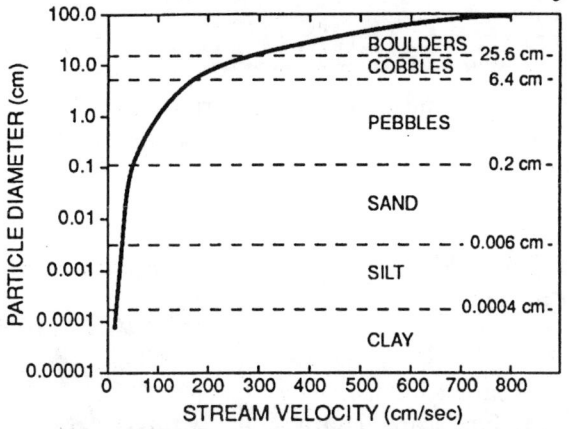

*This generalized graph shows the water velocity needed to maintain, but not start movement. Variations occur due to differences in particle density and shape.

WIND & ICE EROSION

Wind (such as a desert sand storm) and ice (such as a glacier) act as transporting agents for rock materials. The factors involved in wind and ice erosion are similar to the factors which affect running water erosion. The steeper the slope of a mountain, the faster a glacier will move and the more and larger the erosional materials that can be carried. Glaciers can transport the largest sized sediments (boulders). Light winds move only the smallest sediments, but strong winds may carry heavier and larger materials, such as sand, but rarely more than a meter or so above the ground.

EFFECT OF HUMANS

Humans add greatly to the natural processes of land erosion through activities, such as highway and industrial construction, destruction of forests from the careless setting of forest fires and uncontrolled tree cutting without reforestation, strip mining, poor landfill projects, and other such activities.

C. DEPOSITION

Deposition is a part of the **erosional-depositional system**, in which sediments carried by a transporting agent are dropped from the medium. **Deposition** is also called **sedimentation**. Since water is the predominant transporting agent, final deposition often occurs at the end of a stream, where the stream flows into a larger body of water, such as a lake or an ocean. Dissolved ionic minerals and colloids are released by the process of **precipitation**, and larger particles settle out of the transporting medium.

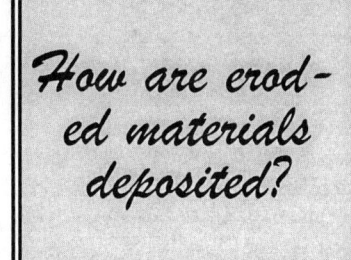

How are eroded materials deposited?

FACTORS AFFECTING DEPOSITION

The major factors that affect the rate of deposition are particle size, shape and density, and the velocity of the transporting medium.

Size. The smaller particles settle more slowly than the larger particles, because as the velocity of a stream slows, the larger and usually heavier particles drop the fastest pulled down by gravity. The smaller particles tend to remain in suspension for longer periods of time, settling more slowly.

Colloids, particles approximately 10^{-4} to 10^{-6} millimeters across, generally remain in suspension indefinitely as long as there is even the slightest movement of the transporting medium.

When there is little or no movement in the transporting medium, sorting in a **quiet medium** (such as still-water or air) produces horizontal layers.

Sorting in a Quiet Medium

Shape. The other factors affecting deposition being equal, a round (spherical) shaped sediment particle will settle out of the transporting medium faster than a flat (disk) shaped particle. The greater the resistance of a flat particle causes it to settle more slowly.

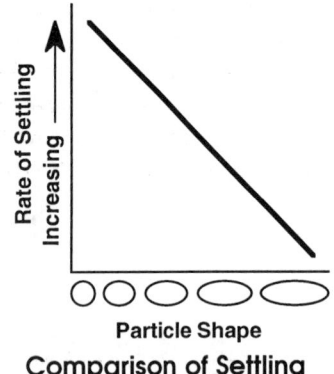

Comparison of Settling Rate & Particle Size

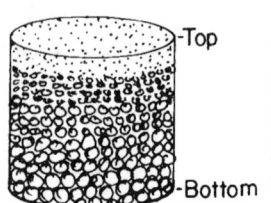

Density. Two particles of the same size and shape, but of different densities will settle at different rates. A high density particle will settle faster than a low density particle, since a high density particle is heavier.

Rapid Deposition in a Quiet Medium

Velocity. As the velocity of a **sediment laden flow** decreases, the particles of greater weight and density settle out first. This decreasing velocity results in **horizontal sorting** (see the illustration above right, 2nd column). When a stream enters another body of water such as the ocean, the heavier particles settle out first, and the lighter particles settle farther out into the ocean..

If deposition is rapid, **graded bedding**, or **vertical sorting** occurs. The heavier, larger, and more dense particles settle first, followed by the smaller, lighter, and less dense particles. In general the bottom of the stream bed will have the larger particles and the particle size will decrease toward the top of the stream bed.

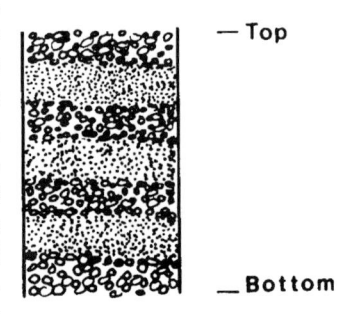

Graded Bedding in a Series of Depositions

The particles in a moving medium do not necessarily move at the same velocity as the transporting medium. Particle movement is usually slower than the transporting medium movement.

Colloidal particles in suspension tend to move at the same velocity as the stream, but particles that are more dense (heavier) or have a greater resistance (larger and flatter) often move at a slower velocity than the fluid transporting medium. Pebbles rolling along a stream bed travel slower than the water.

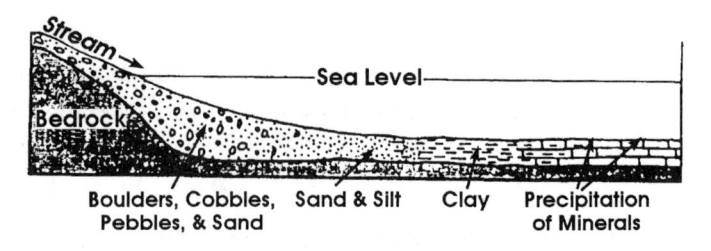

Horizontal Sorting
The cross section of the stream illustrates horizontal sorting showing that larger, rounder, high density particles settle out of decreasing water velocity the fastest.

Deposition made by gravity acting alone or by glacial ice is unsorted, since sediments of all sizes, shapes and densities are deposited together. Outwash, sediment deposited by the meltwaters of a glacier shows characteristics similar to stream deposits.

D. LANDSCAPE CHARACTERISTICS

Earth's landscapes are the results of the interaction of crustal materials, forces, climate, humans, and time.

LANDSCAPE REGIONS

Distinctive landscape regions can be identified by sets of landscape characteristics that seem to occur together. Obviously, the landscape characteristics of Long Island (lowlands) region are expected to be different from the landscape of the Adirondack (mountains) region.

How is the Earth's surface shaped by weathering, erosion, and deposition?

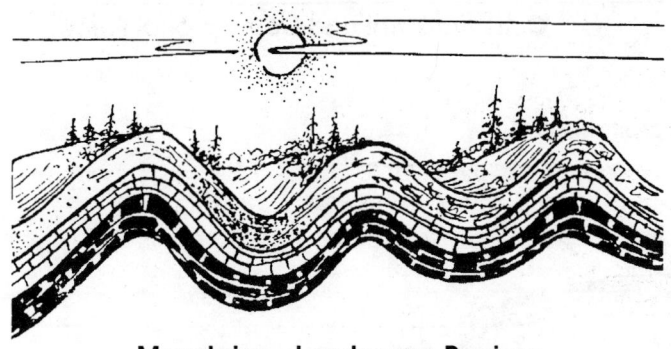

Mountains – Landscape Regions

Landscapes may be classified as **mountains**, **plains** or **plateaus**. They may be distinguished from each other by their relief and internal rock structure. **Mountains** have high relief and deformed rock structures. **Plateaus** have moderate relief, high elevation, and horizontal rock structures. **Plains** are characterized by low relief and horizontal rock layers.

Generally, the boundaries between landscape regions are well defined. Landscape regions tend to be separated by mountains, large bodies of water, and other natural boundaries.

Since the continents are so large and usually bordering on oceans, they have several distinctive types of landscape regions. For example, the North American continent has regions that represent all of the major landscape regions of the Earth.

The continental United States has eight major landscape regions, including (west to east) the Pacific Mountain System, Intermontane Plateaus, Rocky Mountain, Interior Plains, Interior Highlands, Laurentain Upland, Appalachian Highlands, and Atlantic Plain.

The surface of New York State has many distinctive landscape regions because of a large number of bedrock variations. These include (north to south) Saint Lawrence Lowlands, Champlain Lowlands, Adirondack Highlands (mountains), Tug Hill Plateau, Erie-Ontario Lowlands (plains), Appalachian Uplands (plateau), Hudson-Mohawk Lowlands, New England Highlands, and the Atlantic Coastal Lowlands (plains).

ENVIRONMENTAL FACTORS

The environmental factors involved in landscape development, include uplifting and leveling forces (constructive and destructive, respectively), climate, bedrock characteristics, time, and human activities (positive and negative). There is a delicate balance between the multiple environmental factors in all landscapes. Any change in these factors results in a modification of the landscape and the establishment of a new equilibrium.

UPLIFTING & LEVELING FORCES

The two major forces which oppose each other in the formation of landscapes are the forces of **uplifting** and **leveling**, also called the **constructional** and **destructional forces**.

Uplifting forces are referred to as constructional forces because their affect is to build mountains, enlarge the continents, and increase the elevation of some landscapes. The forces originate from within the Earth and include the forces of **diastrophism** (folding and faulting), earthquakes, volcanoes, and isostasy. The forces of destruction, **leveling forces**, include the processes of the erosional-depositional system, weath-

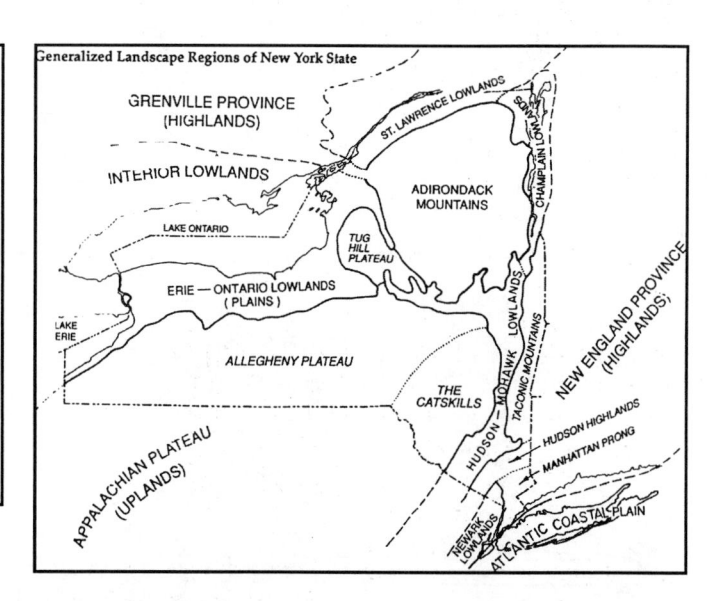

Above: **Landscape Regions of the United States**
Right: **Common Landscape Regions of NYS**

Constructional and Destructional Forces

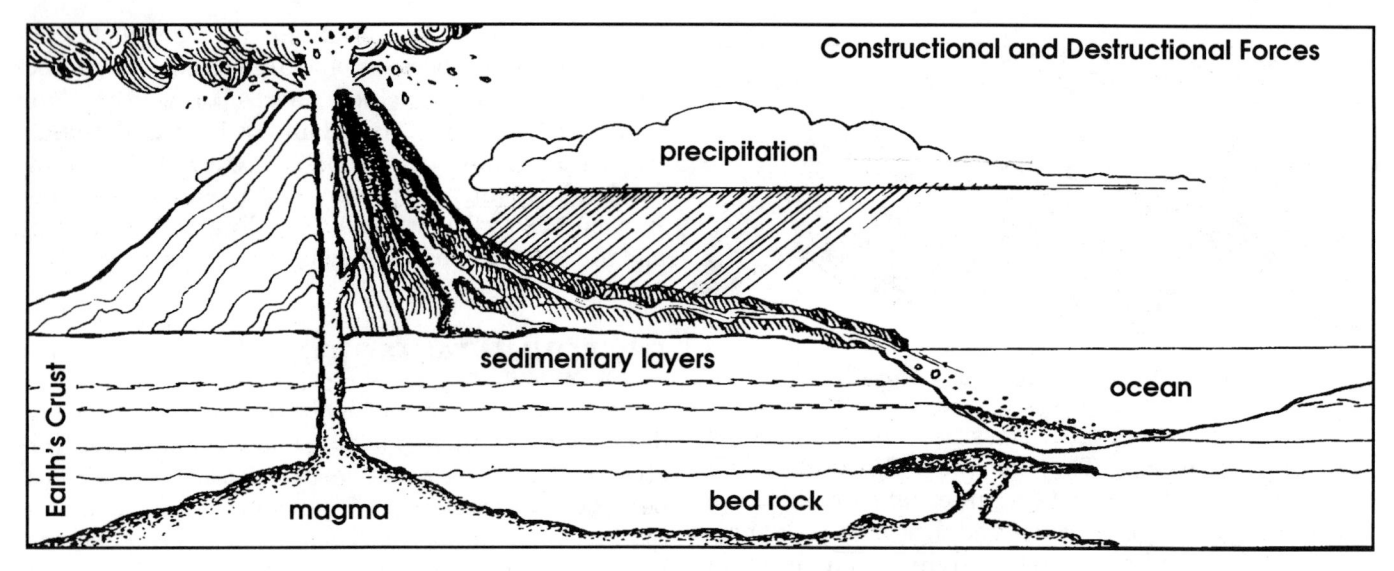

ering, and subsidence. These are forces found working on the surface of the Earth which operate primarily due to gravity, removing, and transporting materials from higher elevations and depositing them at lower elevations.

In any particular landscape or at any specific time in the Earth's history, either the force of leveling or uplifting may be dominant in a region, depending on the rate at which the uplifting or leveling occurs. The rate of uplift or subsidence may result in a modification of landscape by altering hill slopes, the drainage patterns of mountain and river regions, or orographic wind patterns over various landscape regions.

What forces modify landscapes?

During the building of the Appalachian Mountains in the eastern United States, the force of uplifting was dominant. However, today the forces of leveling are dominant in that region since the elevation is decreasing. This change is due to weathering and erosion which are occurring faster than uplifting. In general, the uplifting forces are dominant where landscapes are increasing in elevation, and the forces of leveling are dominant when landscapes are becoming lower.

CLIMATIC EFFECTS ON LANDSCAPE DEVELOPMENT

The climatic factors of temperature and moisture greatly affect the rate of the change in characteristics of landscapes. A change in climate may result in a modification of the landscape. For example, some scientists are concerned over the effects caused by a build up of carbon dioxide in the atmosphere. They suggest that this increase may bring about an increase in the Earth's average temperatures, resulting in a general warming of the Earth's surface. This could cause the melting of the polar ice regions, raising the sea level, thereby flooding low lying continental regions. A change in moisture from arid to humid would increase the rate of weathering and erosion and cause an angular landscape to become more rounded.

The steepness of hill slopes in an area is affected by the balance between weathering and the removal of materials. This can be observed in the hill slopes of arid and humid regions. Since arid regions have less water erosion, the hill slopes are generally more steep and angular, showing fewer signs of erosion. But in humid regions, there is more weathering, erosion, and deposition of materials causing rounded hills and less steep hill slopes The Great Ice Age is an example of an extreme climatic change modifying the landscape.

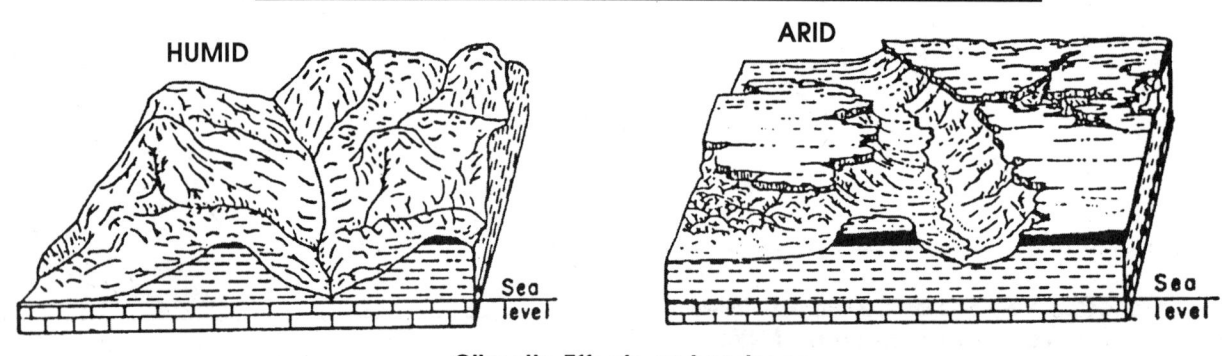

Climatic Effects on Landscapes

Stream characteristics are affected by the climate. If the climate is dry (arid), the streams are intermittent (seasonal) and the water is usually collected in basins, remaining in the region. In wet (humid) climates, the streams are permanent, usually flowing into rivers and eventually into large bodies of water. Therefore, in arid regions most of the weathered sediment remains near the source, but in humid regions the sediment is often transported far from the source of the weathering.

Glaciers have greatly altered the landscapes of New York State. In some areas glacial landscapes have many erosional features. These have been formed by the weight of the moving ice scraping, polishing, rounding, and depressing the land. Other areas are covered with depositional features made of sediment transported and deposited directly by the glacial ice or the meltwaters of the glacier.

River valleys through which glaciers have passed have been rounded into the characteristic **U-shape** of a glacial valley. Rock over which the ice has moved often has grooves and scratches or **striations** gouged out of it by sediment frozen into the bottom of the moving glacier.

Glacial Valley

Depositional features are made of unsorted glacial material called **till** or **outwash** which is sorted and deposited by glacial streams. Features made of till are called **moraines**. Often the surface is covered with **ground moraine** which has been deposited under the glacier. A ridge marking the farthest advance of the ice is called a **terminal moraine**. Winding ridges of glacial material deposited in streams within or below the ice are **eskers**. **Kames** are cone-shaped hills formed along the ice front while **drumlins** are elongated hills shaped like the back of a spoon.

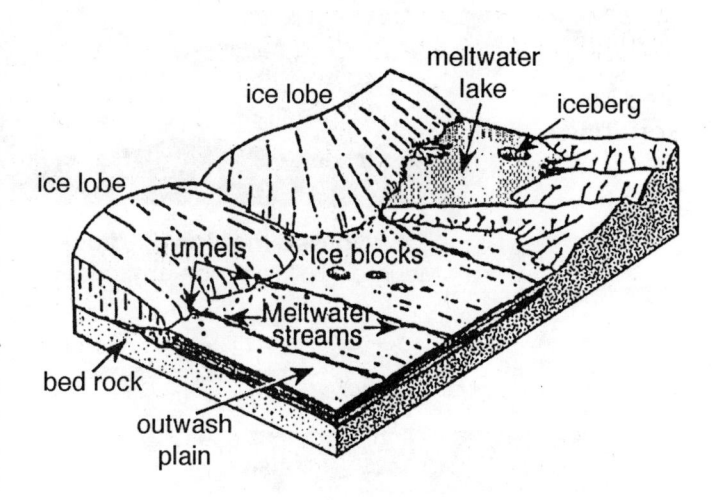

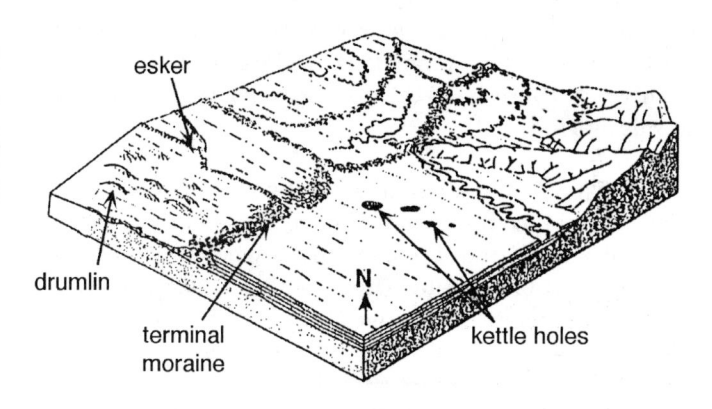

Glacial Features

A broad area in front of the terminal moraine deposited by water from the melting glacier is an **outwash plain**. **Kettle holes** are depressions resulting from melting block of ice that were buried in either the moraine or outwash.

BEDROCK EFFECT ON LANDSCAPE DEVELOPMENT

The rate at which landscape development occurs may be influenced by the bedrock of the region. Different landscape regions have various kinds of bedrock. **Rock resistance** is the ability of different rock types to resist the forces of weathering and erosion. For example, where hard bedrock is exposed to the environment, only a small amount of weathering and erosion occurs. This is because the hard bedrock is more resistant to the forces of weathering and erosion than softer kinds of bedrock.

Where bedrock with varying degrees of rock resistance occur together, the softer bed rocks are weath-

Samples Of Bed Rock Resistance

Nonsedimentary Bedrock
Mountain Building (non-existent leveling)

Bedrock
Varying Resistance (escarpments)

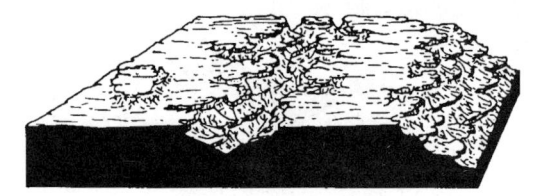

Sedimentary Bedrock
(plateau)

Complex Bedrock
(faulting and variations of leveling)

ered and eroded at a faster rate than the more resistant bedrock. The landscape surface is less regular than when there is a rock with a consistent resistance to weathering.

The shape and steepness of hills are affected by the local bedrock composition. Competent rocks (resistant to weathering and erosion) are responsible for the development of plateaus (high, stable landscapes with little or no distortion), mountains, and escarpments. These areas often have slow rates of change.

Weak or incompetent rocks (poor resistance to weathering and erosion) usually underlie valleys and other low level areas. Meandering streams, thick soil horizons, and deep layering usually characterize weak rock areas.

The difference between competent and weak areas can be observed in an **escarpment**, which is a steep slope separating two gently sloping surfaces. Many escarpments are the result of weathering forces on slopes with rocks of different resistance.

Structural features in bedrock, such as faults, folds, and joints, frequently affect the development of hill slopes. Generally, the more distortion in the rock mass of the hill slope, the greater the variety of changes occurring in the surface. The weathering effect is varied, since faults, folds, and joints expose rocks with different resistance to the forces of change.

Stream characteristics, including gradient (slope), width and depth, and drainage pattern and direction of flow, are controlled by bedrock characteristics. When the resistance of the bedrock over which the stream flows is consistent, a random drainage pattern forms, called **dendritic drainage**. **Trellis drainage** forms in regions of parallel folds and/or faults, whereas **radial drainage** is associated with volcanic cones and

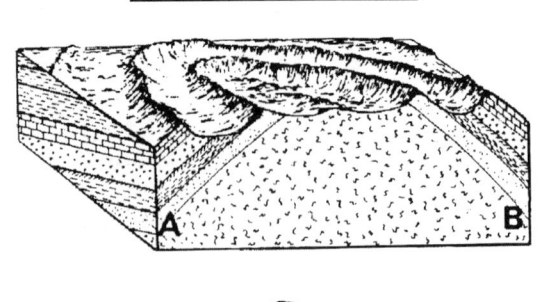

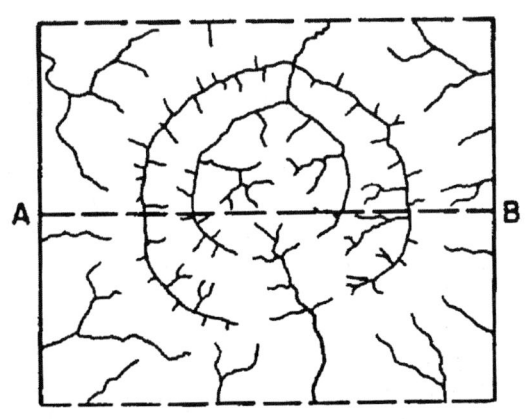

Stream Pattern on Bedrock with Varing Resistance

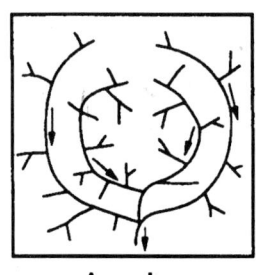

Annular Drainage

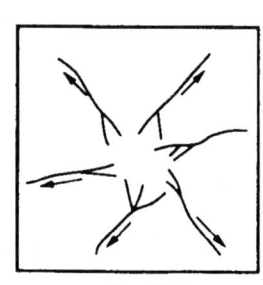

Radial Drainage

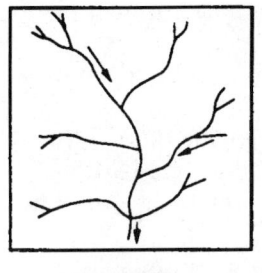

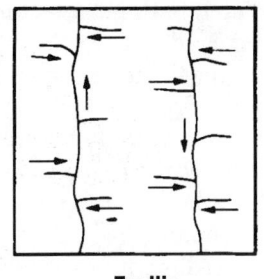

Dendritic Drainage

Trellis Drainage

young domes. **Annular** (ring-shaped) **drainage** develops on domes as they become eroded.

TIME AFFECT ON LANDSCAPE DEVELOPMENT

The stage of development of a landscape is determined by the duration of time during which environmental factors have been active. Older landscapes generally show more effects of weathering and erosion than do younger landscapes.

AFFECT ON LANDSCAPE DEVELOPMENT

The activities of humans have altered the landscapes in many areas. Humans have modified landscapes both in negative and positive ways.

NEGATIVE AFFECTS BY HUMANS ON THE LANDSCAPE

Both intentional and unintentional **technological oversights** have brought about unplanned consequences that have destroyed or reduced the quality of life in many landscape regions. In some cases, technological advances have produced waste products which humans do not know how to dispose of safely. In other cases, industrial wastes are disposed of carelessly and/or criminally.

Agricultural lands, water sources, grazing grasslands for livestock, and forests have been so badly misused as to have resulted in the starvation and extinction of total populations of plants and animals.

High population density areas are the most affected regions because of the concentration of the landscape pollution or the misuse of the landscape.

POSITIVE AFFECTS BY HUMANS ON THE LANDSCAPE

Positive affects of human activities on landscapes have come through an increased awareness of various ecological and landscape interactions. Humans have begun to intervene in the widespread destruction, through the efforts of individuals, community groups, and conservation clubs. There are attempts to clean up toxic waste dumps, reclaim wasted lands, and prevent continued disruption of the environment.

QUESTIONS FOR UNIT 4

1   The principal cause of the chemical weathering of rocks on the Earth's surface is
    1   rock abrasion
    2   the heating and cooling of surface rock
    3   mineral reactions with air and water
    4   the expansion of water as it freezes

2   Which type of climate causes the fastest chemical weathering?
    1   cold and dry
    2   cold and humid
    3   hot and dry
    4   hot and humid

3   Which factor has the least effect on the weathering of a rock?
    1   climatic conditions
    2   composition of the rock
    3   exposure of the rock to the atmosphere
    4   the number of fossils found in the rock

4   Water is a major agent of chemical weathering because water
    1   cools the surroundings when it evaporates
    2   dissolves many of the minerals that make up rocks
    3   has a density of about one gram per cubic centimeter
    4   has the highest specific heat of all common earth materials

UNIT FOUR – SURFACE PROCESSES AND LANDSCAPES – N&N©   Page 81

5 The four limestone samples illustrated below have the same composition, mass, and volume. Under the same climatic conditions, which sample will weather fastest?

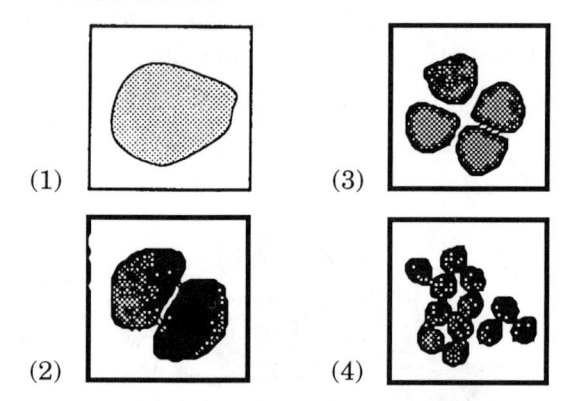

(1)   (3)
(2)   (4)

6 The diagram below shows a process of weathering called frost wedging.

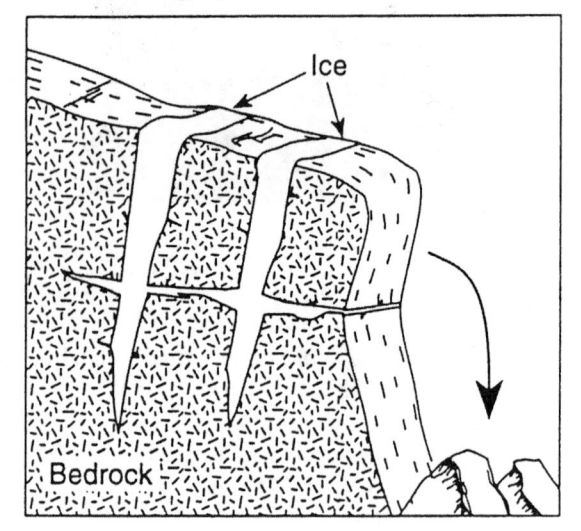

Frost wedging breaks rocks because as water freezes it increases in
1   density   3   mass
2   specific heat 4   volume

7 As the humidity of a region decreases, the amount of weathering taking place usually
1   decreases
2   increases
3   remains the same

8 According to the *Reference Tables*, which stream velocity would transport cobbles, but would *not* transport boulders?
(1) 50 cm/sec   (3) 200 cm/sec
(2) 100 cm/sec   (4) 400 cm/sec

9 A river transports material by suspension, rolling, and
1   solution   3   evaporation
2   sublimation   4   transpiration

10 Which is the best evidence that erosion has occurred?
1   a soil rich in lime on top of a limestone bedrock
2   a layer of basalt found on the floor of the ocean
3   a large number of fossils embedded in limestone
4   sediments found in a sand bar of a river

11 In the diagram below, the arrow shows the direction of stream flow around a bend.

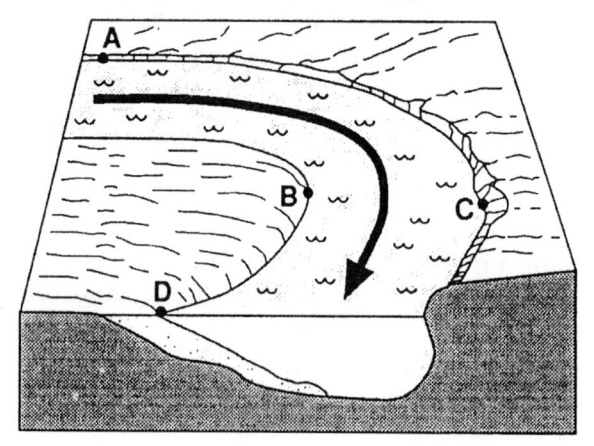

At which point does the greatest stream erosion occur?
(1) A   (3) C
(2) B   (4) D

12 The composition of sediments on the Earth's surface usually is quite different from the composition of the underlying bedrock. This observation suggests that most
1   bedrock is formed from sediments
2   bedrock is resistant to weathering
3   sediments are residual
4   sediments are transported

13 Soil horizons develop as a result of
1   capillary action and solution
2   erosion and ionization
3   leaching and color changes
4   weathering processes and biologic activity

14 The cross section below shows residual soils that developed on rock outcrops of metamorphic quartzite and sedimentary limestone. [Question and answers at top of next page.]

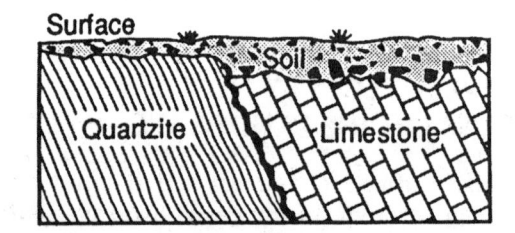

Which statement best explains why the soil is thicker above the limestone than it is above the quartzite?
1. The quartzite formed from molten magma.
2. The limestone is thicker than the quartzite.
3. The quartzite is older than the limestone.
4. The limestone is less resistant to weathering than the quartzite.

15 Which change would cause the topsoil in New York State to increase in thickness?
1. an increase in slope
2. a decrease in rainfall
3. an increase in biologic activity
4. a decrease in air temperature

16 The graph below shows how environmental temperatures affect the amount of organic material (humus) added to and removed from soils in humid regions.

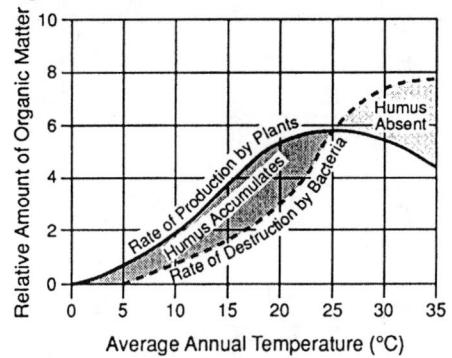

The graph supports the conclusion that soils in regions with average annual temperatures above 25° C have
1. little humus present
2. the highest production of humus
3. a low breakdown of humus
4. the same amount of humus as soils in cooler regions

Base your answers to questions 17 through 19 on your knowledge of Earth science and on the vertical cross section showing a stream profile with reference points A through F within the stream bed.

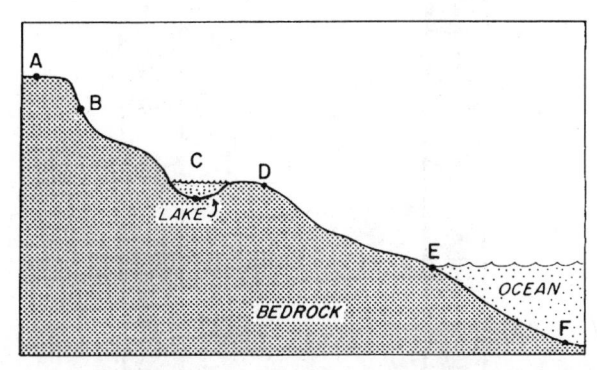

17 The primary force causing the movement of materials from point B to point F is
1. air pressure
2. insolation
3. water
4. gravity

18 At which point would erosion most likely be greatest?
1. A
2. B
3. C
4. D

19 Which would most likely happen if the stream discharge between points D and E were to increase?
1. The average velocity of water would increase.
2. The amount of soil erosion would decrease.
3. The size of the particles carried in suspension would decrease.
4. The length of the stream would decrease.

20 A stream is entering the calm waters of a large lake. Which diagram best illustrates the pattern of sediments being deposited in the lake from the stream flow?

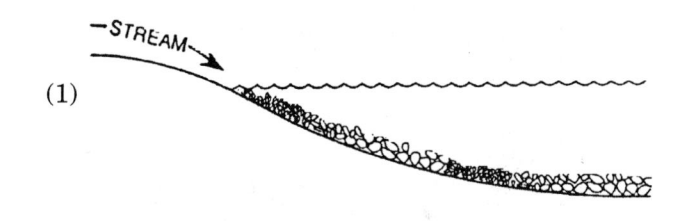

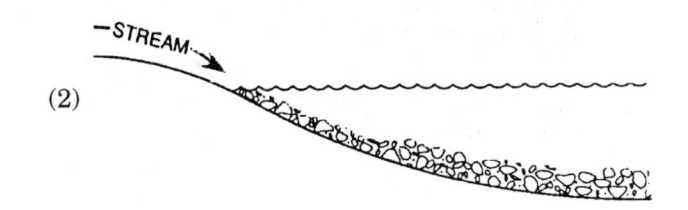

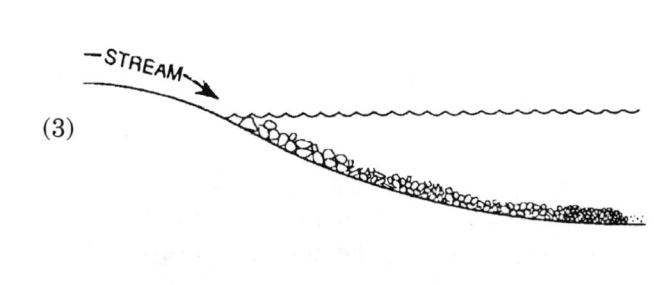

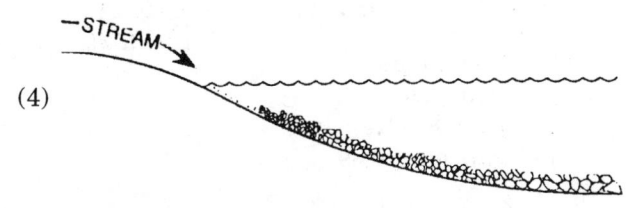

(1)

(2)

(3)

(4)

UNIT FOUR – SURFACE PROCESSES AND LANDSCAPES – N&N© Page 83

21 If all the particles illustrated have the same mass and density, which particle will settle fastest in quiet water? [Assume settling takes place as shown by arrows.]

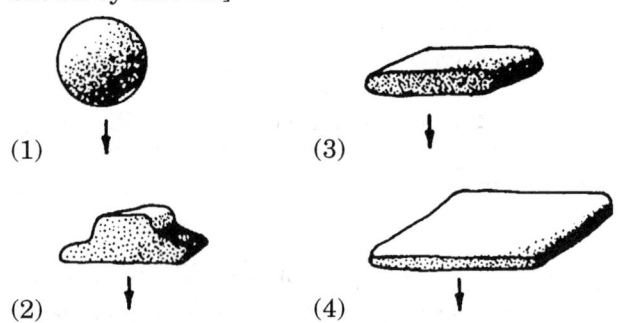

22 Why do particles carried by a river settle to the bottom as the river enters the ocean?
 1 The density of the ocean water is greater than the density of the river water.
 2 The kinetic energy of the particles increases as the particles enter the ocean.
 3 The velocity of the river water decreases as it enters the ocean.
 4 The large particles have a greater surface area than the small particles.

23 Which rock particles will remain suspended in water for the longest time?
 1 pebbles
 2 sand
 3 silt
 4 clay

24 Small spheres that are identical in shape and size are composed of one of four different kinds of substances: A, B, C, or D. The spheres are mixed together and poured into a clear plastic tube filled with water.

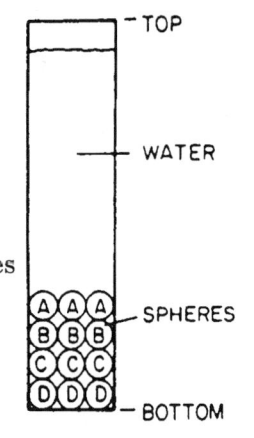

Which property of the spheres caused them to settle in the tube as shown in the diagram?

 1 their shape
 2 their size
 3 their density
 4 their hardness

Base your answers to questions 25 and 26 on the diagram and your knowledge of Earth science.

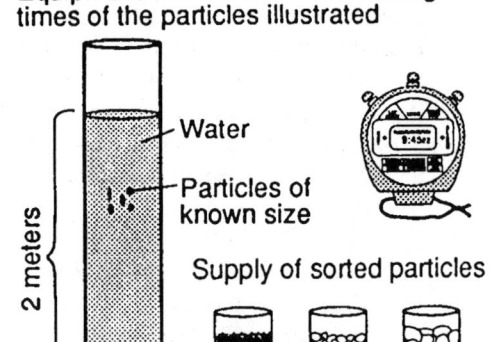

25 Which graph below best represents the effect of particle size on settling time if the particle densities are the same?

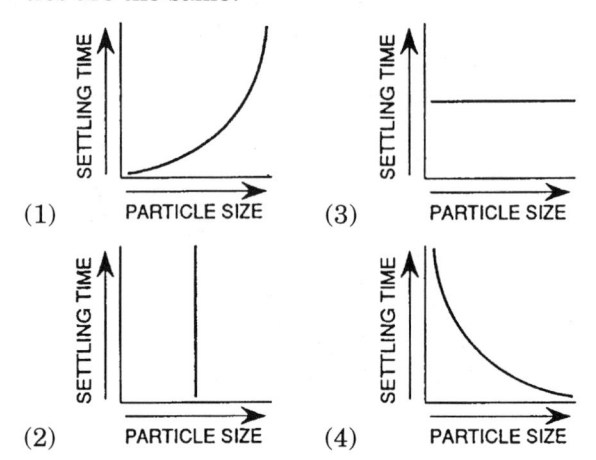

26 All three sediments in the diagram are placed in the cylinder with water, the mixture is shaken, and the particles are allowed to settle. Which diagram best represents the order in which the particles settled?

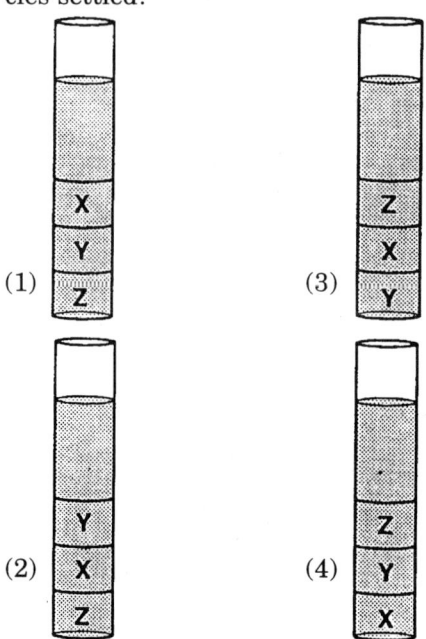

27 A dynamic equilibrium exists in an erosional-depositional system when
1 all sediments are transported to the sea and erosion stops
2 the amounts of the kinetic energy and potential energy both equal zero
3 the rate of erosion exceeds the rate of deposition
4 the rate of erosion is the same as the rate of deposition

28 Which diagram best represents a cross section of a valley which was glaciated and then eroded by a stream?

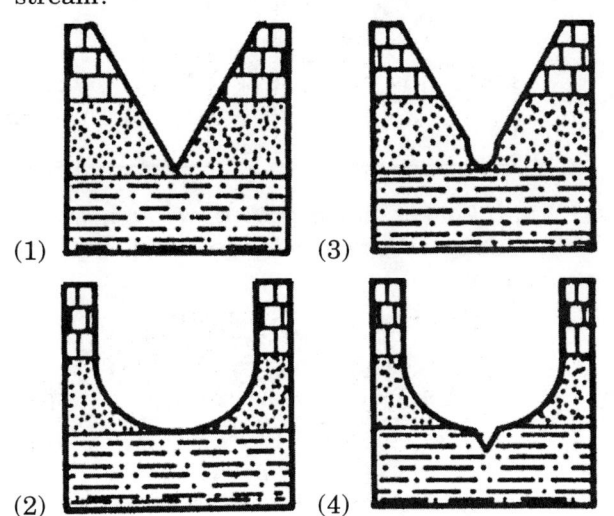

Base your answers to questions 29 through 31 on the diagram, the *Reference Tables*, and your knowledge of Earth science. The diagram represents a glacier moving out of a mountain valley. The water from the melting glacier is flowing into a lake. Letters A through F identify points within the erosional-depositional system.

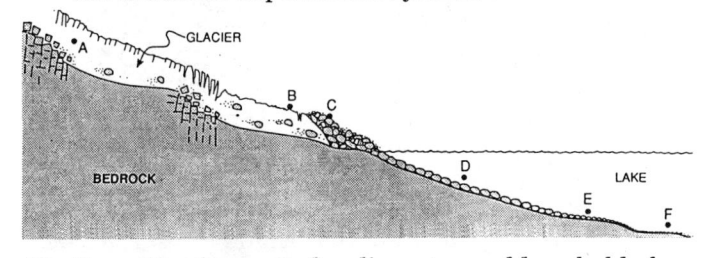

29 Deposits of unsorted sediments would probably be found at location
(1) E   (3) C
(2) F   (4) D

30 An interface between erosion and deposition by the ice is most likely located between points
(1) A and B   (3) C and D
(2) B and C   (4) D and E

31 Colloidal-sized sediment particles carried by water are most probably being deposited at point
(1) F   (3) C
(2) B   (4) D

32 Four samples of aluminum, A, B, C, and D, have identical volumes and densities, but different shapes. Each piece is dropped into a long tube filled with water. The time each sample takes to settle to the bottom of the tube is shown in the table.

Sample	Time to Settle (sec)
A	2.5
B	3.7
C	4.0
D	5.2

Which diagram most likely represents the shape of sample A?

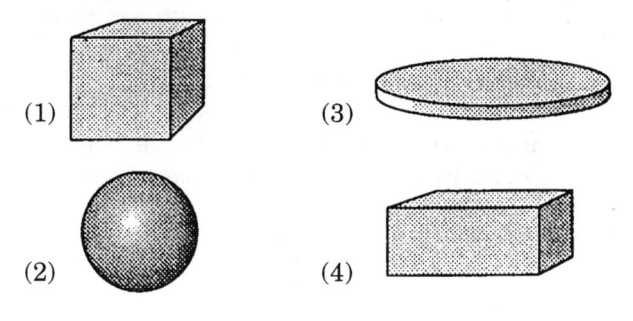

33 One characteristic used to classify landscape regions as plains, plateaus, or mountains is
1 type of soil
2 amount of stream discharge
3 weathering rate
4 underlying bedrock structure

34 Which cross-sectional diagram best represents a landscape region that resulted from faulting?

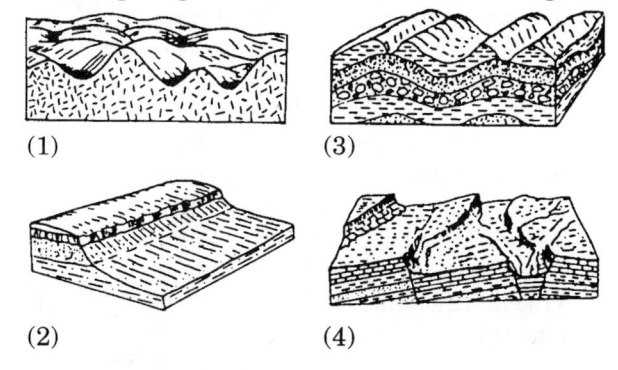

35 An area of gentle slopes and rounded mountaintops is most likely due to
1 climatic conditions
2 earthquakes
3 the age of the bedrock
4 the amount of folding

36 The diagram below represents the surface topography of a mountain valley.

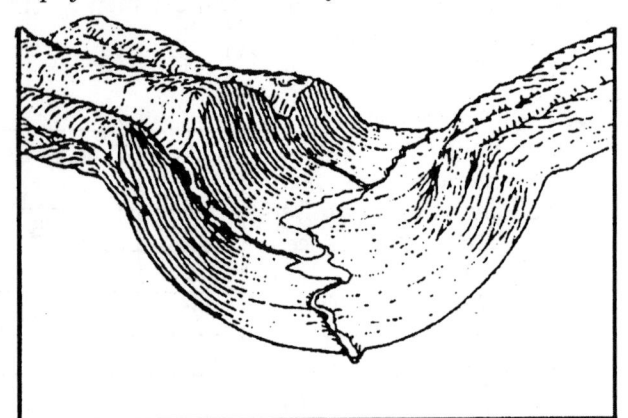

Which agent of erosion most likely created the shape of the valley shown in the diagram?
1 wind
2 glaciers
3 ocean waves
4 running water

Base your answers to questions 37 through 41 on your knowledge of Earth science and the diagram which represents a geologic cross section in which no overturning has occurred. The letters identify regions in the cross section.

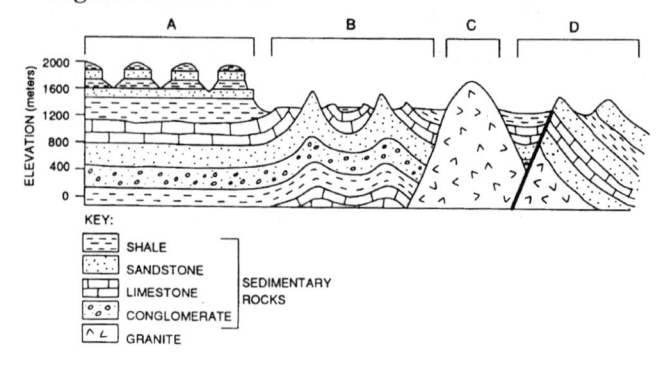

KEY:
- SHALE
- SANDSTONE
- LIMESTONE
- CONGLOMERATE } SEDIMENTARY ROCKS
- GRANITE

37 The surface features in region A were produced primarily as a result of the process of
1 folding
2 faulting
3 erosion
4 glaciation

38 Which type of crustal movement is shown in region B?
1 faulting
2 volcanic eruption
3 jointing
4 folding

39 Which region shows the typical characteristics of plateau
(1) A
(2) B
(3) C
(4) D

40 Which region is least likely to have fossils in the surface bedrock?
(1) A
(2) B
(3) C
(4) D

41 Which stream drainage pattern would most likely be found on the surface of region C?

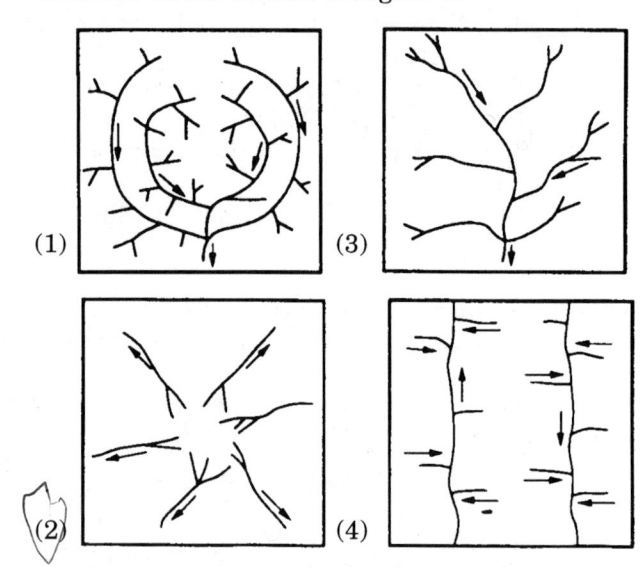

42 Which kind of stream pattern would most likely be found on the type of landscape shown in the diagram?

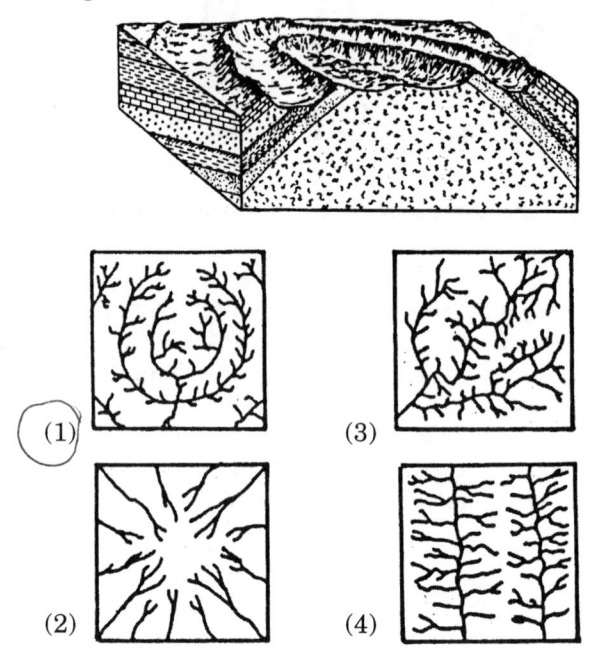

Base your answers to question 43 [on the next page] on the diagram that shows the surface landscape features and the internal rock structure of a cross section of the Earth's crust.

43 How would this region most likely appear immediately after undergoing a period of glaciation?

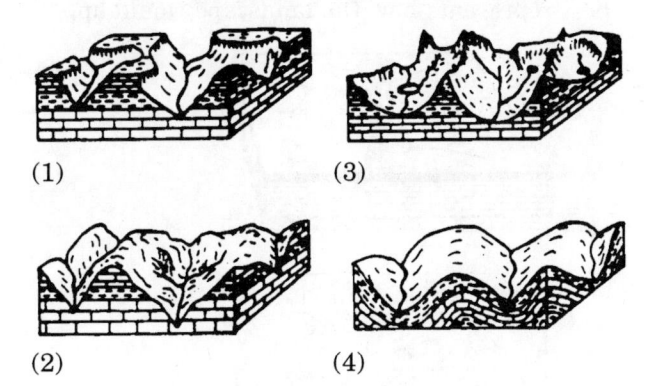

(1)   (3)
(2)   (4)

44 The diagrams below represent geologic cross sections from two widely separated regions.

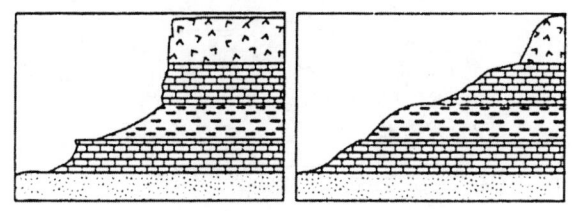

The layers of rock appear very similar, but the hillslopes and shapes are different. These differences are most likely the result of
1   volcanic eruptions
2   earthquake activity
3   soil formation
4   climate variations

45 Which change is most likely to occur in a landscape region that is uplifted rapidly by folding?
1   The climate will become warmer.
2   The stream drainage patterns will change.
3   The composition of the bedrock will change.
4   The hillslopes will become less steep.

46 The diagram below represents a cross section showing glacial deposition on top of solid bedrock.

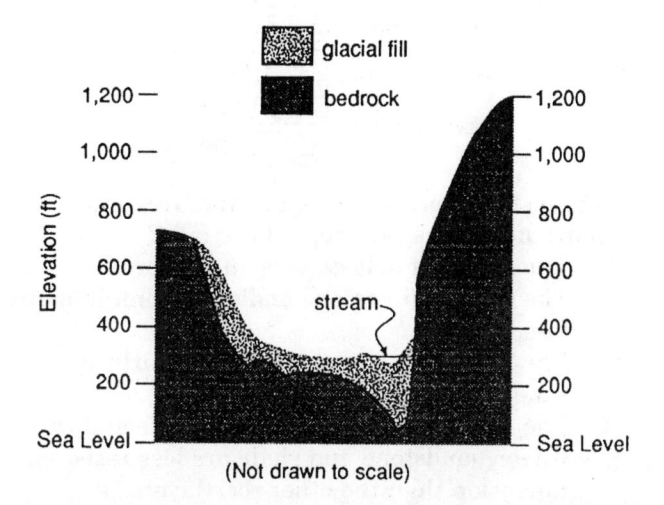

The evidence suggests that the maximum elevation of glacial deposition prior to erosion by the stream
(1)   100 ft   (3)   300 ft
(2)   200 ft   (4)   700 ft

47 The diagrams below represent profiles of four different landscapes, A, B, C, and D.

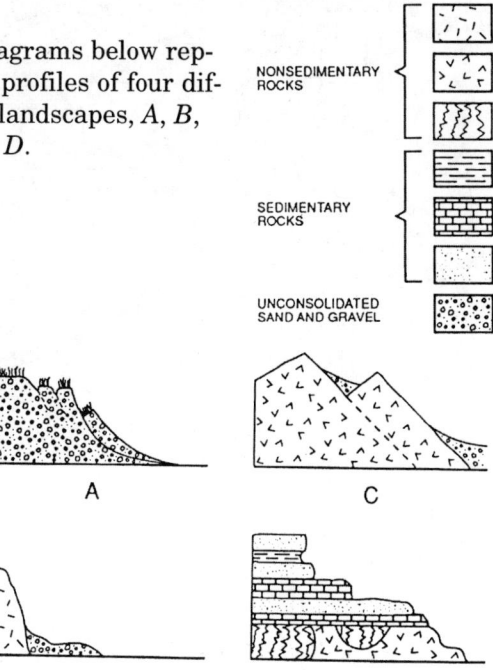

Which landscape is most likely to have a noticeable change in its profile after a heavy rainstorm?
(1)   A   (3)   C
(2)   B   (4)   D

48 The block diagrams below show a river and its landscape during four stages of erosion.

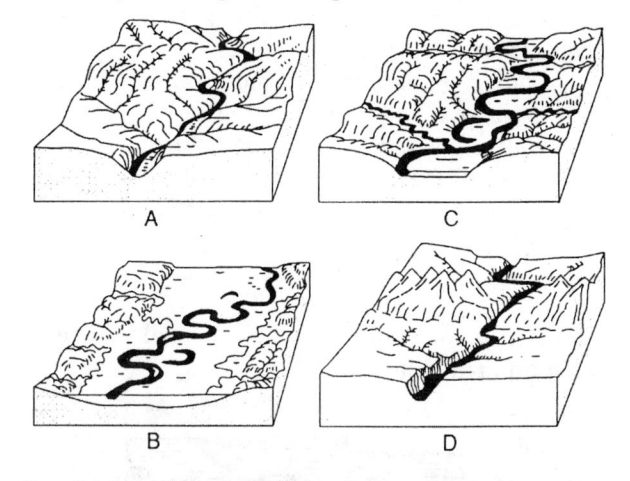

In which order should the diagrams be placed to show the most likely sequence of river and landscape development?
(1)   A, D, B, C   (3)   C, B, A, D
(2)   B, D, C, A   (4)   D, A, C, B

UNIT FOUR – SURFACE PROCESSES AND LANDSCAPES – N&N©   Page 87

Base your answers to questions 49 through 51 on the *Earth Science Reference Tables*, the diagram below, and your knowledge of Earth science. The diagram shows a cross section of the Grand Canyon. The rock type of layer *X* has been purposely left blank.

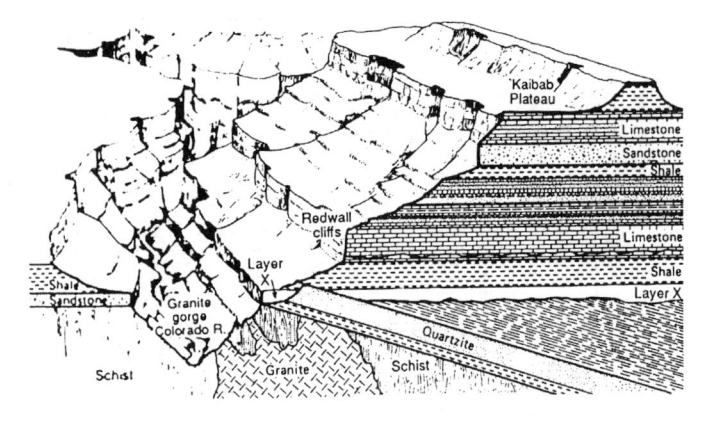

49 The Grand Canyon is primarily the result of erosion due to
1 wind
2 volcanic activity
3 glaciers
4 running water

50 This landscape area developed in an arid climate. If the climate of this region were to become humid, the hills would eventually
1 become more angular in shape
2 become more rounded in shape
3 erode at a slower rate
4 continue to erode in the same way

51 The most probable explanation for the development of the steep Redwall cliffs is that this limestone layer
1 is more resistant to erosion than the rock layers above and below the cliffs
2 was deposited in a marine environment
3 was uplifted by the granite intrusion
4 is younger than the rock layers above the cliffs

52 The landscape shown below developed in a region with an arid climate.

If the erosion of this plateau had taken place in a much more humid climate, which diagram below best represents how the landscape would appear?

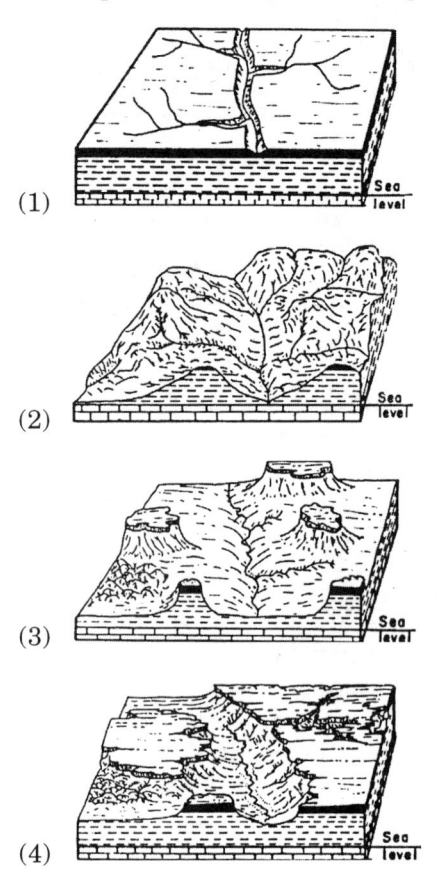

(1)
(2)
(3)
(4)

53 The diagram shows a geologic cross section of the rock layers in the vicinity of Niagara Falls in western New York State.

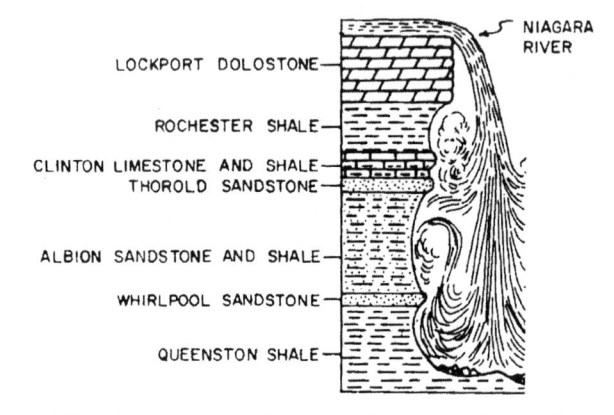

Which statement best explains the irregular shape of the rock face behind the falls?
1 The Lockport dolostone is an evaporite
2 The Clinton limestone and shale contain many fossils.
3 The Thorold sandstone and the whirlpool sandstone dissolve easily in water.
4 The Rochester and Queenston shale and the Albion sandstone and shale are less resistant to erosion than the other rock layers.

54 Although the Adirondacks are classified as a mountain landscape, the Catskills are classified as a plateau landscape because of a major difference in their
1 amount of rainfall
2 bedrock structure
3 index fossils
4 glacial deposits

55 If an area became more arid, the steepness of the slopes and sharpness of the landscape features would
1 decrease
2 increase
3 remain the same

SKILL ASSESSMENTS

Base your answers to questions 1 through 6 on the data table below. Samples of three different rock materials, A, B, and C, were placed in three containers of water and shaken vigorously for 20 minutes. At 5-minute intervals, the contents of each container were strained through a sieve. The mass of the materials remaining in the sieve was measured and recorded as shown in the data table below.

Mass of Material Remaining in Sieve

Shaking Time (minutes)	Rock Material A (grams)	Rock Material B (grams)	Rock Material C (grams)
0	25.0	25.0	25.0
5	24.5	20.0	17.5
10	24.0	18.5	12.5
15	23.5	17.0	7.5
20	23.5	12.5	5.0

Directions: Using the information in the data table, construct a line graph on the grid, following the directions below.

1 Plot the data for rock sample A for the 20 minutes of the investigation. Surround each point with a small circle and connect the points.

2 Plot the data for rock sample B for the 20 minutes of the investigation. Surround each point with a small triangle and connect the points.

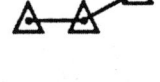

3 Plot the data for rock sample C for the 20 minutes of the investigation. Surround each point with a small square and connect the points.

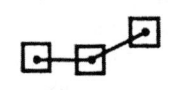

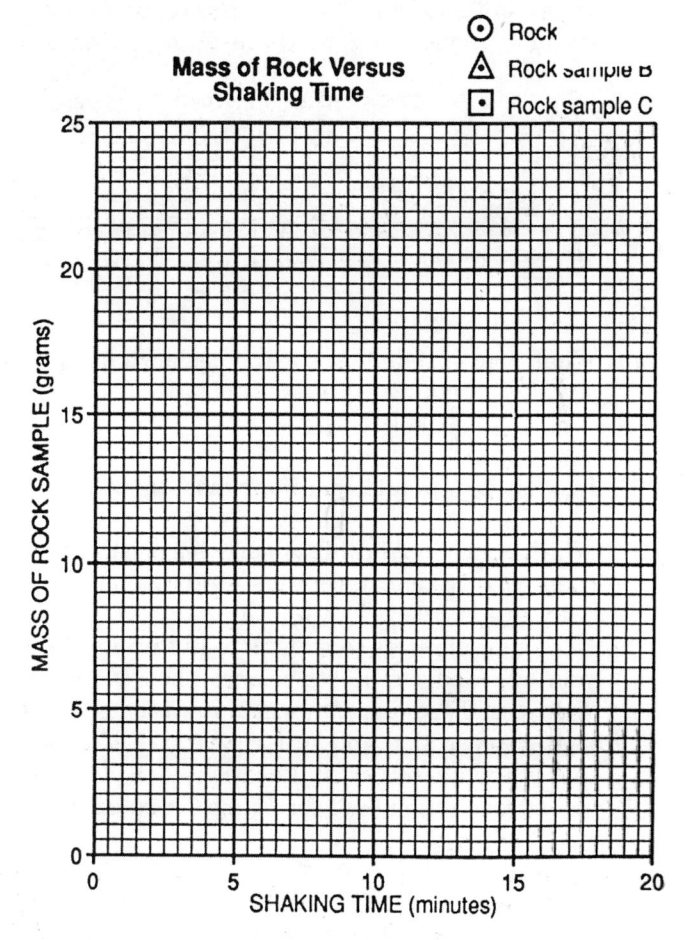

Mass of Rock Versus Shaking Time

Legend:
⊙ Rock
△ Rock sample B
☐ Rock sample C

4 Using one or more complete sentences, state the most likely reason for the differences in the weathering rate of the three rock materials.

5 Using the directions in parts *a* through *c* below, calculate the average rate of change in the mass of rock material C for the 20 minutes of shaking.

   a Write the equation for rate of change in mass.

   b Substitute data into the equation.

   c Calculate the rate of change in mass and label your answer with proper units.

UNIT FOUR – SURFACE PROCESSES AND LANDSCAPES – N&N©   Page 89

6. Using one or more complete sentences, describe the most likely appearance of the corners and edges of rock material *C* at the end of the 20 minutes.

Base your answers to questions 7 through 9 on your knowledge of Earth science, the *Reference Tables*, and the cross-sectional diagram. The diagram shows a sediment-laden stream entering the ocean. Points *X* and *Y* are in the stream and the ocean is divided into four zones *A*, *B*, *C*, and *D*.

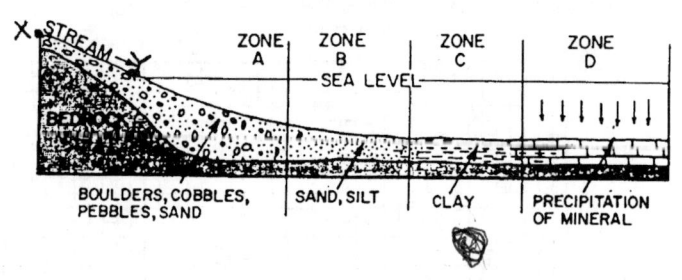

7. What type of sediment sorting is shown in the diagram? Explain what causes this type of sorting.

   horizontal sorting

8. In what zone would limestone most likely form?

   D

9. Which zone would contain particles mostly in the range of 0.05 to 0.10 centimeter in diameter?

   B

Base your answers to questions 10 through 14 on your knowledge of Earth science, the *Reference Tables*, and the diagram which represents the geologic cross section and surface features of a portion of the Earth's crust. Locations *A*, *B*, *C*, and *D* are reference points of the Earth's surface.

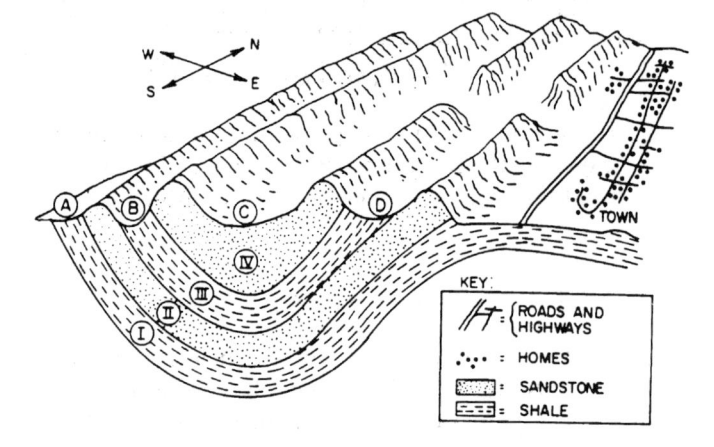

10. In a sentence or two, describe the type of rock structure underlying the regions located west of the town.

    folding

11. Index fossils are found in the surface bedrock under the town. Where else are fossils of the same type likely to be found?

    A

12. What type of rock appears to be the most resistant to weathering?

    Sandstone

13. Draw a diagram of the stream pattern that would most likely form in the region west of the town. What is the name of this stream pattern?

14. Both of the landscapes shown in the diagrams below are eroded plateaus. In a sentence or two, infer why their surfaces differ in appearance.

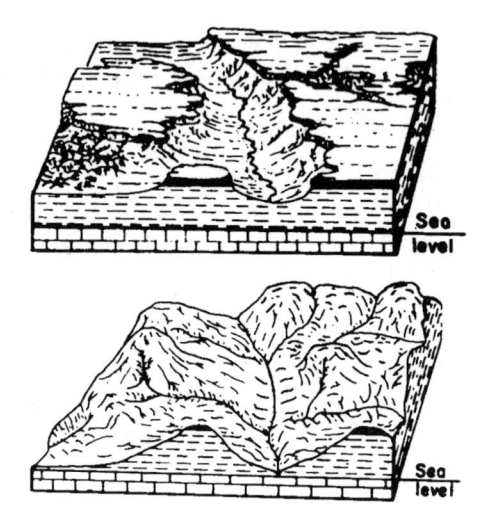

Topic C
Oceanography

Vocabulary To Be Understood

Abyssal Plain	Groin	Sea Cliff
Backwash	Gyre	Seamount
Baymouth Bar	Headland	Spit
Breaker	Hook	Surf
Continental Margin	Jetty	Swash
Continental Rise	Lagoon	Trough
Continental Shelf	Longshore Current	Tsunami
Continental Slope	Organic Sediments	Turbidity Current
Crest	Salinity	Wave Height
Density Current	Salinity Current	Wavelength
Fetch	Sea Arch	Wave Refraction

A. OCEAN WATER

Most of the water on our planet is found in the oceans. Look at a globe and you see that the oceans cover approximately 71% of the surface of the Earth.

COMPOSITION

The ocean water tastes salty, because it contains a variety of dissolved minerals. Much of the rain that falls on the land runs off into streams and eventually makes its way to the ocean. Water is a good **solvent**, so it dissolves soluble minerals in its path and carries them into the ocean. A measure of the dissolved solids in sea water is called the **salinity**.

Sea Water Composition Dissolved Ion	%
Chloride (Cl–)	55.04
Sulfate (SO$_4$$^{2-}$)	7.68
Bicarbonate (HCO$_3$–)	0.41
Bromide (Br–)	0.19
Sodium (Na+)	30.61
Magnesium (Mg^{2+})	3.69
Calcium (Ca^{2+}–)	1.16
Potassium (K+)	1.10
All others	0.12
	100.00

Put one kilogram of seawater into an open container. Let the water evaporate. Approximately 35 grams of solid material will be left in the container. Most of this solid material will be white crystals of sodium chloride or table salt (NaCl). Sodium and chloride ions accumulate in sea water to a greater degree than other ions because of their high solubility.

Since ancient times people have evaporated ocean water to get the salt to sell and to use for cooking and preserving food. Sodium chloride is not the only mineral in the ocean water. Technology has developed economical methods of removing bromine and magnesium from the ocean. There are many other mineral resources that someday may be "mined" from the ocean as well.

Why is it important to investigate the oceans?

B. OCEAN FLOOR

The floor of the ocean is not smooth but has mountains and valleys like the continents. The areas that make up the **continental margin** are the continental shelf, continental slope and continental rise.

The **continental shelf** is the shallow submerged land sloping gently out from the shoreline. At the seaward edge of the continental shelf is the **continental slope** which leads more steeply to the deeper ocean. In some places submarine canyons are cut into the continental slope. Some of these canyons appear to be extensions of river valleys cut into the continental shelf when sea level was lower, others are associated with **turbidity currents**, movement of thick sediment-laden water.

The Representative Ocean Floor

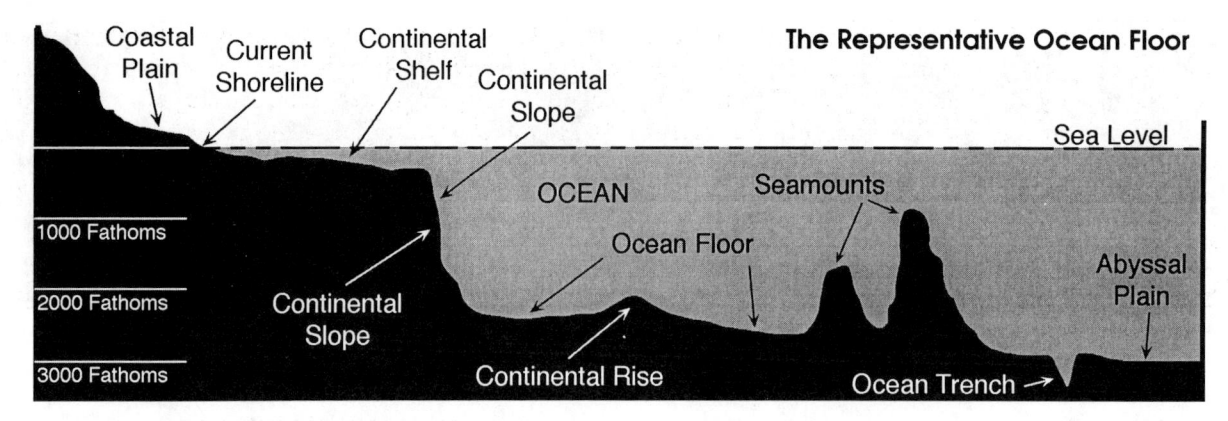

The **continental rise** is a gently sloping surface of sediments at the base of the continental slope. Out past the continental rise is the flat **ocean basin** or **abyssal plain** which extends to a mid ocean ridge. Found along coastlines associated with colliding plates are deep **ocean trenches**.

Core samples of the basaltic ocean crust show that it is much younger than the crust of the continents. Zones of magnetic reversals recorded in the rocks of the ocean basins indicate how the ocean basins are formed. New ocean crust forms at mid-ocean ridges where lithospheric plates are pulled apart and magma wells up from below to cool and solidify into new crust. Where the ocean crust collides with a continental plate, the ocean crust subducts down into the mantle forming a trench on the surface. Thus the ocean basins form and reform in a conveyor belt-like motion.

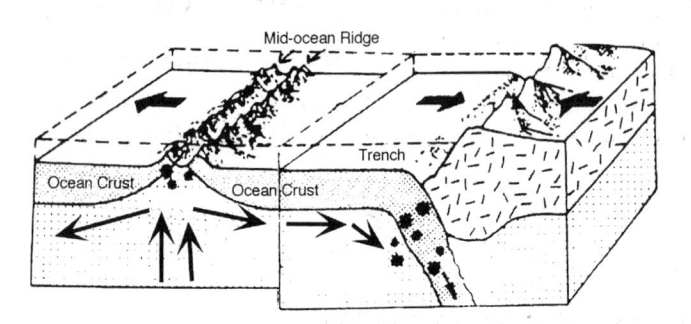

Ocean Basis Reformation

The direction and rate of formation of chains of volcanic islands (like the Hawaiian Islands) enable scientists to calculate the rate and direction of ocean plate motion (refer to Topic B for more information).

OCEAN SEDIMENTS

Along with dissolved minerals, sediments are eroded from the land and deposited in the ocean. These **land-derived sediments** are often found as graded deposits left by turbidity currents. The thickness of these sediments decreases from the continental rise to the mid-ocean ridge.

Organic sediments, formed from the skeletal remains and shells of microscopic organisms, are sediments of marine origin. These organic sediments settle out of the water onto the ocean floor.

C. OCEAN CURRENTS

A map of global winds and ocean currents look similar. This is because the winds blowing over the water surface create a frictional drag on the water resulting in **surface currents**. Currents are like huge "rivers" in the ocean.

What is the nature of ocean waves and currents?

The Coriolis Effect, due to the Earth's rotation, causes surface currents to be deflected clockwise in the Northern Hemisphere and counterclockwise in the Southern Hemisphere, circulating in large cells called **gyres**.

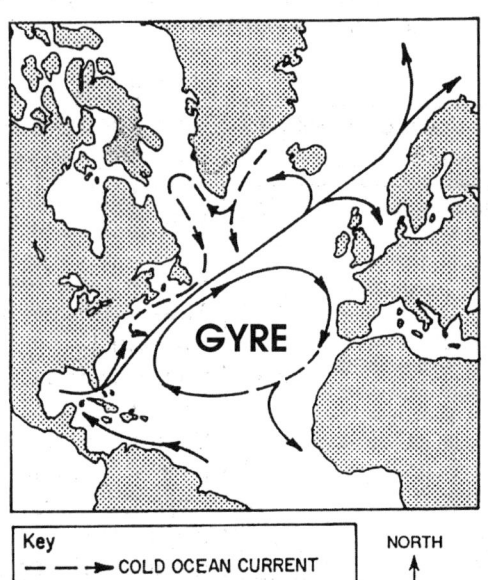

Page 92 — N&N© SCIENCE SERIES – EARTH SCIENCE – MODIFIED PROGRAM

Surface currents may be either warm or cold depending upon the temperature of the water through which they are passing and their direction. **Warm currents** flow away from the Equator and **cold currents** flow toward the Equator. As the warm water flows away from the Equator it helps to distribute the solar energy it has absorbed in the low latitudes to the higher, cooler latitudes.

The currents that flow in the opposite direction from the wind-driven currents are called **countercurrents**. They help return some of the water the winds have moved.

Currents under the ocean surface are called **density currents**, because they have a density greater than that of the surface water. The water may be denser because it is more saline (saltier), colder, or filled with sediment.

Extreme cooling and freezing of sea water in the high latitudes results in the formation of cold, saline currents in the polar regions. These currents sink to the ocean floor and creep very slowly toward the Equator.

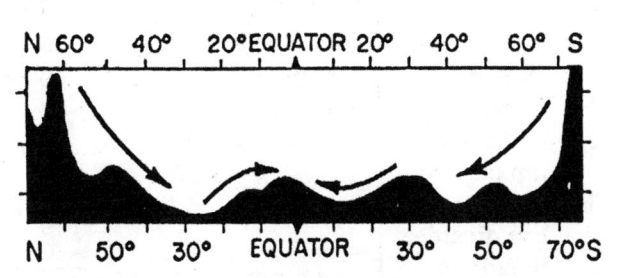

Currents Move From Cold Poles toward Warm Equator

Salinity currents are the result of evaporation, removing water leaving a higher concentration of dissolved salts behind. A salinity current flows from the Mediterranean Sea out into the Atlantic Ocean.

Currents filled with sediment are called **turbidity currents**. Turbidity currents may form where sediment is dislodged (perhaps by an earthquake) along the continental shelf and flow rapidly down the continental slope taking land-derived sediments out into the ocean basin.

OCEAN WAVES

Most of the ocean's surface waves are caused by the wind. The size of the waves is determined by wind speed, the length of time the wind blows in one direction and the distance the wind has blown across the water. This distance is known as the **fetch**.

Submarine earthquakes or **landslides** may generate **tsunamis**. Although these giant waves are often called "tidal waves," they are unrelated to tidal effects. Tsunamis can cause catastrophic loss of life and property damage.

D. OCEAN & POLLUTION

The constant motion of ocean water in waves, currents and tides acts to distribute pollutants throughout the ocean. Deliberate dumping of garbage and waste, as well as, accidental spills of oil and other toxic materials contaminate the ocean. Damage to the marine ecosystem kills and injures marine life, decreasing both the amount and quality of food from the sea and hurting coastal recreation.

COASTAL PROCESSES

The top of a wave in open water is known as its **crest**, and the low point is the **trough**. **Wavelength** is the distance from one crest to the next and **wave height** is the vertical distance from the crest to the trough. As a wave moves through water, the water particles rise and fall in **circular paths** with little forward movement of the water until the wave gets close to shore.

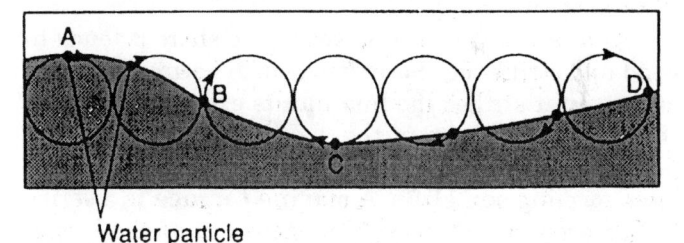

Water Particles Rotate in Circular Paths in Waves

When a wave reaches shallow water along the shore, friction causes the bottom of the wave to move more slowly than the crest, which falls over on the shore forming a **breaker** resulting in the foaming water called **surf**. The water from the breaking wave washes up on the shore as **swash** moving sand forward. As gravity pulls the water back, the **backwash** rolls beach materials toward the ocean.

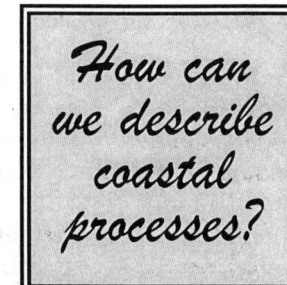

How can we describe coastal processes?

As waves strike the shoreline at an angle, they form a **longshore current** in the surf zone which moves sediment parallel to the coast. The sands transported by these currents are deposited farther along the coast. Coastal communities may build walls, called **groins**, perpendicular to the shore to prevent the longshore currents from removing sand from their beaches. **Jetties**, much like groins, are rock barriers built on both sides of a harbor entrance to slow down sedimentation which would clog the entrance to the harbor.

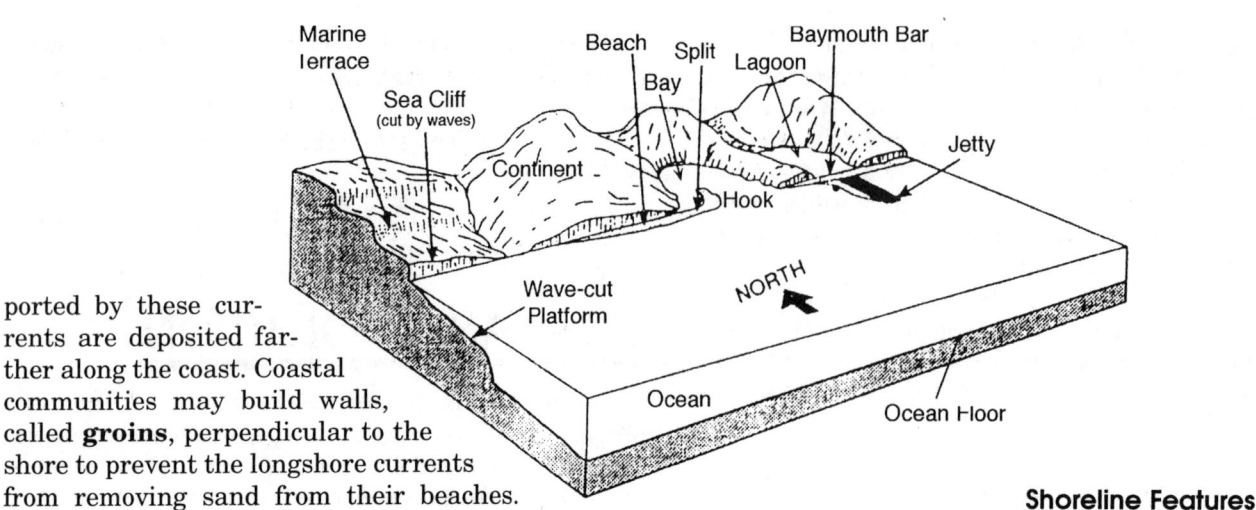

Beaches vary in size, shape, and composition. Although a majority of beaches are made of sand, there are gravel, pebble, cobble and boulder beaches, named for the size of the particles. Beaches vary in color depending upon the parent material forming the sand. Black sand beaches may form and are made from basaltic lava. Pink and very white sand beaches are formed from shells and coral.

WAVE REFRACTION

As a line of waves approaches the shore it tends to bend and strike the shore "head on." The greatest force of the water strikes the **headlands** extending out into the water. The force of the waves striking the headlands breaks the rocks straightening the coastline and often forming **sea cliffs**. A **marine terrace** is a series of flat areas which look likes steps out of the ocean. They are cut by the action of waves or streams.

SHORELINE FEATURES

Erosion and deposition are constantly changing coastlines. Several meters of material may be removed by one violent storm. Pounding waves may cut through a headland and form a **sea arch**.

Sand moved by currents may be deposited forming beaches or **spits** that extend out from the shoreline. Currents may curve the end of the spit into a **hook**. Occasionally spits form across the entire opening of a bay forming a **baymouth bar** with a **lagoon** behind it.

Shoreline Features

QUESTIONS FOR TOPIC C

1 The table shows the percentage by mass of dissolved ions in seawater.

Which mineral is most abundant in seawater?
1. calcite (calcium carbonate)
2. halite (sodium chloride)
3. quartz (silicon dioxide)
4. magnetite (iron oxide

Dissolved Ion	Percentage
Chloride (Cl–)	55.04
Sulfate (SO4 2–)	7.68
Bicarbonate (HCO3–)	0.41
Bromide (Br–)	0.19
Sodium (Na+)	30.61
Magnesium (Mg2+)	3.69
Calcium (Ca2+–)	1.16
Potassium (K+)	1.10
All others	0.12
	100.00

2 Historically, political boundaries have been associated with land areas. Most of the ocean basin, however, has not been divided politically. Why are people becoming more concerned about the political status of the ocean basins?
1. Worldwide, rising sea level is changing the shapes of the continents.
2. The salinity of ocean water is changing.
3. The size of ocean basins is decreasing.
4. Some ocean basins contain important mineral resources.

3 Most of the sediments deposited on the deep ocean bottoms consist of
1. volcanic dust and materials dissolved by ocean currents
2. windblown particles and meteorites
3. materials directly deposited by streams and glaciers
4. the remains of marine organisms and particles from land areas

4 What is the primary cause of the major surface currents found in the North Atlantic Ocean?
1 planetary winds
2 mantle plumes beneath ocean plates
3 undersea earthquakes
4 high winds from storms

Base your answers to questions 5 and 6 on the chart below. The chart shows the salinity of three different water samples and their densities. Salinity is a measure of total amount of dissolved minerals in the sea water expressed as parts per thousand (o/oo). The temperature of all three water masses is 20°C.

Water Sample	Salinity (percent)	Density (g/cm3)
A	36.3	1.026
B	34.0	1.024
C	35.3	1.025

5 The three water samples were found in the same calm regions of the ocean. What are their relative vertical positions?
(1) C would be above A but below B
(2) A would be above both B and C
(3) B would be below both A and C
(4) C would be below A but above B

6 What is the major source of the dissolved minerals that affects the salinity of ocean water?
1 submarine volcanoes
2 sediments eroded from land
3 deep ocean sediments
4 shells of sea animals

7 The usual direction of movement of major surface ocean currents is best described as
1 clockwise in the Northern Hemisphere and counterclockwise in the Southern Hemisphere
2 clockwise in the Southern Hemisphere and counterclockwise in the Northern Hemisphere
3 clockwise in both hemispheres
4 counterclockwise in both hemispheres

8 Which is the most common source of energy for surface ocean waves?
1 the Earth's rotation
2 disturbances of the ocean bottom
3 planetary winds
4 heat flow and seafloor spreading

9 Which is a characteristic of water that helps the oceans to moderate the climates of the Earth?
1 Water is a fluid with a high specific heat.
2 Water can exist as a high-density solid.
3 Water can dissolve and transport materials.
4 Water can flow into loose sediments to deposit mineral cements.

Base your answer to question 10 on the diagram and the *Earth Science Reference Tables*.

10 Which conditions describe the ocean current that would be found at position X in the diagram?

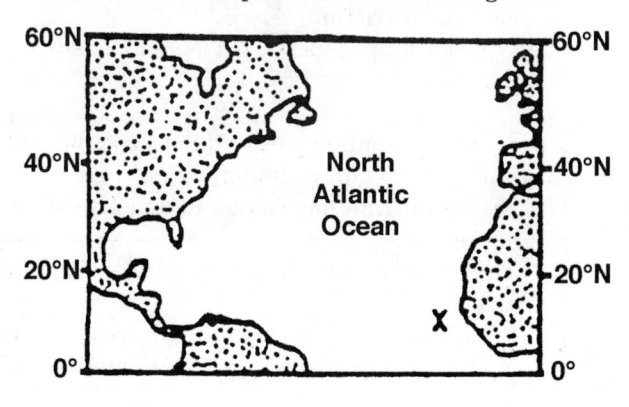

1 warm and flowing towards the north
2 warm and flowing towards the south
3 cold and flowing towards the north
4 cold and flowing towards the south

11 The major source of sediments found on the deep ocean bottom is
1 erosion of continental rocks
2 submarine landslides from the mid-ocean ridges
3 icebergs that have broken off glaciers
4 submarine volcanic eruptions

Base your answers to questions 12-14 on the diagram which shows the movement of water particles in ocean waves. The particles are represented by black dots. Letters A, B, C, and D are points of reference.

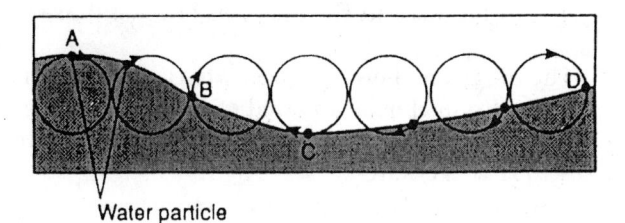

12 The passage of a wave will cause a particle of water at the surface to
1 move horizontally, only
2 move vertically, only
3 move in a circular path
4 remain stationary

13 The wave pattern shown in the diagram would occur most often in
1 a longshore current parallel to a coastline
2 shallow water in the breaker zone of a beach
3 a turbidity current
4 deep water

14 Why do waves form breakers as they move from deep water into shallow water?
1 The speed of the waves increases.
2 The wave crests are slowed by the air.
3 Waves collapse as the bottom of the waves is slowed by friction.
4 Waves collapse as the bottom of the wave moves faster.

15 What do mid-ocean ridges and hot spots beneath ocean plates have in common?
1 Rising magma moves due to density differences
2 They are located along crustal plate boundaries.
3 Local earthquakes originate at great depths.
4 Neither is associated with plate motions.

16 The diagram represents the magnetic fields of rocks in the oceanic crust at the mid-ocean ridge.

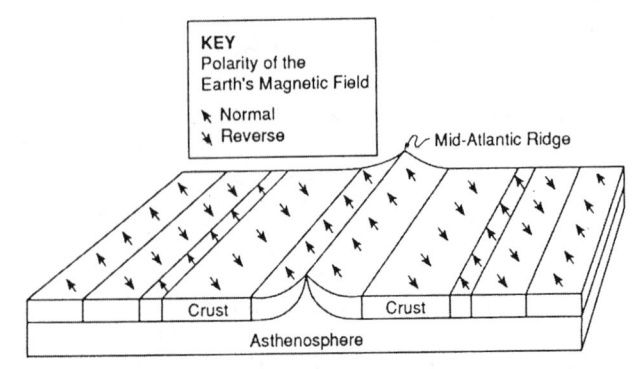

The mapping of these magnetic fields provides information about
1 the origin of ocean basins
2 climatic patterns over the oceans
3 the circulation of ocean currents
4 differences in the climates of coastal areas

17 The diagram shows an area where sea level gradually dropped over a period of thousands of years. A continuous sandy beach deposit stretching from A to B was created.

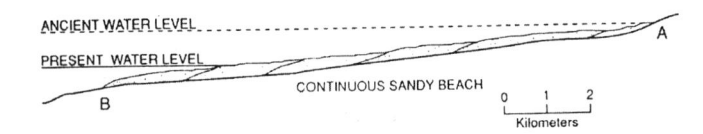

Which statement about the beach deposit would most likely be true?
1 It is older at A than at B.
2 It is older at B than at A.
3 It is the same age at A and B.

18 Points A through E on the map below represent locations in the Atlantic Ocean between the United States and Africa. Assume that a complete core sample of sediment and sedimentary rock could be taken from the ocean bottom at each of these five locations.

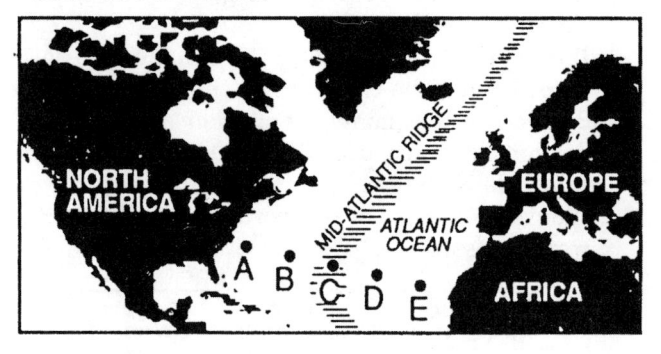

Which diagram of core samples best represents the relative thickness of the sediments at the five locations?

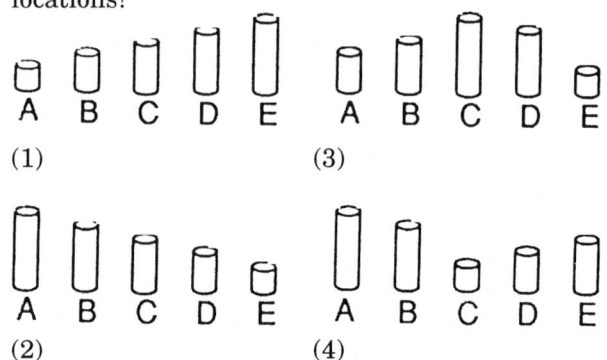

Base your answers to questions 19 through 23 on the topographic map showing a peninsula in the coastal area of North Carolina.

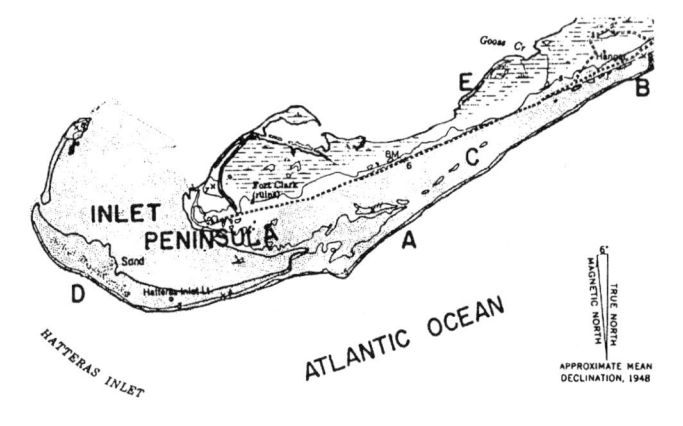

19 What is the elevation of the shoreline?
(1) 0 meters   (3) 5 meters
(2) 1 meter   (4) 10 meters

20 Inlet Peninsula is an example of
1 a spit   3 an atoll
2 a sea cliff   4 a marine terrace

21 The small hills shown by contour lines near C are
   1 drumlins   3 dunes
   2 buttes   4 kames

22 Along the coastline between points A and B, the longshore currents are probably moving toward the
   1 northeast
   2 northwest
   3 southeast
   4 southwest

23 The least amount of erosion of the shoreline by wave action is taking place near
   (1) A   (3) D
   (2) B   (4) E

Base your answers to questions 24 through 27 on the profile below. Four zones within the profile are labeled. The profile shows ocean waves approaching a beach. The circles show the motion of water beneath the wave.

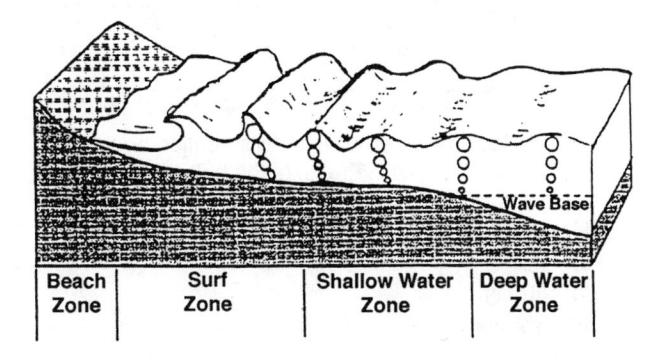

24 In which zone is sand most likely to be moving along the shore?
   1 deep water zone
   2 shallow water zone
   3 surf zone
   4 beach zone

25 Which statement best describes the waves in the surf zone?
   1 The waves are unaffected by the ocean bottom.
   2 The wave height is decreasing.
   3 The speed of the wave bottom is decreasing.
   4 The waves collapse as their wave heights increase.

26 The sand on this beach has its origin from the weathering of granite bedrock. According to the *Earth Science Reference Tables*, a mineral that is likely to be found in the sand is
   1 quartz   3 halite
   2 calcite   4 olivine

27 If a strong coastal storm with high wind speed moves into the region, the wave heights are likely to
   1 increase
   2 decrease
   3 remains the same

28 Structures such as jetties and breakwaters are built along the shore of the ocean in order to
   1 prevent ice formation along the shore
   2 increase the speed of longshore currents
   3 absorb and deflect the energy of the water
   4 lower the salinity of the water

29 At which location is ocean water most likely to be clean and unpolluted?
   1 near the west coast of North America
   2 near the east coast of South America
   3 in the mid-Pacific
   4 around Australia

Base your answers to questions 30 through 35 on the diagram which represents a shoreline in New York State along which several general features have been labeled. Letter B identifies a location on the shoreline.

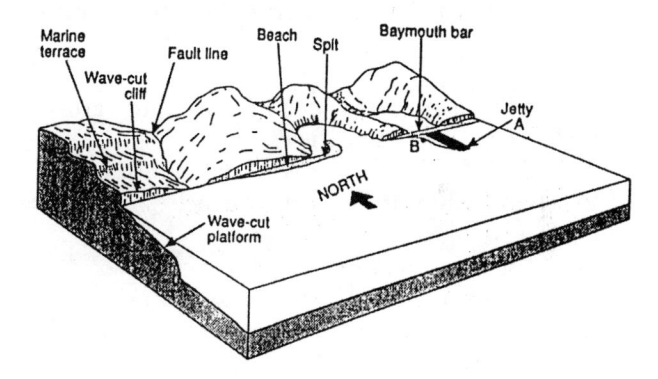

30 What is the most likely source of the waves approaching this coastline?
   1 variations in water temperature
   2 density differences within the water
   3 the rotation of the earth
   4 surface winds

31 Which statement best describes the longshore current that is modifying the coastline?
   1 The current is flowing northward at a right angle to the shoreline.
   2 The current is flowing southward at a right angle away from the shoreline.
   3 The current is flowing eastward parallel to the shoreline.
   4 The current is flowing westward parallel to the shoreline.

32 Past movements along the fault line most likely caused the formation of
1 the baymouth bar
2 tsunamis
3 longshore currents
4 tidal currents

33 After the formation of the baymouth bar, the jetty (a structure made of rocks that extends into the water) labeled *A* was constructed perpendicular to the shoreline. Which statement best describes the result of the construction of this jetty?
1 Water current velocity at location *B* decreased.
2 Water current velocity at location *B* increased.
3 Sand deposition at location *B* decreased.
4 Sand deposition was not affected.

34 The marine terraces represent former positions of the wave-cut platform. These terraces indicate that
1 crustal uplift has occurred
2 crustal sinking has occurred
3 the coastline was subjected to many tidal waves
4 the coastline was affected by strong winter storms

35 Which feature was formed because more erosion took place than deposition?
1 spit   3 baymouth bar
2 wave-cut cliff 4 beach

36 Often, sediments that are carried down submarine canyons by turbidity currents settle out over the ocean floor. Which cross section represents the pattern of deposition of these sediments?

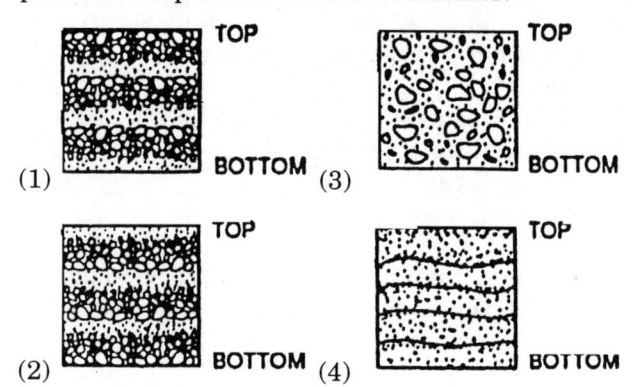

37 Based on the graph below, at which latitude is evaporation equal to precipitation in the ocean?

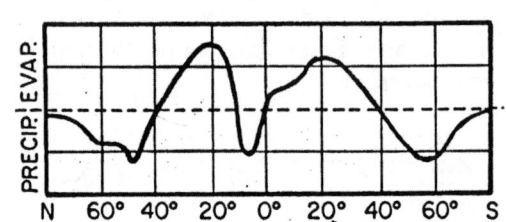

Note: Graph represents Evaporation above the dashed line and Precipitation below dashed line.

(1) 8° N
(2) 25° N
(3) 40° N
(4) 55° S

Topic D
Glacial Geology

Vocabulary To Be Understood

Ablating	Drumlins	Outwash Plain
Alpine Glacier	Erratic	Pleistocene Epoch
Arête	Finger Lakes	Snowfield
Centers of Accumulation	Horn	Striations
Cirque	Kettle Lake	Terminal Moraine
Continental Glacier	Interglacial Period	Till

A. GLACIAL GEOLOGY

What are glaciers?

Glaciers are found at high altitudes and in the high latitudes, because these are the locations where more snow falls during the winter season than melts during the summer season. The snow accumulates and compaction and recrystallization changes it to **glacial ice**. The ice flows slowly out of these **centers of accumulation** or **snow fields** due to the force of gravity. The ice always has a forward motion. The fastest ice flow occurs in the center of the glacier where there is the least friction with the valley walls. If the front of the ice appears to move backward or retreat, it is because the glacier is **ablating** or melting faster than it is moving forward.

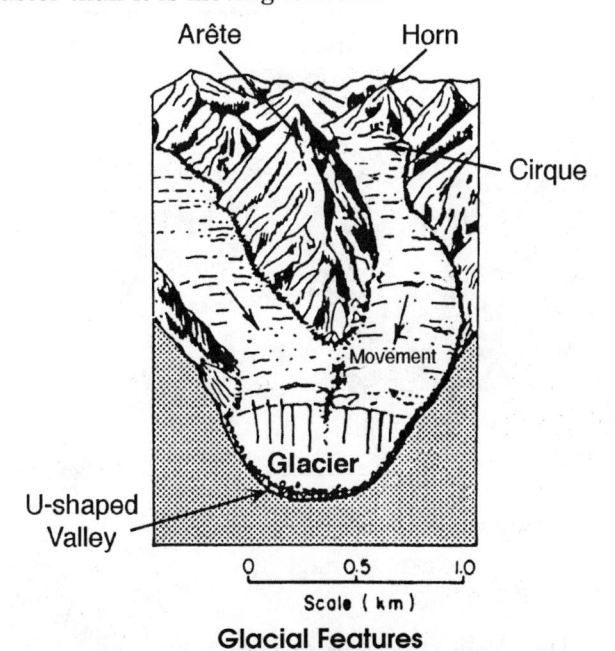

Glacial Features

Glaciers found in high mountains are called **Alpine glaciers**. These glaciers tend to make a rough, jagged landscape gouging out bowl-shaped **cirques** and **u-shaped valleys** and leaving **arêtes** (sharp ridges) and **horns** (pyramid-shaped mountains), as they move out of a snow field.

Continental glaciers are large **ice sheets** like those found in Antarctica and Greenland. As they move out from the center of accumulation, the tremendous weight of the thick ice tends to depress, scour, and round the land surface (see Unit 4 for additional information and explanation).

As glaciers move, they carry, push, and drag huge amounts of **glacial till** (unsorted sediment). The sediment ranges in size from enormous boulders to the finest clay. Sediment frozen in the bottom of the glacial ice (the cutting tools) abrades the surface over which the glacier passes. Large fragments form **striations (grooves)** in the surface indicating the direction the glacier was moving. Finer sediment polishes the rock surface. The rounding of rock surfaces and glacial striations in both the Catskills and Adirondacks shows that the glacial ice that covered New York State buried the land at least one mile deep.

The melting ice left large boulders, called **erratics** scattered over the landscape. The rock type of an erratic

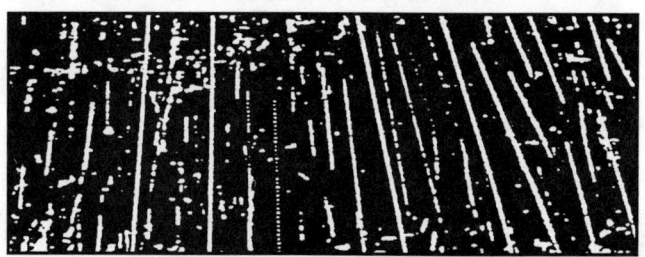

Striations on Limestone Bedrock
(Exposed in New York State)

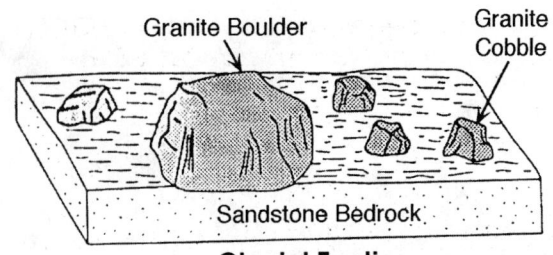

may be so distinct as to enable scientists to trace the rock to its point of origin giving evidence of the direction of ice flow.

Clues to prehistoric conditions on Earth are gathered by studying cores of glacial ice collected from deep within glaciers. Some of this ice formed centuries ago trapping air, dust, and pollen from the distant past.

GLACIAL HISTORY

Layers of weathered till, buried under other layers of weathered till, show that there have been several periods of glaciation during the **Pleistocene Epoch**. New York State experienced a series of **glacial** and **interglacial periods**.

In addition to changing the landscape, the ice age in New York State caused the migration of animals and changes in the animal and plant communities. Most of the soils found in New York State are thin, rocky and poorly developed, made of weathered glacial till.

As glaciers grew and the ice advanced, sea level dropped because of the increased amount of water being held as ice. Pleistocene fossil evidence shows that, during this time, the continental shelf off Long Island was exposed and was inhabited by terrestrial (land) organisms. As the ice gradually melted during the interglacial periods, sea level rose again returning the continental shelf to a marine environment.

B. NEW YORK GLACIATION

Features resulting from glacial erosion and deposition may be seen over most of New York State. The north-south trending valleys of the Finger Lakes show river valleys widened by glacial erosion and dammed up by the deposition of moraine.

Swarms of **drumlins**, hills shaped like the back of a spoon, are found on the Lake Ontario Plain. The alignment of these drumlins indicates the direction of ice movement.

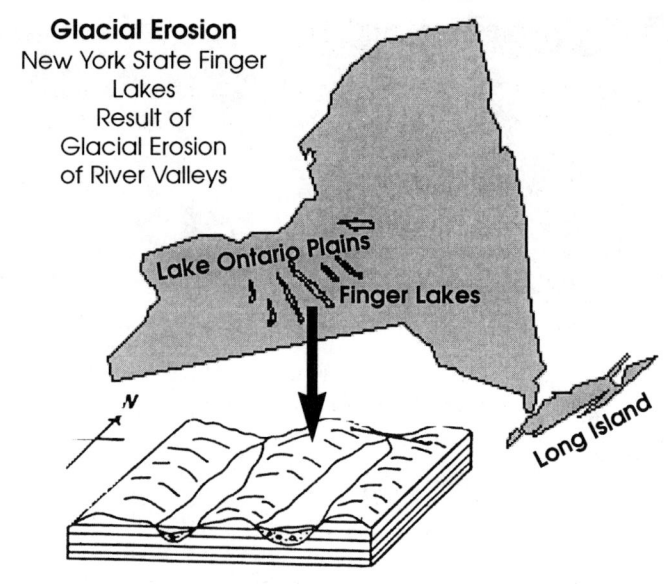

The northern ridge forming Long Island is **terminal moraine**, marking the farthest advance of the glacier. The southern part of Long Island is composed of **outwash plain** made from layers of horizontally deposited sorted material left behind by the water from the melting ice front. Lake Ronkonkoma is an example of a **kettle lake** left by a large mass of melting ice buried in the glacial sediment.

Almost all of New York State was covered by the ice sheet originating in Labrador. The water from this melting ice deposited gravel and sand in large sorted and layered deposits. These deposits are an important geological resource in the economy of New York State.

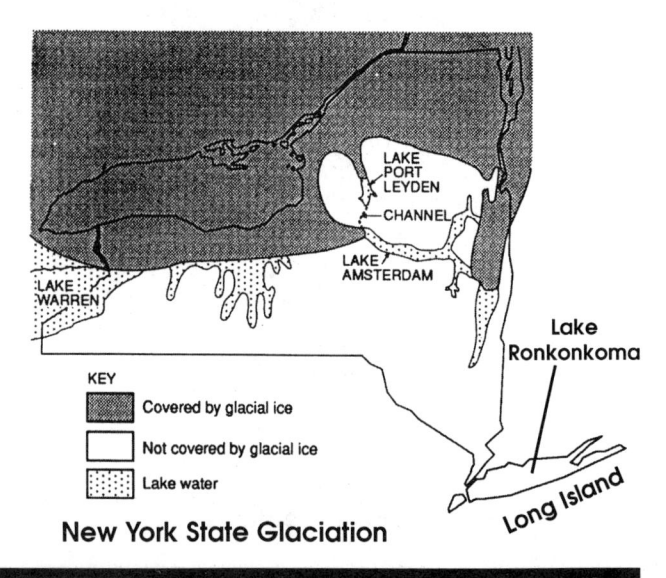

How did glaciers influence the landscape and economy of New York?

QUESTIONS FOR TOPIC D

1. Over the past 2 million years, which erosional agent has been most responsible for producing the present landscape surface features of New York State?
   1. groundwater
   2. wind
   3. glaciation
   4. human activities

2. What was one of the major effects of the continental glaciers on the landscapes of New York State?
   1. They formed numerous sharp mountain peaks and knife-edged ridges.
   2. They folded many of the rock layers.
   3. They deposited a covering of transported rock material over most of the state.
   4. They carved the wide U-shaped valleys into narrow V-shaped valleys.

3. The Adirondack Highlands landscape region was formed primarily by
   1. changes in the water levels of the Great Lakes
   2. erosion by the Hudson and Mohawk Rivers
   3. mountain building and glacial erosion
   4. wind erosion in an arid climate

4. Which is the best evidence that more than one glacial advance occurred in a region?
   1. ancient forests covered by glacial deposits
   2. river valleys buried deeply in glacial deposits
   3. scratches in bedrock that is buried by glacial deposits
   4. glacial deposits that overlay soils formed from glacial deposits

5. Many elongated hills, each having a long axis with a mostly north-south direction, are found scattered across New York State. These hills contain unsorted soils, pebbles, and boulders. Which process most likely formed these hills?
   1. stream deposition
   2. wind deposition
   3. wave deposition
   4. glacial deposition

6. The formation of the Finger Lakes of central New York State and the formation of Long Island are both examples of
   1. climatic changes resulting in a modification of the landscape
   2. uplifting and leveling forces being in dynamic equilibrium
   3. soils differing in composition depending upon the bedrock composition
   4. activities of man altering the landscape

7. The diagram represents a landscape and bedrock section of an area of Central New York State

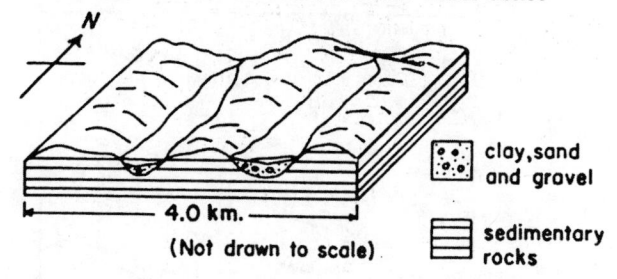

   The present surface landscape of this area was produced chiefly by
   1. faulting
   2. folding
   3. volcanic activity
   4. glaciation

8. The diagram illustrates the formation of a
   1. plunge pool
   2. kettle hole
   3. sinkhole lake
   4. finger lake

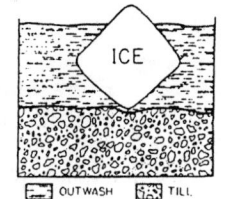

9. While studying the movement of alpine glaciers that are advancing from the north, geologists placed metal stakes extending in a straight line from one valley wall to the other. Which sketch would best illustrate the position of the stakes one year later?

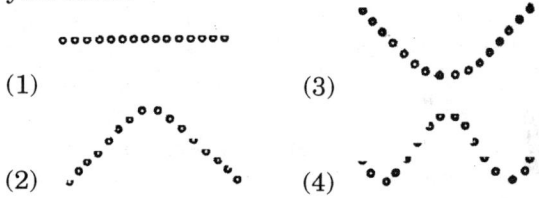

10. Which observation LEAST supports inferences that New York State was covered in the recent past by great sheets of ice?
    1. boulders on hilltops
    2. increased average annual temperatures
    3. stream valleys that have been widened and deepened
    4. soils foreign to the bedrock beneath them

11. Which diagram best represents a cross section of a valley which was glaciated and then eroded by a stream?

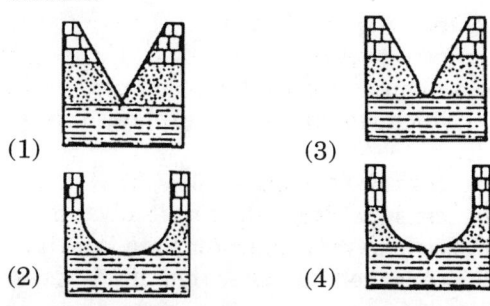

12 This section of Long Island shows considerable evidence of

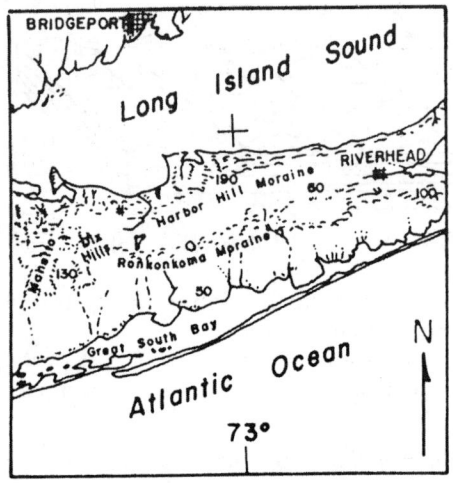

1 Precambrian metamorphism
2 Pleistocene glaciation
3 Triassic intrusions
4 Permian mountain building

13 The graph below shows the average temperature of the Earth during the past 250,000 years and the beginning and end of the most recent glacial and interglacial stages.

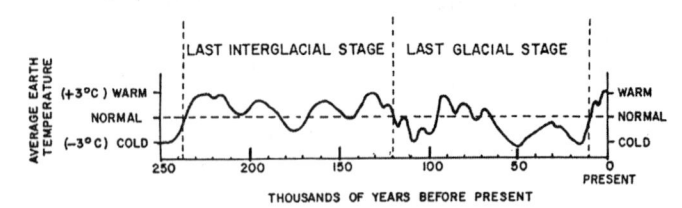

According to the graph, the duration of the last glacial stage was from
(1) 250,000 years ago to 240,000 years ago
(2) 240,000 years ago to 120,000 years ago
(3) 120,000 years ago to 10,000 years ago
(4) 10,000 years ago to the present

14 Which force is primarily responsible for the movement of a glacier?
1 ground water
2 running water
3 wind
4 gravity

15 The front of an active glacier is observed to be stationary. Which statement best explains this observation?
1 The ice is not moving at all.
2 The ice is melting as fast as it advances.
3 The ice is advancing faster than it melts.
4 The ice is melting faster than it advances.

16 A glacial deposit would most likely consist of
1 particles in a wide range of sizes
2 particles the size of pebbles and larger
3 sediments in flat, horizontal layers
4 sediments found only in the bottoms of stream valleys

17 When were large parts of North America covered by ice sheets?
1 only once, early in the geologic history of the Earth
2 only once, in the recent geologic past
3 once early in the geologic history of the Earth, and once in the recent geologic past
4 many times during the geologic history of the Earth

18 Which event would most likely cause a new ice age in North America?
1 a decrease in the energy produced by the Sun
2 a decrease in the light reflected by the surface of the Earth
3 an increase of carbon dioxide in the Earth's atmosphere
4 an increase in the westward drift of the North American continent

19 Which statement presents the best evidence that a boulder-sized rock is an erratic?
1 The boulder has a rounded shape.
2 The boulder is larger than surrounding rocks.
3 The boulder differs in composition from the underlying bedrock.
4 The boulder is located near potholes.

20 The direction of movement of a glacier is best indicated by the
1 elevation of erratics
2 alignment of grooves in bedrock
3 size of kettle lakes
4 amount of deposited sediments

21 The general direction of continental glacial advance in New York State was from
1 south to north
2 north to south
3 west to east
4 east to west

22 Another ice age would probably result in a change in
1 sea level
2 Moon phases
3 the speed of the Earth in its orbit
4 the time between high tides

Base your answers to questions 23 through 25 on the diagrams below. The diagrams represent glacial events in the geologic history of the Yosemite Valley region in California.

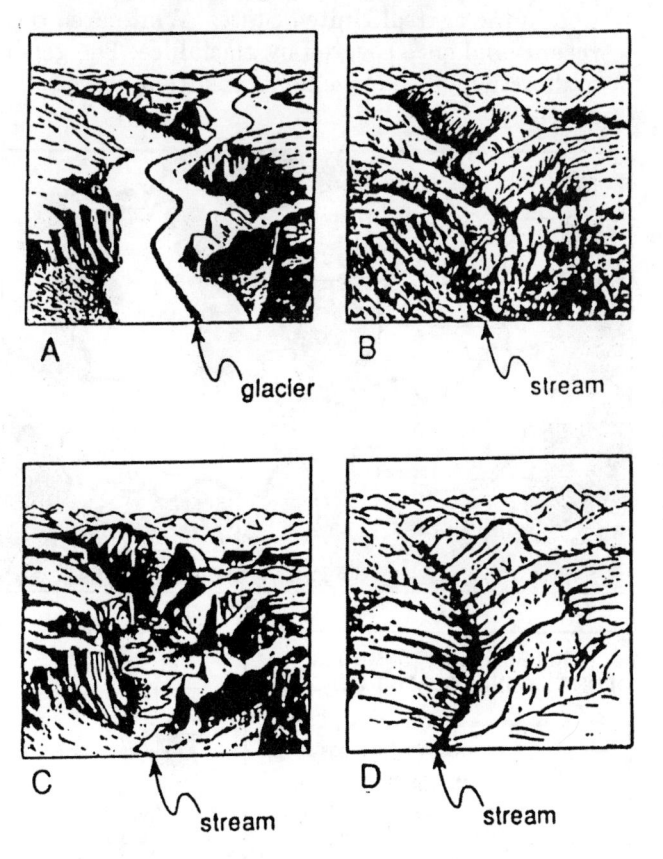

23 Which diagram most likely represents the most recent stage of landscape development?
(1) A
(2) B
(3) C
(4) D

24 Which features currently found in this region are the result of glaciation?
1 intrusions of volcanic rock
2 deep U-shaped canyons with steep sides
3 fossils of marine organisms
4 thick soils covering all rock surfaces

25 Which natural resource of economic value would most likely be found in this region?
1 rock salt
2 petroleum
3 natural gas
4 sand and gravel

Base your answers to questions 26 through 28 on the diagrams below. Diagram I represents a section of the northeastern United States and Canada. Five different source regions, A through E, are shown along with the pattern of glacial deposits containing boulders which originated from each location. Diagram II represents the appearance of the surface of a typical boulder from any of the deposit locations.

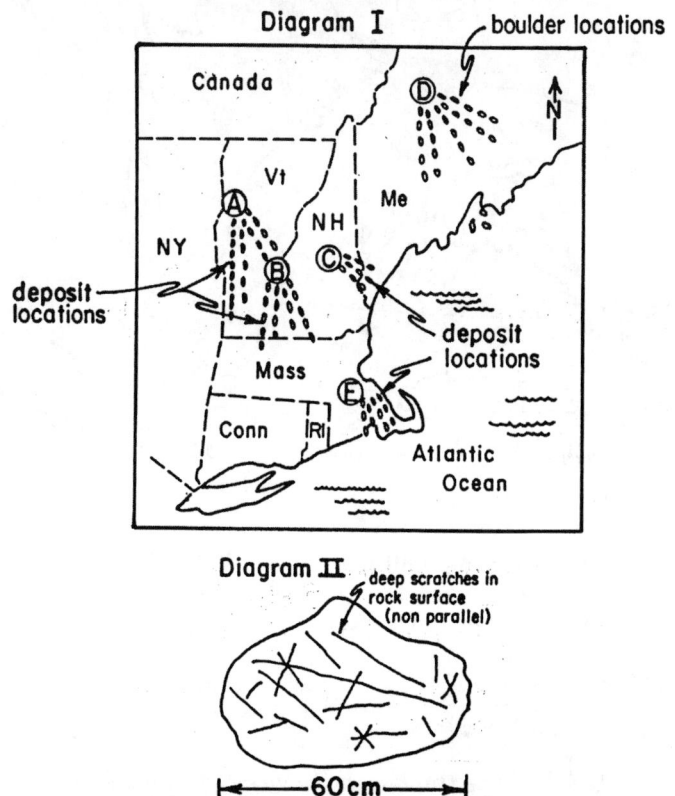

26 The force that caused the deposits to be distributed in the pattern shown in Diagram I most likely came from which general direction?
1 northwest
2 northeast
3 southwest
4 southeast

27 Which characteristic do all of the deposits most likely have in common?
1 They have the same chemical composition.
2 They were eroded from source region A.
3 They are composed of unsorted sediments.
4 They are found at the ends of large rivers.

28 The scratches in the boulder shown in Diagram II were most likely caused by the
1 internal arrangement of the minerals in the boulder
2 splitting of a large boulder into two smaller boulders
3 erosion of the boulder by running water
4 movements of the boulder over bedrock

Base your answers to questions 29 through 32 on the map of New York State and surrounding areas shown below. The short lines indicate the location and direction of streamline features such as drumlins and glacial striations. The moraine deposits and drumlins consist of glacial till.

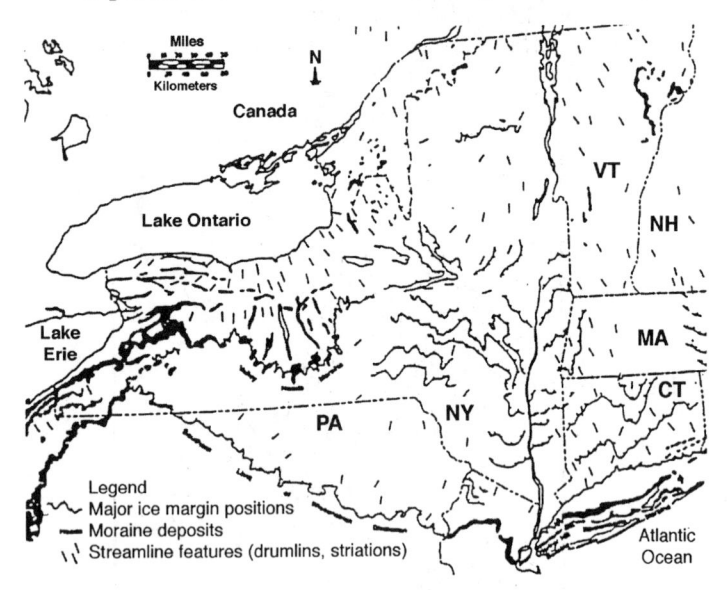

Base your answers to questions 33 through 36 on the map below and on your knowledge of Earth science. The map shows the southernmost advance of four major stages of continental glaciation in the central United States. White area represent land once covered by glacial ice. The general direction of ice movement was from north to south.

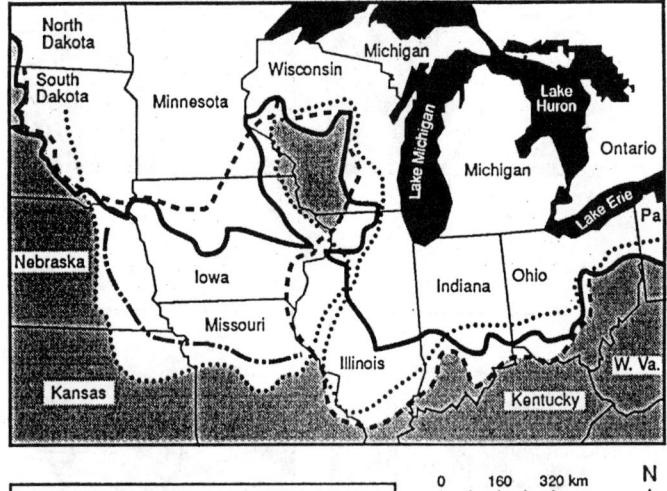

29 The glacial ice that caused the deposits in the region south of Lake Ontario and north of the Finger Lakes most likely came from which general direction
1 north
2 southeast
3 southwest
4 west

30 According to the *Earth Science Reference Tables*, during which geologic time interval were the glacial deposits in New York State most likely formed?
1 Mesozoic
2 Jurassic
3 Pleistocene
4 Paleocene

31 Which state would show the smallest percentage of its surface area covered by glacial deposits?
1 Connecticut (CT)
2 Vermont (VT)
3 Pennsylvania (PA)
4 Massachusetts (MA)

32 In which two landscape regions are most of the moraine deposits shown on the map located?
1 Adirondack Mountains and the Catskill Mountains
2 Allegheny Plateau and the Atlantic Coastal Plain
3 Tug Hill Plateau and the Erie-Ontario Lowlands
4 The Atlantic Coastal Plain and the St. Lawrence Lowlands

33 The landforms that mark the terminal glacial boundaries are made up of
1 residual soil particles resting on a flat plain
2 rounded grains in a sand dune
3 layered clay particles on a flat plain
4 unsorted gravel in low hills

34 Which state was partly or completely covered by glacial ice during all four stages of ice advance?
1 Iowa
2 Kansas
3 Kentucky
4 Missouri

35 Which map best represents the southernmost advance of the continental ice sheet during the Wisconsinan Stage?

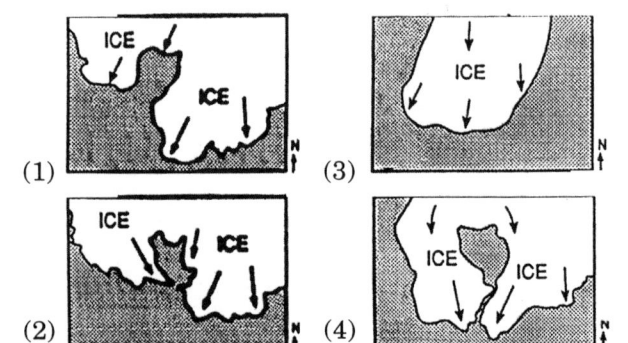

36 What evidence found on the former ice-covered areas would best show the direction of continental glacial movement?
1 resistant folded metamorphic bedrock
2 high-temperature igneous and volcanic bedrock
3 parallel scratches and grooves in the bedrock
4 bedrock containing fossils of animals that lived in cold water

Base your answers to questions 37 through 40 on the map below and on your knowledge of Earth science. The map shows the inferred position of the continental ice sheet in New York State approximately 12,000 years ago.

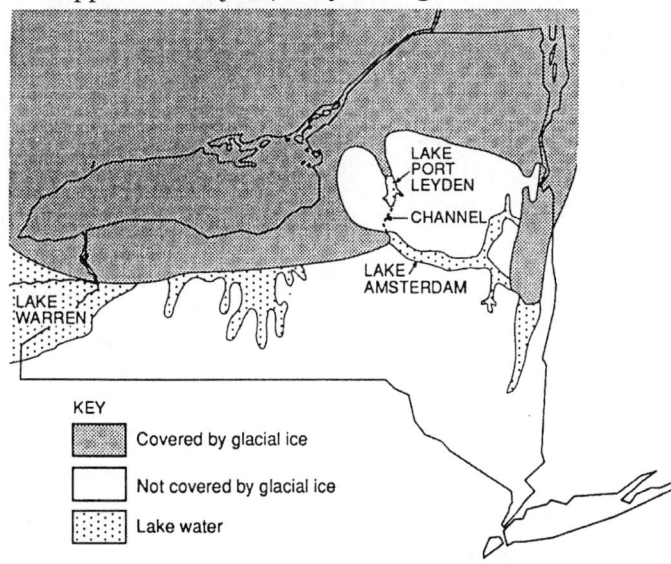

37 Which evidence would have been most useful to geologists for locating the edges of the ice sheet shown on the map?
1 flat, thick deposits of impermeable clay
2 piles of unlayered, unsorted sediment
3 folded layers of bedrock
4 formations of rocks with interlocking crystals

38 A steep-walled channel was cut by meltwater pouring from Lake Port Leyden into Lake Amsterdam about 12,000 years ago. What condition probably existed in the channel at that time?
1 The channel contained a small volume of water.
2 The channel had a very gentle slope.
3 The water flowed at a high velocity in the channel.
4 The water in the channel was very warm.

39 Which map best represents the inferred position of the ice sheet at the time of glacial deposition on Long Island? [The shaded portion represents the areas covered by glacial ice.]

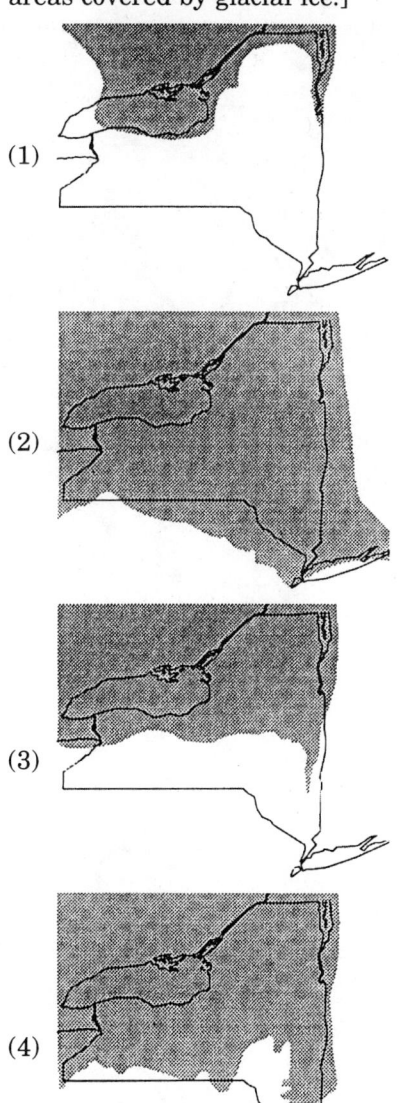

40 Which fossil found in sediments in New York State support the inference that these sediments were deposited during the Pleistocene Epoch?
1 eurypterid fossils
2 stromatolite mounds
3 mastodont bones
4 coelophysis footprints

Unit Five
Earth's History

Vocabulary To Be Understood

Absolute Age (Dates)	Isotope (Radioactive)	Rock Formation
Bedrock (Local Rock)	Joint	Species
Carbon-14 Dating	Organic Evolution	Unconformity
Correlation	Orogeny	Uniformitarianism
Extrusion, Intrusion	Outcrop	Uranium-238
Fossil, Geologic Time Scale	Principle of Superposition	Vein
Half-life	Radioactive Dating, Decay	Volcanic Ash
Index Fossil	Relative Age (Dates)	Walking the Outcrop

A. GEOLOGIC EVENTS

The history of the changing Earth is told in the Earth's **geologic events**. The analysis, synthesis, and interpretation of these geologic events is a form of puzzle solving.

SEQUENCE OF GEOLOGIC EVENTS

Knowing the sequence of the geologic events that took place during the formation of the Earth's crust makes it possible to develop a **geologic history** of the Earth and to better understand the forces that have and still are changing the Earth's crust.

Relative age is concerned with the sequence of geologic events that have occurred in an area as shown by the appearance of the rock layers. Relative age is not concerned with the actual ages of the rocks. This method uses sedimentary rock layers, igneous extrusions and intrusions, faults, folds, continuity, similarity of rock, fossil evidence, and volcanic time markers as clues to determine the probable order and conditions under which rock layers formed.

The actual age of a rock or fossil is called **absolute age**. The most accurate method of determining the absolute age of geologic events and rock is by techniques such as **radioactive dating**. Every radioactive element decays. A "parent" element emits radiation and particles until it is transformed by **decay** (breakdown) into a stable "daughter" element. Sometimes the "changing" element passes through a series of transformations into other radioactive elements before reaching this stability. Each radioactive element also has its own identifiable pattern and **rate of decay**.

CHRONOLOGY OF LAYERS

The bottom layer in a series of horizontal sedimentary rock layers is the oldest, unless the series has been overturned or has had older rock thrust over it. This concept, called the **principle of superposition**, is used to determine the sequence in which a series of sedimentary layers was formed.

How can the order in which geologic events occurred be determined?

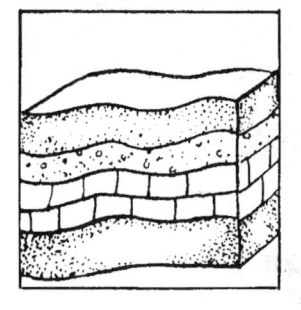

Principle of Superposition
with "youngest" on top and "oldest" on the bottom

IGNEOUS INTRUSIONS AND EXTRUSIONS

The rock layers through which **igneous intrusions** or extrusions cut are older than the intrusions or extrusions themselves, since the rock layers must be formed prior to the intrusion of magma or extrusion of lava.

Contact metamorphism of the rocks, through which the magma has moved, provides an additional clue to their relative age. Contact metamorphism is

Contact Metamorphism

more recent than the rock layers that were metamorphosed.

FAULTS, JOINTS, AND FOLDS

Faults (cracks in the rock along which movement has occurred), **joints** (immovable cracks), and **folds** (bends in the rock strata) are younger than the rocks in which they appear. These distortions in rock occur due to changes in temperature and pressure.

INTERNAL CHARACTERISTICS

Fragments that occur within a rock are older than the rocks in which they are found, since they were formed previously from other rocks. However, cracks and **veins** (mineral deposits that have filled a rock crack or permeable zone) are younger than the rocks in which they occur.

Sedimentary rocks are younger than the sediments and the cements that formed them. Another kind of internal characteristic of rock layers is an **unconformity**, which is a zone where rocks of different ages meet. An unconformity is a "gap" in the geologic rock record due to erosion or nondeposition. It is usually seen as a buried erosional surface.

B. RELATIVE AGE

The determination of the relative age of rock in geologic history can be accomplished through the use of correlation techniques. But, it is very important for the observer to distinguish actual *evidence* from *inferences*.

Correlation is the process of determining that the rock layers or geologic events in two separate areas are the same. Correlation involves observing the similarity and continuity of rock layers in different locations, comparing fossil evidence and using volcanic time markers.

CONTINUITY

When bedrock is exposed at the Earth's surface, it is called an **outcrop**, and correlation can be accomplished directly by "**walking the outcrop**." Over many years of destructional action, the landscape of a particular region may change greatly. For example, thick layers of level sedimentary rock could be cut into a wide valley by the action of streams. After many years, the valley will not resemble the original sedimentary formation. However, by careful examination of the rock strata (layers) exposed on opposite sides of the valley, the geologist may be able to reconstruct the geologic history of the valley.

How can rocks and geologic events in one place be matched to another?

SIMILARITY OF ROCKS

Rocks can often be tentatively matched on the basis of similarity in **appearance**, **color**, and **composition**. Referring back to the previous example, the geologist can use these rock characteristics to help figure out the puzzle of the valley's geologic history.

FOSSIL EVIDENCE

The remains or traces of many once-living organisms, found almost exclusively in sedimentary rock, are called **fossils**. These fossils provide clues to the environment in which the organisms once lived. If a geologist finds fossils of a marine organism, evidence is provided that this sedimentary rock was formed in the sea.

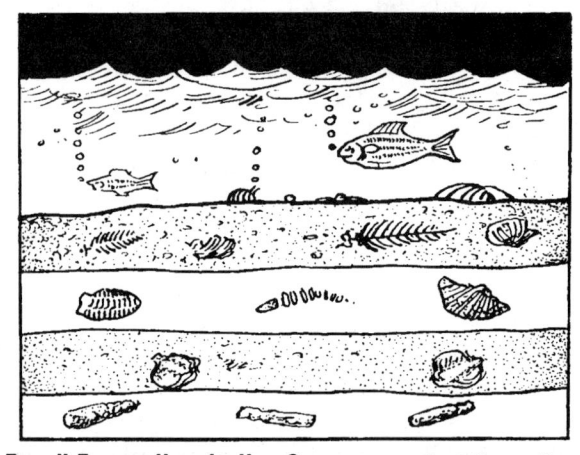

Fossil Formation in the Sea generally follows the same **Principle of Superposition** with the oldest fossils in the lower levels and younger above.

The geologist may then infer that this region was submerged at some time during geologic history.

VOLCANIC TIME MARKERS

A volcanic eruption is relatively short in duration when compared to the many years required for other constructional forces to build up the Earth's surface. When a volcano erupts, a layer of **volcanic ash** (fine particles of igneous rock ejected during the eruption) is rapidly deposited over a large area.

A layer of volcanic ash occurring between other layers of rock may serve as a time marker. Should a geologist discover a layer of volcanic ash buried between other layers of the sedimentary rock, and the actual date of the volcanic eruption is known, this time marker will provide very important information in determining the relative age of the rock layers above and below it.

PROBLEMS WITH CORRELATION

The process of "solving the puzzle" of geologic history appears fairly easy according to our discussion of rocks and geologic events and the correlation evidences. However, this oversimplification of the processes involved in determining geologic history may lead to misconceptions.

The geologist must exercise cautious interpretation to minimize this problem, since the very careful study of two similar rock formations may show that the rock formations are actually of different ages. Also, it is possible to find within a single formation, areas of different ages.

C. GEOLOGIC TIME SCALE

THE ROCK RECORD

A close study of the **rock record**, using fossil evidence to develop a geologic time scale and erosional evidence to help fill in any gaps in the fossil record, can lead to an inferred geologic history of an area.

Fossil Evidence. Fossils provide direct (e.g., shells, bones) and indirect evidence (e.g., footprints, burrows) of organisms that had lived on Earth. Events in geologic history can often be placed in order according to relative age by using evidence provided by certain fossils.

The fossils used to correlate rock layers are called **index fossils** or **guide fossils**. Index fossils are used because of their wide-spread horizontal distribution (geographical) in sedimentary rocks and their relatively short period of existence on the Earth (narrow vertical distribution). By comparing these fossils in various locations on the Earth, it is possible to correlate the relative ages of the rock in which they appear.

How can Earth's geologic history be sequenced from the fossil and rock record?

Scale of Geologic Time. Geologists have subdivided geologic time into units, called **eons** (e.g., Phanerozoic, Proterozoic, Archean), **eras** (e.g., in the Phanerozoic Eon, Cenozoic, Mesozoic, and Paleozoic), **periods** (e.g., in the Mesozoic Era, Cretaceous, Jurassic, and Triassic), and **epochs**, based on the fossil evidence. However, note that most of the geologic past is devoid of a fossil record (see the Geologic Time Scale from the *Reference Tables* on the next page).

A review of the **Geologic Time Scale** suggests the following sequence in the geologic history of Earth.

- The *Precambrian* or *Pre-Paleozoic Era* makes up about 85 percent of the total geologic time of Earth history. There is very little fossil evidence from this era, since the organisms that existed at this time did not lend themselves to making good fossils. They were small, simple, and soft bodied (such as algae and bacteria).

- The *Paleozoic Era* was much shorter, covering about 8 or 9 percent of the geologic history of the Earth. This era, which began the abundant fossil record, progressed from the Age of Invertebrates to the Age of Fishes and ended with the Age of Amphibians. The first vertebrates and the land plants and animals developed during this era. The Periods of the Paleozoic Era began with the Cambrian, passing through the Ordovician, Silurian, Devonian, Carboniferous, and ending with the Permian.

- The *Mesozoic Era* was even shorter, about 3 or 4 percent of the geologic history. This is the era in which the fossils of dinosaurs and the earliest birds and mammals were formed. The Mesozoic Era included the Triassic, Jurassic, and Cretaceous Periods.

UNIT FIVE – EARTH'S HISTORY – N&N©

Geologic Time Scale

EON	ERA	PERIOD	EPOCH	Life on Earth
PHANEROZOIC	CENOZOIC	QUATERNARY	HOLOCENE .01	Humans, mastodonts, mammoths
			PLEISTOCENE 1.6	Large carnivores
			PLIOCENE 5	Abundant grazing mammals
		TERTIARY	MIOCENE 24	Earliest grasses
			OLIGOCENE 37	Large running mammals
			EOCENE 57	Many modern groups of mammals
			PALEOCENE 66	Last of dinosaurs — Earliest placental mammals
	MESOZOIC	CRETACEOUS	LATE 97	Climax of dinosaurs and ammonoids -- followed by extinction. Earliest flowering plants
			EARLY 144	Great decline of brachiopods. Great development of bony fishes
		JURASSIC	LATE 163	Earliest birds and mammals
			MIDDLE 187	Abundant dinosaurs and ammonoids
			EARLY 190	
		TRIASSIC	LATE 230	Modern coral groups appear. Earliest dinosaurs, flying reptiles
			MIDDLE 240	Abundant cycads and conifers
			EARLY 245	Extinction of many kinds of marine animals, including trilobites
	PALEOZOIC	PERMIAN	LATE 256	
			EARLY 286	Little change in land animals
		CARBONIFEROUS	Pennsylvanian 320	Earliest reptiles. Great coal-forming forests
			Mississippian 360	Abundant sharks and amphibians. Large and numerous scale trees and seed ferns
		DEVONIAN	LATE 374	Earliest amphibians, ammonoids, sharks
			MIDDLE 387	Extinction of armored fishes, other fishes abundant
			EARLY 408	Diverse brachiopods
		SILURIAN	LATE 421	Earliest insects. Earliest land plants and animals
			EARLY 438	Peak development of eurypterids
		ORDOVICIAN	LATE 458	First corals
			MIDDLE 478	Invertebrates dominant -- mollusks become abundant
			EARLY 505	Echinoderms expand in numbers and kinds. Graptolites abundant
		CAMBRIAN	LATE	Earliest fish. Algal reefs
			MIDDLE	Earliest chordates
			EARLY 540	Diverse trilobites dominant. Earliest marine animals with shells
PRECAMBRIAN (PROTEROZOIC / ARCHEAN)				Soft-bodied animals. Stromatolites 1300

Precambrian notes (left side of chart):
- Oldest marine invertebrates
- First appearance of sexually reproducing organisms
- Transition to atmosphere containing oxygen
- Oldest microfossils
- Geochemical evidence for oldest biological fixing of carbon
- Oldest known rocks
- Estimated time of origin of earth and solar system

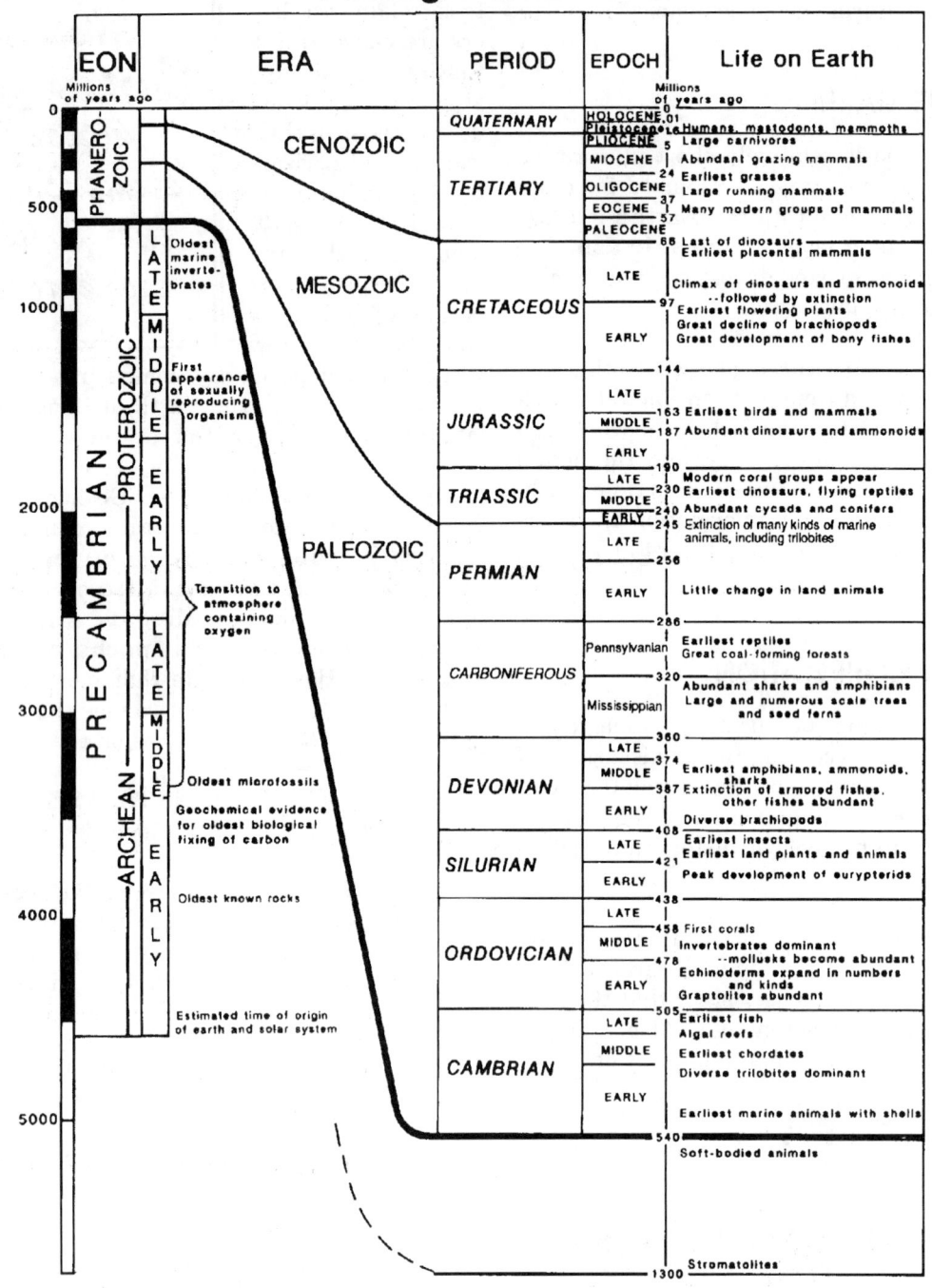

- The most recent era, 2 or 3 percent of the geologic time scale, is the **Cenozoic Era**, which includes the fossils of many modern plants and mammals, including the appearance of humans.

Human existence is infinitesimal (0.04% of geologic time) in comparison to the entire geologic time of the Earth (4.6 billion years).

Plate motions and mountain building (**orogeny**) events may also be identified and placed in this time sequence by using the information in the *Earth Science Reference Tables*.

THE EROSIONAL RECORD

Buried erosional surfaces indicate some gaps in the time record of the rock. These gaps represent periods of destruction (erosion) of the geologic record, or nondeposition. It has been suggested that if the principle of superposition held absolutely true, and no forces of destruction (weathering and erosion) occurred after the origin of life forms, that the fossil record would show a complete time scale and history of the Earth. But, it is clear that destructional forces have worked in the past and are still changing the Earth's surface, since there is no known location on Earth in which the entire rock record has been preserved.

One of the key principles that geologists use to interpret the Earth's geologic history is the **principle of uniformitarianism**. In general, this principle implies that the geologic processes currently changing the Earth also have changed the Earth in the past. There may have been different rates of change at various times, but the patterns and agents of change remain the same.

THE GEOLOGIC HISTORY OF AN AREA

Using the evidence discovered in the rock record, the geologist can infer the geologic history of an area. The New York State geologic map and the geologic time scale may be used to illustrate the various portions of the rock record that have been preserved in New York State (see the *Earth Science Reference Tables*).

RADIOACTIVE DECAY

The evidences of the rock record discussed in this topic allow the geologist to make relative age estimates in the scale of geologic time and Earth history. To obtain **absolute** or **actual age**, the process of radioactive dating is used.

DECAY RATES

How can the absolute age of a rock be determined?

When the spontaneous and natural nuclear breakdown of unstable atoms occurs, particles and energy are released. This constant, predictable process is called **radioactive decay**. By measuring the amount of radioactive isotope, compared to the amount of decay product (more stable form of the element) present in a rock sample, absolute age can be determined. The decay rate is unaffected by external factors, such as pressure and temperature, that would normally affect chemical reactions.

Some, but not all, rocks contain atoms whose nuclei undergo radioactive decay. This decay occurs as a random event and is not influenced by other changes occurring in the rock at the same time.

HALF-LIVES OF RADIOACTIVE SUBSTANCES

The **half-life** of a radioactive substance is the time taken for the activity of decay to reduce the total amount of radioactive substance in a material to half of its original amount. Therefore, by knowing the original content of a radioactive material and comparing it to the present content of the same radioactive material, the age of the material can be determined.

The half-lives of **radioactive isotopes** are different for different substances. Some radioactive isotopes, such as **Carbon–14**, have short half-lives and are good for dating recent organic remains (between 1,000 and 50,000 years).

Other radioactive substances, such as Uranium–238 which decays to stable Lead–206, have very long half-lives and are good for dating older rock formations (more than 10 million years). Uranium–238 has a half-life of about 4.5 billions years. The Earth itself is estimated to be 4.6 billion years old. Therefore, Uranium–238 that formed at the same time the Earth was formed has had time to undergo only one half-life.

DETERMINING AGE BY RADIOACTIVE DECAY

The age of a rock or fossil can be inferred from the relative amounts of the undecayed (radioactive) substance and the decay product. For example, the actual age of a fossil can be inferred by the following method.

A sample piece of a fossil is taken and the amount of the Carbon–14 remaining in that piece is measured at 0.5 grams. An equal sample of an existing organism, shows that the amount of original Carbon–14 was 2.0 grams. Since the radioactive decay rate of Carbon–14 is the loss of half of the total amount in the sample piece every 5,700 years, the sample was formed 11,400 years ago.

Explanation: In the first 5,700 years of decay, the 2.0 grams were reduced to 1.0 gram; then, in the second 5,700 years of decay, the 1.0 gram was reduced in half again to 0.5 gram. Since one quarter of the original radioactive substance remained, the fossil had undergone two half-lives.

Radioactive Decay Data

Radioactive Element	Disintegration	Half-life
Carbon–14	C-14 → N-14	5.7×10^3 years
Potassium–40	K-40 → Ar-40	1.3×10^9 years
Uranium–238	U-238 → Pb-206	4.5×10^9 years
Rubidium–87	Rb-87 → Sr-87	4.9×10^{10} years

NYS Fossils Through Geologic Time

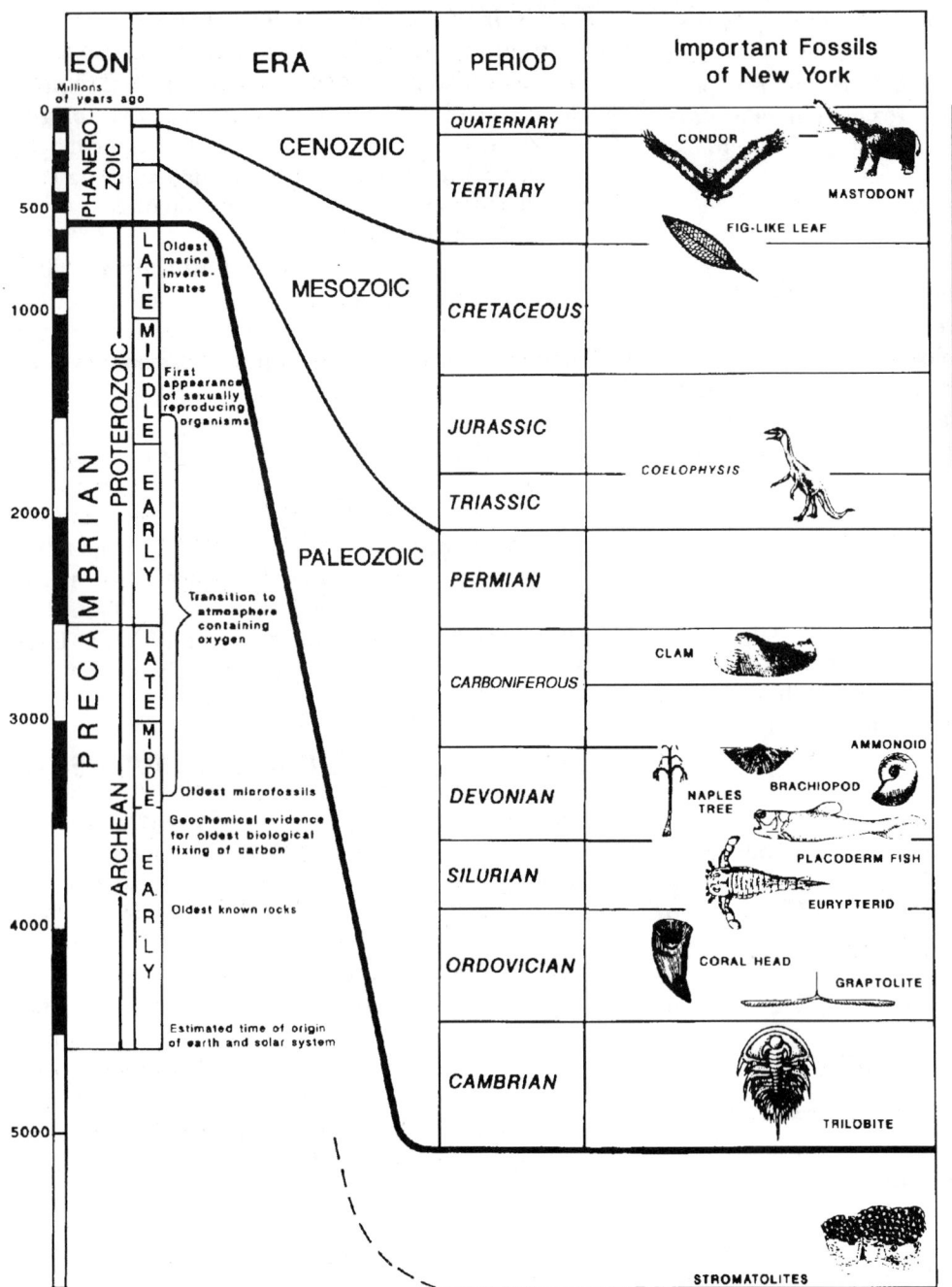

EON	ERA	PERIOD	Important Fossils of New York
PHANEROZOIC	CENOZOIC	QUATERNARY	CONDOR, MASTODONT
		TERTIARY	FIG-LIKE LEAF
	MESOZOIC	CRETACEOUS	
		JURASSIC	COELOPHYSIS
		TRIASSIC	
	PALEOZOIC	PERMIAN	
		CARBONIFEROUS	CLAM
		DEVONIAN	NAPLES TREE, AMMONOID, BRACHIOPOD
		SILURIAN	PLACODERM FISH, EURYPTERID
		ORDOVICIAN	CORAL HEAD, GRAPTOLITE
		CAMBRIAN	TRILOBITE
PRECAMBRIAN (PROTEROZOIC / ARCHEAN)			STROMATOLITES

Precambrian timeline markers (Millions of years ago):
- Oldest marine invertebrates
- First appearance of sexually reproducing organisms
- Transition to atmosphere containing oxygen
- Oldest microfossils
- Geochemical evidence for oldest biological fixing of carbon
- Oldest known rocks
- Estimated time of origin of earth and solar system (~5000)

D. THE FOSSIL RECORD

Ancient Life. The study of the fossils found in sedimentary rock provides clues to the life that existed on the Earth in past eras and to the environments in which the organisms (e.g., plants, animals) lived.

The Variety of Life Forms. Fossils give evidence that a great many kinds of animals and plants have lived in the past on the Earth in a great variety of environmental conditions. Most of these life forms have become **extinct** (do not exist on the Earth today). It is highly probable that in addition to the fossil types that have been found, there existed an even greater number of life forms that have left no traces (fossils) in the rock.

How can the rock record and fossil evidence reveal changes in past life and environments?

EVOLUTIONARY DEVELOPMENT

A **species** is defined as organisms that are able to mate and produce offspring capable of continuing the same species. Within a species there are a great number of variations which can be observed, measured, and described.

Theories of **organic evolution** (how change occurs) have suggested that the variations within a species may provide some members of that species with a higher probability of survival. For example, a variation in the color of rabbits living in a snow covered Arctic region, such as white and a brown color difference, provide the white rabbit with a better chance to survive. The white rabbit may be able to better hide in the snow from predators. But, the brown rabbit would be easily seen. As a result, the white rabbit will have a higher probability of producing more offspring.

The similarity among some fossil forms of various time periods suggests a transition that may be a result of evolutionary development. Generally, the older rock formations contain more simple and marine life forms. The younger rock formations have the fossils of more complex land dwelling organisms.

QUESTIONS FOR UNIT 5

1. Volcanic ash layers may serve as excellent time markers in the geologic rock record because most volcanic ash
   1. contains fine-textured particles
   2. contains many minerals
   3. has a very low resistant to weathering
   4. is rapidly deposited over a wide geographic area

2. Unconformities (buried erosional surfaces) are good evidence that
   1. many life-forms have become extinct
   2. the earliest life-forms lived in the sea
   3. part of the geologic rock record is missing
   4. metamorphic rocks have formed from sedimentary rocks

3. Where is metamorphic rock frequently found?
   1. along the boundary between igneous intrusions and sedimentary bedrock
   2. as a thin surface layer covering huge areas of the continents
   3. on mountaintops that have horizontal layers containing marine fossils
   4. within large lava flows

Base your answers to questions 4 and 5 on your knowledge of Earth science and on the diagram of rock structure below.

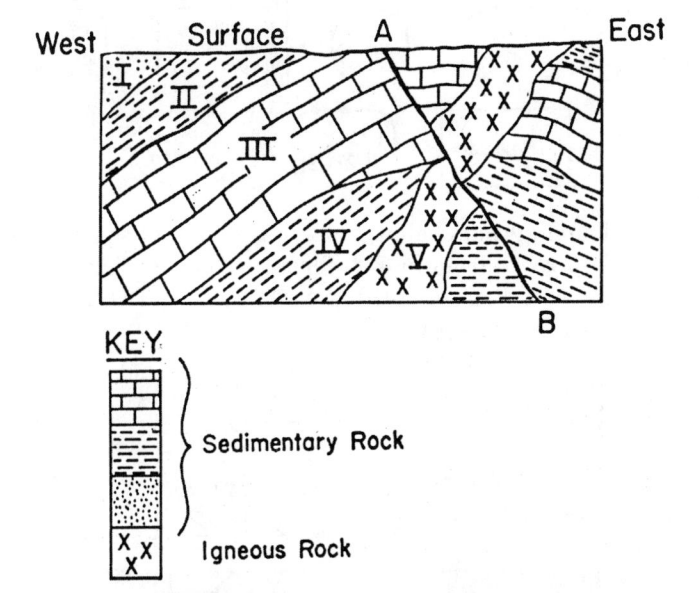

4. Based on the diagram of rock structure, which probably occurred *last*?
   1. folding of the region
   2. deposition of layer III
   3. intrusion of layer V
   4. faulting along line *AB*

5. Rock layer II west of the fault *AB* could be matched with the proper rock unit east of the fault by
   1. chemical analysis of the rocks at the fault
   2. determining the absolute ages of the mineral grains of the various rock layers on both sides of the fault
   3. matching the similarity in appearance, color, and composition of the various rock layers on both sides of the fault
   4. walking the outcrop directly from one side of the fault to the other

6. The map shows the relative age of the bedrock in the continental United States.

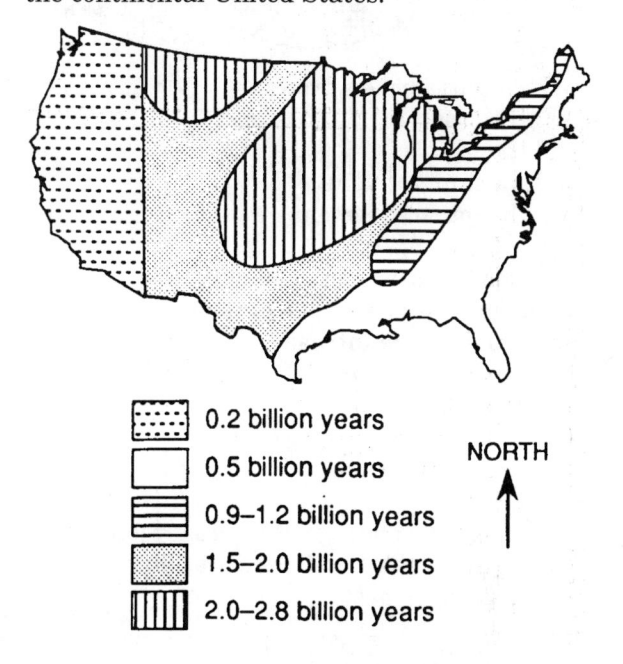

   The general age of the bedrock is found to be progressively
   1. older as an observer moves from the east and west coasts toward the center of the United States
   2. younger as an observer moves from the east and west coasts toward the center of the United States
   3. older as an observer moves across the United States from the east coast to the west coast
   4. younger as an observer moves across the United States from the east coast to the west coast

UNIT FIVE – EARTH'S HISTORY

7 The diagram represents a sample of a sedimentary rock viewed under a microscope.

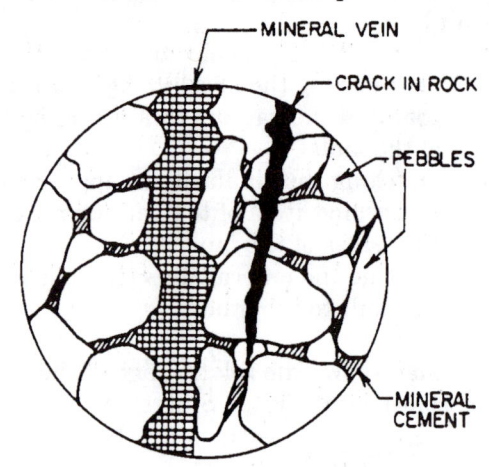

Which part formed first?
1. the crack
2. the pebbles
3. the mineral vein
4. the mineral cement

8 An igneous intrusion is 50 million years old. What is the most probable age of the rock immediately surrounding the intrusion?
(1) 10 million years
(2) 25 million years
(3) 40 million years
(4) 60 million years

9 The diagrams show geologic cross sections of the same part of the Earth's crust at different times in the geologic past.

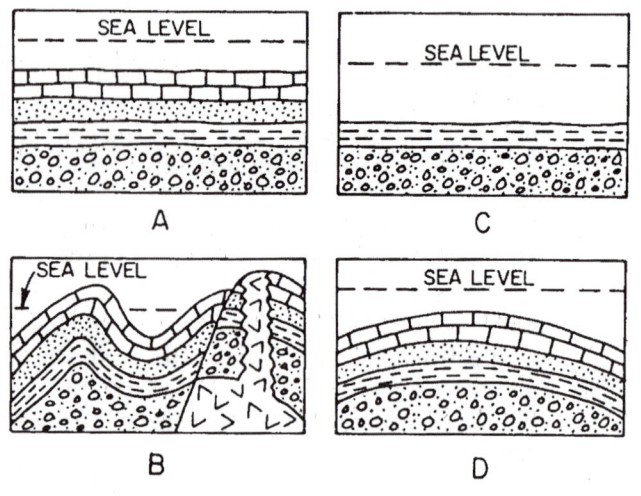

Which sequence shows the order in which this part of the crust probably formed?
(1) $A \to B \to C \to D$
(2) $C \to D \to A \to B$
(3) $C \to A \to D \to B$
(4) $A \to C \to B \to D$

10 The diagram below shows a cross section of the Earth's crust. Line XY is a fault.

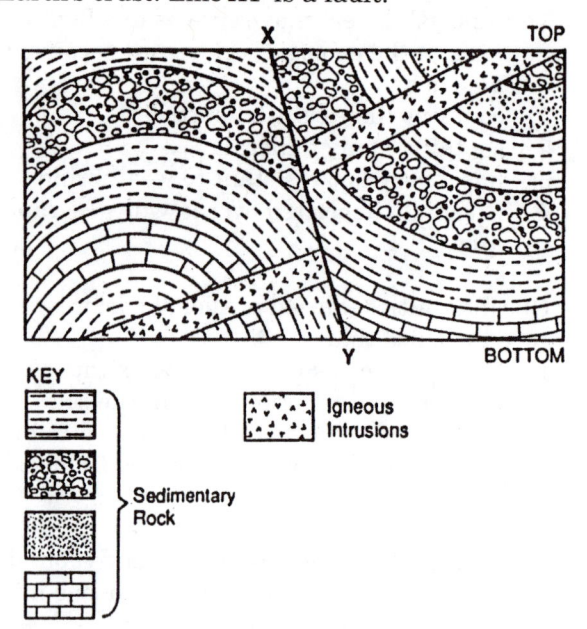

Which sequence of events, from oldest to youngest, has occurred in this outcrop?
(1) formation of sedimentary layers → igneous intrusion → folding of layers → faulting
(2) igneous intrusion → formation of sedimentary layers → folding of layers → faulting
(3) igneous intrusion → faulting → formation of sedimentary layers → folding of layers
(4) formation of sedimentary layers → folding of layers → igneous intrusion → faulting

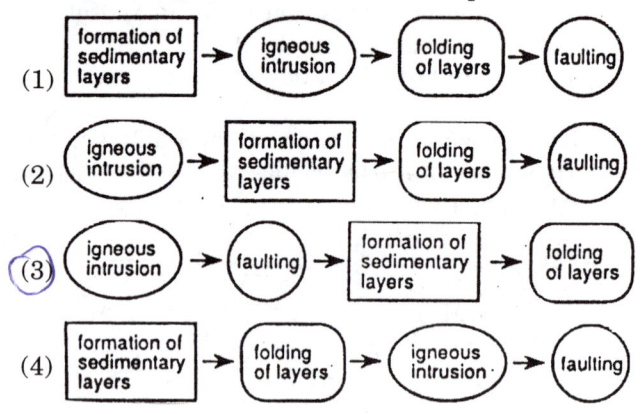

11 The diagrams below represent layers of sedimentary rocks from four different locations. Four of the layers are identified as A, B, C, and D. No layers have been overturned.

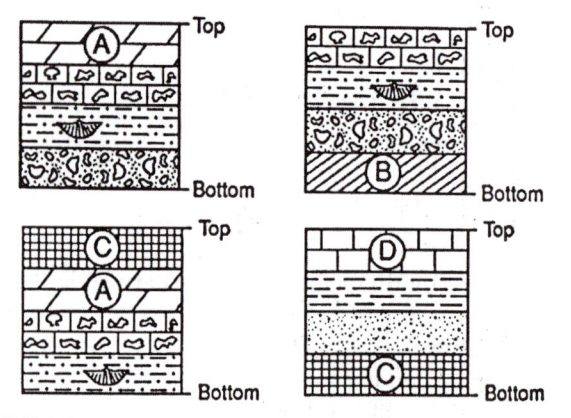

Which rock layer is youngest?
(1) A   (3) C
(2) B   (4) D

Base your answers to questions 12 and 13 on the geologic cross section diagram.

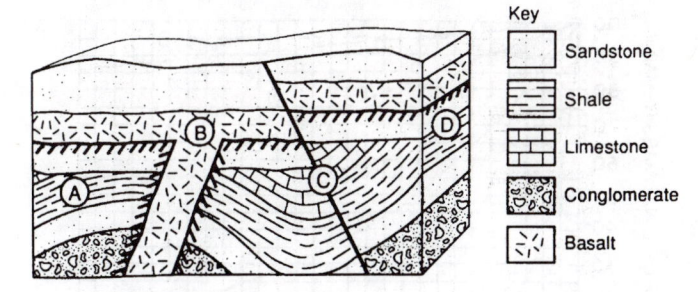

12 Which geologic event occurred most recently?
1   folding at A
2   the intrusion at B
3   faulting at C
4   the unconformity at D

13 The symbol ⊤⊤⊤⊤⊤⊤ in the diagram most likely represents a
1   metamorphic rock in contact with an igneous rock
2   depression caused by underground erosion.
3   convection cell caused by unequal heating
4   large fault from an earthquake or crustal movement

Base your answers to questions 14 through 17 on your knowledge of Earth science, the *Earth Science Reference Tables*, and the block diagram below. The diagram represents a geologic cross section in which overturning has not occurred.

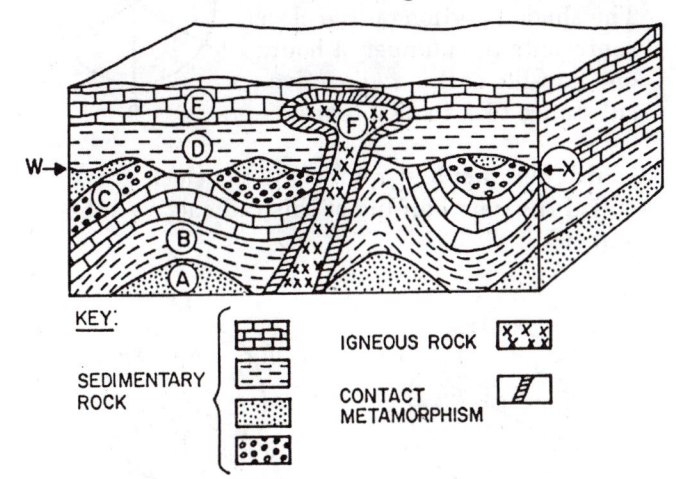

14 Which rock most likely is the oldest?
(1) A   (3) F
(2) B   (4) D

15 When did the folding of rock layer B most likely occur?
1   before the deposition of rock layer A
2   after the deposition of rock layer E
3   after the deposition of rock layer C
4   after the deposition of rock layer D

16 Fossils are *least* likely to be found in which rock?
(1) E   (3) C
(2) F   (4) D

17 What evidence in the rock layers indicates that the formation of igneous rock F occurred after rock layer E was in place?
1   the presence of radioactive minerals in rock F
2   the presence of extrusive igneous rock below rock layer E
3   the unconformity between rock F and rock layer E
4   the zone of contact metamorphism between rock F and rock layer E

18 The Geologic Time Scale has been subdivided into a number of time units called periods based upon
1   fossil evidence
2   rock thickness
3   rock types
4   radioactive dating

19 The diagram represents a cross section of a series of rock layers of different geologic ages.

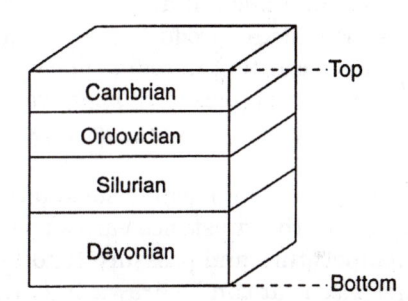

Which statement provides the best explanation for the order of these rock layers?
1   The oldest layer is on the bottom.
2   A buried erosional surface exists between layers.
3   The layers have been overturned.
4   The Permian layer has been totally eroded.

20 According to the *Earth Science Reference Tables*, when did the Jurassic Period end?
(1) 66 million years ago
(2) 144 million years ago
(3) 163 million years ago
(4) 190 million years ago

21 Using the information in the *Reference Tables*, students plan to construct a geologic time line of the Earth's history from its origin to the present time. They will use a scale of 1 meter equals 1 billion years. What should be the total length of the students' time line?
(1) 10.0 m   (3) 3.8 m
(2) 2.5 m   (4) 4.6 m

22 The geologic cross section below shows an unconformity between gneiss and the Cambrium-age Potsdam sandstone in northern New York State.

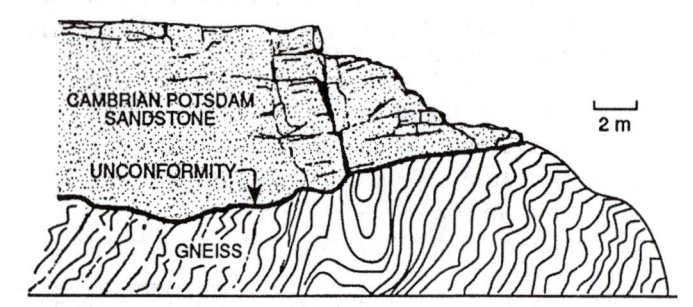

According to the *Earth Science Reference Tables*, what is the most probable age of the gneiss at this location?
1   Precambrian
2   Silurian
3   Ordovician
4   Cretaceous

23 Why are radioactive materials useful for measuring geologic time?
1   The disintegration of radioactive materials occurs at a predictable rate.
2   The half-lives of most radioactive materials are less than five minutes.
3   The ratio of decay products to undecayed material remains constant in sedimentary rocks.
4   Measurable samples of radioactive materials are easily collected from most rock types.

24 The decay rates of radioactive substances remain constant when the substances are subjected to different temperature and pressure conditions. The best inference that can be drawn from this statement is that decay rates are
1   independent of external factors
2   independent of the isotope's composition
3   affected by the mass of the isotope
4   affected by pressure, but not by temperature

25 The table at the right gives information about the radioactive decay of Carbon–14. [Part of the table has been left blank for student use.]

Half-Life	Mass of Original C-14 Remaining (grams)	Number of Years
0	1	0
1	$\frac{1}{2}$	5,700
2	$\frac{1}{4}$	11,400
3	$\frac{1}{8}$	17,100
4		
5		
6		

What is the amount of original Carbon–14 remaining after 34,200 years?
(1) $1/8$ g
(2) $1/16$ g
(3) $1/32$ g
(4) $1/64$ g

26 Which radioactive substance shown on the graph below has the longest half-life?

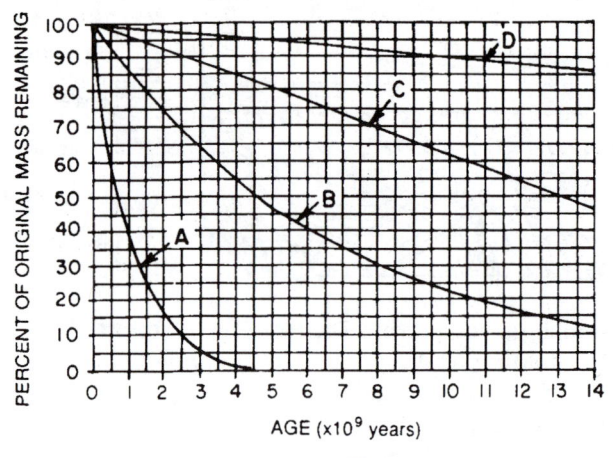

(1) A
(2) B
(3) C
(4) D

27 According to the *Earth Science Reference Tables*, which element is used by Earth scientists for radioactive dating of rocks?
1   Cobalt–60
2   Plutonium–244
3   Potassium–40
4   Silicon–28

28 The diagram represents a clock used to time the half-life of a particular radioactive substance. The clock was started at 12:00. The shaded portion on the clock represents the number of hours one-half-life of this radioactive substance took to disintegrate.

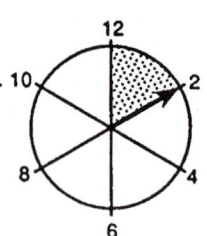

Which diagram best represents the clock at the end of the next half-life of this radioactive substance?

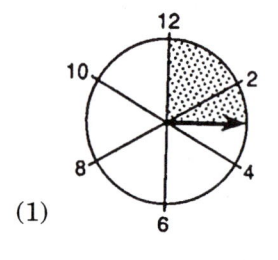

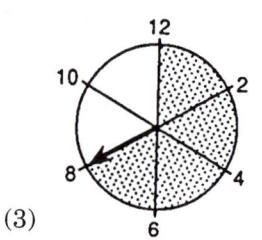

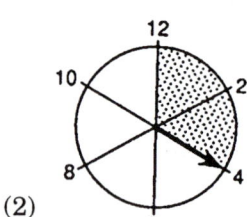

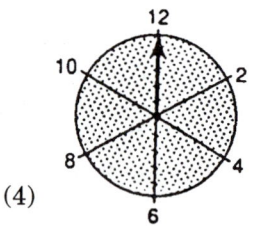

29 A geologist uses Carbon–14 to measure the age of some material found in a sedimentary deposit. If the half life of Carbon–14 is 5.7 x 10³ years and the sample shows that only ¼ of the original carbon–14 is left, the age of the sample is about
(1) 5,700 years
(2) 11,400 years
(3) 17,100 years
(4) 22,800 years

30 According to the *Reference Tables*, which radioactive element would be most useful for determining the age of clothing that is thought to have been worn 2,000 years age?
1 Carbon–14
2 Potassium–40
3 Uranium–238
4 Rubidium–87

31 Which characteristic of a fossil would make it useful as an index fossil in determining the relative age of widely separated rock layers?
1 a wide time range and a narrow geographic range
2 a wide time range and a wide geographic range
3 a narrow time range and a wide geographic range
4 a narrow time range and a narrow geographic range

32 In the Earth's geologic past there were long warm periods which were much warmer than the present climate. What is the primary evidence that these long warm periods existed?
1 U.S. National Weather Service records
2 polar magnetic directions preserved in the rock record
3 radioactive decay rates
4 plant and animal fossils

33 Why are fossils rarely found in Precambrian rock layers?
1 Few Precambrian rock layers have been discovered.
2 Nearly all fossils from this era have been destroyed by glaciers.
3 Few rock layers were formed during the Precambrian Era.
4 Life that would produce fossils was not abundant during the Precambrian Era.

34 From the study of fossils, what can be inferred about most species of plants and animals that have lived on the Earth?
1 They are still living today.
2 They are unrelated to modern life forms.
3 They existed during the Cambrian Period.
4 They have become extinct.

35 The geologic columns *A*, *B*, and *C* in the diagrams represent widely spaced outcrops of sedimentary rocks. Symbols are used to indicate fossils found within each rock layer. Each rock layer represents the fossil record of a different geologic period.

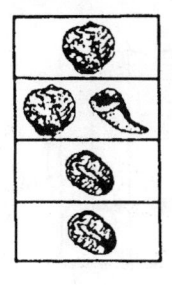

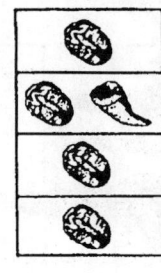

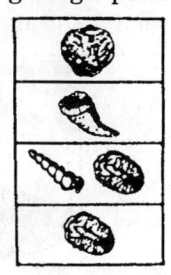

According to the diagrams for all three columns, which would be the best index fossil?

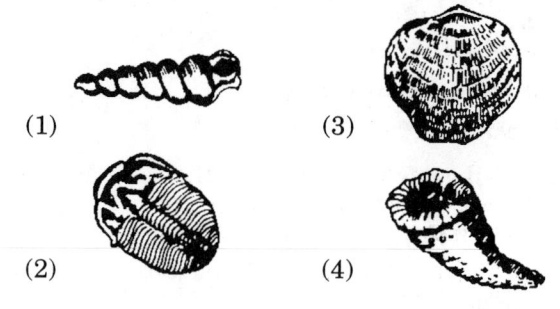

(1)
(2)
(3)
(4)

36 The chart below shows index fossils found in rocks of various ages.

BEDROCK AGE	INDEX FOSSIL
MISSISSIPPIAN	SPIRIFER
DEVONIAN	MUCROSPIRIFER
SILURIAN	EOSPIRIFER
ORDOVICIAN	MICHELINOCERAS

According to the *Earth Science Reference Tables*, which fossil could be found in the same rock as fossils of the first corals?
1 *Spirifer*
2 *Mucrospirifer*
3 *Eospirifer*
4 *Michelinoceras*

37 Trilobite fossils from different time periods show small changes in appearance. These observations suggest that the changes may be the result of
1 evolutionary development
2 a variety of geologic processes
3 periods of destruction of the geologic record
4 the gradual disintegration of radioactive substances

Base your answers to questions 38 and 39 on the *Earth Science Reference Tables*, the graph below, and your knowledge of Earth science. The graph shows the development, growth in population, and extinction of the six major groups of trilobites, labeled A through F.

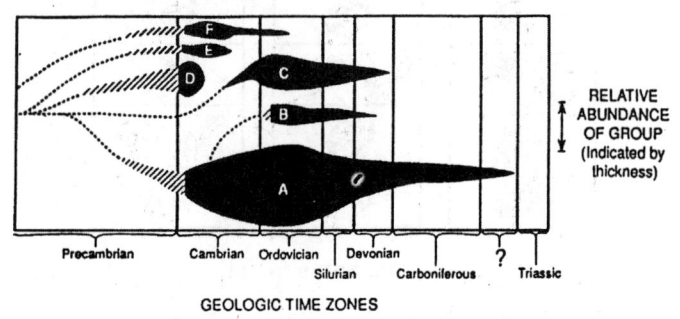

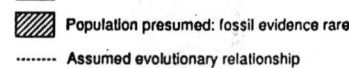

38 The fossil evidence that forms the basis for this graph was most likely found in
 1 lava flows of ancient volcanoes
 2 sedimentary rock that formed from ocean sediment
 3 granite rock that formed from former sedimentary rocks
 4 metamorphic rock that formed from volcanic rocks

39 Which group of trilobites became the most abundant?
 (1) A   (3) C
 (2) B   (4) D

40 The cartoon at the right represents the time of the last dinosaurs and the earliest mammals.

According to the *Earth Science Reference Tables*, the cartoon could represent the boundary between which two units of geologic history?
 1 Archean and Proterozoic
 2 Precambrian and Paleozoic
 3 Ordovician and Silurian
 4 Mesozoic and Cenozoic

41 The diagram shows the probable arrangement of some Earth continents during the Mesozoic Era and the present distribution of four fossils.

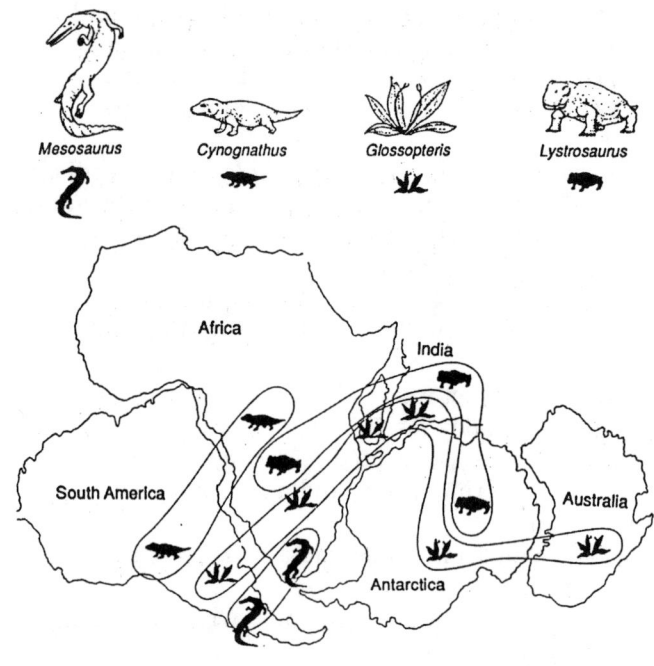

Which fossil shown in the diagram would be most useful for correlating rocks among all the landmasses shown?
 1 *Mesosaurus*   3 *Glossopteris*
 2 *Cynognathus*   4 *Lystrosaurus*

42 The diagram shows a possible sequence of evolutionary development of some vertebrates.

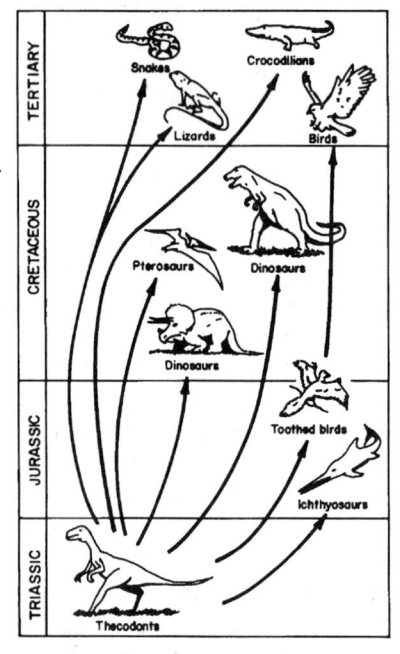

Which statement can best be inferred from the diagram?
 1 The lizard and the bird are both reptiles.
 2 The *pterosaurs* and the *ichthyosaurs* became extinct at the same time.
 3 The *pterosaurs* evolved into modern birds.
 4 The *thecodont* is the ancestor of several different types of animals.

SKILL ASSESSMENTS

Base your answers to questions 1 through 4 on your knowledge of Earth science, the *Reference Tables*, and the diagrams which represent the bedrock geology of an area outside New York State.

TOP VIEW

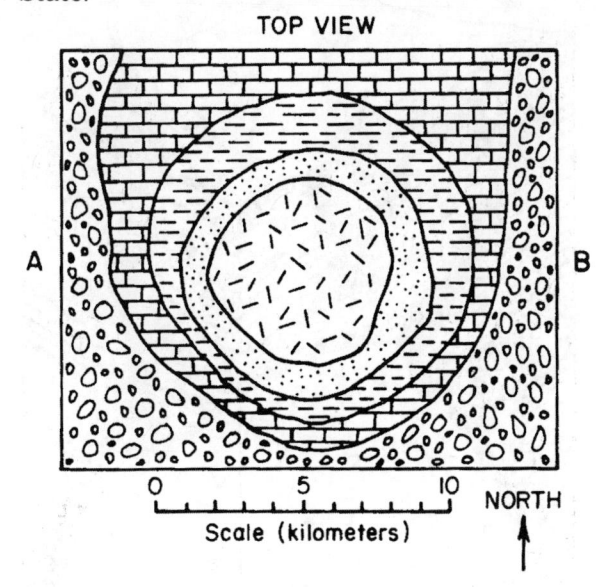

CROSS SECTION (not drawn to scale)

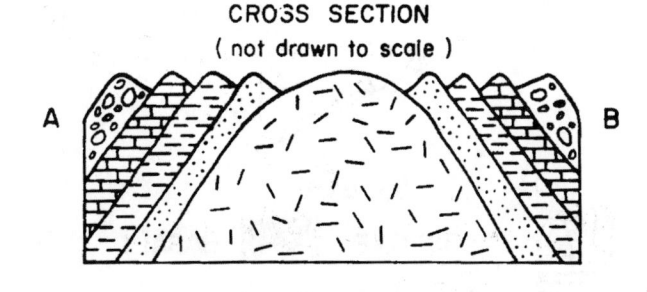

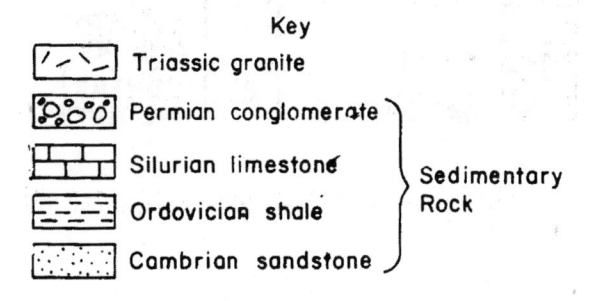

Key
- Triassic granite
- Permian conglomerate
- Silurian limestone } Sedimentary Rock
- Ordovician shale
- Cambrian sandstone

1. In which bedrock unit are fossils *least* likely to be found? Why?

2. Which bedrock could be 460 million years old?

3. What kind of fossils might be found in the Silurian limestone? [See *Reference Tables*]

4. Which type of bedrock shown in the diagrams is *not* found in New York State?

Base your answers to questions 5 and 6 on your knowledge of Earth science and the diagram below. The diagram shows a profile view of a bedrock outcropping. None of the layers has been overturned.

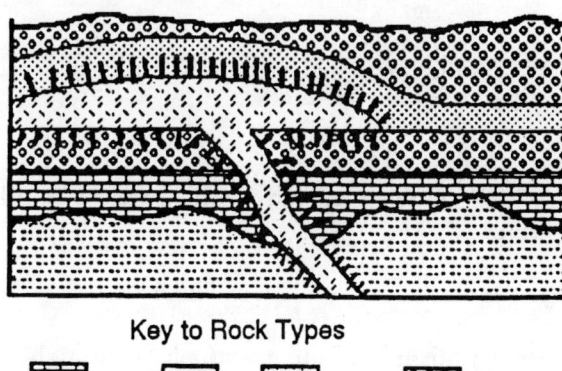

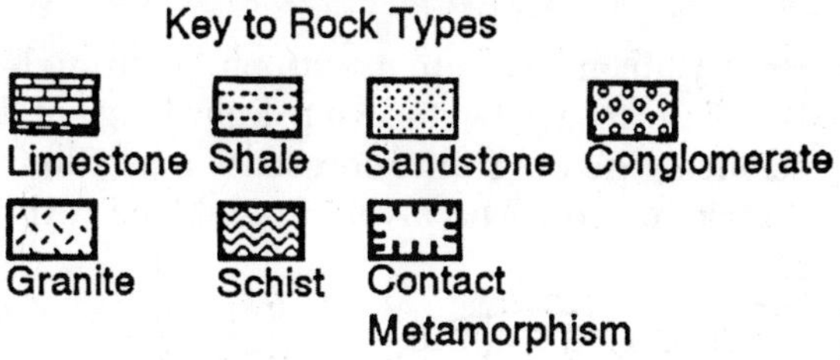

Key to Rock Types
- Limestone
- Shale
- Sandstone
- Conglomerate
- Granite
- Schist
- Contact Metamorphism

5. How can one tell that the sandstone is older than the granite below it? Answer in one or more compete sentences.

6. List the names of the rocks in the order of their age from oldest to youngest using the number one for the oldest rock.

   1 _____   4 _____
   2 _____   5 _____
   3 _____   6 _____

Base your answers to questions 7 through 10 (on the next page) on your knowledge of Earth science, the *Reference Tables*, and the chart which illustrates the geologic timespan and the relative abundance of species of ten types of animals and plants.

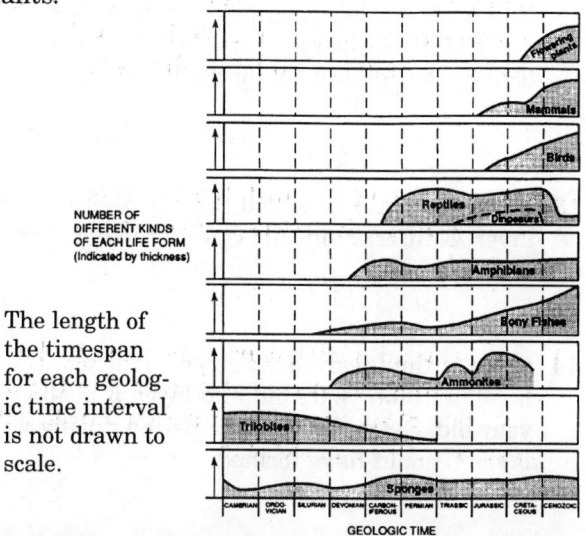

The length of the timespan for each geologic time interval is not drawn to scale.

UNIT FIVE — EARTH'S HISTORY — N&N©   Page 119

7   Which life form reached a peak three separate times in its existence on Earth?

8   When did the extinction of the dinosaurs occur?

9   Which life form appeared on Earth most recently?

10 Which life form existed for the greatest length of time?

Base your answers to questions 11 through 14 on the diagram below and your knowledge of Earth science. The diagram represents a profile view of a rock outcrop. The layers are labeled A through H.

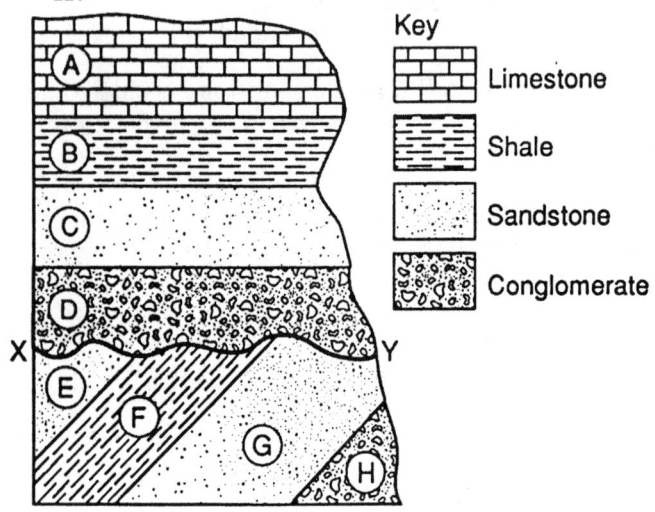

Key
- Limestone
- Shale
- Sandstone
- Conglomerate

11 State the range of particle sizes of the sediment that formed rock layer C.

12 Using one or more complete sentences, briefly describe the geologic process that resulted in the boundary represented by the line XY.

13 State two ways in which the composition of rock layer A differs from the composition of rock layer B.

14 None of the layers has been overturned. Layer D is 505 million years old and layer B is 438 million year old. State the geologic period during which layer C could have formed.

Base your answers to questions 15 and 16 on your knowledge of Earth science, the *Earth Science eference Tables*, and the diagram which represents a geologic cross section of a volcanic area where overturning of rock layers has not occurred.

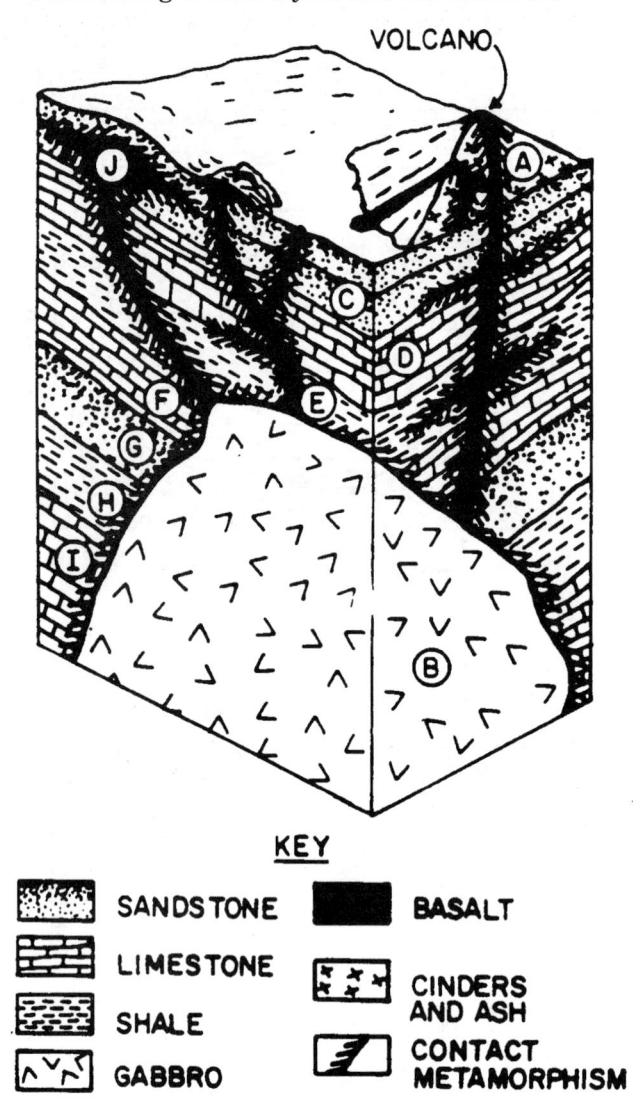

KEY
- SANDSTONE
- LIMESTONE
- SHALE
- GABBRO
- BASALT
- CINDERS AND ASH
- CONTACT METAMORPHISM

15 At which location is the oldest rock most likely to be found?

16 In a sentence, explain what most likely happened to rock layers E, F, G, H, and I where they came in contact with the molten rock that formed B.

Page 120   N&N© SCIENCE SERIES – EARTH SCIENCE – MODIFIED PROGRAM

Unit Six
METEOROLOGY

Vocabulary To Be Understood

Air Mass	Isobar & Isotherm	Source Region
Anemometer	Jet Stream	Station Model
Cloud	Land & Sea Breezes	Stationary Front
Cloud Base	Millibar	Storm Track
Cold Front	Moisture Capacity	Synoptic Map
Condensation	Occluded Front	Tornado
Condensation Nuclei, Surface	Orographic Effect	Vortex
Coriolis Effect	Precipitation	Warm Front
Cyclone, Anticyclone	Pressure Gradient	Water Vapor
Deposition	Probability of Occurrence	Weather
Dew, Dew Point Temperature	Psychrometer	Weather Forecasting
HIGH & LOW Pressure	Relative Humidity	Wind
Hurricane	Saturation Point	Wind Vane

A. WEATHER

Weather is the short term condition of the atmosphere at any location. It is the result of the interrelationship of temperature, humidity, air pressure, and winds.

How can weather be described?

TEMPERATURE

Temperature usually varies in a daily cycle. The temperature at Earth's surface is greatly affected by the amount (intensity and duration) of sunlight. Other factors such as different surface materials (land and water), altitude, cloud cover, and movements of air masses contribute to a variety of temperature conditions.

HUMIDITY

The **dew point** temperature is the temperature at which the air is holding the maximum amount of water vapor. At this temperature the air is said to be **saturated** with water vapor. The water holding capacity of air is directly related to the temperature of the air so air of different temperatures will have different dew points.

Relative humidity is the percent of saturation of the air. It is the expression of the ratio between the actual amount of water vapor in the atmosphere and the air's capacity at that temperature. When the relative humidity is 50%, the air is holding half of the water vapor it is capable of holding.

dew point and relative humidity are determined by using an instrument called a **psychrometer**. The psychrometer has two thermometers, a **dry bulb thermometer** and a **wet bulb thermometer** with a wet wick around the bulb. When whirled in the air, the wet bulb temperature usually drops due to the evaporation of the water, cooling the bulb of the thermometer. The amount of evaporation depends on the moisture content of the air. The lower the moisture content of the air, the more evaporation will occur from the wet bulb, and the lower the wet bulb temperature will be. The bigger the difference between the wet and dry bulb temperatures, the drier the air.

Unit Six – Meteorology – N&N© Page 121

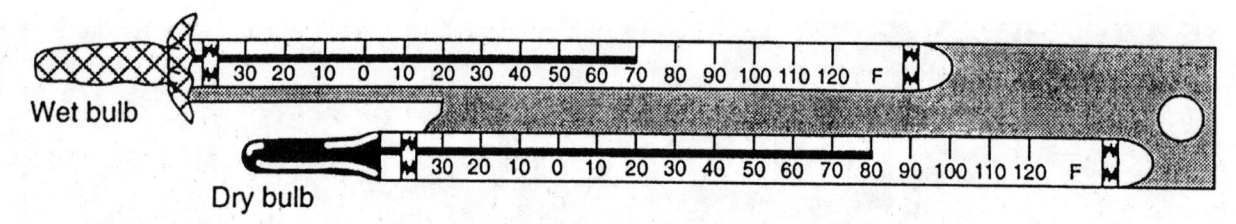

Dew Point and Relative Humidity Psychrometer

Compare the difference between the wet and dry bulb readings. Convert the Fahrenheit (°F) to Celsium (°C) using the Temperature Chart in the *Earth Science Reference Tables*, and use the Dew Point Temperature Chart to determine the Dew Point. Relative Humidity is also determined as above, but with the use of the Relative Humidity Chart.

Dewpoint Temperatures

Dry-Bulb Temperature (°C)	1	2	3	4	5	6	7	8	9	10	11	12	13	14	15
-20	-33														
-18	-28														
-16	-24														
-14	-21	-36													
-12	-18	-28													
-10	-14	-22													
-8	-12	-18	-29												
-6	-10	-14	-22												
-4	-7	-12	-17	-29											
-2	-5	-8	-13	-20											
0	-3	-6	-9	-15	-24										
2	-1	-3	-6	-11	-17										
4	1	-1	-4	-7	-11	-19									
6	4	1	-1	-4	-7	-13	-21								
8	6	3	1	-2	-5	-9	-14								
10	8	6	4	1	-2	-5	-9	-14	-28						
12	10	8	6	4	1	-2	-5	-9	-16						
14	12	11	9	6	4	1	-2	-5	-10	-17					
16	14	13	11	9	7	4	1	-1	-6	-10	-17				
18	16	15	13	11	9	7	4	2	-2	-5	-10	-19			
20	19	17	15	14	12	10	7	4	2	-2	-5	-10	-19		
22	21	19	17	16	14	12	10	8	5	3	-1	-5	-10	-19	
24	23	21	20	18	16	14	12	10	8	6	2	-1	-5	-10	-18
26	25	23	22	20	18	17	15	13	11	9	6	3	0	-4	-9
28	27	25	24	22	21	19	17	16	14	11	9	7	4	1	-3
30	29	27	26	24	23	21	19	18	16	14	12	10	8	5	1

WIND

One major effect of differences in air pressure is the movement of air. The horizontal movement of air is called **wind**. Note: Winds are described in terms of both speed and direction and are named for the direction from which they blow. A north wind blows from the north toward the south and an ocean wind blows from the ocean toward the land.

A **wind vane** is used to indicate the direction the wind is blowing and an **anemometer** with a series of rotating cups measures wind speed. Wind direction is an important weather variable used to predict the weather.

By using the dry bulb temperature and the difference between the wet bulb and dry bulb temperatures (as found of the *Dew point Temperature Chart* in the *Reference Tables*) the dew point may be determined. The same data may be applied to the *Relative Humidity Chart* to determine the relative humidity.

AIR PRESSURE

Air pressure is caused by the weight of the air. Air pressure is measured by a **barometer in inches of mercury** or **millibars**. Standard air pressure, also known as one atmosphere, at sea level is equal to 1013.2 millibars or 29.92 inches of mercury. **Isobars**, lines of equal pressure, are used on a weather map to indicate air pressure variations. On U.S. Weather Bureau maps, the interval between isobars is 4 millibars.

Relative Humidity (%)

Dry-Bulb Temperature (°C)	1	2	3	4	5	6	7	8	9	10	11	12	13	14	15
-20	28														
-18	40														
-16	48	0													
-14	55	11													
-12	61	23													
-10	66	33	0												
-8	71	41	13												
-6	73	48	20	0											
-4	77	54	32	11											
-2	79	58	37	20	1										
0	81	63	45	28	11										
2	83	67	51	36	20	6									
4	85	70	56	42	27	14									
6	86	72	59	46	35	22	10	0							
8	87	74	62	51	39	28	17	6							
10	88	76	65	54	43	33	24	13	4						
12	88	78	67	57	48	38	28	19	10	2					
14	89	79	69	60	50	41	33	25	16	8	1				
16	90	80	71	62	54	45	37	29	21	14	7	1			
18	91	81	72	64	56	48	40	33	26	19	12	6	0		
20	91	82	74	66	58	51	44	36	30	23	17	11	5	0	
22	92	83	75	68	60	53	46	40	33	27	21	15	10	4	0
24	92	84	76	69	62	55	49	42	36	30	25	20	14	9	4
26	92	85	77	70	64	57	51	45	39	34	28	23	18	13	9
28	93	86	78	71	65	59	53	47	42	36	31	26	21	17	12
30	93	86	79	72	66	61	55	49	44	39	34	29	25	20	16

B. WEATHER VARIABLES

Warm air has a greater water holding capacity than cool air. Warm air is less dense and has more space between the molecules allowing for a greater capacity to hold water. For every 10°C rise in temperature, the air can hold approximately twice as much water vapor.

As the air temperature changes, so does the dew point and relative humidity. As air is cooled to the temperature at which it becomes saturated (the dew point temperature) the relative humidity approaches 100%. Cooling the air causes it to contract and become more dense, decreasing the space Available between the molecules for water vapor.

As the difference between the dew point temperature and the air temperature decreases, there is a greater probability of water vapor condensing and forming precipitation. **Condensation** is the change of phase of water vapor (gas) to liquid water.

What are the relationships among the various weather conditions?

Variations in air pressure may be caused by temperature, altitude, or moisture content. A change in temperature brings about a change in pressure. Since warm air is less dense than cool air, warm air has a lower mass per unit volume (weight). Being less dense, warm air exerts less pressure and is lighter. As air cools it contracts increasing its density and so is heavier and exerts more pressure on Earth's surface. Note: Since temperature changes produce pressure variations, pressure changes also produce temperature variations.

Air pressure also decreases with an increase in altitude, because there are fewer gas molecules per unit volume as you go up in the atmosphere. Moist or humid air also exerts less pressure than dry air. This is due to small, light water vapor molecules replacing the heavier gas molecules.

There is an inverse relationship between air temperature, altitude, and moisture content and the air pressure.

Uneven heating of the Earth's surface causes differences in temperature from one location to another. These temperature variations cause pressure variations which make the wind blow. Wind always move from areas of high pressure to areas of low pressure.

Along the coast on a summer afternoon the land becomes much warmer than the adjacent water surface. The air over the warm land heats, expands, and becomes less dense, forming lower pressure over the land. This results in cooler, denser air moving on to the land from the sea and pushing the warm air over the land upward. This results in a **sea breeze**.

At night, energy radiates more rapidly from the land surface than from the water surface. This results in higher air pressure over the land. As cooler air moves from the land toward the sea, a **land breeze** results.

Winds do not blow in a straight path from high to low pressure because wind direction is also influenced by the rotation of the Earth. In the Northern Hemisphere winds are deflected to their right and in the Southern Hemisphere to their left. This is known as the **Coriolis Effect**. Often satellite photographs of cloud patterns show this curvature.

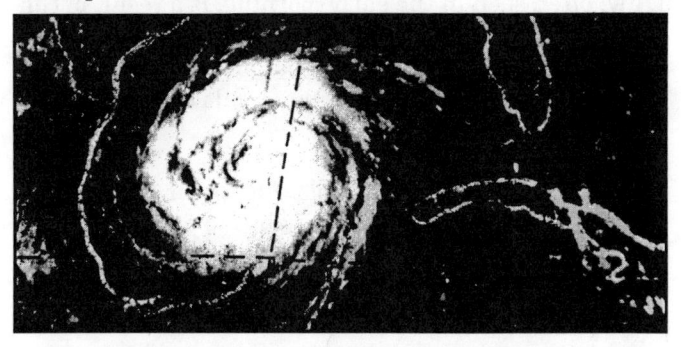

Counterclockwise Rotation of a Hurricane in the Gulf of Mexico (Northern Hemisphere).

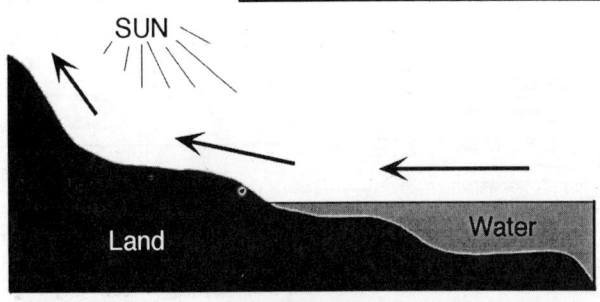

Sea Breeze
When air over land is warmer than over water, cooler air over water moves over land to replace rising air.

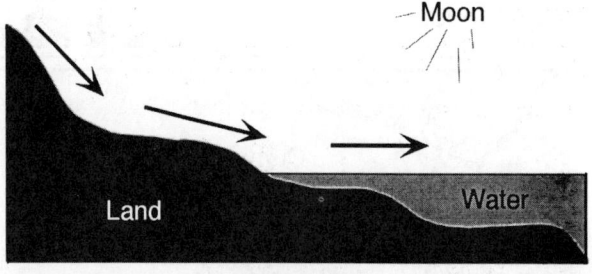

Land Breeze
When air over water is warmer than over land, cooler air over land moves over water to replace rising air.

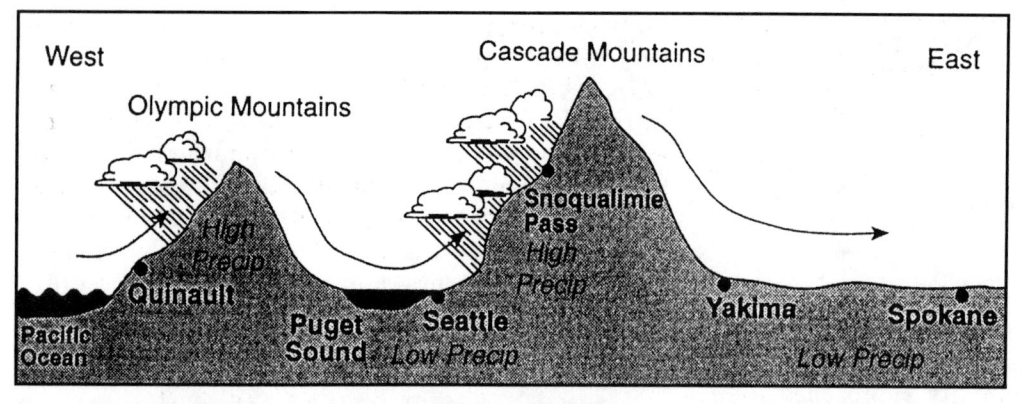

Orographic Effect
As air mass moves over the mountains, the rising air cools – producing clouds and precipitation on the windward side of the obstruction (mountain).

C. CLOUDS - PRECIPITATION

When air is cooled to the dew point and **condensation nuclei** (surfaces) are available, **condensation** will occur in the air. Microscopic particles of dust, salt, or smoke are common condensation nuclei. When the dew point temperature is above freezing, tiny droplets of liquid water form. If the dew point temperature is below freezing, water vapor solidifies directly into solid ice crystals (**deposition**).

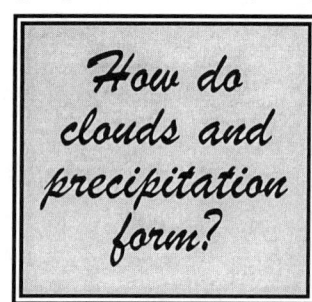

How do clouds and precipitation form?

The same process occurs at the Earth's surface. **Dew**, often seen in the early morning, is a result of condensation of water vapor from the atmosphere onto the Earth's surface. **Frost**, often seen in the fall and early spring after a cold night, is the result of deposition of water vapor below 0°C.

Wherever air rises, it cools. Rising air results from heating, convection currents, air moving up over a mountain (**orographic effect**) or being forced up at a frontal boundary. When the rising air cools to the dew point, condensation or deposition begins resulting in clouds made of tiny water droplets and/or ice crystals.

Where air is pushed up vertically, tall **cumuliform clouds** (cumulus, cumulonimbus) form. **Stratiform clouds** (cirrus and stratus) layer across the sky where air drifts up at a low angle.

The height of the base of the clouds may be determined by using the *Lapse Rate Chart* like the one in the *Reference Tables*. Since condensation begins when the air temperature and dew point temperature are equal, the intersection of the air temperature and dew point temperature lines indicates the altitude of the **cloud base**.

Precipitation does not fall from all clouds because the cloud droplets are so small that the motion of the air keeps them suspended in the air. When condensation droplets and/or ice crystals coalesce (grow together) and become large enough to fall, **precipitation** occurs. As precipitation falls, it acts to clean the atmosphere by bringing down condensation nuclei and other materials suspended in the air.

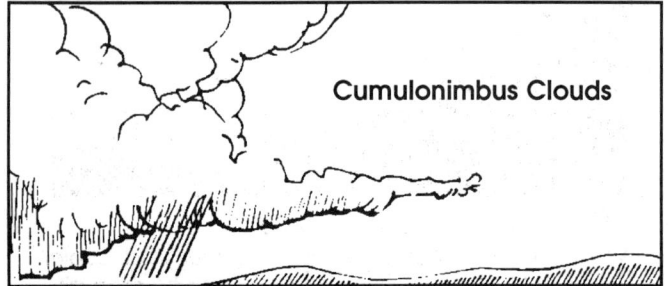

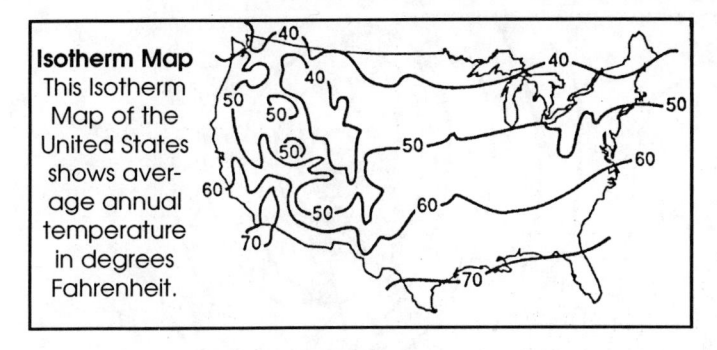

Isotherm Map
This Isotherm Map of the United States shows average annual temperature in degrees Fahrenheit.

D. WEATHER MAPS

Field values such as temperature and pressure may be shown on weather maps with the use of **isolines**, lines connecting places with the same value. A map may show the temperature field by using **isotherms** or the air pressure field by using **isobars**.

What information is shown by weather maps?

Proximity of the isolines indicates the **gradient** of the field. Where the isolines are close together, the gradient is steep; where the isolines are far apart, the gradient is gentle.

Temperature or pressure gradient may be determined numerically by determining the difference in numerical value between two points and dividing that difference by the distance between the points.

The speed of the wind is directly related to the **pressure field gradient**. The greater the difference between a high and low pressure area and the less distance between the pressure centers, the greater the wind speed. Where the isobars on a weather map are closest together, the wind speed is the greatest.

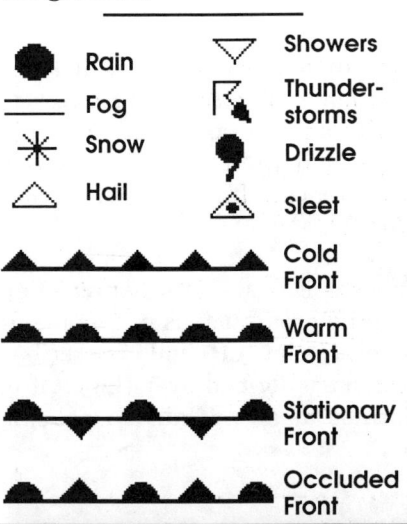

STATION MODEL

On the daily United States Weather Bureau *Weather Map*, surface data is plotted at many stations using numbers and symbols. The following is a sample station model for a weather map (for further information, see the Weather Map Information section of the *Reference Tables*).

WEATHER MAP INFORMATION
STATION MODEL

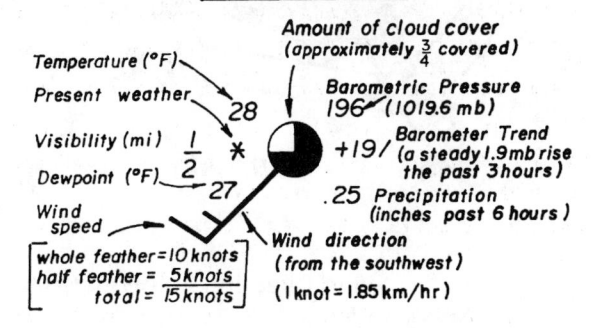

Temperature (°F)
Present weather
Visibility (mi)
Dewpoint (°F)
Wind speed
whole feather = 10 knots
half feather = 5 knots
total = 15 knots

Amount of cloud cover (approximately 3/4 covered)
Barometric Pressure 196 (1019.6 mb)
Barometer Trend +19 (a steady 1.9 mb rise the past 3 hours)
.25 Precipitation (inches past 6 hours)
Wind direction (from the southwest)
(1 knot = 1.85 km/hr)

The **weather station model** describes:
- the sky around the weather observation station, including the amount of cloud cover;
- the present weather conditions, including precipitation type and amount;
- the visibility (measured in miles);
- the air temperature and dew point in °F;
- the wind speed and direction; and,
- the barometric pressure (measured in millibars) and the pressure trend.

Weather Map Example
The symbols shown on the map identify the many aspects of weather and weather forecasting.

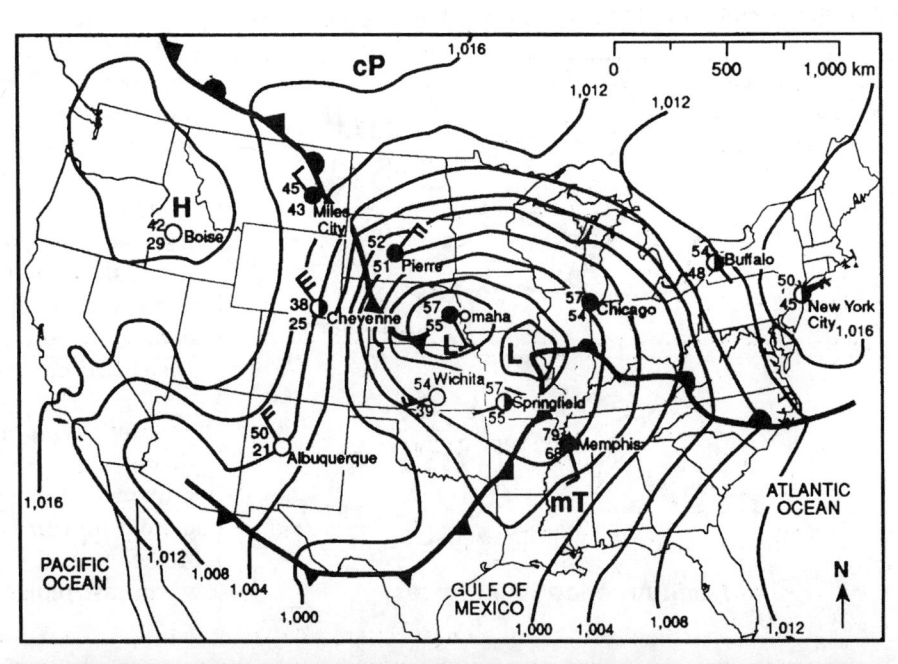

Symbols:
- ● Rain
- ▽ Showers
- ═ Fog
- Thunderstorms
- ✴ Snow
- Drizzle
- △ Hail
- Sleet
- ▲▲▲ Cold Front
- ⌒⌒⌒ Warm Front
- Stationary Front
- Occluded Front

UNIT SIX – METEOROLOGY – N&N© — Page 125

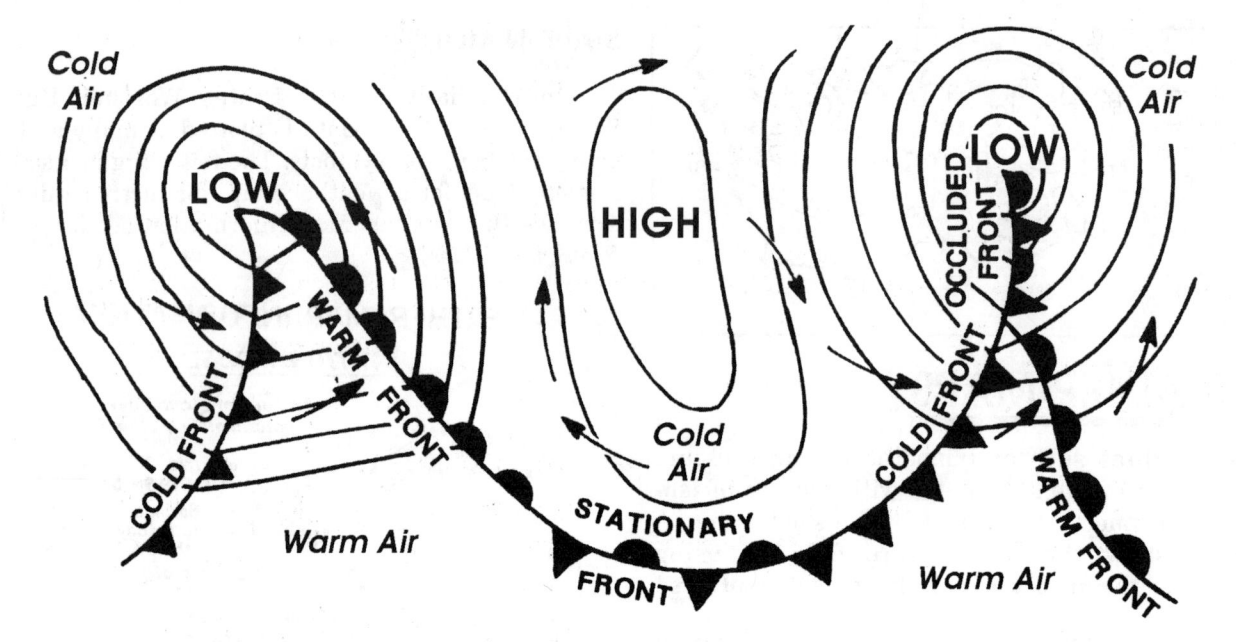

Wind Direction Relationship Between LOW And HIGH Pressure Areas
In the Northern Hemisphere, air currents move out of a HIGH in a clockwise direction and into a LOW in a counterclockwise direction. On a Weather Map, similar symbols are used to represent various fronts.

SYNOPTIC WEATHER MAPS

A **synoptic weather map** offers a "bird's eye view" of the weather. By studying the atmospheric conditions as they exist simultaneously over a broad area (synoptic observations), it is possible to make short term predictions of future weather conditions.

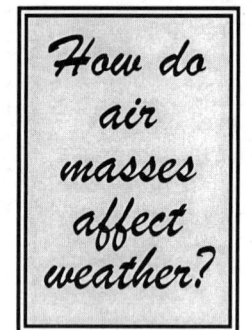

AIR MASS CHARACTERISTICS

In the atmosphere, an **air mass** is a large body of air having characteristics, including temperature and moisture, that are fairly uniform at any given level.

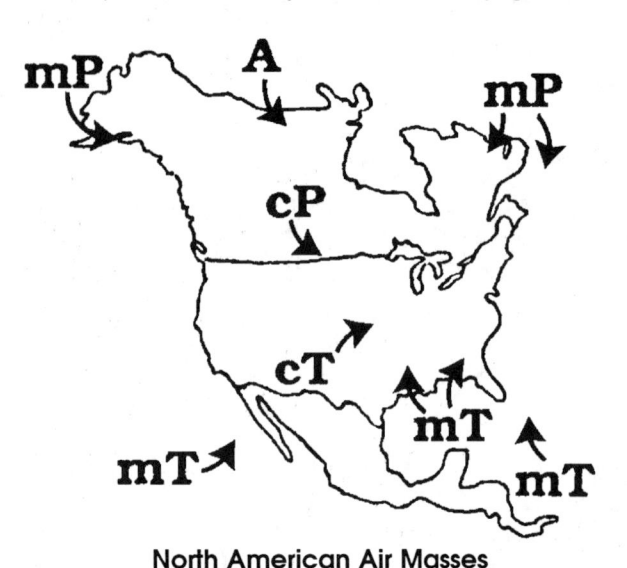

North American Air Masses

How do air masses affect weather?

These air masses are identified on the basis of their **average air pressure**, **moisture content**, **winds**, and **temperature**.

AIR MASS SOURCE REGIONS

Air masses have definite characteristics which depend upon their geographic region of origin, called the **source region**. In general, if the origin of an air mass is at high latitude, it is cold. If the source region is at low latitude, it is warm. For example, in the Northern Hemisphere, cool air masses come mainly from the north, and warm air masses come mainly from the south. An air mass formed over water is moist, but if the air mass forms over land, it is dry. As an air mass remains stationary in its source region, it becomes more intense and generally larger.

On a weather map, air masses are identified and described according to their source regions. An **arctic** air mass is identified with an **A**, a **polar** with a **P**, and a **tropical** with a **T**. In addition, if the source is over *water, it is called **maritime** (**m**). If it forms over land, it is called **continental** (**c**).

Identifying an air mass is easy. If the source of an air mass is a northern land area, such as in Canada, it is identified as continental polar (**cP**) and is described as cold and dry. An air mass formed over the Gulf of Mexico would be identified as an **mT** air mass, being warm and moist.

CYCLONES & ANTICYCLONES

Cyclone (Low Pressure System). In the Northern Hemisphere, air circulates *into* a low pressure or **cyclone** in a counterclockwise direction. This pattern of air motion in a cyclone may be hundreds of kilometers in diameter. Converging air *rises* near the center of the low. Cyclones are associated with clouds and precipitation. Decreasing air pressure often brings warm, rainy, or unsettled weather.

Anticyclone (High Pressure System). In the Northern Hemisphere, the circulation of air in a high pressure air mass or **anticyclone** is clockwise, spiraling *outwards* at the surface. The *descending* air in anticyclones is associated with cool and clear weather.

FRONTS

A **front** is the boundary or interface between two air masses on the ground. In the air this boundary is called the **frontal surface**. A front forms between two air masses having differing temperatures. Along the frontal surface, atmospheric conditions are usually unstable, and precipitation is most probable.

NORTHERN HEMISPHERE CYCLONIC WEATHER

A **warm front** forms when the warm air meets and slowly glides up and over the back of cold air. A **cold front** is the result of cold air pushing into a region of warm air. As the fronts are forming, the entire cyclonic storm (low pressure from which the warm and cold fronts extend) generally moves toward the northeast directed by the strong winds high in the troposphere.

Both fronts are associated with clouds and precipitation, because warm air is moving upward, resulting in cooling and condensation. However, at warm fronts stratiform clouds produce steady precipitation, usually of longer duration than cold fronts. The cumuliform clouds associated with the faster moving cold fronts, usually result in brief but heavy precipitation, often accompanied by gusty winds and thunderstorms.

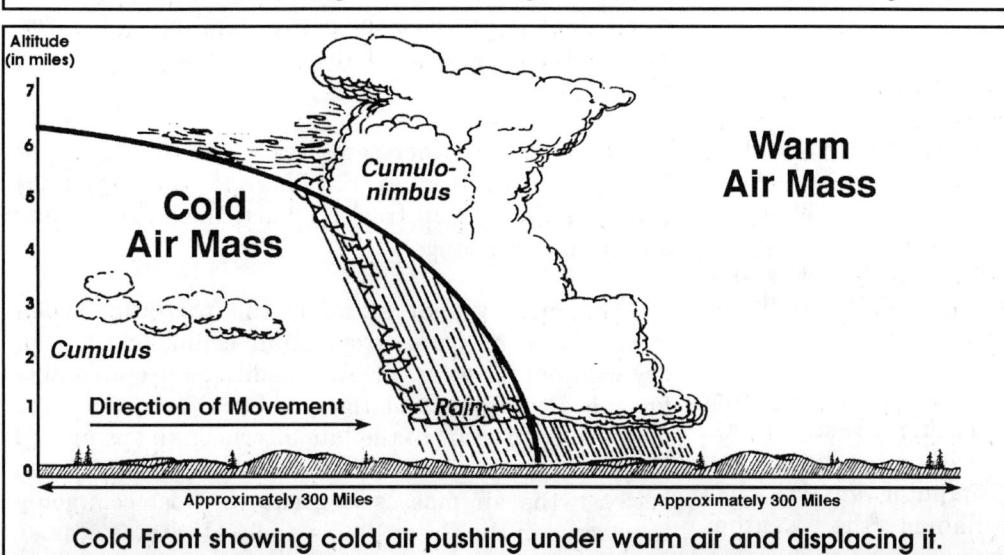

When a cyclone is first developing, it moves rapidly, but as it becomes older, the cyclone slows. Over time, the wedge of warm air tends to narrow, because the cold air moves faster than the warm air, forcing the less dense warm air up. The cold air front eventually overtakes the warm front, resulting in an **occlusion**.

An **occluded front** is the interface between a cool and a cold air mass. When the trailing cold front overtakes the warm front, the two cooler air masses converge and force all of the warm air to rise. Occluded fronts may produce some type of heavy precipitation.

A **stationary front** is produced when the boundary between two air masses of differing characteristics is not moving. The weather of

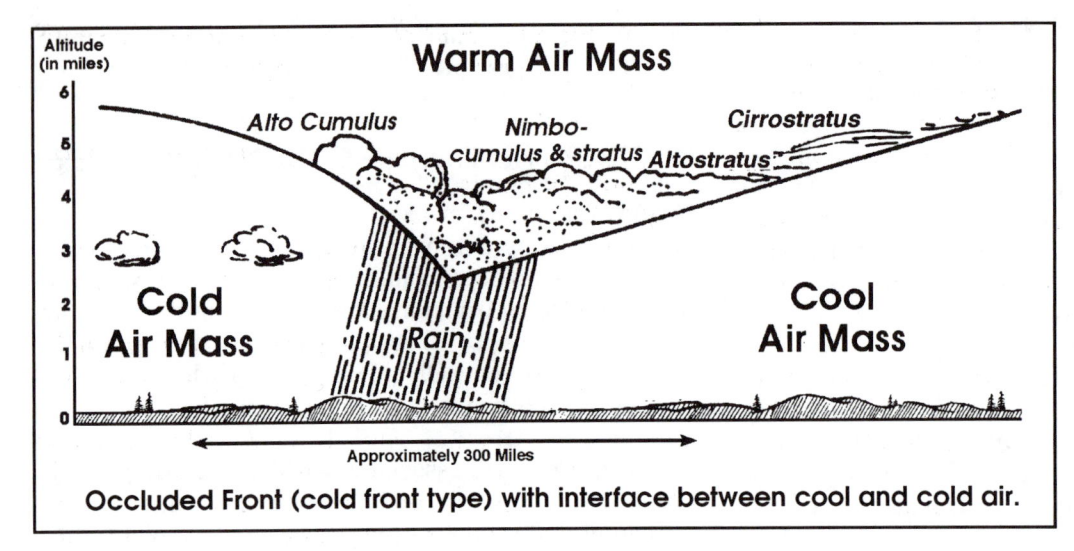

Warm Air Mass

Altitude (in miles)

Alto Cumulus, Nimbocumulus & stratus, Altostratus, Cirrostratus

Cold Air Mass — Rain — Cool Air Mass

Approximately 300 Miles

Occluded Front (cold front type) with interface between cool and cold air.

these fronts most closely resembles warm front weather conditions

WEATHER FORECASTING

The path that the center of a low or storm travels is called the **storm track**. The rate of movement of a storm can usually be determined and is generally predictable in the continental United States. Tropical air masses most often move northeast, and polar air masses usually move southeast.

In the northern middle latitudes, the prevailing westerly winds blow from west to east and tend to direct the weather patterns. High in the troposphere, there are strong winds capable of slowing down or speeding up a jet airplane as much as 300 kilometers per hour. These wave-like currents with high winds, called the **jet stream**, move toward the east.

Most weather forecasts are based upon movement of air masses. Using data to determine the path and speed of a storm center, forecasters can predict the weather changes ahead of the storm.

Meteorologists use records of past weather data and computer analysis to predict weather conditions as a probability of occurrence. If it rained 4 out of 5 days in a region when the wind was from the east, the pressure was falling, and the relative humidity was rising, when these conditions occur again, the probability of rain would be 80%.

Using data from weather instruments, weather satellites, and computer analysis has enabled the weather forecaster to make short term forecasts (1-3 days) with good accuracy, 80 to 90 percent of the time. It is far more difficult to make forecasts extending a week or more in advance.

How can weather information be used to make forecasts?

SEVERE STORMS

From June through September tropical depressions which may develop into **hurricanes** form in the low latitudes over warm, tropical water. Energy released by condensation over the warm water surface intensifies these huge rotating storms. With much lower pressure and greater intensity than a middle-latitude cyclone, the hurricane has a central area of calm with descending air currents called the "**eye**." Around the eye there are violent winds (in excess of 119 km/hr) and heavy precipitation.

Hurricanes are the most destructive of the severe storms because of their large size having diameters from 200 to 400 miles and paths which may cover a thousand miles. The most severe damage is done by water along low-lying coastal areas.

Where and how are the most hazardous weather conditions likely to happen?

Because hurricanes pick up moisture and gain energy when they are over warm water, they die down when their path takes them over land or cool water. The average life span of these destructive storms is nine days.

Data from weather satellites assist the **National Hurricane Center** in Florida and the **National Weather Service** in tracking hurricanes and issuing appropriate warnings.

The most violent storm is the **tornado**. When warm, moist maritime tropical air collides with cold, dry continental polar air, the conditions are favorable for the development of these storms. Tornadoes are most likely to occur in the late afternoon in spring and early summer when the temperature differences between the air masses is greatest. Most commonly they occur in the Great Plains and Gulf States area.

Their funnel-shaped cloud extends down out of the cumulonimbus cloud of a thunderstorm. The center of the funnel is called the **vortex**. Winds move counter-clockwise around the vortex in speeds in excess of 600 km/hr (370 mph).

The "twisters" are fortunately small in size with a maximum diameter approximately 100 meters. These storms travel at 40-60 km/hr as they cause devastating damage in their unpredictable paths which vary from 25-65 kilometers in length. The **National Severe Storm Forecast Center** in Kansas City predicts and tracks tornadoes.

QUESTIONS FOR UNIT 6

1. A balloon carrying weather instruments is released at the Earth's surface and rises through the troposphere. As the balloon rises, what will the instruments generally indicate? [Refer to the *Earth Science Reference Tables*.]
   1. a decrease in both air temperature and air pressure
   2. an increase in both air temperature and air pressure
   3. an increase in air temperature and a decrease in air pressure
   4. a decrease in air temperature and an increase in air pressure

2. The graph shows air temperature for an area near the Earth's surface during a 12-hour period. Which graph best illustrates the probable change in air pressure during the same time period?

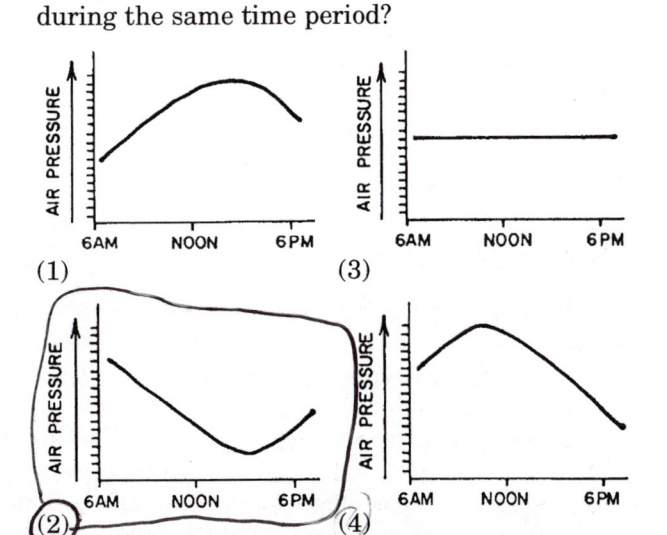

3. According to the *Reference Tables*, an air pressure of 29.65 inches of mercury is equal to
   (1) 984.0 mb
   (2) 999.0 mb
   (3) 1001.0 mb
   (4) 1004.0 mb

4. Wind moves from regions of
   1. high temperature toward regions of low temperature
   2. high pressure toward regions of low pressure
   3. high precipitation toward regions of low precipitation
   4. high humidity toward regions of low humidity

5. The wind speed between two nearby locations is affected most directly by differences in the
   1. latitude between the location
   2. longitude between the locations
   3. air pressure between the locations
   4. Coriolis effect between the locations

6. The Coriolis effect is caused by the
   1. rotation of the Earth on its axis
   2. revolution of the Earth around the Sun
   3. movement of the Earth in relation to the Moon
   4. movement of the Earth in relation to the Milky Way

7. In the Northern Hemisphere, a wind blowing from the north will be deflected toward the
   1. northwest
   2. northeast
   3. southwest
   4. southeast

8. The map represents a portion of an air-pressure field at the Earth's surface.

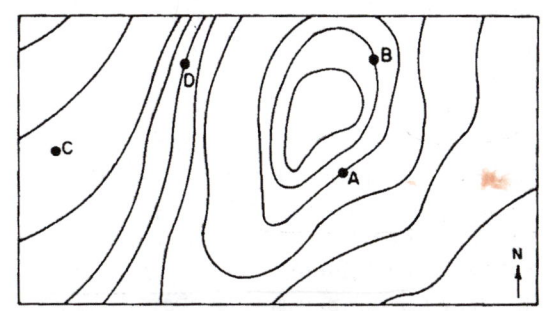

   At which position is wind speed *lowest*?
   (1) A   (3) C
   (2) B   (4) D

9. As the amount of moisture in the air increases, the atmospheric pressure will probably
   1. decrease
   2. increase
   3. remains the same

10 The air temperature and the wet bulb temperature were measured and both were found to be 18°C. Two hours later, measurements were taken again and the air temperature was 20°C, while the wet bulb temperature remained at 18°C. The relative humidity of the air during those two hours
1 decreased
2 increased
3 remained the same

11 The two thermometers show the dry-bulb and wet-bulb temperatures of the air. According to the *Reference Tables*, what is the approximate dew point temperature of the air? 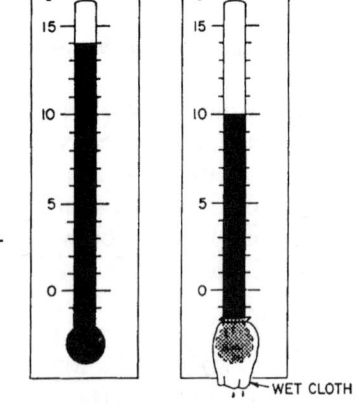
(1) -25°C
(2) 6°C
(3) 3°C
(4) 4°C

12 The graph below shows changes in air temperature and dew point temperature over a 24-hour period at a particular location.

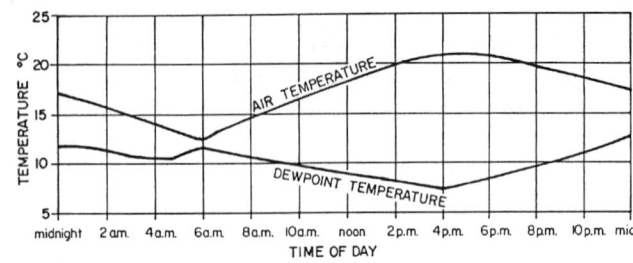

At what time was the relative humidity *lowest*?
(1) midnight
(2) 6 a.m.
(3) 10 a.m.
(4) 4 p.m.

13 Which conditions must exist for condensation to occur in the atmosphere?
1 The air is saturated and a condensation surface is available.
2 The air temperature is above the dew point and the air pressure is high.
3 The air is calm and the relative humidity is low.
4 The relative humidity is low and the air pressure is high.

14 What is the approximate relative humidity if the dry-bulb temperature is 12°C and the wet-bulb temperature is 7°C?
(1) 28%
(2) 35%
(3) 48%
(4) 65%

15 Which event will most likely occur in rising air?
1 clearing skies
2 cloud formation
3 decreasing relative humidity
4 increasing temperature

16 Which statement best explains why a cloud is forming as shown in the diagram?

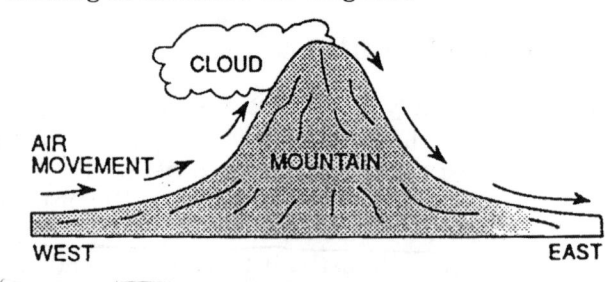

1 Water vapor is condensing.
2 Moisture is evaporating.
3 Cold air rises and compresses.
4 Warm air sinks and expands.

17 On a clear, dry day an air mass has a temperature of 20°C and a dew point temperature of 10°C.

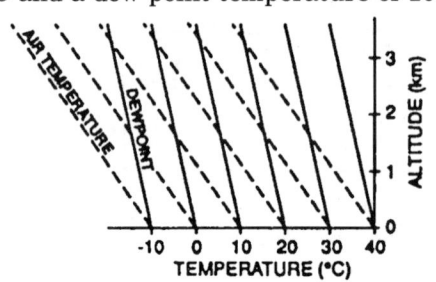

According to the graph, about how high must this air mass rise before a cloud can form?
(1) 1.6 km
(2) 2.4 km
(3) 3.0 km
(4) 2.8 km

18 Which is a form of precipitation?
1 frost
2 snow
3 dew
4 fog

19 Why is it possible for no rain to be falling from a cloud?
1 The water droplets are too small to fall.
2 The cloud is water vapor.
3 The dew point has not yet been reached in the cloud.
4 There are no condensation nuclei in the cloud.

20 The air temperature is 10°C. Which dew point temperature would result in the highest probability of precipitation?
(1) 8°C
(2) 6°C
(3) 0°C
(4) -4°C

21 Which letter on the map at the right represents the area closest to the source region of a cT air mass? 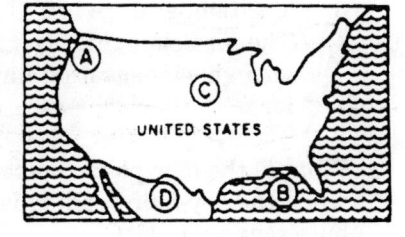
(1) A
(2) B
(3) C
(4) D

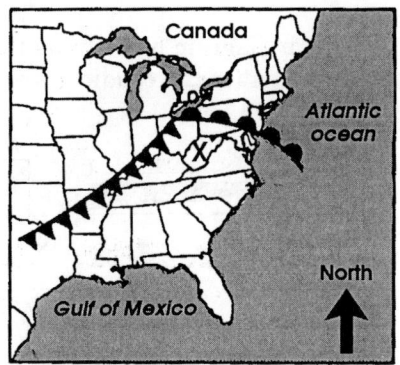

22 The weather map shows a frontal system that has followed a typical storm track. The air mass located over point X most likely originated over the
1 northern Atlantic Ocean
2 central part of Canada
3 Gulf of Mexico
4 Pacific Northwest

23 An air mass located over central United States will most likely move toward the
1 northeast
2 southeast
3 northwest
4 southwest

Base your answers to questions 24 and 25 on the *Reference Tables* and the diagram of the station model.

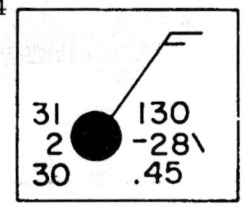

24 The barometric pressure is
(1) 1013.0 mb
(2) 913.0 mb
(3) 130.0 mb
(4) 10.28 mb

25 The weather forecast for the next six hours at this station most likely would be
1 overcast, hot, unlimited visibility
2 overcast, hot, poor visibility
3 overcast, cold, probable snow
4 sunny, cold, probable rain

26 Which diagram below best represents the air circulation around a Northern Hemisphere low-pressure center?

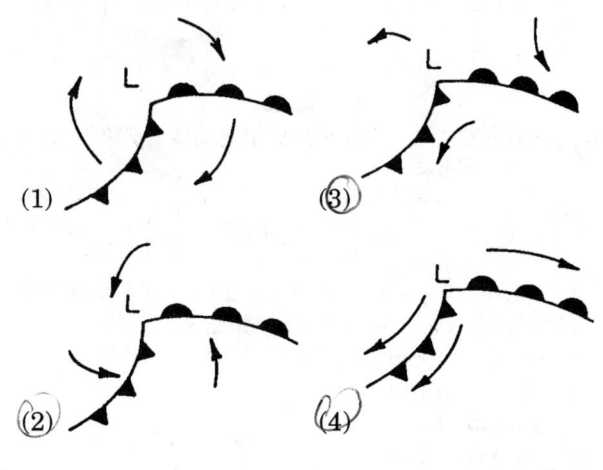

27 Cities A, B, C, and D on the weather map are being affected by a low-pressure system (cyclone).

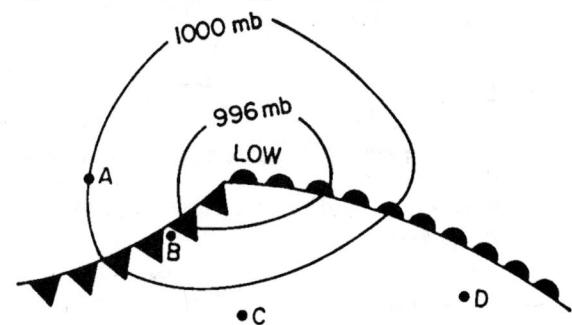

Which city would have the most unstable atmospheric conditions and the greatest chance of precipitation?
(1) A
(2) B
(3) C
(4) D

28 Which map best represents the normal air circulation around a high-pressure air mass located over central New York State?

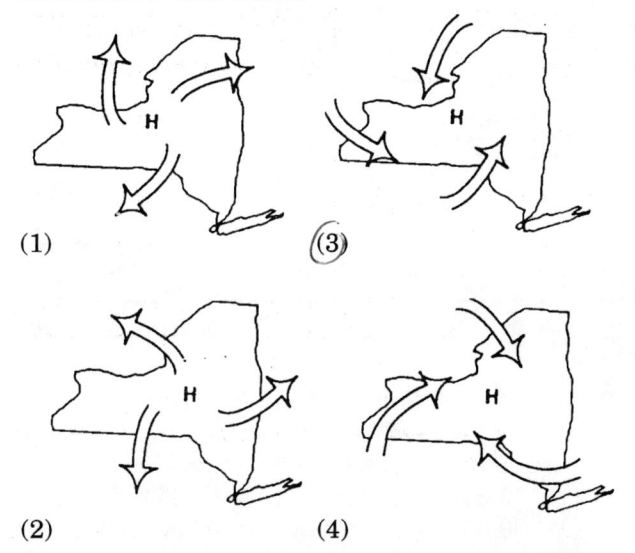

29 Which weather station model indicates the greatest probability of precipitation?

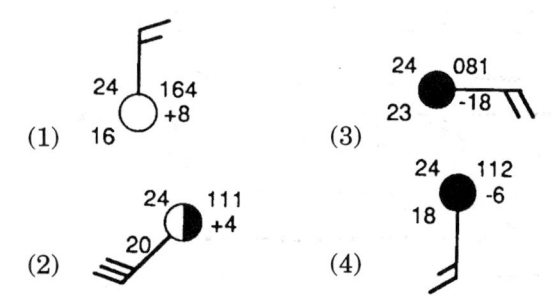

(1) 24 ⦿ 164 +8 / 16
(2) 24 ◐ 111 +4 / 20
(3) 24 ● 081 -18 / 23
(4) 24 ● 112 -6 / 18

30 A weather station reporting clear, cold weather with little wind is probably located
1 in the center of a high
2 in the center of a low
3 ahead of a warm front
4 at a cold front

31 An observer reports the following data for a location in New York State:

Air temperature = 35°C
Pressure = 996 mb
Relative humidity = 84%

The weather conditions at this location would best be described as
1 hot and dry
2 hot and humid
3 cool and dry
4 cool and humid

Base your answers to questions 32 through 35 on the *Earth Science Reference Tables*, the diagram below, and your knowledge of Earth science. The diagram shows a section of the shore of Lake Ontario. Surface air-pressure readings are shown for three of the locations.

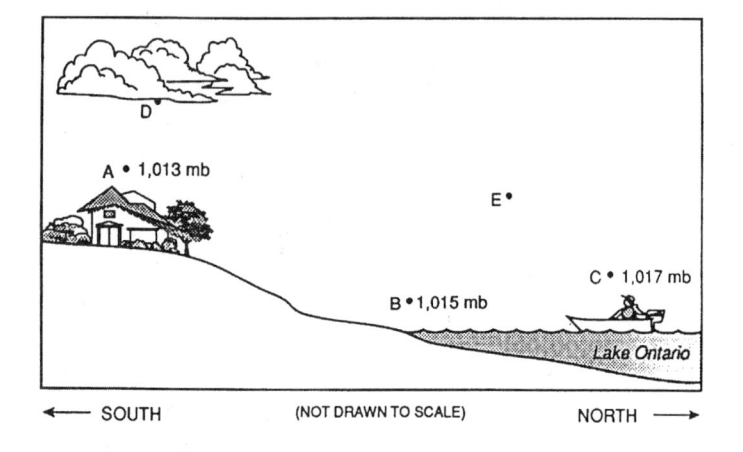

[Diagram showing D (clouds), A • 1,013 mb, E •, B • 1,015 mb, C • 1,017 mb on Lake Ontario. SOUTH ← (NOT DRAWN TO SCALE) → NORTH]

32 [Refer to the *Earth Science Reference Tables*.] When converted to inches of mercury, the air pressure reading of 1,017 millibars at C is equal to
(1) 33.0 in
(2) 30.30 in
(3) 30.03 in
(4) 30.00 in

33 Why do the clouds begin to form at the elevation of D?
1 The air has cooled to the dew point temperature at this elevation.
2 The water droplets are too small to be seen below this elevation.
3 The temperature is 0°C at this elevation.
4 The air below this elevation does not have enough condensation nuclei for clouds to form.

34 What is the dew point temperature at location E when the dry-bulb reading is 18°C and the wet-bulb reading is 11°C?
(1) 1°C
(2) –10°C
(3) 7°C
(4) 4°C

35 Which diagram best shows the probable wind direction for the conditions shown?

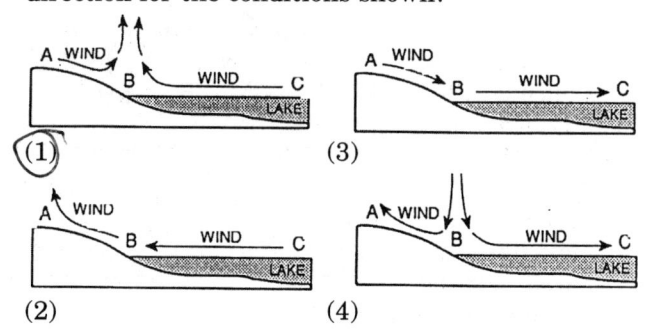

(1) (2) (3) (4) [diagrams showing wind directions over land and lake]

Base your answers to questions 36 through 38 on the *Earth Science Reference Tables*, the map below, and your knowledge of Earth science. The map shows the source regions for various types of air masses affecting the weather of the continental United States. Regions labeled with the same letter produce air masses with similar characteristics. Point X represents a location in the central United States.

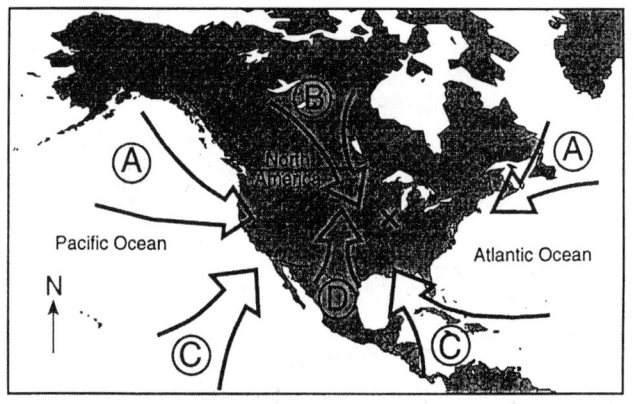

[Map of North America showing air mass source regions labeled A, B, C, D with Pacific Ocean and Atlantic Ocean]

36 On a weather map, which symbol would be used to represent an air mass that formed in region B?
(1) mP
(2) mT
(3) cP
(4) cT

Page 132   N&N© SCIENCE SERIES – EARTH SCIENCE – MODIFIED PROGRAM

37 Which atmospheric conditions will most likely exist when air masses from source regions *B* and *C* meet at point *X*?
1. clearing skies and little wind
2. cloudiness and precipitation *(circled)*
3. decreasing relative humidity and rising temperature
4. appearance of condensation nuclei and constant dew point temperature

38 Which map symbol represents a stationary front that formed when an air mass from source region *B* met an air mass from source region *D*?

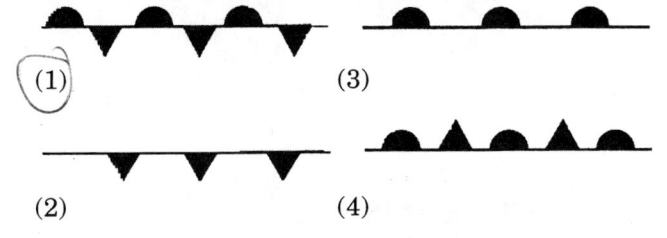

(1) *(circled)*   (2)   (3)   (4)

Base your answers to questions 39 through 41 on the *Earth Science Reference Tables*, the information and diagram below, and your knowledge of Earth science.

The diagram represents a model that shows how air density is affected by the addition of water vapor to the air. Marbles with different masses, representing nitrogen, oxygen, and carbon dioxide, were used to fill a container to show a certain volume of dry air. The container was placed on a scale to find the mass of this volume of dry air.

A few marbles representing nitrogen (N_2) and oxygen (O_2) were removed and replaced with marbles representing water vapor (H_2O) to show the same volume of air with water vapor present. The relative mass of each gas, as represented by the marbles, is shown in the data table.

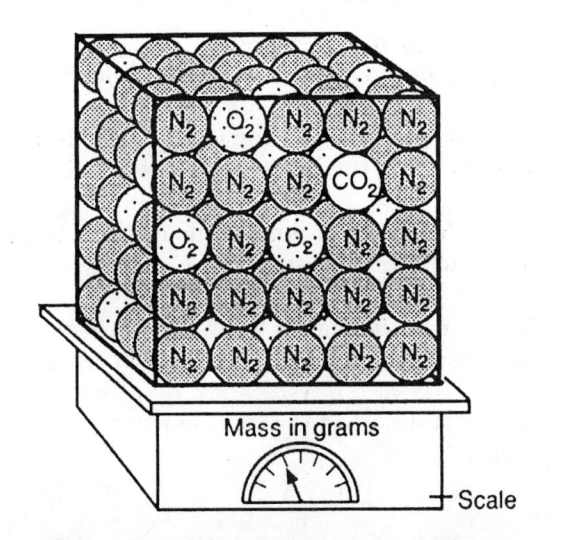

DATA TABLE

Molecule Symbol	Gas	Mass
N_2	Nitrogen	28 g
O_2	Oxygen	32 g
CO_2	Carbon Dioxide	44 g
H_2O	Water Vapor	18 g

39 According to the data table, which gas molecule has the *least* mass?
1. nitrogen
2. oxygen
3. carbon dioxide
4. water vapor *(circled)*

40 When a few of the marbles representing nitrogen and oxygen are replaced with marbles representing water vapor, the air model will become
1. lighter and less dense *(circled)*
2. lighter and more dense
3. heavier and less dense
4. heavier and more dense

41 After water vapor molecules enter the Earth's atmosphere, what conditions must occur before they can become liquid?
1. warming temperatures and condensation
2. warming temperatures and evaporation
3. cooling temperatures and condensation *(circled)*
4. cooling temperatures and evaporation

Base your answers to questions 42 through 44 (found on the next page) on the *Earth Science Reference Tables*, the weather map below, and your knowledge of Earth science. The map shows part of the southern United States and northern Mexico.

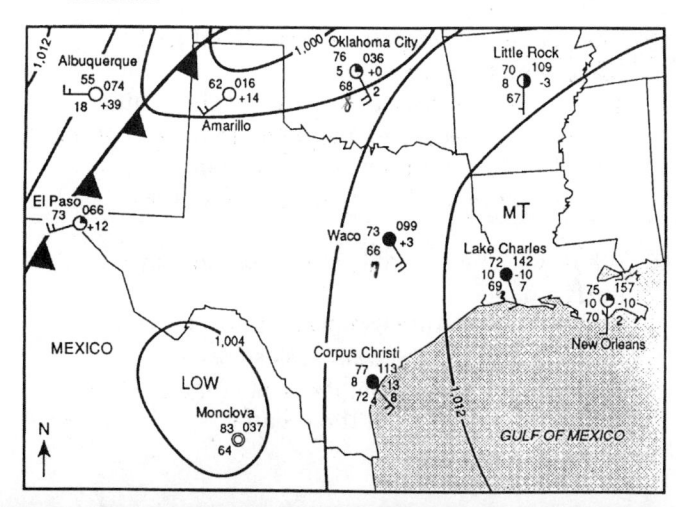

42 At which city is the visibility 8 miles?
1. Little Rock, Arkansas
2. Lake Charles, Louisiana
3. Oklahoma City, Oklahoma
4. New Orleans, Louisiana

43 The isolines on this map connect locations that have the same
1. dew point temperature
2. air temperature
3. barometric pressure
4. relative humidity

44 Which city has the *least* chance of precipitation during the next 3 hours?
1. Oklahoma City, Oklahoma
2. Waco, Texas
3. Lake Charles, Louisiana
4. Albuquerque, New Mexico

Base your answers to questions 45 through 48 on the *Earth Science Reference Tables*, the diagram below, and your knowledge of Earth science. The diagram represents a satellite image of Hurricane Gilbert in the Gulf of Mexico. Each **X** represents the position of the eye of the storm on the date indicated.

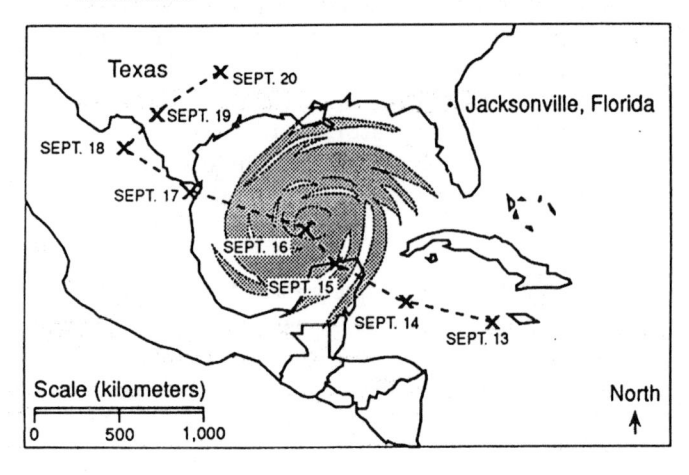

45 The surface wind pattern associated with Hurricane Gilbert was
1. counterclockwise and toward the center
2. counterclockwise and away from the center
3. clockwise and toward the center
4. clockwise and away from the center

46 What was the probable source of moisture for this hurricane?
1. carbon dioxide from the atmosphere
2. winds from the coastal deserts
3. transpiration from tropical jungles
4. evaporation from the ocean

47 On September 18, Hurricane Gilbert changed direction. Which statement provides the most probable reason for this change?
1. The air mass was cooled by the land surface.
2. The storm entered the prevailing westerlies wind belt.
3. The amount of precipitation released by the storm changed suddenly.
4. The amount of insolation received by the air mass decreased.

48 The air mass that gave rise to Hurricane Gilbert would be identified as
(1) cP   (3) mT
(2) cT   (4) mP

SKILL ASSESSMENTS

Base your answers to questions 1 through 9 on your knowledge of Earth science, the *Earth Science Reference Tables*, and the diagram which represents a section of a weather map for locations in the central United States. The letters *A* through *I* identify reporting weather stations.

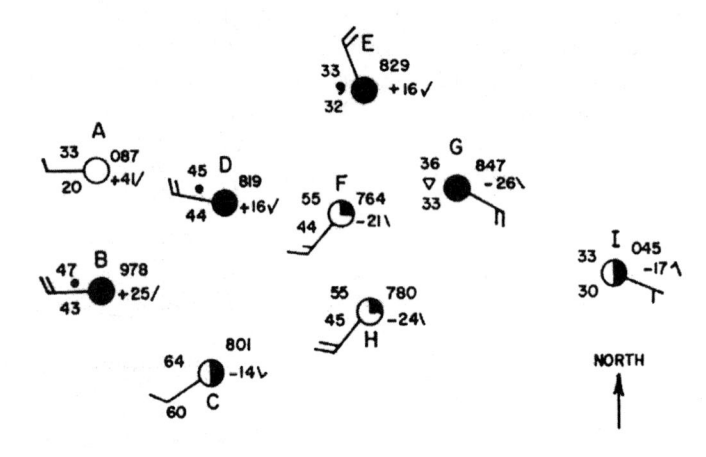

1 On the map draw isolines with an interval of 10°F, beginning with the 40°F isoline.

2 Find the station with the lowest barometric pressure. What is the pressure at that station?

3 Place the letter "L" just to the north of the station with the lowest pressure.

4 Draw and label a cold front and a warm front on the appropriate places extending out of the Low.

5 Which station has the least amount of cloud cover?

6 Which station has a wind from the southeast at 5 knots?

7 Which station shows the pressure has dropped 2.6 mb in the past three hours?

8 What is the air pressure at station *D*?

9 In order to test the rate of evaporation, equal amounts of water are exposed to the open air outside weather stations *B, E, H,* and *I*. In a sentence explain at which station the water will probably evaporate the fastest.

Base your answers to questions 10 through 15 on your knowledge of Earth science and the satellite photograph of a tropical storm centered in the Gulf of Mexico. An outline of the southeastern United States and the latitude-longitude system have been drawn on the photograph.

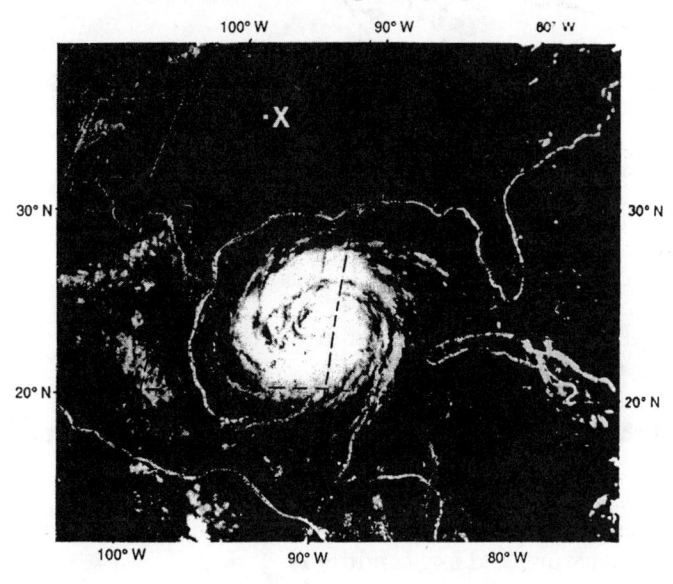

Use one or two sentences to answer the following questions.

10 What is the approximate latitude and longitude of the center or eye of the tropical storm on the satellite photograph?

11 What type of air mass would most likely be associated with the storm?

12 Describe the weather conditions at point *X* at the time this photograph was taken.

13 What will happen to barometric pressure along the coast of Texas as the storm approaches?

14 What is the source of energy for this storm?

15 Describe the general direction of movement of the surface winds associated with this tropical storm.

Base your answers to questions 16 and 17 on the data table below. The data table shows the air temperature and dew point over a 24-hour period for a particular location in New York State.

TIME OF DAY	AIR TEMPERATURE (°C)	DEWPOINT (°C)
12:00 MIDNIGHT	19	12
2:00 a.m.	17	11
4:00 a.m.	14	10
6:00 a.m.	13	12
8:00 a.m.	15	11
10:00 a.m.	17	10
12:00 NOON	18	9
2:00 p.m.	21	7
4:00 p.m.	23	6
6:00 p.m.	21	8
8:00 p.m.	19	10
10:00 p.m.	18	12
12:00 MIDNIGHT	17	13

16 Use the data to construct a graph following the directions below.
   a Mark an appropriate scale on the axis labeled "Temperature."
   b Plot a line graph for air temperature and label the line "Air Temperature."
   c Plot a line graph for dew point and label the line "Dew Point."

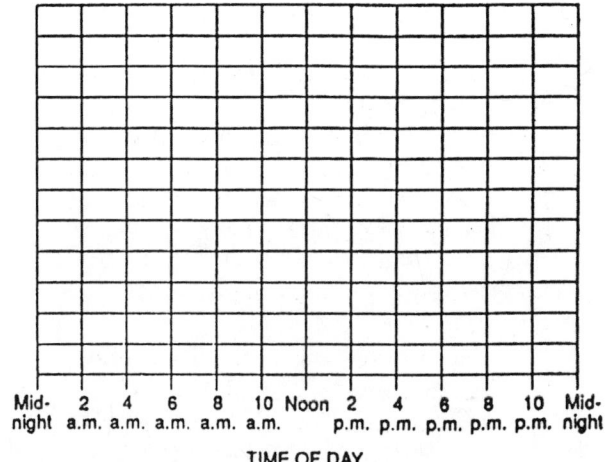

UNIT SIX – METEOROLOGY – N&N©   Page 135

17 Based on your graph, state the hour of the day when the relative humidity was *lowest*.

Base your answers to questions 18 and 20 on the *Earth Science Reference Tables*, the diagram below, and your knowledge of Earth science. The diagram represents a weather map of the United States. On the map, station models for selected cities indicate weather conditions and isobars indicate the air pressure pattern. Two fronts, A and B, have been identified.

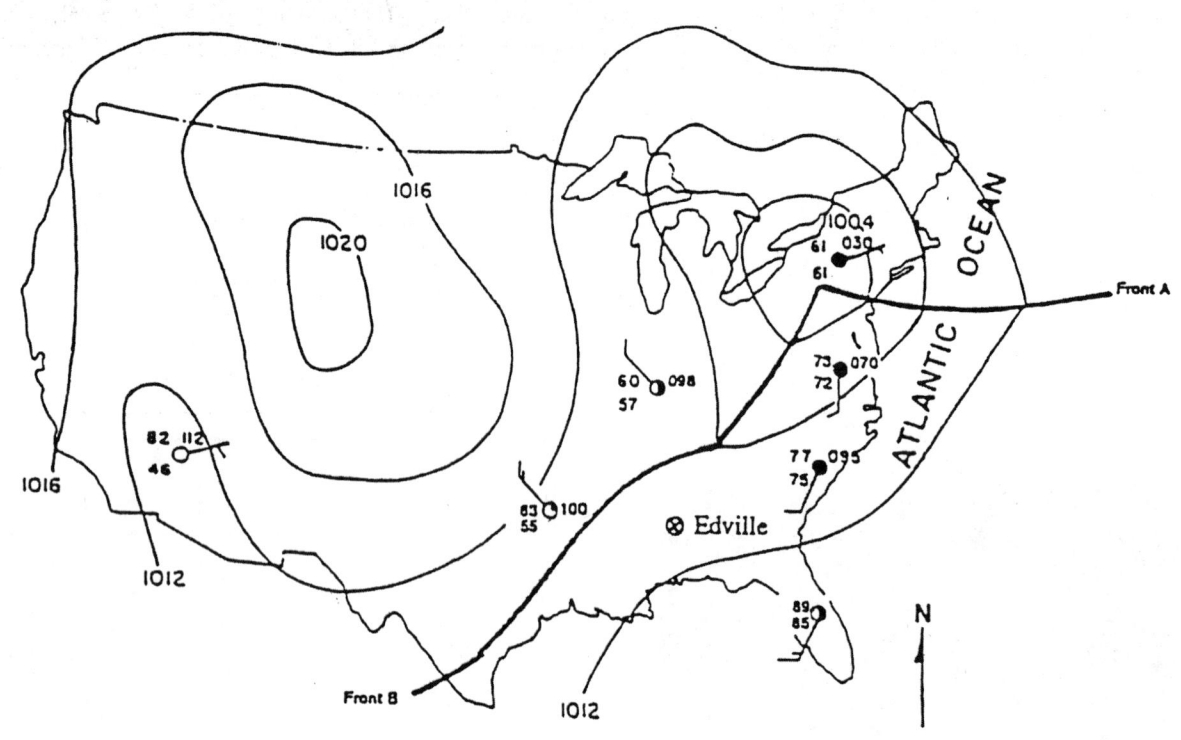

18 On the map above, write the words "**HIGH**" and "**LOW**" directly on the map to indicate the high and low pressure centers.

19 On the map above, label both front **A** and front **B** with the correct symbol to indicate the type of front and its direction of motion.

20 In one or more complete sentences, write a short weather forecast for the next twenty-four hours at Edville. Include in your forecast any anticipated changes in: temperature and sky conditions.

Topic E
Atmospheric Energy

Vocabulary To Be Understood

Adiabatic Temperature Change	Heat of Fusion	Solar Electromagnetic Spectrum
Change of Phase	Heat of Vaporization	Solar Radiation
Condensation	Kinetic Energy	Specific Heat
Conduction	Latent Heat	Transpiration
Convection	Potential Energy	Vapor Pressure
Convection Cell	Pressure Belt	Wavelength
Electromagnetic Energy	Radiation	Zone of Convergence
Evaporation	Radiational Cooling	Zone of Divergence

A. EARTH'S ENERGY

The primary source of energy for the Earth is **solar radiation** which is responsible for driving the surface systems which change the Earth such as weather and climate patterns. Solar radiation has its greatest intensity occurring in the visible wavelength (about 50%) of the solar electromagnetic spectrum.

SOLAR ELECTROMAGNETIC SPECTRUM

There are many forms of electromagnetic energy, such as heat and light. They all have the capacity (ability) to do work. Each form is distinguished from the others by its **wavelength** – the distance between the crests (peaks) of successive waves.

The **solar electromagnetic spectrum** is similar to the general electromagnetic spectrum, since the Sun radiates almost every kind of wave energy. The principal solar electromagnetic radiations includes visible light as well as X-rays, ultraviolet rays, and infrared rays.

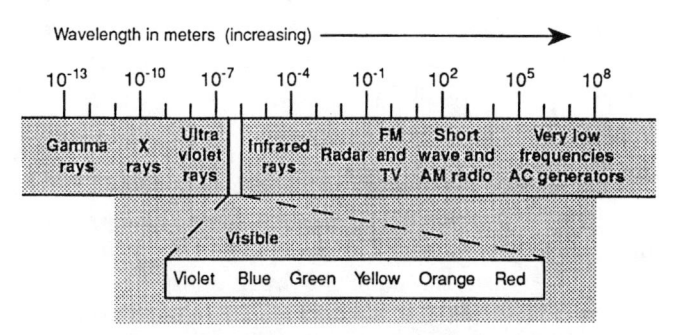

Electromagnetic Spectrum

ENERGY TRANSFER

Energy may be transferred in three ways: **conduction**, **convection**, and **radiation**. In matter (solids, liquids, and gases), heat energy may be transferred by either conduction or convection. In space, energy is transferred through electromagnetic wave radiation.

How is energy transferred in solids, liquids, gases, and space?

Conduction of thermal (heat) energy occurs as an interaction of matter at the molecular or atomic level. When atoms or molecules collide at an interface, thermal energy is transferred from one atom (or molecule) to another. The efficiency of the energy transfer depends upon the densities of the substances involved. For example, the molecules within a gas are relatively far apart; therefore, the energy transferred is less than within a liquid, where the molecules are relatively close together. Solids have the most compact molecules and have the greatest energy transfer.

Conduction in a Pan
Heat energy is transfered (conducted) through the metal spoon from the hot pan to the hand.

Convection is the transfer of thermal energy by the movement of molecules within liquids and gases. The

Convection Cell In a Room
Warm air rises, then cools and falls, maintaining the cycle.

movement is from regions of higher density molecules to lower density molecules. Since warm air is less dense than cool air, the warm air rises as it is displaced by the cool air that settles due to the cool air's greater density.

For example, heated air rises, because it is displaced by cooler air. As the cool air comes in contact with the heat source, it is warmed and also rises. The warm air moves away from the heat source, then cools and falls, maintaining the cycle. The resulting circulation is referred to as a **convection current** or a **convection cell**.

MOISTURE & ENERGY INPUT

Atmospheric moisture content is constantly changing due to evaporation and transpiration. **Evaporation** is the conversion of liquid water to vapor (gas) from the Earth's surface, such as soils, lakes, and streams, but primarily from the oceans. Moisture enters the air by means of evaporation and transpiration.

Transpiration is the loss of water from plants through their leaves. Plants absorb water from the ground through their roots, carry it up through the stems and release the water into the air from their leaves. Combined, these moisture input processes are known as **evapotranspiration**.

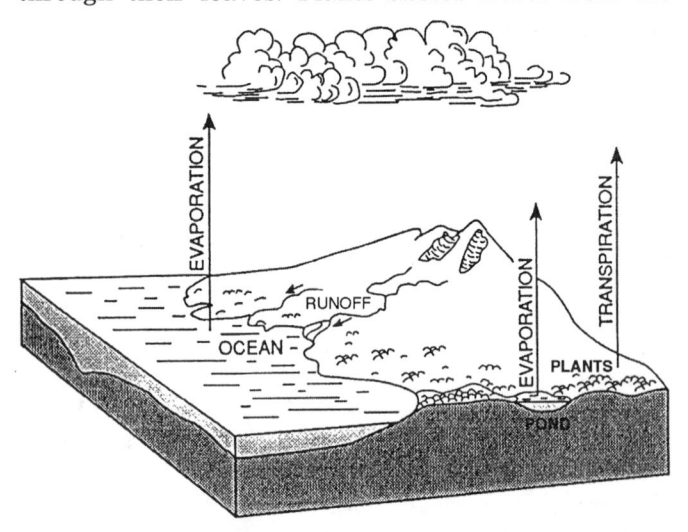

Evaporation and Transpiration

Both evaporation and transpiration require the absorption of energy and so constitute an energy input into the atmosphere in the form of more energetic water molecules.

VAPOR PRESSURE

The rate of evaporation is dependent upon:

- the amount of energy available (temperature),
- the surface area of the water source, and
- the moisture content of the air (vapor pressure).

The pressure exerted by the water vapor within the atmosphere at a given location is called the **vapor pressure**. Near the surface of water, vapor pressure is greatest, decreasing with altitude above the water's surface.

When there is very little water vapor in the air, evapotranspiration can take place rapidly, but as the amount of vapor pressure increases, the rate of evaporation decreases until the air is saturated and cannot hold any additional moisture.

How does the atmosphere store and release energy?

B. LATENT HEAT

Water exists in our environment as solid (ice), liquid, and water vapor (gas). **Latent heat** is a form of **potential energy** which is absorbed or released when a change of phase occurs. When heat is transferred within the same phase (solid, liquid, or gas), the temperature of the material is changed.

However, when heat is applied to different states of matter and there is a **change of phase** (such as solid to liquid), the temperature of the material remains the same, since there is no increase in the **kinetic energy** of the molecules. Instead, the heat energy is either **absorbed** (solid to liquid to gas) or **released** (gas to liquid to solid), increasing or decreasing the potential energy of the molecules.

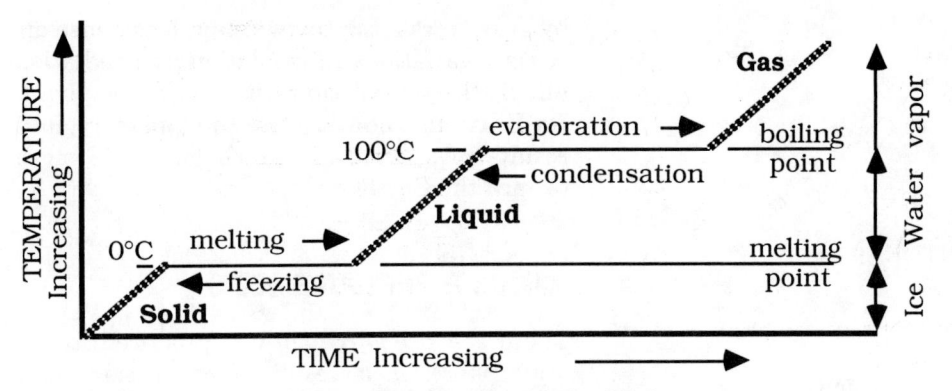

Phase Change Graph for Water

Heat is gained or lost in the phase change. The amount of latent heat is different for various substances and the type of phase change. The amount of heat either gained or lost is equal to the product of the mass of the matter times the latent heat per unit of mass.

LATENT HEAT OF WATER

Melting occurs when water changes phase from a solid to a liquid. **Freezing** occurs when water changes from a liquid into a solid. The latent heat for these changes is 80 calories per gram of water and is called the **heat of fusion**. In other words, one gram of ice absorbs (gains) 80 calories of heat energy when changed to a liquid. When the phase change is reversed, liquid to ice, 80 calories per gram of heat is released (lost) by the water.

A significantly greater amount of energy is required to change a given mass of liquid water to water vapor than is required to change ice to liquid. Evaporation, requiring the addition of 540 calories per gram of water (called the **heat of vaporization**) occurs when water in the liquid phase is changed to water vapor (gas).

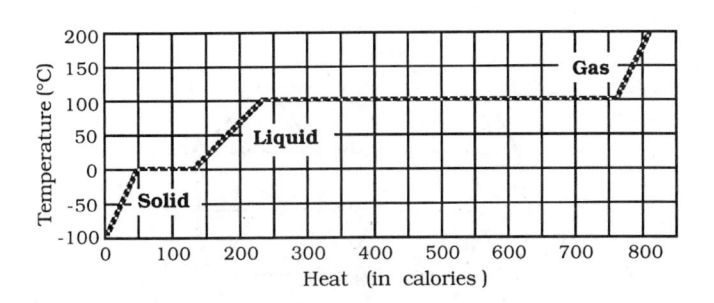

Heating Curve for Water

When **condensation** (vapor to liquid) occurs, latent heat of 540 calories per gram of water is released. When condensation is taking place in the atmosphere, solar energy that was stored in the process of evaporation is released into the environment. This latent heat is the energy that intensifies and sustains violent storms.

The amount of energy needed to produce an equal temperature change in equal masses of different materials varies with the materials. The amount of heat energy needed to raise the temperature of one gram of a material one degree on the Celsius scale is called **specific heat**. The higher the specific heat a material has, the more energy it takes to heat it and the longer the time it takes for it to cool.

Liquid water has the highest specific heat capacity among all of the naturally occurring materials on Earth (1.0 calories/gram/°C). (See the *Earth Science Reference Tables* for the specific heats of other materials.)

HEAT LOST OR GAINED

The heat (measured in calories) lost or gained by a material is proportional to the product of the mass and the temperature change in the material times the specific heat. To determine the amount of heat lost or gained, use the following formula: Heat (cal) = Mass **x** Temperature Change **x** Specific Heat (see below).

The heat lost or gained in a phase change is equal to the product of the mass times the change in potential energy per unit mass.

Latent Heat – Solid ↔ liquid:

HEAT (CAL) = MASS x HEAT OF FUSION

Latent Heat – Liquid ↔ gas:

HEAT (CAL) = MASS x HEAT OF VAPORIZATION

Formula for the Determination of Heat Lost or Gained –

HEAT (CAL) = MASS x TEMPERATURE CHANGE x SPECIFIC HEAT

TOPIC E – ATMOSPHERIC ENERGY – N&N©

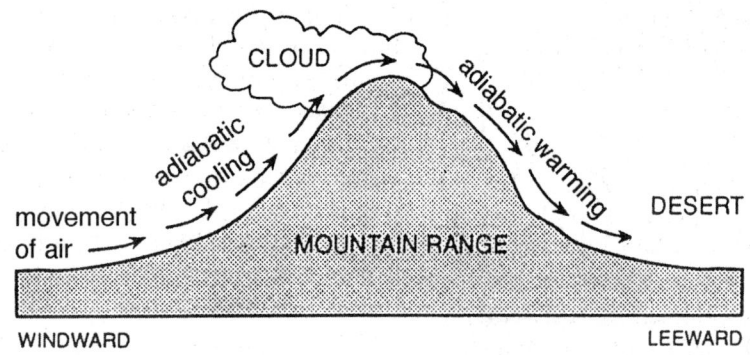

Adiabatic Cooling & Heating
Resultant of air movement to higher and lower elevations.

C. MOISTURE & ENERGY TRANSFER

Moisture and energy transfer is accomplished in the atmosphere through three primary means: (1) the adiabatic changes, (2) the density differences, and (3) the wind speed and direction.

ADIABATIC TEMPERATURE CHANGE

Most clouds are formed by adiabatic cooling in rising air. An **adiabatic temperature change** is the change in the temperature of air, due to the expansion or compression of the air mass. As air rises in the atmosphere, expansion causes cooling as potential energy is released. When the dewpoint temperature is reached in the rising air, condensation (or deposition) begins to occur. Since the condensation process releases heat energy into the air, the cooling rate decreases as the air continues to rise.

Conversely, as air descends, it compresses (due to an increase in pressure), and its temperature rises.

DENSITY DIFFERENCES

As the moisture content and/or the temperature of the air increases, the density of that volume decreases. Therefore, density increase is expressed as the inverse (opposite) of temperature and moisture increase.

One of the results of density differences and the effect of the gravity field is the formation of convection cells. A **convection cell** is formed when the air circulates by rising in one place and sinking at another. The effect may be observed in the ocean and Earth's mantle as well as the atmosphere.

Convection cells, or currents, occur because cooler, more dense air sinks toward the Earth's surface, pulled by gravity, causing the warmer, less dense air to rise. Variations in insolation and radiation affect the convection in the atmosphere. **Radiational cooling** in the polar regions results in sinking air which begins to move toward the Equator.

WINDS & PRESSURE BELTS

There are **pressure "belts"** produced in the atmosphere as a result of convection. Low pressure belts are found at the equator and at the 60° North and South Latitudes. High pressure belts are found at the 30° North and South Latitudes as well as at the Poles.

How does the atmosphere transfer energy?

In the low pressure belts, air converges (moves together) and rises; whereas, belts of high pressure are associated with descending air that diverges (moves apart) at the surface. Therefore, regions of low pressure are often called **zones of convergence**, while regions of high pressure are called **zones of divergence**.

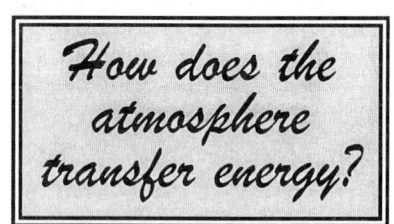

Wind is a result of the movement of the air from regions of divergence to regions of convergence. In this way the atmosphere distributes solar energy over the whole Earth. However, the cold, dense, sinking air in the polar regions does not move in a straight path to replace the warm, light air rising in the tropical regions. As the Earth rotates, the wind direction is modified as it flows between the pressure belts. The Coriolis Effect deflects the winds to their right in the Northern Hemisphere and their left in the Southern Hemisphere.

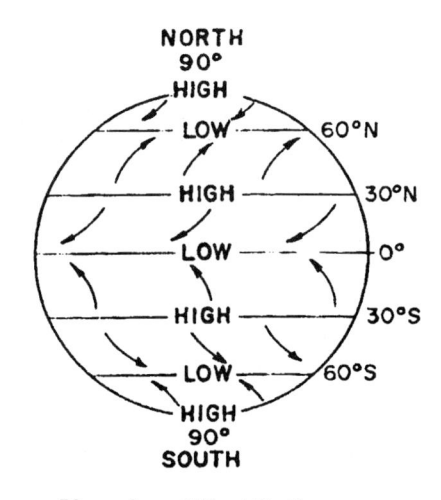

Planetary Wind Patterns

D. MODERN FORECASTING

Since weather variables are related in a complex way, accurate weather prediction is also complex. The National Weather Service in the United States uses some of the largest computers in the world to analyze the huge amount of data that streams in every hour of the day and night. Ground observations, radar and satellite imagery all contribute to the increasing reliability of the day-to-day weather forecast.

It is, however, unlikely that accurate weather predictions will be made for a week or more in advance, because a small, unexpected disturbance may have world wide unexpected effects on the weather.

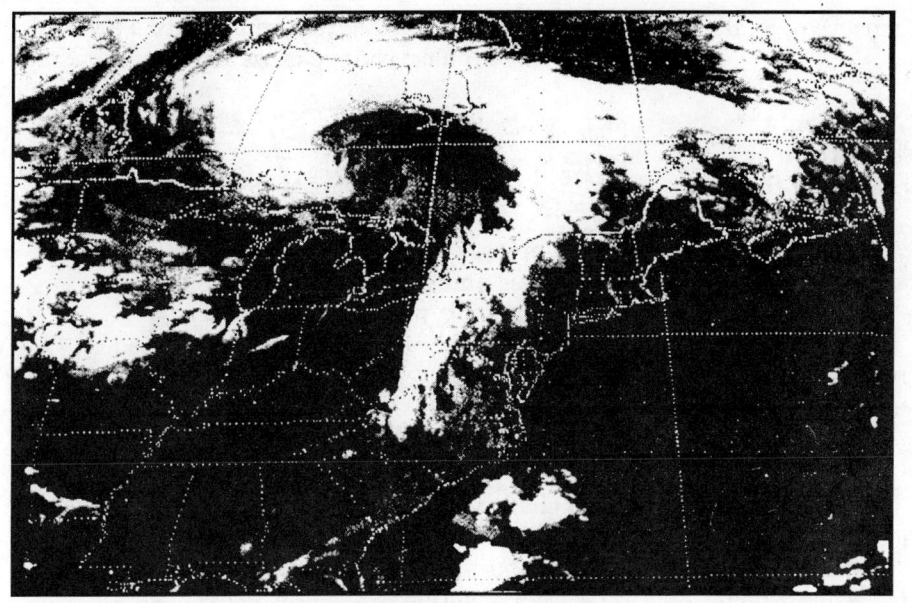

Satellite Weather Map

The location of the Earth's pressure belts and the affect of the Earth's rotation determines the general position and direction of planetary wind circulation. Factors such as altitude and position relative to mountains and large bodies of water may modify the wind pattern. (See Topic 7 for more information).

How can we forecast the weather?

QUESTIONS FOR TOPIC E

1   Most moisture enters the atmosphere by the processes of
    1   convection and conduction
    2   condensation and radiation
    3   reflection and absorption
    4   transpiration and evaporation

Base your answers to questions 2 through 6 on the *Earth Science Reference Tables*, the graph below and your knowledge of Earth science. The graph shows the results of a laboratory activity in which a 200-gram sample of ice at –50°C was heated in an open beaker at a uniform rate for 70 minutes and was stirred continually.

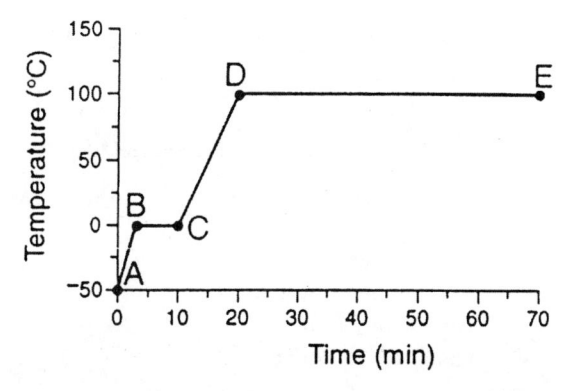

2   What was the temperature of the water 17 minutes after the heating began?
    (1) 0°C   (3) 75°C
    (2) 12°C   (4) 100°C

3   Which change occurred between point *A* and point *B*?
    1   Ice melted.   3   Water froze.
    2   Ice warmed.   4   Water condensed.

4   What was the total amount of energy absorbed by the sample during the time between points *B* and *C* on the graph?
    (1) 200 calories   (3) 10,800 calories
    (2) 800 calories   (4) 16,000 calories

5   During which time interval was the greatest amount of energy added to the water?
    (1) *A* to *B*   (3) *C* to *D*
    (2) *B* to *C*   (4) *D* to *E*

6   Which change could shorten the time needed to melt the ice completely?
    1   using colder ice
    2   stirring the sample more slowly
    3   reducing the initial sample to 100 grams of ice
    4   reducing the number of temperature readings taken

7 The rate of evaporation from the surface of a lake would be increased by
   1 a decrease in wind velocity
   2 a decrease in the amount of insolation
   3 an increase in the surface area of the lake
   4 an increase in the moisture content of the air

8 In the diagram, at which location would the vapor pressure of the air most likely be greatest? 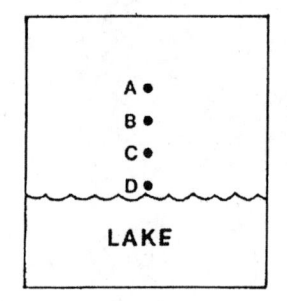
   (1) A
   (2) B
   (3) C
   (4) D

9 Equal quantities of water are placed in four uncovered containers with different shapes and left on a table at room temperature. From which container will the water evaporate most rapidly?

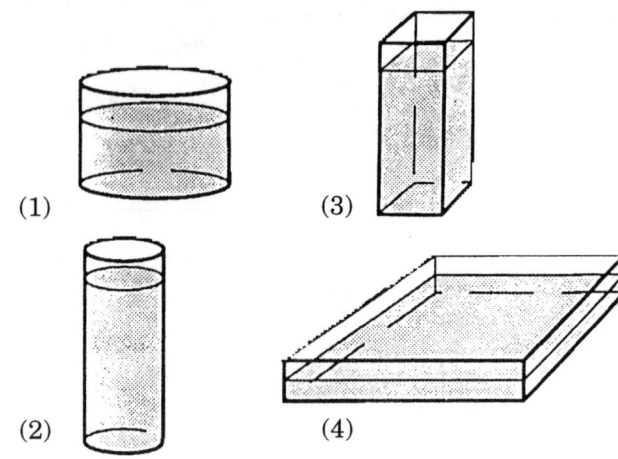

10 At what temperature would ice crystals form from air that has a dew point temperature of -6°C?
    (1) 6°C
    (2) 0°C
    (3) –2°C
    (4) –6°C

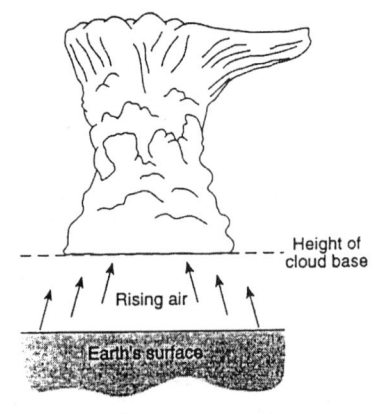

11 The diagram shows air rising from the Earth's surface to form a thunderstorm cloud. According to the *Lapse Rate Chart*, what is the height of the base of the thunderstorm cloud when the air at the Earth's surface has a temperature of 30°C and a dewpoint of 22°C? 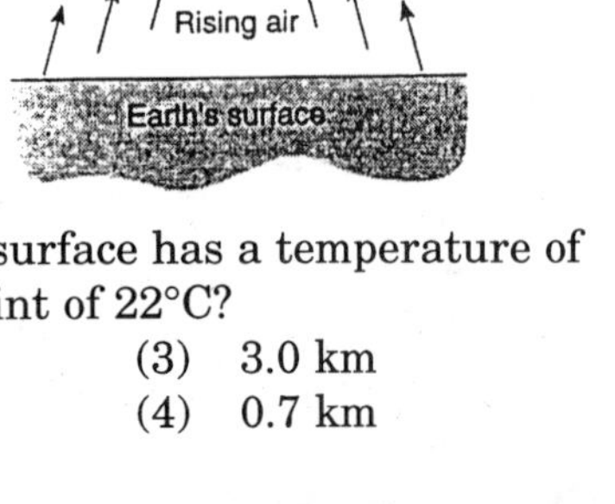
    (1) 1.0 km
    (2) 1.5 km
    (3) 3.0 km
    (4) 0.7 km

12 Each arrow in the diagram represents a process involving a phase change of water.

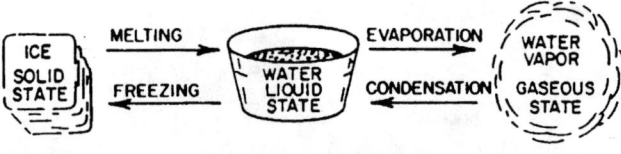

    Each process can take place only if
    1 mass and volume remains the same
    2 heat energy is added or released
    3 the number of molecules is increased or decreased
    4 the dewpoint temperature is increased or decreased

13 The change from vapor phase to liquid phase is called
    1 evaporation
    2 condensation
    3 precipitation
    4 transpiration

14 Which process results in a release of latent heat energy?
    1 melting of ice
    2 heating of liquid water
    3 condensation of water vapor
    4 evaporation of water

15 How many calories of latent heat would have to be absorbed by 100 grams of liquid water at 100°C in order to change all of the liquid water into water vapor at 100°C?
    (1) 100 calories
    (2) 8,000 calories
    (3) 1,000 calories
    (4) 54,000 calories

16 The diagram shows a container of water that is being heated.

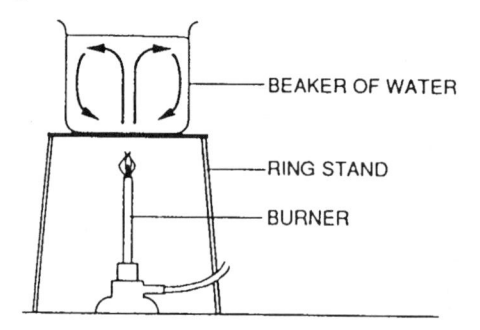

    The movement of water shown by the arrows is most likely caused by
    1 density differences
    2 insolation
    3 the Coriolis Effect
    4 the Earth's rotation

Base your answers to questions 17 through 19 on the *Reference Tables* and the diagram which represents the water cycle.

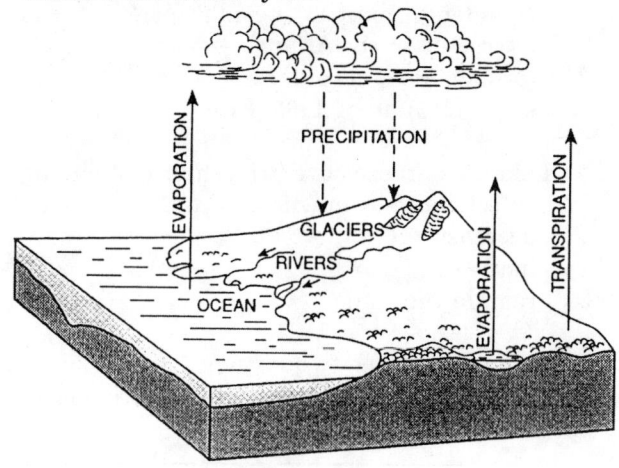

17 By which process does most water vapor enter the atmosphere?
1 evaporation from lakes and rivers
2 evaporation from ocean surfaces
3 evapotranspiration from land areas
4 sublimation from polar ice and snow

18 During which process does water vapor release 540 calories of latent heat per gram?
1 condensation   3 transpiration
2 evaporation   4 precipitation

19 Precipitation is most likely occurring at the time represented in the diagram because
1 the air has been warmed due to expansion
2 no condensation nuclei are present in the air
3 the relative humidity of the air is low
4 the water droplets are heavy enough to fall

Base your answers to questions 20 through 22 on the diagrams and the *Earth Science Reference Tables*. The diagrams represent weather conditions on **two** *consecutive days* (Day 1 below, Day 2 2nd column). The sea level air temperature for both days at the time shown is 20°C, and the arrows in each diagram represents the wind direction.

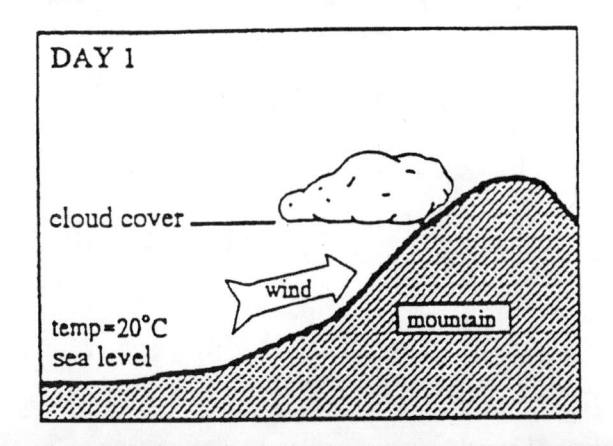

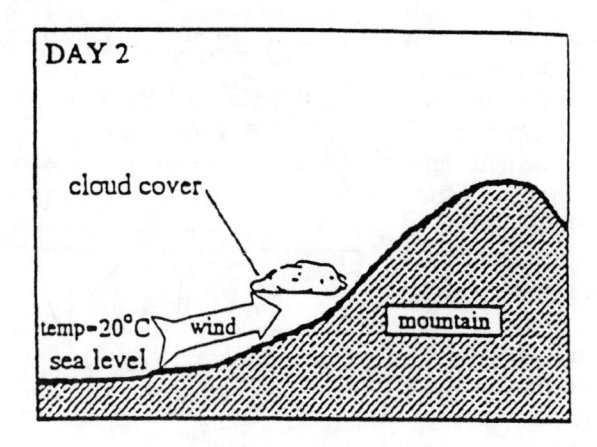

20 On both days the clouds formed as the rising air was
1 warmed by compression
2 cooled by expansion
3 warmed by expansion
4 cooled by compression

21 On day 1 at sea level, the difference between the wet-bulb and dry-bulb temperatures of a sling psychrometer was 8°C. The cloud base formed at an altitude of approximately
(1) 1.0 km   (3) 2.0 km
(2) 1.5 km   (4) 3.0 km

22 Which statement best explains why the cloud base altitude was lower on day 2?
1 The air had lower humidity on day 2.
2 The sea level air temperature was warmer on day 1.
3 The dewpoint temperature was lower on day 2.
4 The sea level air temperature was closer to the dewpoint on day 2.

23 Radiant energy from the sun is the primary energy source for
1 condensation nuclei
2 weather changes
3 station models
4 adiabatic cooling

24 Wind is part of heat flow by
1 radiation
2 conduction
3 convection
4 reflection

25 Which of the following statements best describes the planetary wind pattern at the equator?
1 Winds converge from higher latitudes.
2 Winds diverge from the equator.
3 Winds sink from higher altitudes.
4 Winds flow as high altitude jet streams.

TOPIC E – ATMOSPHERIC ENERGY – N&N©

Base your answers to questions 26 through 30 on the diagram which represents the general circulation of the Earth's atmosphere and the Earth's planetary wind and pressure belts. Points A through F represent locations on the Earth's surface.

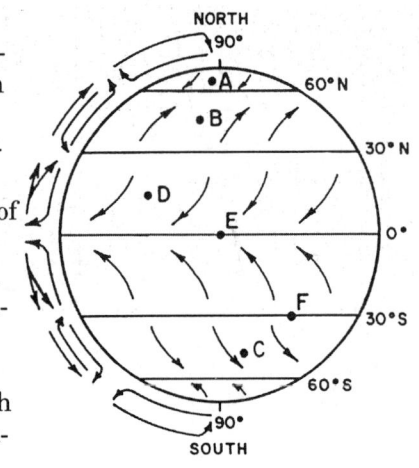

26 The curving path of the surface winds shown in the diagram are caused by the Earth's
1   gravitational field   3   rotation
2   magnetic field   4   revolution

27 Which location might be in New York State?
(1) A   (3) C
(2) B   (4) F

28 Which location is experiencing a southwest planetary wind?
(1) A   (3) C
(2) B   (4) F

29 Which location is near the center of a low-pressure belt where daily rains are common?
(1) E   (3) F
(2) B   (4) D

30 The arrows in the diagram represent energy transfer by which process?
1   conduction   3   convection
2   radiation   4   absorption

31 Hurricanes accounted for an average of 17 deaths per year from 1972 until 1991. However, the hurricane of 1938 was responsible for at least 600 deaths as it moved across Long Island into New England. What is the best explanation for the decrease in hurricane related deaths in recent years?
1   Hurricanes prior to 1970 were of greater intensity than all recent hurricanes.
2   Recent hurricanes have not struck populated areas.
3   Recent forecasts are more accurate due to the use of satellites.
4   Recent forecasts are more accurate because fewer weather instruments are used.

32 The most difficult type of weather event to forecast several hours in advance is the development of a
1   thunderstorm along a cold front
2   hurricane in the Atlantic Ocean
3   tornado along a cold front
4   snowstorm along Lake Erie

33 As Lake Erie freezes over, the chance of having a lake-effect snowstorm will
1   decrease
2   increase
3   remain the same

Base your questions to questions 34 and 35 on the diagram which shows part of the electromagnetic spectrum

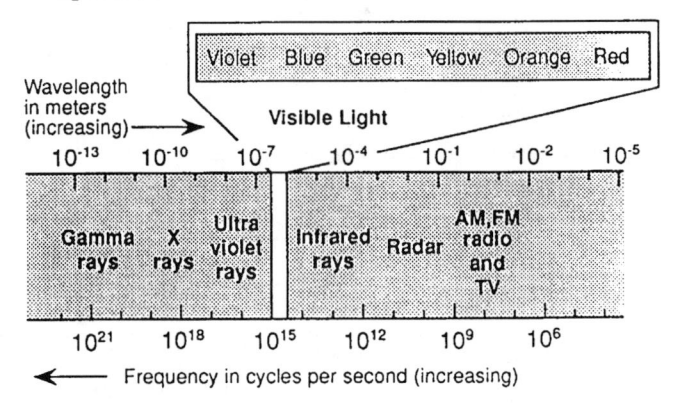

34 Which form of electromagnetic energy shown on the diagram has the lowest frequency and longest wavelength?
1   AM radio   3   red light
2   infrared rays   4   gamma rays

35 Which statement about electromagnetic energy is correct?
1   Violet light has a longer wavelength than red light.
2   Gamma rays have a shorter wavelength than visible light.
3   X rays have a longer wavelength than infrared rays.
4   Radar waves have a shorter wavelength than ultraviolet rays.

UNIT SEVEN
WATER CYCLE & CLIMATE

VOCABULARY TO BE UNDERSTOOD

Actual Evapotranspiration	Hydrologic or Water Cycle	Recharge
Aerosols	Incident Insolation	Runoff
Angle of Insolation	Infiltration	Scattering
Arid Climate	Insolation	Seasonal Lag
Base Flow	Intensity of Insolation	Stream Discharge
Capillarity	Marine Climate	Storage
Continental Climate	Moisture Deficit	Surplus
Deficit	Orographic Effect	Temperature Lag
Duration of Insolation	Parallelism of Axis	Terrestrial Radiaton
Evapotranspiration	Permeability	Usage
Global Warming	Porosity	Water Budget
Greenhouse Effect	Potential Evapotranspiration	Water Table
Ground Water	Radiative Balance	Zone of Aeration
Humid Climate	Random Reflection	Zone of Saturation

A. EARTH'S WATER

Where does water come from?

There is only one water supply. The water supply of the Earth is continually moving from the atmosphere to the Earth and from the Earth to the atmosphere. This is known as the **hydrologic** or **water cycle**. The water cycle includes the phase changes of water and the movements of water above, on, and below the Earth's surface.

The water cycle is dependent upon the atmosphere to provide water through precipitation to the land and oceans of the Earth. Precipitation amounts vary year to year and from place to place.

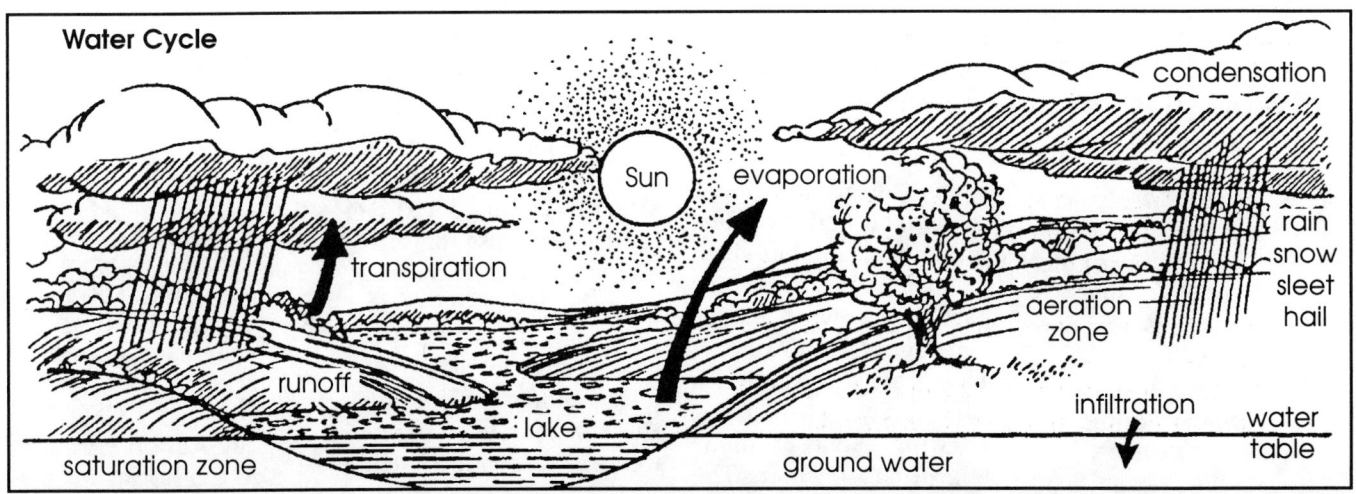

This precipitation can:

- infiltrate (sink into) the Earth's surface and become **ground water**,

- runoff from the surface into streams, lakes and the ocean,

- be stored in the form of ice and snow on the Earth's surface, and/or

- be evapotranspired back into the atmosphere from large bodies of water, soil, plants and animals.

Water Cycle

The oceans and large lakes act as the temporary storage areas for the majority of water within the water cycle and act as the major stabilizing factor in the Earth's climates.

Water is essential to life, so availability of water is an important key to the habitability of a location.

GROUND WATER POROSITY

How permeable a material is depends on the **porosity**, which is defined as the percentage of open space between the particles. The porosity of loose material is largely dependent upon the particle's shape, how tightly the material is packed, and the degree to which the particles are **sorted**.

How does water move into the Earth?

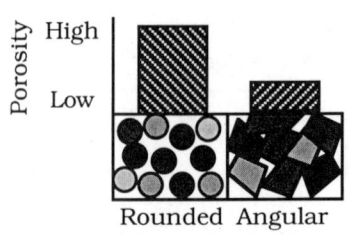

For example, water will more easily pass through a cylinder full of round beads, than a cylinder full of square blocks of the same size, since there would be more space between the round beads.

Particle Shape Comparison

When a test hole is being dug to determine whether or not a septic (waste) field can be placed in a yard, the hole must be dug in "undisturbed" earth, since water will pass through loose soil much faster than the hard packed soil of "undisturbed" ground.

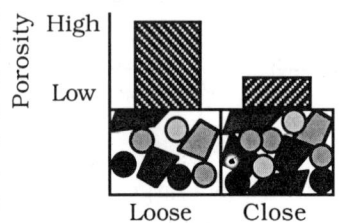

Particle Packing Comparison

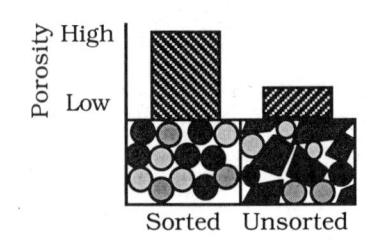

When a septic field is made, the fill used must be gravel and pebbles, rather than smaller sized particles, since the larger particles provide bigger pores for better drainage. The gravel and pebbles will pack less with time, allowing the continuous migration of water. The smaller particles, like clay and silt, will pack and become almost impermeable to water flow.

Particle Sorting Comparison

INFILTRATION

The precipitation from the atmosphere can **infiltrate** (sink into), runoff, or evaporate from the Earth's surface.

Before runoff and evapotranspiration, water will usually infiltrate the Earth's surface and become part of the **groundwater**. In order for water to move into the surface materials of the Earth, these materials must be permeable and unsaturated. Therefore, loose rock materials such as sand and gravel will allow for greater infiltration than the more dense, closely packed particles or the solid rock on the Earth's surface.

PERMEABILITY

A material is said to be **permeable** if it allows water to pass though the connecting pore spaces in the material.

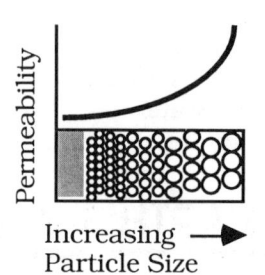

Water that has infiltrated loose material continues downward through the **Zone of Aeration** until the water reaches the **zone of saturation**. The top of the zone of saturation is called the **water table**. The depth of the water table is dependent upon the type of earth materials, thickness of those materials, the amount of water infiltrating the ground, and the characteristics of the surrounding materials.

Permeability Particle Size Comparison

For example, the larger the pore spaces between the particles that make up a material, the greater the permeability of that material.

CAPILLARITY

Water can also move upward within a material. This is called **capillarity**. In loose materials, capillarity increases with the decrease in particle size. The finer the loose particle size, the faster and the farther the water can move upward through the

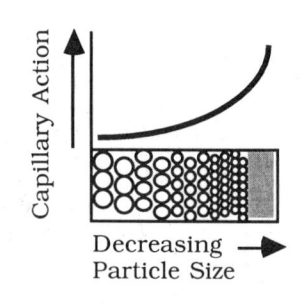

Capillary Action Particle Size Comparison

material. Along a beach composed of sand, the water will seep up along the shoreline, moving inward from the water line. However, along a rocky beach, the water will not seep very far inward from the water line.

SURFACE WATER & RUNOFF

When water moves over the surface of the Earth, it is referred to as **runoff**. Surface runoff can occur when rainfall exceeds the permeability rate of the material. For example, a specific sample of soil has a permeability rate of 0.3 liters per hour, but the rate of rain fall is 0.4 liters per hour. The rainfall is greater than the permeability rate by 0.1 liters per hour, which is the amount of runoff from that specific soil sample.

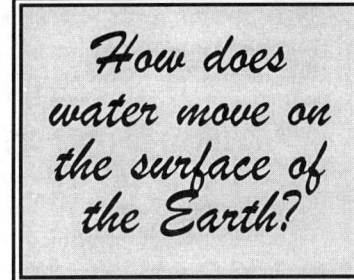

How does water move on the surface of the Earth?

Surface runoff can also occur when the slope of the surface is too steep to allow infiltration to occur. Even though the permeability rate of the slope materials may be greater than the rate of rainfall, the water may not have sufficient "standing" time to allow for the infiltration, before it runs off the surface. Water is less likely to soak into a hill side, than to soak into the soil of a flat valley.

If the saturation level of the soil type has reached the surface, such as after long and heavy rains, the added rainfall will run over the surface. This is often the case in severe lowland flooding associated with hurricanes and large maritime tropical storms. Runoff will also occur when rain falls on a frozen surface. The **capillary water** retained at the points of contact between the soil particles freezes, preventing infiltration of the water which then runs off the surface.

WATER & STREAM DISCHARGE

The rate of streamflow in volume is called **stream discharge** and is measured by determining the amount of water that flows past a specific part of the stream during a specific amount of time. Often the stream discharge is a measurement of the **surplus water** that drains from the area around the stream. However, if there is no surplus water in the area of the stream, then stream discharge may be related to the depletion of groundwater from soil storage. This is the case during dry seasons when streams are being fed by the local water table. This is called **base flow**.

B. FACTORS AFFECTING INSOLATION

The primary source of energy for the Earth is **solar radiation**. Solar radiation has its greatest intensity occurring in the visible wavelength (about 50%) of the solar electromagnetic spectrum. The other forms of insolation include ultraviolet, infrared, x-rays, and other longer and shorter wavelengths.

The term **Insolation** comes from **In**coming **So**lar **Radiation**. It is that portion of the Sun's radiation which is received at the Earth's surface.

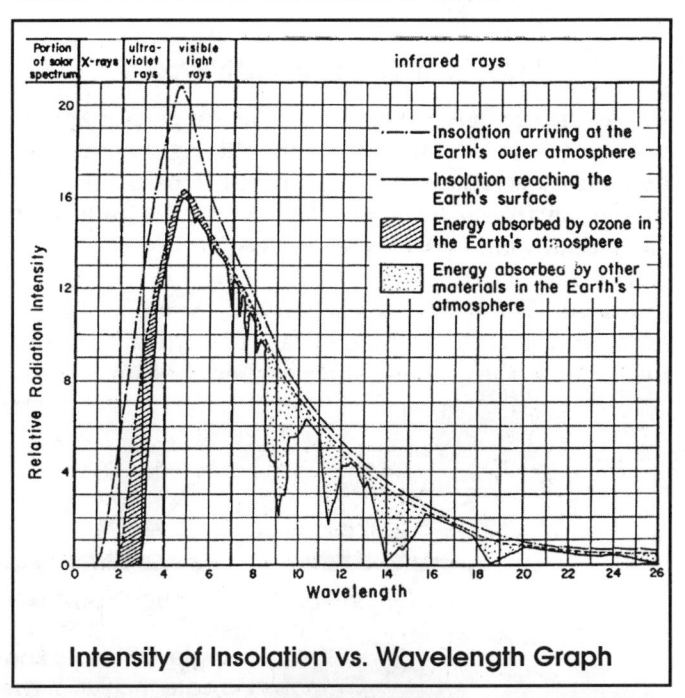

Intensity of Insolation vs. Wavelength Graph

INSOLATION

Angle of Insolation. The **intensity of insolation** in any area of the Earth's surface increases as the angle of insolation approaches perpendicular. At an angle of 90° to the Earth's surface, the solar radiation is concentrated in a relatively small area of the surface. Therefore, where the Sun's rays are vertical, the maximum amount of solar energy is received at the surface.

What factors determine the amount of solar energy an area receives?

The angle of insolation changes on any particular date with either an increase or decrease in latitude. For example (in the diagram below), at 23½° North, the Sun's rays are vertical and therefore have maximum intensity. The intensity of insolation per unit area will decrease with increasing latitude towards the North and South Poles, because the insolation received is spread over a greater area.

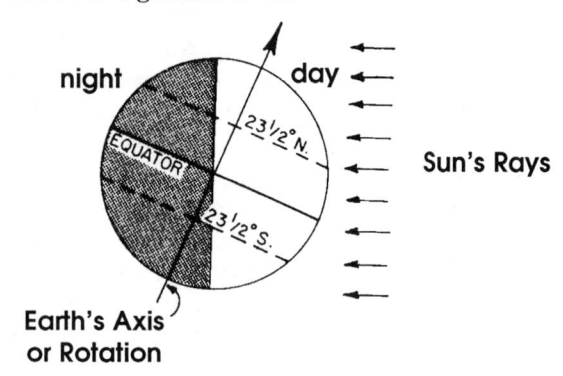

Angle of Insolation to Earth's Axis

At the equinoxes, the Sun's rays are perpendicular to the equator (0° latitude), while during the summer and winter solstices, the Sun's rays are perpendicular to the Earth's surface at 23½° North Latitude and 23½° South Latitude, respectively.

In addition to changes in latitude, the angle of insolation varies with the time of day. The intensity of the insolation is the greatest at noon when the Sun is highest in the sky and is least when the Sun is very low in the sky. The greater the angle of insolation, the greater the intensity of insolation.

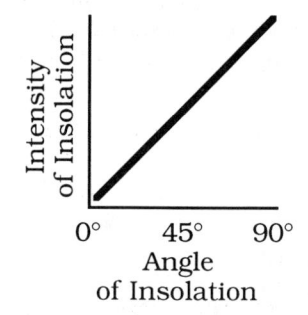

DURATION OF INSOLATION

Duration of insolation refers to the number of daylight hours. It is determined by the length of the Sun's path across the sky. For every 15° of arc in the Sun's path, there is one hour of daylight. The longer the Sun's path, the greater the intensity and duration of insolation.

Relationship of Intensity and Angle of Insolation

DURATION VARIES WITH LATITUDE AND SEASONS

Because of the 23½° tilt of the Earth's axis and the revolution of the Earth around the Sun, the rays of the Sun hit the Earth's surface with varying angles, and the length of the Sun's path across the sky also varies. For every 15° of arc in the Sun's path, the surface receives one hour of insolation. The greater the **angle of insolation** and the longer the duration, the more total energy received.

The seasons are the result of yearly cyclic changes in the duration and intensity of solar radiations or insolation. These cyclic changes result from:

- **Inclination (tilt) of the Earth's axis**, at 23½°, allows the vertical rays of the Sun to fall on different latitudes of the Earth (between the Tropics of Cancer — 23½° North and Capricorn — 23½° South).

- **Parallelism of the Earth's axis.** Since the Earth's axis is always pointing into space in the same direction, the axis of the Earth at any given point in the Earth's orbit around the Sun remains parallel to the axis at any other given point of the orbit.

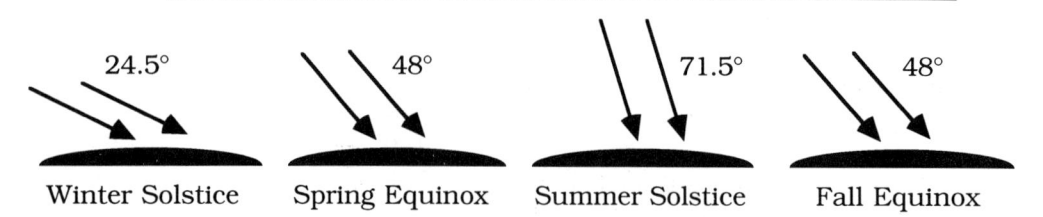

24.5°	48°	71.5°	48°
Winter Solstice	Spring Equinox	Summer Solstice	Fall Equinox

Angle Of Insolation Related To Seasons
(Arrows indicate the Sun's rays at noon for 42° North latitude.)

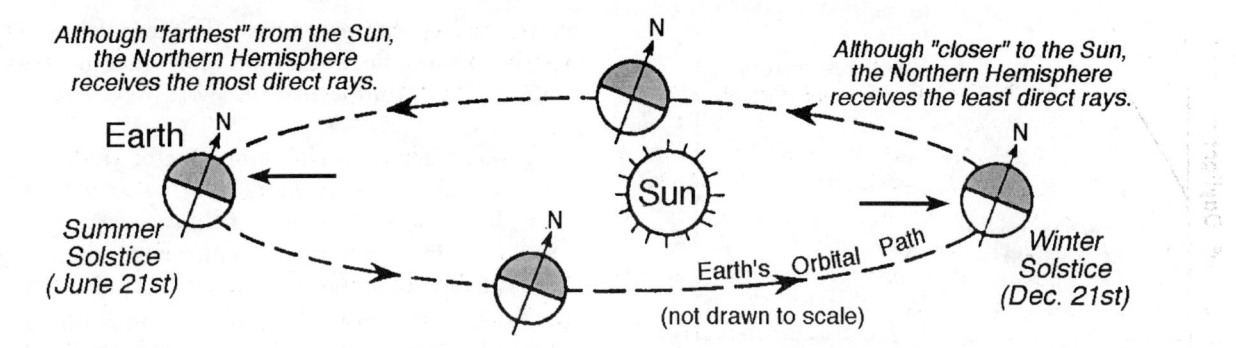

Summer and Winter Solstice (in Northern Hemisphere), Showing Relative Earth – Sun Positions

- **Revolution of the Earth** causes the Sun's perpendicular rays to fall on different Earth latitudes between 23½° North and South Latitude.

- **Rotation of the Earth** causes the alternation of day and night.

In the northern mid-latitudes (42° North), the **maximum insolation** occurs about June 21st (**summer solstice**). The angle of insolation at noon is 71½°, and the duration of insolation is 15 hours.

Average insolation occurs on or about March 21st and September 23rd (**equinoxes**). The angle of insolation at noon is 48°, and the duration of insolation is 12 hours.

The **minimum insolation** occurs about December 21st (**winter solstice**). The angle of insolation at noon is 24½°, and the duration of insolation is 9 hours.

SUMMER & WINTER SOLSTICE

In the Northern Hemisphere, the Sun's path is the longest on the summer solstice (June 21st). Therefore, New York State has the greatest duration of insolation. The shortest duration of insolation occurs on the winter solstice (December 21st), because the Sun is the lowest in the sky and has the shortest path over New York State.

TEMPERATURE & INSOLATION

The surface temperature of the Earth is directly related to insolation received at the surface.

Incoming solar radiation raises the Earth's surface temperature (heat gain). However, **terrestrial radiation** causes a cooling effect on the Earth's surface (heat loss).

Surface temperature is a result of the relationship between heat gained and lost. Where insolation exceeds radiation, the temperature rises. If more heat is lost (radiation) than is gained (insolation), the temperature at the Earth's surface decreases.

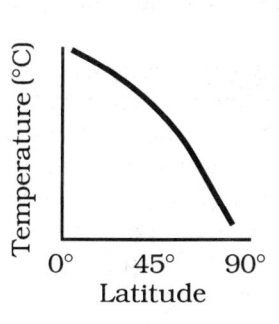

Relationship of Average Annual Surface Temperature & Latitude

DAILY TEMPERATURES

During the day, the **maximum surface temperature** usually occurs sometime **after maximum insolation** – usually during the early afternoon. The minimum surface temperature usually occurs approximately an hour before sunrise, due to the continuous loss of heat during nighttime (radiation). This is called **temperature lag**.

There are other factors which affect the maximum and minimum temperatures, such as cloud cover. Clouds tend to reduce the amount of heat lost due to surface radiation, but also reduce the amount of insolation reaching the Earth's surface.

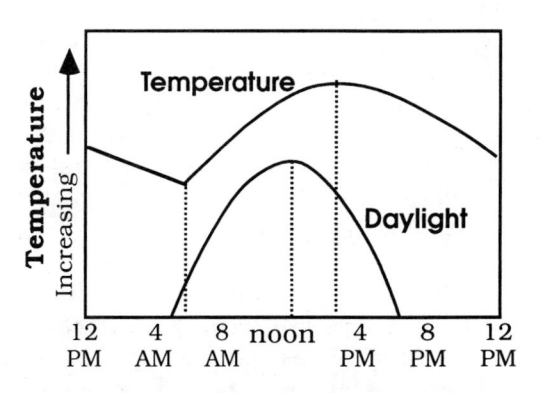

Relationship of Average Daylight Hours & Daily Temperatures

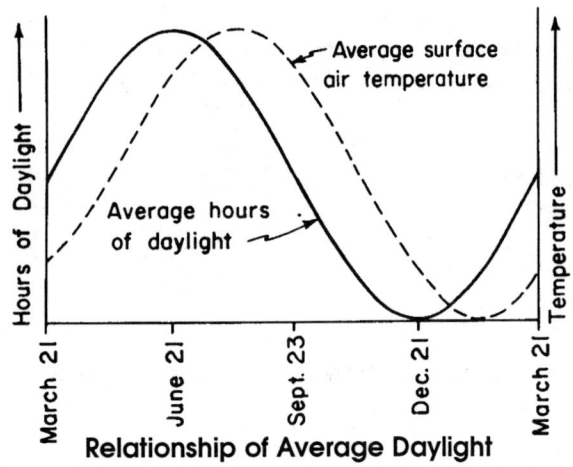

Relationship of Average Daylight Hours & Annual Temperatures (in No. Hemisphere)

YEARLY TEMPERATURES

During the year the maximum surface temperature most often occurs sometime after the maximum insolation, generally midsummer, following the summer solstice. This is the case, because temperatures continue to rise as long as insolation received during the long days exceeds radiation during the shorter nights.

The minimum surface temperature usually occurs sometime after the minimum angle of insolation, soon after the winter solstice. This is called **seasonal lag**. Again there are factors (such as weather systems) which affect the yearly maximum and minimum temperatures. As long as radiation during the long nights exceeds insolation received during the shorter days, temperature will continue to decrease.

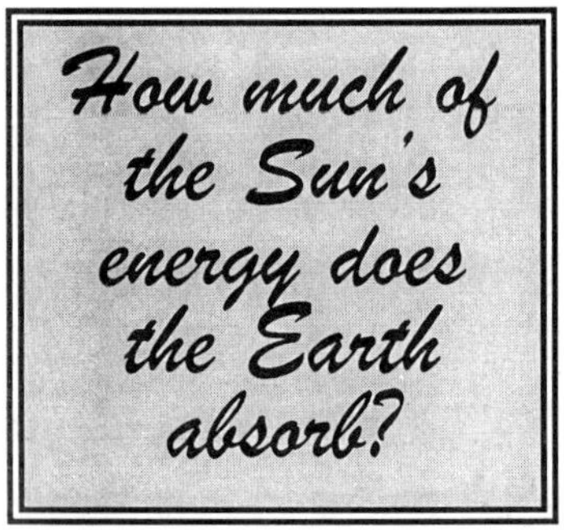

How much of the Sun's energy does the Earth absorb?

ABSORPTION

Since the atmosphere is largely transparent to visible radiation, most of the Sun's visible (light) radiation reaches the Earth's surface. But, because the atmosphere selectively absorbs solar radiation, most of the Sun's ultraviolet radiation is absorbed by the atmosphere's ozone, and much of the infrared radiation is absorbed by the atmosphere's carbon dioxide and water vapor.

To a large extent, the surface of the Earth itself tends to control temperature changes through absorption and radiation. Water surfaces heat more slowly than land surfaces, and water also tends to hold heat longer. Therefore, land surface temperatures change more rapidly and change to a greater degree than water surface temperatures. The resulting temperature differences between land and water masses greatly affect Earth temperatures.

The surface material is one factor that determines how much of the electromagnetic energy will be reflected or absorbed. In the case of light energy, if a surface is rough or dark, it is likely to absorb more energy than if the surface is smooth or light. In the south, houses are built with white or light colored shingles on the roofs to reduce the solar heat absorption. In the north, darker shingles are used, allowing more solar heat absorption during the winter months.

WAVELENGTH ABSORPTION & RADIATION

As previously discussed, a good absorber of energy is also a good radiator. The characteristics of the surface of the material receiving the electromagnetic energy determine the quantity and type of energy absorbed or reflected.

The Earth's surface often absorbs solar electromagnetic energy (generally strong, short wavelengths) and converts the energy to longer wavelengths (having less strength) which are reradiated. For example, solar visible light rays (shorter wavelengths) are absorbed by the Earth and reradiated as infrared or heat (long wavelengths). Gases in the Earth's atmosphere absorb much of the infrared energy. This is referred to as the **greenhouse effect**.

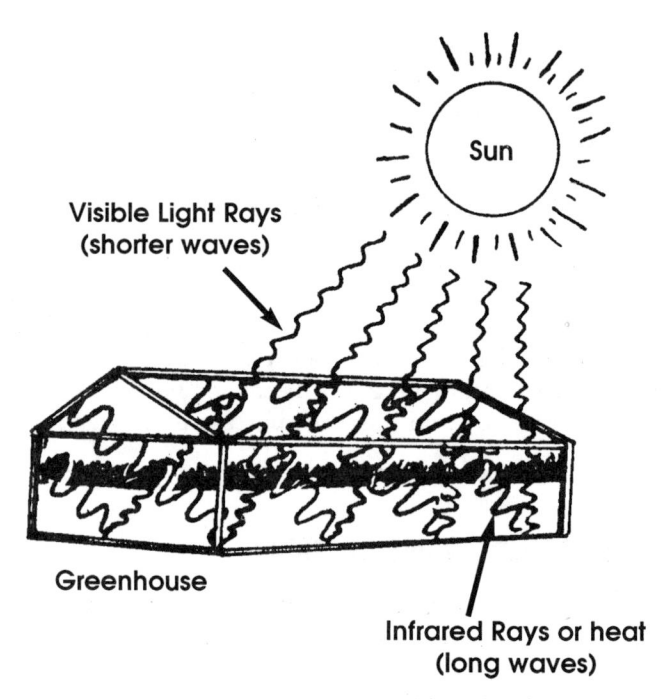

Greenhouse Effect
Solar visible light rays (shorter wavelengths) are absorbed by the Earth and reradiated as infrared or heat (long wavelengths).

REFLECTION

Particles in the atmosphere and materials on the surface of the Earth greatly affect the amount of insolation that reaches the Earth's surface and is absorbed. For example, clouds may reflect approximately 25 percent of the incident insolation.

The reflectivity of the Earth also depends upon both the surface and the angle of insolation. The greater the angle of insolation, the greater the absorption, but as the angle of insolation decreases, the reflection becomes greater. A surface of ice and/or snow may reflect almost all of the **incident insolation**.

SCATTERING

Aerosols (finely dispersed particles, such as water droplets and dust) in the atmosphere cause a **random reflection** or **scattering** of insolation. The amount of insolation reaching the Earth's surface decreases as the amount of random reflection increases.

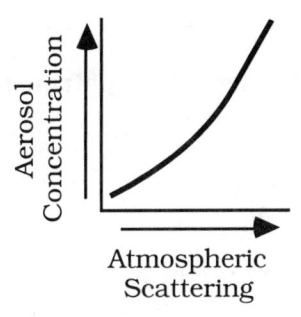

Scattering Effect of Aerosols Iin the Atmosphere

For example, a volcanic eruption may put sufficient ash into the air to decrease the amount of insolation reaching the Earth's surface, resulting in a temperature decrease.

ENERGY CONVERSION

Not all of the insolation is directly reflected or radiated as heat energy. Instead, some of the insolation is converted into **potential energy** by the evaporation of water and the melting of ice. This energy conversion does not change the surface temperature.

GREENHOUSE EFFECT

As previously mentioned, most of the radiant energy from the Sun 1) passes through the Earth's atmosphere, 2) is absorbed by the atmosphere or Earth's surface, or 3) is reradiated by the Earth's surface back into the atmosphere. Visible light is the most intense energy (from the Sun) to be absorbed by the Earth's surface. After it is absorbed and heats the surface, it is reradiated back into the Earth's atmosphere as infrared energy and is absorbed by water vapor and carbon dioxide. This process which causes the atmosphere to be heated is known as the **greenhouse effect**. The warmed atmosphere acts as a "thermal blanket," reducing the loss of energy to space and raising the temperature of the Earth's surface.

The greenhouse effect is the result of the conversion of shortwave energy (insolation, such as visible radiation) to longer wave radiation, infrared, which is prevented from escaping the Earth's atmosphere.

C. TERRESTRIAL RADIATION

Terrestrial radiation refers to the electromagnetic energy that the Earth gives off to its atmosphere and space. This radiation from the surface is nearly all in the infrared region of the electromagnetic spectrum.

What are some factors that affect terrestrial radiation?

The Greenhouse Effect

Sun Primary Radiant Energy Source for Earth

The layers of gas, surrounding the planet, filter sunlight and maintain temperature levels.

Deforestation, modern industries, and urbanized life produce too many waste gases they have radically altered the composition of the planet's protective atmosphere.

Rising air (i.e. global warming) and water (i.e. El Nino) temperature levels may alter weather patterns, shift seasons, change ocean levels and currents, and cause droughts and great hurricanes.

RADIATION

The maximum intensity of outgoing radiation from the Earth's surface, **terrestrial radiation**, is in the infrared region of the electromagnetic spectrum. As previously discussed, some of this infrared (heat) energy is absorbed by the atmosphere, and the balance of the energy escapes into space. The **greenhouse gases** (i.e., water vapor and carbon dioxide) are the primary absorbers of infrared energy in the Earth's atmosphere. They are responsible for the greenhouse effect. Since carbon dioxide is a good absorber of infrared energy, scientists today are concerned about **global warming** as a long term effect of increasing amounts of carbon dioxide in the atmosphere.

RADIATIVE BALANCE

The Earth's temperature depends upon the relationship between incoming and outgoing energy. When the average Earth temperature remains stable, the Earth is said to be in **radiative balance**, gaining as much energy as it gives off.

Comparing the radiative balance of the Earth over long and short time periods indicates the following:

- **Long-term Measurements** (thousands of years) of worldwide surface temperatures indicate that the Earth *is not* in radiative balance. For example, during the Pleistocene Epoch Age, the estimated average temperatures of the Earth varied approximately 5°C to 10°C cooler, resulting in the Ice Age.

- **Annual Measurement** of worldwide surface temperatures indicates that the Earth *is not* in radiative balance. Daily, weekly, monthly, and seasonal temperatures are constantly changing, resulting in variations in the yearly average.

D. CLIMATE

Unlike weather conditions which are short term affects, a **climate** is the average weather conditions over much longer periods of time. Climate is primarily concerned with temperature and moisture conditions.

WATER BUDGET

Local water budget characteristics can be used to distinguish climatic regions. A **water budget** is a system of accounting for moisture income, storage, and outgo for the soil in a specific area. The variables involved in the water budget are precipitation, potential and actual evapotranspiration, storage, usage, deficit, and recharge.

The moisture source for the water budget is **precipitation** regardless of the form it takes (i.e. snow, rain, sleet, hail). This moisture is measured in millimeters of water.

Potential evapotranspiration refers to the maximum amount of water that can be evaporated, if the water is available. It is directly proportional to the energy available. Therefore, the greatest potential evapotranspiration occurs in the summer when the temperatures are the highest. The warmer the temperature, the more water can evaporate.

The **actual evapotranspiration** is the amount of water that is really evaporated and/or transpired. It is equal to the potential evapotranspiration if sufficient water is available. The actual can never exceed the potential.

The amount of water stored in the top of the soil (root zone) is called the **storage**. When water evaporates out of the soil the storage decreases. This process is called **usage**. When precipitation infiltrates (sinks into) the soil, the storage increases. This process is called **recharge**.

When the soil moisture is depleted and there is not enough precipitation to meet the potential evapotranspiration there is a **moisture deficit** or **drought**. **Surplus** exists when ground storage is at maximum and there is more precipitation than can be evaporated. This surplus water becomes runoff and contributes water to local streams.

Local water budget characteristics can be used to distinguish climatic regions. **Humid climates** are those in which the total amount of precipitation is greater than the total amount of potential evapotranspiration. The water budget graphs of humid climates show months with surplus water.

Arid climates have significantly more total potential evapotranspiration than precipitation. The water budget graphs of arid climates show months of deficit.

CLIMATE PATTERN FACTORS

The variables which affect climate patterns are latitude, altitude, and proximity to large bodies of water, ocean currents, mountain barriers, and prevailing winds.

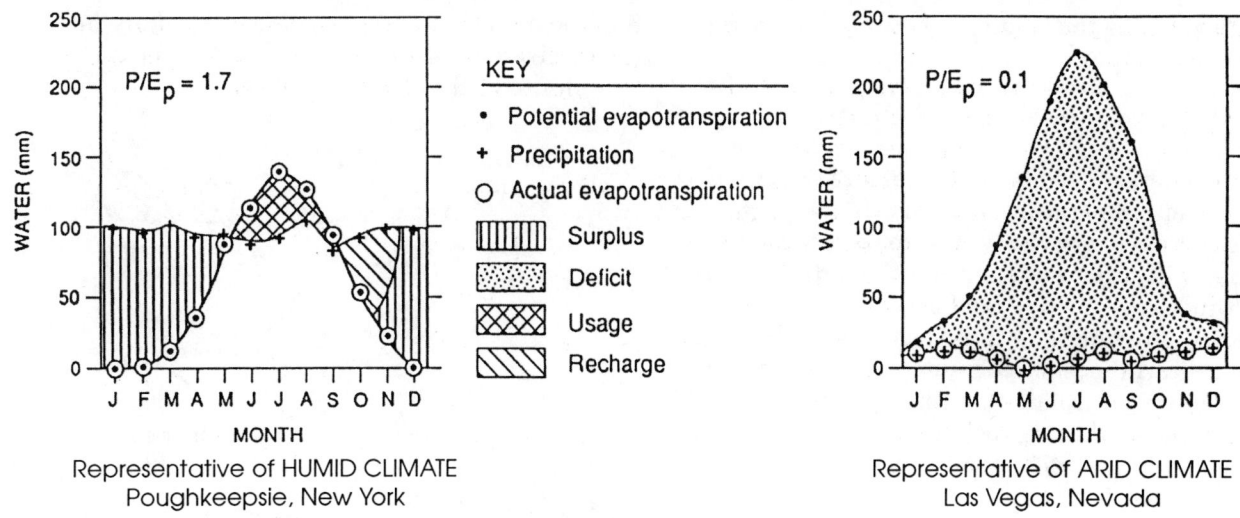

Water Budget Graph Examples

Representative of HUMID CLIMATE — Poughkeepsie, New York ($P/E_p = 1.7$)

Representative of ARID CLIMATE — Las Vegas, Nevada ($P/E_p = 0.1$)

Key: Potential evapotranspiration, Precipitation, Actual evapotranspiration, Surplus, Deficit, Usage, Recharge

What factors determine the climate of an area?

LATITUDE

Latitude is the most important factor in determining climate, especially influencing temperature patterns. Since the duration of insolation at low latitudes is fairly constant, about 12 hours per day, temperature variance is small. Also, at low latitudes, the angle of insolation is always quite high. Therefore, the temperatures remain relatively high. At high latitudes, temperatures vary but remain relatively low due to the generally low angle of insolation. The duration of insolation varies between 0 hours and 24 hours per day causing a great seasonal variation in temperatures.

In general, as the duration of insolation increases, the temperature increases. As latitude increases, average yearly temperature decreases, but the annual temperature range increases.

ELEVATION

The elevation (altitude) influences the temperature and moisture patterns of a region. The effects of elevation are very similar to those of latitude. Lower elevations are generally more stable in temperature and moisture, and higher elevations have more varied conditions. As the altitude or elevation increases, the average yearly temperature decreases, and the precipitation generally increases.

LARGE BODIES OF WATER

Large bodies of water (large lakes and oceans), ocean currents, and prevailing winds modify the latitudinal climate patterns of their shoreline areas. The slow heating and cooling of large bodies of water cause the land masses near them to have modified temperatures. For example, the **marine climate**, of the northwest coast of the United States, is characterized by cooler summers and warmer winters, than would normally be expected for that latitude. Areas of marine climate have small temperature ranges.

Inland regions are not directly affected by large bodies of water, and have **continental climates**, that are characterized by large yearly temperature ranges with hot summers and cold winters.

MOUNTAIN (OROGRAPHIC) BARRIERS

The overall effect of mountains on climatic patterns is called the **orographic effect**. Latitudinal climate patterns are modified by mountains that act as barriers to local weather systems by interrupting the normal path of a prevailing wind. The windward side of a mountain is the side facing the prevailing wind. As the wind hits the windward side of a mountain the air is forced upward and cools adiabatically until the dew point is reached. Then condensation occurs and cooling slows. These conditions cause the water in the air to condense, forming clouds and precipitation on the windward side of the mountain.

The other side of the mountain, the leeward side, will be drier and warmer than the windward side. It is said to be in the *rain shadow*. As the air rises over the mountain and begins to descend on the leeward side, it warms adiabatically (due to the compression), increasing its ability to hold moisture.

When a mountain is high enough, it can act as a barrier to air masses and prevent warm or cold air from getting past the mountain to the other side. In this case a mountain can be the boundary between two very different climatic areas.

WIND BELTS

The planetary winds and pressure belts affect moisture and temperature patterns. If the prevailing winds first cross a large body of water before coming on land, they will bring moisture to the land. If the prevailing winds cross a large land mass, the effect will be more arid. Prevailing winds from tropical areas, including warm ocean currents, bring warm air. Conversely, winds from polar regions, including cold ocean currents, will bring cool air.

QUESTIONS FOR UNIT 7

1. Electromagnetic energy that reaches the Earth from the Sun is called
   1. insolation
   2. conduction
   3. specific heat
   4. terrestrial radiation

2. Short waves of electromagnetic energy are absorbed by the Earth's surface during the day. They are later reradiated into space as
   1. visible light rays
   2. infrared rays
   3. ultraviolet rays
   4. x-rays

3. An object that is a good absorber of electromagnetic energy is also a good
   1. reflector of electromagnetic energy
   2. refractor of electromagnetic energy
   3. radiator of electromagnetic energy
   4. convector of electromagnetic energy

4. Which type of surface would most likely be the best reflector of electromagnetic energy?
   1. dark-colored and rough
   2. dark-colored and smooth
   3. light-colored and rough
   4. light-colored and smooth

5. The factor that contributes most to the seasonal temperature changes during one year in New York State is the changing
   1. speed at which the Earth travels in its orbit around the Sun
   2. angle at which the Sun's rays strike the Earth's surface
   3. distance between the Earth and the Sun
   4. energy given of by the Sun

6. The map shows isolines of average daily insolation received in calories per square centimeter per minute at the Earth's surface.

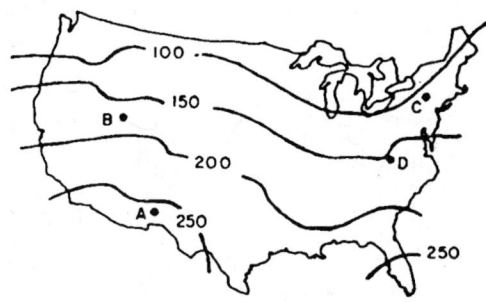

   If identical solar collectors are placed at the lettered locations, which collector would receive the least insolation?
   (1) A
   (2) B
   (3) C
   (4) D

7. The intensity of the insolation that reaches the Earth is affected most by
   1. the angle at which the insolation strikes the Earth's surface
   2. changes in the length of the Earth's rotational period
   3. the temperature of the Earth's surface
   4. the distance between the Earth and the Sun

8. In which diagram would the incoming solar radiation reaching Earth's surface heat the ground the most?
   (1) Location 1
   (2) Location 2
   (3) Location 3
   (4) Location 4

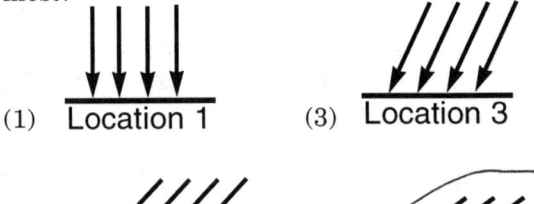

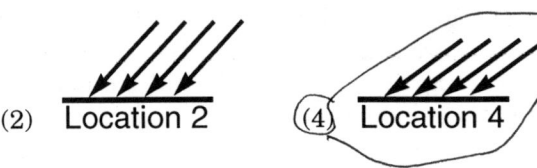

9. Which graph best illustrates the relationship between the angle of insolation and the time of day at a location in New York State?

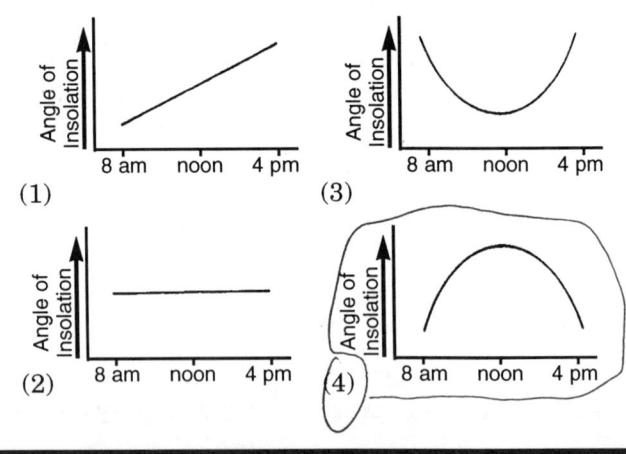

10 The tilt of the Earth on its axis is a cause of the Earth's
   (1) uniform daylight hours
   (2) changing length of day and night
   (3) 24-hour day
   (4) 365½-day year

11 Over a period of one year, which location would probably have the greatest average intensity of insolation per unit area? [Assume equal atmospheric transparency at each location.]
   (1) Tropic of Cancer (23½°N)
   (2) New York City (41°N)
   (3) the Arctic Circle (66½°N)
   (4) the North Pole (90°N)

12 Compared to the polar areas, why are equatorial areas of equal size heated much more intensely by the Sun?
   1 The Sun's rays are more nearly perpendicular at the Equator than at the poles.
   2 The equatorial areas contain more water than the polar areas do.
   3 More hours of daylight occur at the Equator than at the poles.
   4 The equatorial areas are nearer to the Sun than the polar areas are.

13 What happens to the angle of insolation between solar noon and 6 p.m. in New York State?
   1 It decreases steadily.
   2 It increases steadily
   3 It remains the same.
   4 It first increases and then decreases.

14 On which date does the maximum duration of insolation occur in the Northern Hemisphere?
   1 March 21   3 September 23
   2 June 21   4 December 21

15 The diagram represents a model of the Sun's apparent path across the sky in New York State for selected dates.

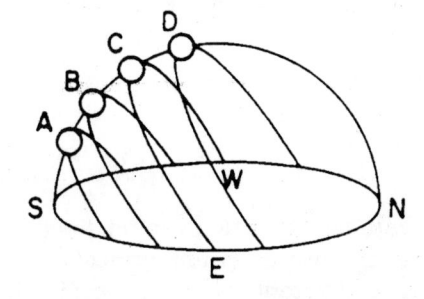

For which path would the duration of insolation be greatest?
   (1) A   (3) C
   (2) B   (4) D

16 For which date and location will the longest duration of insolation normally occur?
   1 June 21, at 60°N.
   2 June 21, at 23½°N.
   3 December 21, at 60°N.
   4 December 21, AT 23½°N.

17 The diagram represents four positions of the Earth as it revolves around the Sun.

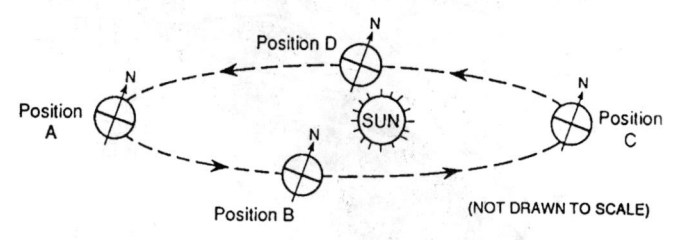

At which position is the Earth located on December 21?
   (1) A   (3) C
   (2) B   (4) D

18 In New York State, it is observed that the north-facing slopes of mountains usually retain their snow later in the spring than the south-facing slopes. This is caused by the fact that the north slopes of the mountains
   1 are protected from the prevailing south winds
   2 receive greater rainfall
   3 usually are steeper
   4 receive less insolation than the south slopes

19 The graph shows the average daily temperatures and the duration of insolation for a location in the mid-latitudes of the Northern Hemisphere during a year.

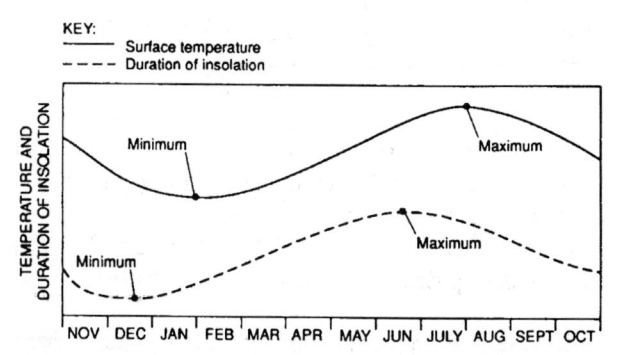

Compared to the date of maximum duration of insolation, the date of maximum surface temperature for this location is
   1 earlier in the year
   2 later in the year
   3 the same day of the year

Unit Seven – The Water Cycle and Climate – N&N©

Base your answers to questions 20 through 22 on your knowledge of Earth science, the *Earth Science Reference Tables*, and the diagram below. The diagram represents a hot-air solar collector consisting of a wooden box frame, an absorber plate, a glass cover, and insulation.

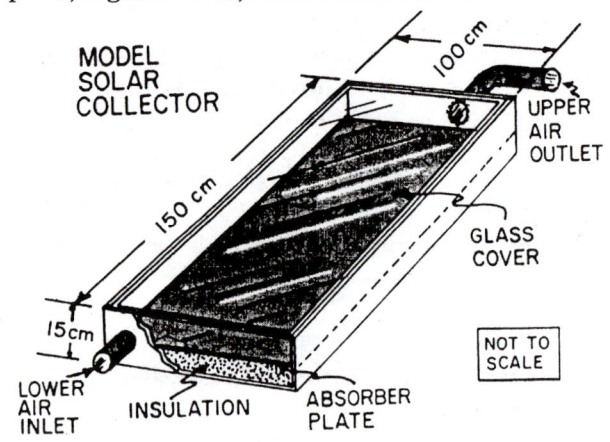

20 The solar collector is placed outside in sunlight, facing south and tilted 40° from the horizontal.

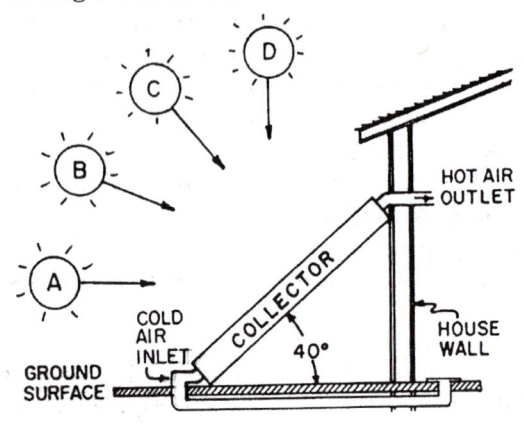

At which position of the Sun would the collector receive the most intense solar radiation?
(1) A
(2) B
(3) C
(4) D

21 Which paint should be used on the absorber plate if it is designed to absorb the greatest possible amount of insolation?
1 red paint
2 white paint
3 yellow paint
4 black paint

22 What is the primary function of the glass cover?
1 It reduces the amount of insolation entering the collector.
2 It increases the heat that is lost by convection to the outside air.
3 It allows short wavelengths of radiation to enter, but reduces the amount of long-wavelength radiation that escapes.
4 It allows all wavelengths of radiation to pass through in either direction.

23 Which form of radiation given off by the Earth causes heating of the Earth's atmosphere?
1 infrared
2 ultraviolet
3 visible
4 x-ray

24 Which gases in the atmosphere best absorb infrared radiation?
1 hydrogen and nitrogen
2 hydrogen and carbon dioxide
3 water vapor and carbon dioxide
4 water vapor and nitrogen

25 Which model best represents how a greenhouse remains warm as a result of insolation from the Sun?

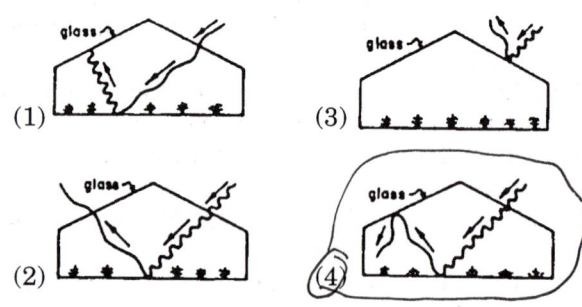

26 What is the primary reason New York State is warmer in July than in February?
1 The Earth is traveling faster in its orbit in February.
2 The altitude of the noon Sun is greater in February.
3 The insolation in New York is greater in July.
4 The Earth is closer to the Sun in July.

27 If an object with a constant source of energy is in radiative balance, the temperature of the object will
1 decrease
2 increase
3 remains the same

28 Stream discharge would normally be highest during a period of
1 recharge
2 deficit
3 usage
4 surplus

29 The type of climate for a location can be determined by comparing the yearly amounts of
1 precipitation and potential evapotranspiration
2 soil storage and potential evapotranspiration
3 precipitation and infiltration
4 change in soil storage and stream discharge

30 The table shows the precipitation / potential evapotranspiration ratio (P/Ep ratio) for different types of climates.

Climate Type	P/E_p Ratio
Humid	Greater than 1.2
Subhumid	0.8 to 1.2
Semiarid	0.4 to 0.8
Arid	Less than 0.4

The total annual precipitation (P) for a city in California is 420 millimeters. The total annual potential evapotranspiration (Ep) is 840 millimeters. What type of climate does this city have?
1. humid
2. sub-humid
3. semi-arid
4. arid

31 Which type of climate would most likely be found in an area that has a high potential for evaporation of water but a low actual evaporation of water?
1. polar
2. rain forest
3. desert
4. temperate

32 What is the best explanation for the two statements below?
- Some mountains located near the Earth's Equator have snow-covered peaks
- Icecaps exist at the Earth's poles.

1. High elevation and high latitude have a similar effect on climate.
2. Both mountain and polar regions have arid climates.
3. Mountain and polar regions receive more energy from the Sun than other regions do.
4. An increase in snowfall and an increase in temperature have a similar effect on climate.

33 The diagram represents several locations on the surface of the Earth. Each location is at sea level and is surrounded by ocean water. The average annual air temperature at point P is most likely higher than the average annual air temperature at point
(1) A
(2) B
(3) C
(4) D

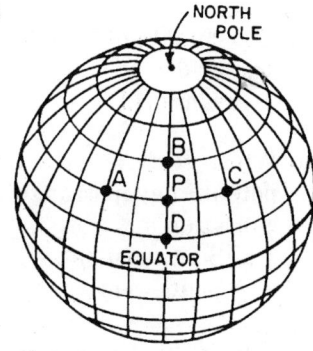

34 Compared to a coastal location of the same elevation and latitude, an inland location is likely to have
1. warmer summers and cooler winters
2. warmer summers and warmer winters
3. cooler summers and cooler winters
4. cooler summers and warmer winters

35 The diagram below shows the positions of the cities of Seattle and Spokane, Washington. Both cities are located at approximately 48° North latitude, and they are separated by the Cascade Mountains.

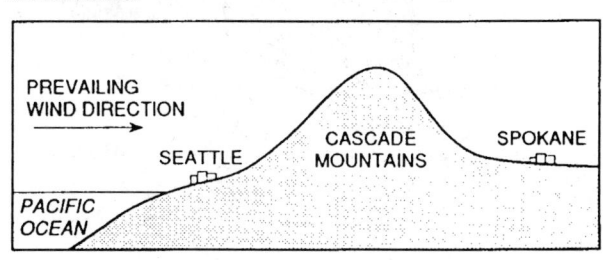

How does the climate of Seattle compare with the climate of Spokane?
1. Seattle – hot and dry
   Spokane – cool and humid
2. Seattle – hot and humid
   Spokane – cool and dry
3. Seattle – cool and humid
   Spokane – hot and dry
4. Seattle – cool and dry
   Spokane – warm and humid

36 The climates of densely populated industrial areas tend to be warmer than similarly located sparsely populated rural areas. From this observation, what can be inferred about the human influence on local climate?

1. Local climates are not affected by increases in population density.
2. The local climate in densely populated areas can be changed by human activities.
3. In densely populated areas, human activities increase the amount of natural pollutants.
4. In sparsely populated areas, human activities have stabilized the rate of energy absorption.

Base your answers to questions 37 through 41 on your knowledge of Earth science and on the diagram below. The diagram represents an imaginary continent on the Earth surrounded by water. The arrows indicate the direction of the prevailing winds. Two large mountain regions are also indicated. Points A, B, E, and H are located at sea level; C, D, and F are in the foothills of the mountains; G is high in the mountains.

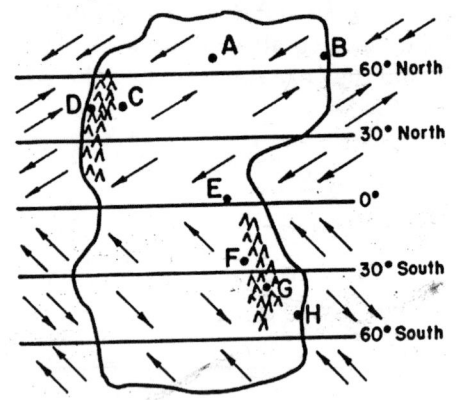

37 Which graph best represents the average monthly temperatures that would be recorded during one year at location E?

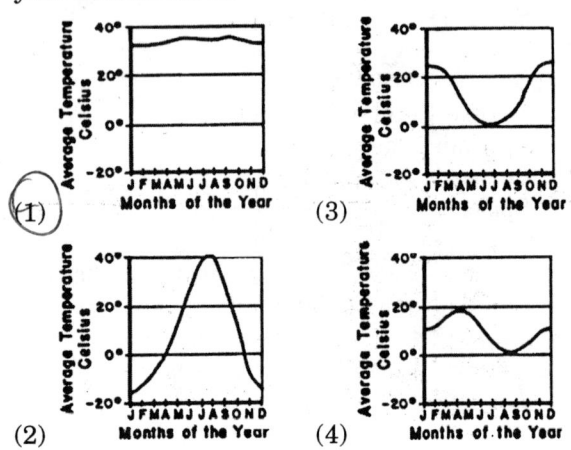

(1) [circled]
(2)
(3)
(4)

38 Which physical characteristic would cause location G to have a colder yearly climate than any other location?
1. The nearness of location G to a large ocean
2. the location of G with respect to the prevailing winds
3. the elevation of location G above sea level [circled]
4. the distance of location G from the Equator

39 Which location probably has the greatest annual rainfall?
(1) A (3) C
(2) F (4) D [circled]

40 Which location probably has the greatest range in temperature during the year?
(1) A [circled] (3) H
(2) B (4) D

41 Which location will probably record its highest potential evapotranspiration values for the year during January?
(1) A (3) C
(2) F [circled] (4) D

Base your answers to questions 42 through 46 on the *Earth Science Reference Tables*, your knowledge of Earth science and on the maps below. Map A shows the generalized climatic zones of the United States based on the P/E_p ratio (average yearly precipitation divided by average yearly potential evapotranspiration). Map B shows the average annual precipitation for sections of the United States.

MAP A – CLIMATE ZONES
(Numbers represent P/E_p ratios)

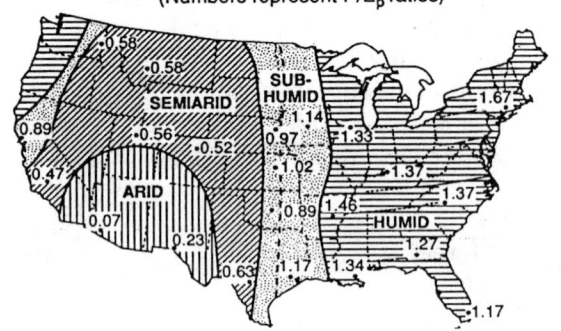

MAP B – AVERAGE ANNUAL PRECIPITATION

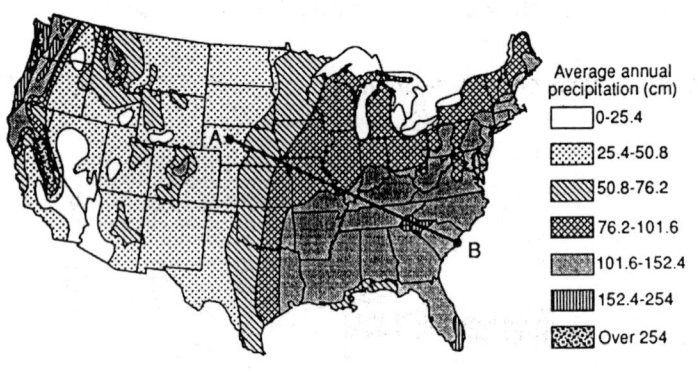

42 A P/E_p ratio 0.59 is classified as
1. arid 3. humid
2. semiarid [circled] 4. subhumid

43 The climate of New York State is classified as
1. arid 3. humid [circled]
2. semiarid 4. subhumid

44 Which section of the United States has high potential evapotranspiration with very little precipitation?
1. Southeast 3. Northeast
2. Southwest [circled] 4. Northwest

45 Which climate zone would have the greatest number of months of moisture deficit?
1. arid [circled] 3. humid
2. semiarid 4. subhumid [circled]

46 Which graph best represents the annual precipitation received along line AB on map B?

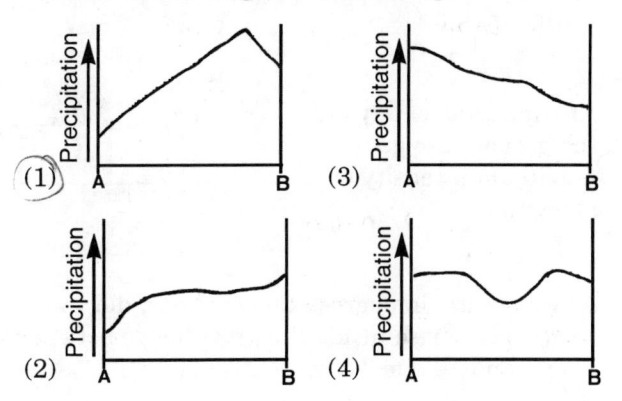

49 Compared to the amount of insolation reflected by the roof of house A, the amount of insolation reflected by the roof of house B is
1 usually less
2 usually more
3 always the same
4 less in summer and greater in winter

Base your answers to question 50 on the maps below of California. Landscape areas are shown on map I, and yearly average rainfall in inches is shown on map II. (Isoline intervals vary.)

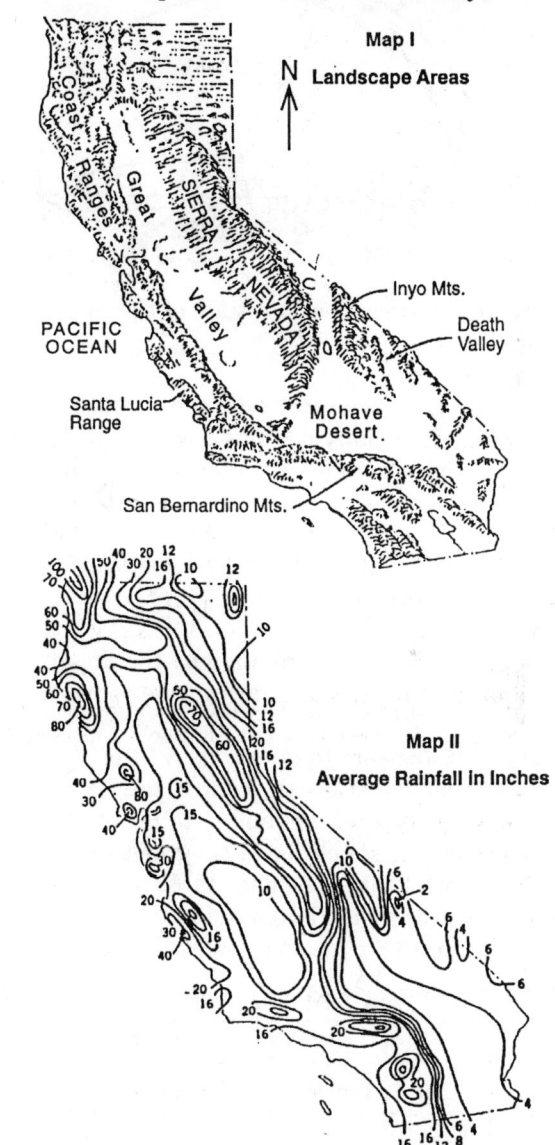

Base your answers to questions 47 through 50 on your knowledge of Earth science and on the diagram below. Two identical houses, A and B, were built in a city in New York State. One house was built on the east side of a factory, and the other house was built on the west side of the factory. Both houses originally had white roofs, but the roof on house has been blackened by factory soot falling on it over the years.

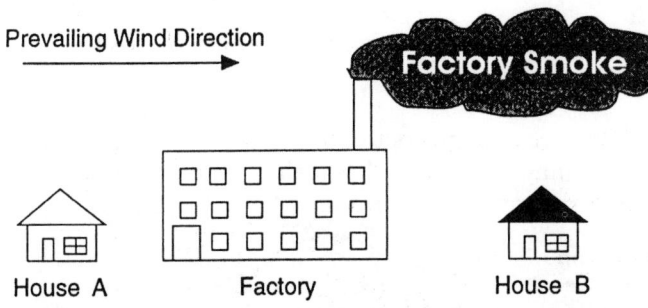

47 Which graph best shows the amount of insolation received at house A during the daylight insolation hours?

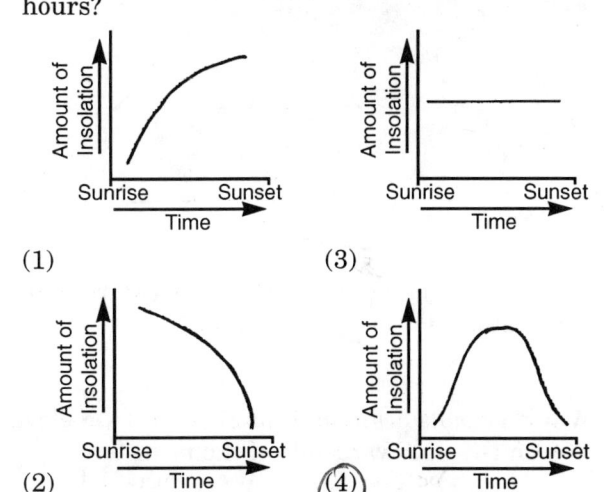

48 On which date would the greatest amount of insolation normally be received at house A?
1 December 21   3 June 21
2 March 21   4 September 21

50 Why are rainfall amounts greater in some regions of California than in others?
1 More evaporation occurs in rainy areas.
2 Areas closer to the Equator receive more rainfall.
3 Moist air is cooled as it rises and moves over mountains.
4 Desert regions produce condensation nuclei for cloud formation.

UNIT SEVEN – THE WATER CYCLE AND CLIMATE – N&N©   Page 159

SKILL ASSESSMENTS

Base your answers to questions 1 through 4 on your knowledge of Earth science and the diagram which represents a plastic hemisphere upon which lines have been drawn to show the apparent path of the Sun on four days at a location in New York State. Two of the days are December 21 and June 21. The protractor is placed over the North–South line.

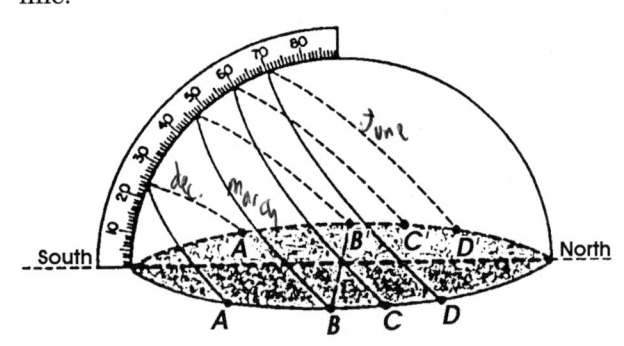

1. What is the solar noon altitude of the Sun for path C–C'? **60°**

2. Label the December 21 and June 21 paths on the diagram. How many degrees does the altitude of the Sun change from December 21 to June 21? **A–D**

3. Which path was recorded on a day that had twelve hours of daylight and twelve hours of darkness? How can you tell? **47°**

4. In a sentence or two explain why the apparent path of the Sun changes during the year. *due to a revolution and a tilt on its axis cause the apparent path of sun changes during the year.*

Base your answers to questions 5 through 10 on your knowledge of Earth science, the *Reference Tables* and the diagram and data below. The diagram represents a closed glass greenhouse located in New York State. The data table shows the air temperatures inside and outside the greenhouse from 6 a.m. to 6 p.m. on a particular day.

AIR TEMPERATURE

Time	Average Outside Temperature	Average Inside Temperature
6 a.m.	10°C	13°C
8 a.m.	11°C	14°C
10 a.m.	12°C	16°C
12 noon	15°C	20°C
2 p.m.	19°C	25°C
4 p.m.	17°C	24°C
6 p.m.	15°C	23°C

Use one or more complete sentences to answer each of the following questions:

5. Where and when did the highest temperature occur? **Inside 2 p.m.**

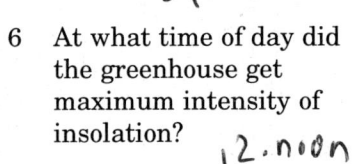

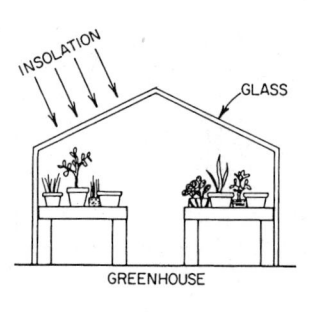

6. At what time of day did the greenhouse get maximum intensity of insolation? **12 noon**

7. At what rate, in degrees Celsius/hour, did the temperature rise inside the greenhouse between 8 a.m. and 10 a.m.? **2/2 = 1°C/hr**

8. What happens to the intensity of insolation received by the greenhouse from February 1 to April 1? *increase because it gets warmer and summer is abt to come up.*

9. Explain why the inside of the greenhouse heats up between 6 a.m. and 2 p.m. *absorbs more than it radiates*

10. Explain how the "greenhouse effect" heats the atmosphere.

Base your answers to questions 11 through 13 on the diagram which shows the prevailing wind moving over a mountain.

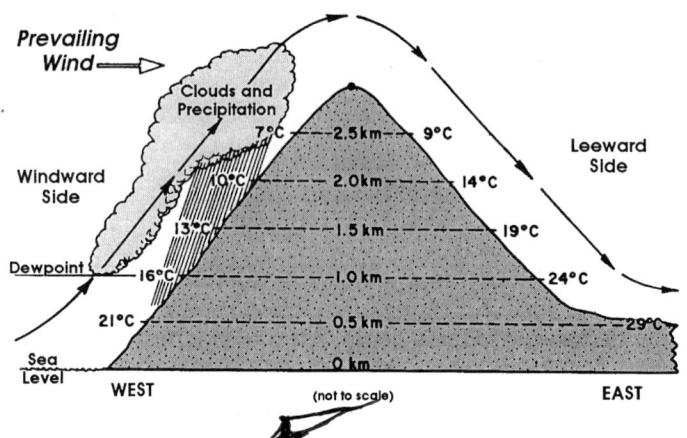

11. What would the approximate temperature of the air be at the top of the mountain? **4°C**

12. Why do clouds begin to form at the 1.0 km elevation on the windward side of the mountain? *because the dew pt air reaches*

13. The air temperature on the leeward side of the mountain at the 1.5 km level is higher than the temperature at the same level on the windward side. Explain why. *The air on the windward side has moisture in it so the temp. does change much but the leeward side is very dry.*

Page 160

UNIT EIGHT
EARTH IN SPACE

VOCABULARY TO BE UNDERSTOOD

Angular Diameter
Annular Eclipse
Aphelion
Apparent Daily Motion
Apparent Solar Day
Cluster
Constellation
Cyclic Energy Transformation
Eccentricity of Ellipse
Equinox
Focus

Galaxy
Gravitation
Kepler's Harmonic Law
Light-year
Lunar Eclipse
Milky Way Galaxy
Neap Tide
Orbit
Orbital Speed
Penumbra
Perihelion

Phases of the Moon
Satellite
Sidereal, Synodic Months
Solar Eclipse
Solar System
Solstice
Speed of Light
Spring Tide
Umbra
Vertical or Direct Rays
Zenith

A. CELESTIAL OBSERVATIONS

Collectively, the objects observed in the sky during the day or night are called **celestial objects**. These objects include the other planets, our Sun, Moon, the stars, and comets. For the most part, these celestial objects *appear to move daily in a path from east to west* in the sky as observed from the Earth's surface. This apparent motion appears as a part of a circle around the Earth, called an **arc**.

How can we account for our observations of celestial objects?

As viewed from Earth, the apparent center of the arcs of the stars or constellations is very near to Polaris, the North Star. A constellation is a group of stars that make an identifiable pattern in the sky. Since Polaris is located above the Earth's axis of rotation, the stars and planets seem to rotate counterclockwise around Polaris at approximately 15° per hour.

Actually, this apparent daily motion of the stars, Moon, and planets is due to the Earth's rotation toward the east, at a rate of 15° per hour or 360° in 24 hours.

SOLAR DAY

The **apparent solar day** is the amount of time required for the Earth to rotate from one noon time to the next. It is measured by two successive appearances of the Sun at a given meridian. Because the Earth is revolving as well as rotating, the Earth must actually rotate slightly more than 360° in order to return to noon on successive days.

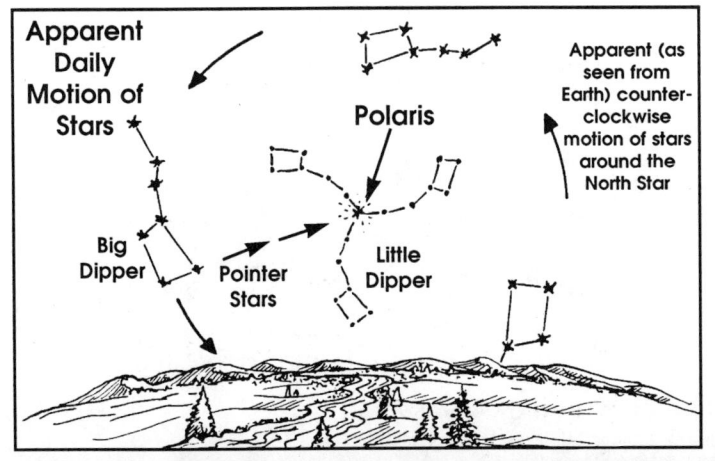

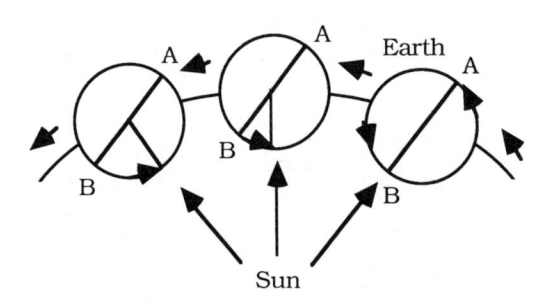

Apparent Solar Day
Earth's rotation and revolution are counterclockwise. As the Earth revolves, the Earth must rotate more than 360° for the Sun's zenith to return to the same Earth longitude.

UNIT EIGHT – EARTH IN SPACE – N&N©

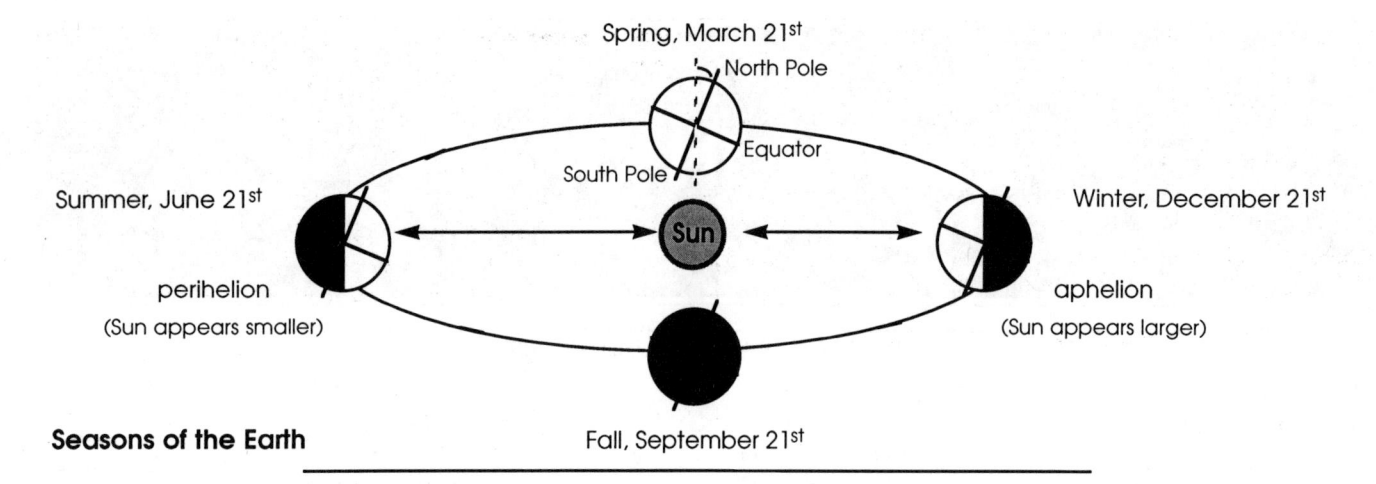

Seasons of the Earth

Spring, March 21st
North Pole
Equator
South Pole
Summer, June 21st
perihelion
(Sun appears smaller)
Sun
Winter, December 21st
aphelion
(Sun appears larger)
Fall, September 21st

APPARENT MOTION OF THE SUN

The Sun appears to move the same as the other celestial objects, in an arc from east to west, sunrise to sunset. For every 15° of arc in the Sun's path, there is one hour of daylight. This apparent path varies with the seasons and with latitude.

SEASONS

The **seasons** on Earth are the direct result of the relative position of the Sun in the sky at different times of the year. Because of the tilt of the Earth's axis ($23\frac{1}{2}°$ from a perpendicular drawn to the plane of the orbit), the Sun's rays are only perpendicular (directly overhead) at noon, between $23\frac{1}{2}°$ North latitude and $23\frac{1}{2}°$ South latitude during the year. As the Earth revolves around the Sun the **perpendicular or vertical rays** of the Sun move between the tropics. This motion causes the points on the horizon of sunrise and sunset to vary during the year.

During summer in the Northern Hemisphere, when the Sun has its highest arc, the Sun rises north of east and sets the same distance north of west. Conversely, during our winter season, the Sun's path is lower, causing shorter daylight hours, and sunrise and sunset occur south of east and south of west, respectively.

The solar diameter varies in a cyclic pattern during the year, due to the change in the Earth's distance from the Sun. The Sun appears largest at **perihelion** (shortest distance between the Earth and Sun), which occurs during the winter in the Northern Hemisphere (about January 3) and the smallest at **aphelion**, during our summer (approximately July 4).

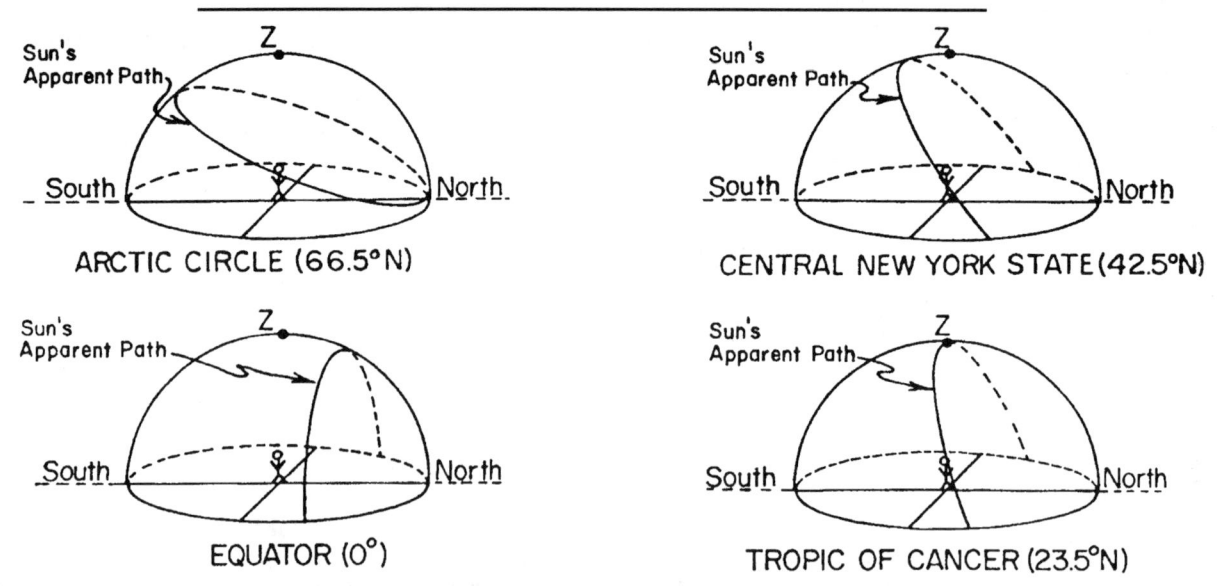

How does a revolving sphere with a tilted spin axis, model yearly celestial observations?

ARCTIC CIRCLE (66.5°N)

CENTRAL NEW YORK STATE (42.5°N)

EQUATOR (0°)

TROPIC OF CANCER (23.5°N)

Apparent Paths of Sun on June 21 as seen by 4 Observers
The zenith (Z) is the point in the sky directly over the observer.

NOON SUN

Local noon is when the Sun reaches its maximum altitude at the observer's longitude. When the Sun is in the **zenith position** (directly overhead) at the equator (Sun rays are vertical), daylight and dark hours are equal on the Earth. This is known as an **equinox** (generally March 21st and September 23rd). For the Northern Hemisphere, the first day of summer occurs when the Sun's vertical rays are at $23\frac{1}{2}°$ N, over the **Tropic of Cancer**. During the **summer solstice** (usually, June 21st), the Northern Hemisphere has its longest daylight hours. The **winter solstice** occurs when the Sun's rays are vertical at $23\frac{1}{2}°$S, over the **Tropic of Capricorn** (generally December 21st). At this time the Northern Hemisphere has its shortest daylight hours. Because the Sun's vertical rays are never seen in the continental United States, an observer in New York State will never see the Sun directly overhead.

B. GEOMETRY OF ORBITS

A **revolution** is defined as the movement of one body around another body. This motion is also referred to as orbiting. An **orbit** is the path taken by one body as it revolves around the other body. For example, the *Earth orbits the Sun* and the *Moon orbits the Earth*.

A circle has one fixed central point called the **focus**, but an ellipse has two focuses or foci. Ellipses vary in their eccentricity. The **eccentricity** or "out of roundness" of an ellipse may be determine by using the following formula:

$$\text{ECCENTRICITY} = \frac{\text{DISTANCE BETWEEN FOCI}}{\text{LENGTH OF MAJOR AXIS}}$$

The Earth orbits the Sun in a slightly elliptical path with the Sun at one of the foci. The Earth's orbit is so slightly eccentric that, drawn to scale on a piece of paper, it appears to be a circle. It is only careful measurement that shows that the Earth's orbit is elliptical. This small variation in the Earth's yearly distance from the Sun shows why distance from the Sun is not a cause of seasons on the Earth. The other planets of the solar system also orbit the Sun in much the same manner with orbits of varying eccentricities (see chart in *Reference Tables*).

ORBITAL SPEED

The speed of a satellite is related to the distance from the body around which it orbits. Changes in a planet's **orbital velocity** (time per distance) are based on the following principle:

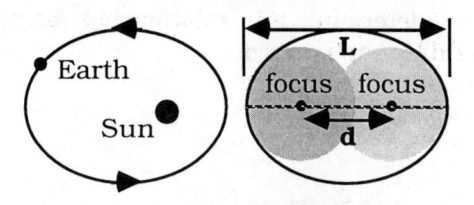

The areas swept out by an imaginary line connecting the Sun and a planet are equal for equal intervals of time.

In an elliptical orbit, the point at which the planet is closest to the Sun is called its **perihelion**, and the point at which the planet is the farthest from the Sun is its **aphelion**. A planet's fastest orbital velocity occurs at perihelion, and its slowest orbital velocity occurs at aphelion.

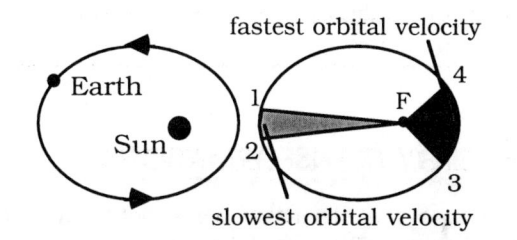

Sun – Earth Orbit
L = major axis length
d = distance between foci

Note: Eccentricity of Earth's orbit is greatly exaggerated.

Orbital Velocity Comparison
The Earth orbit time between 1-2 and 3-4 is the same. The areas within 1-2-F and 3-4-F are equal.

Note: Eccentricity of Earth's orbit is greatly exaggerated.

PERIOD OF A PLANET

The "period of a planet" is explained by Kepler's *Harmonic Law Of Planetary Motion*. A planet's **period** is the length of time required for the planet to orbit (revolve) once around the Sun. Therefore, the **period** of a planet **equals one year** for that planet.

The period of any planet is related to the mean radius of its orbit. The further a planet is from the Sun, the longer its period and time to revolve.

This formula illustrates this relationship:

$$T^2 \propto R^3$$

The square of the planet's period (T^2) is proportional to the cube of its mean distance from the Sun (R^3).

In order to determine this relationship for any planet, this formula can be used.

$$T^2 = R^3$$

Where:
T is expressed in <u>Earth years</u>, and

R is expressed in <u>Astronomical Units</u>.
(1 Astronomical Unit is the mean distance of the Earth from the Sun – about 150,000,000 kilometers.)

The result should be: **1 = 1**

Sample Problem:
Planet "X" has a period of 2.33 Earth years and a distance from the Sun of 1.76, Earth's average distance.

Solution:
$T^2 = R^3$
$T^2 = (2.33)^2 = 5.43$
$R^3 = (1.76)^2 = 5.45$

relationship verification: 1 = 1 (rounded off)

FORCE & ENERGY TRANSFORMATIONS

Gravitation is the attractive force between any two objects anywhere. The principle that governs the gravitational attraction in the simple celestial model is:

Gravitational force is directly proportional to the product of the masses of the objects and inversely proportional to the distance between their centers squared.

This proportion can be expressed in the following formula:

$$F \propto \frac{M_1 M_2}{d^2}$$

Where:
F = gravitational force
M_1 = mass of the first object
M_2 = mass of the second object
d^2 = distance between the centers of the two objects

Gravity is the force that keeps all satellites, both natural and human-made, moving in their curved orbits.

A cyclic energy transformation between **kinetic energy** (energy of motion) and **potential energy** (stored energy) takes place as the Earth orbits the Sun, resulting in a change in the Earth's speed (see illustration below).

What keeps satellites in their orbits?

Since the gravitational force acting on the Earth is greatest when the Earth is closest to the Sun, the orbital velocity of the Earth is increased as it moves closer to the Sun in its orbit. As the Earth speeds up, some of the Earth's **potential energy is converted to kinetic energy**. However, the Earth is *not pulled* into the Sun, because of inertia. **Inertia** is the tendency of an object in motion to continue in motion in a straight path. It is the balance between gravity and inertia that keeps all satellites in their orbits.

As the Earth moves away from the Sun, its orbital speed will decrease as the Sun's gravitational force opposes the Earth's motion (away from the Sun). The gained kinetic energy is reconverted back into potential energy, causing the **cyclic energy transformation**.

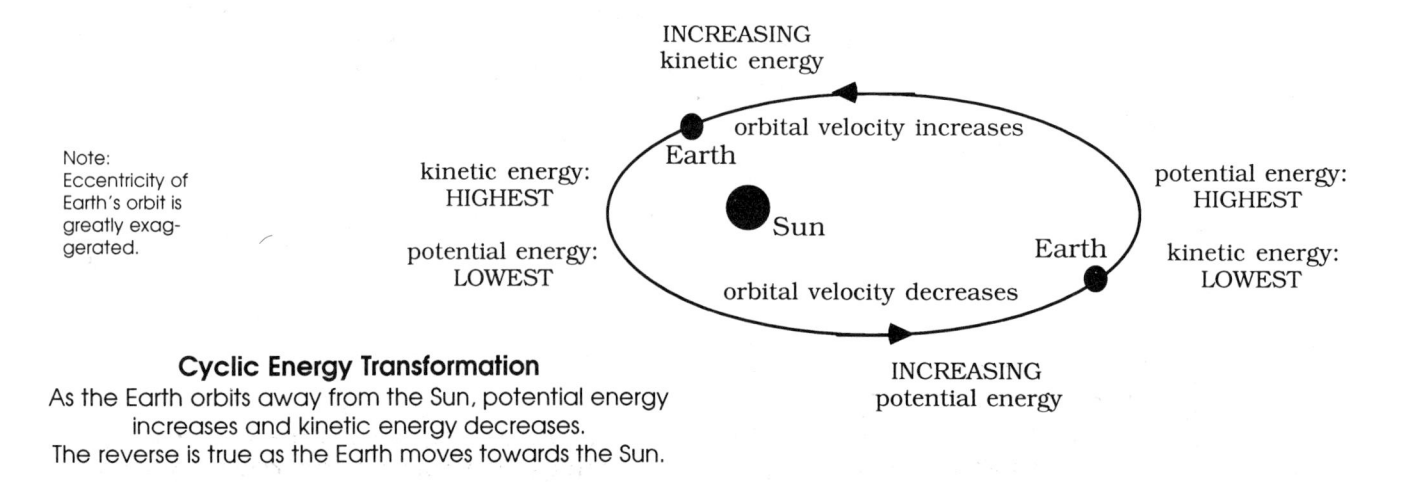

Note: Eccentricity of Earth's orbit is greatly exaggerated.

INCREASING kinetic energy
orbital velocity increases
Earth
kinetic energy: HIGHEST
potential energy: LOWEST
Sun
potential energy: HIGHEST
Earth
kinetic energy: LOWEST
orbital velocity decreases
INCREASING potential energy

Cyclic Energy Transformation
As the Earth orbits away from the Sun, potential energy increases and kinetic energy decreases. The reverse is true as the Earth moves towards the Sun.

In addition to the cyclic energy transformation, the change in the orbital speed of the Earth causes the length of the day to vary slightly.

ANGULAR DIAMETER

Unlike the apparent motion of stars, the movement of planets through the star field is not uniform. The difference in the uniform motion of the stars and the non-uniform motion of the planets is explained by the orbital velocities of the planets as they move around the Sun.

From Earth observation, the angular diameter of the Moon, Sun, and planets appears to change in size in a cyclic manner. This change in **angular diameter** is due to change in the distance of the Moon, Sun, and planets from the Earth. The closer they are to the Earth, the larger they appear and the faster they appear to move in their orbits.

SATELLITE MOTIONS

The **Moon** is the primary satellite of the Earth, since it revolves around the Earth. (The Earth can be considered a satellite of the Sun for the same reason.) As the Moon revolves around the Earth, it seems to have the same motion characteristics as the other celestial objects. Based on Earth time, one complete revolution of the Moon takes $27\frac{1}{3}$ days (**sidereal month**).

The angular diameter of the Moon changes as it orbits the Earth. This apparent change in the Moon's diameter is due to the actual orbit of the Moon being slightly elliptical, not circular (see Section D of this Topic for a further explanation).

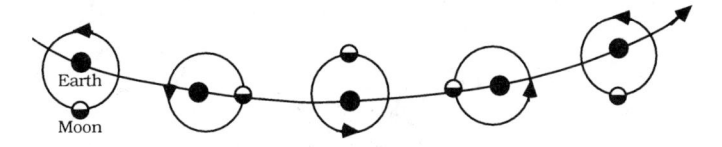

Moon (satellite) Motion Around The Earth
Earth and Moon revolve around their common mass center (barycenter). Note that the Moon does not revolve around the Earth's center.

PHASES OF THE MOON

As the Moon revolves it passes through a cyclic series of **phases**. The Sun's light rays are always illuminating one half of the Moon. As the Moon revolves around the Earth, the Earth observer sees varying amounts of the illuminated portion of the Moon.

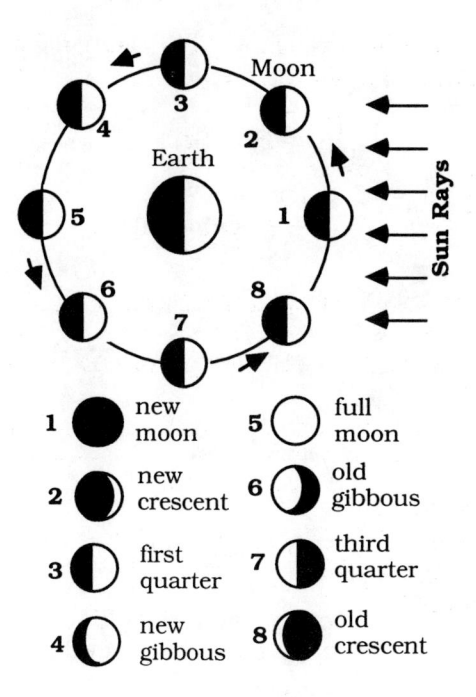

1. new moon
2. new crescent
3. first quarter
4. new gibbous
5. full moon
6. old gibbous
7. third quarter
8. old crescent

The Moon makes a complete phase cycle around the Earth in $27\frac{1}{3}$ days. However, as the Moon revolves around the Earth, the Earth is moving in its orbit around the Sun, constantly changing the relative positions of the Sun, Earth, and Moon.

Therefore, it takes longer (approximately 2 days) for the Moon to complete its full revolution cycle from new Moon to new Moon. The time for this cycle is $29\frac{1}{2}$ days, known as a **synodic month**.

C. AFFECTS ON EARTH

Along the coastline, the water level is continually changing caused by the gravitational attraction of the Moon and the Sun. Although the Sun is much more massive than the Moon, the Moon is much closer to the Earth. Thus, its gravitation has a greater impact on the tides. As the Earth rotates, the tides move around the Earth alternating between high and low tide approximately every six hours.

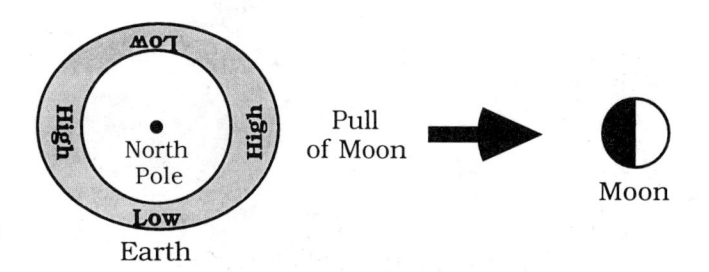

High & Low Tides
Tides are primarily the result of the gravitational attraction between the Earth and the Moon.

UNIT EIGHT – EARTH IN SPACE – N&N©

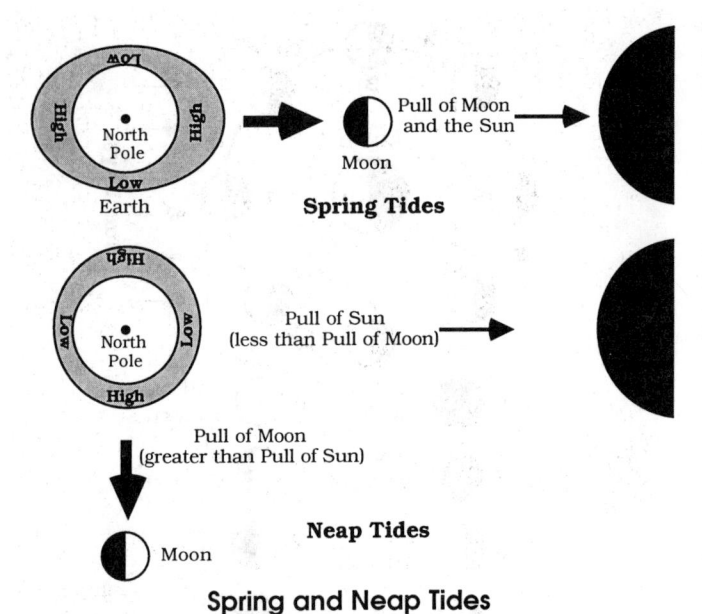

Spring and Neap Tides

Spring Tides of large range are produced when the gravitational attraction of both the Moon and the Sun is lined on the same axis. Neap Tides of small range are produced when the gravitational attraction of the Sun and the Moon are at right angles (90°) to the Earth.

Tidal range (difference between high and low tide levels) varies during the month. At new and full Moon phases, the Sun's gravitation reinforces that of the Moon's resulting in tides with the greatest range. These tides are called **spring tides**.

Neap tides, those with the smallest tidal range, also occur twice a month but at the first and third quarter phases. At the quarter phases the gravitational attraction of the Sun is pulling at a right angle to that of the Moon.

ECLIPSES

Objects like the Earth and the Moon cast shadows into space. These shadows have a part of total darkness called the **umbra** and a part of partial darkness called the **penumbra**. When the entire Moon passes into the Earth's umbra, a **total lunar eclipse** occurs. This is only possible at full moon phase. Since the plane of the Moon's orbit is inclined approximately 5° to that of the Earth, lunar eclipses do not occur every month.

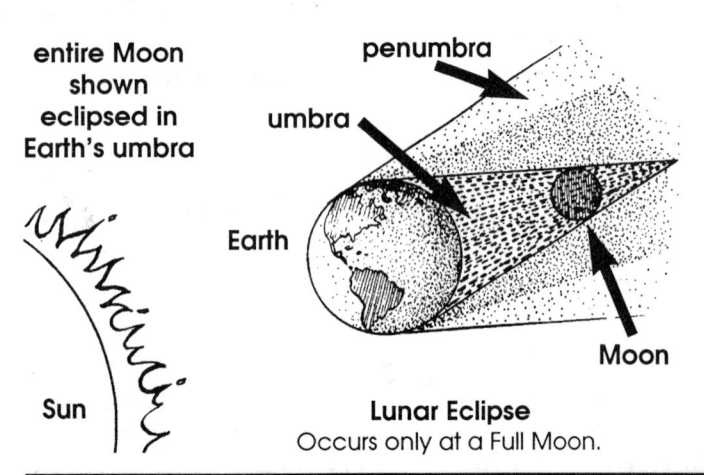

When the Moon is close enough so that the umbra reaches the Earth's surface, a **total solar eclipse** occurs. This is only possible during the new moon phase. A partial eclipse is seen by areas in the penumbra.

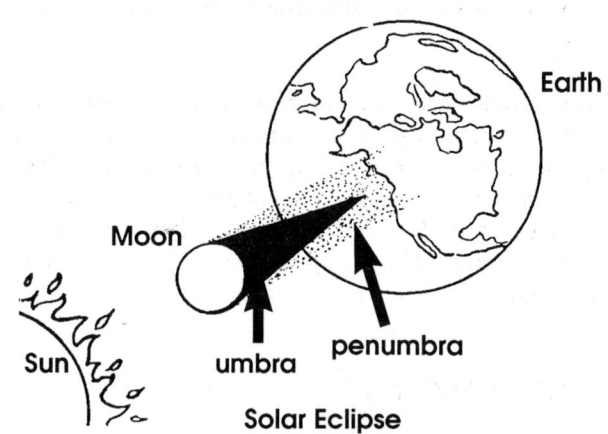

Lunar Eclipse
Occurs only at a Full Moon.

Solar Eclipse
Occurs only at a New Moon. Within the umbra, a total eclipse of the Sun is seen, whereas within the penumbra, only a partial eclipse is seen.

When the Moon is too far from the Earth for its umbra to reach the surface, observers see an **annular eclipse**, a bright ring of solar surface seen behind the dark disk of the moon.

D. EARTH IN THE UNIVERSE

Outside the solar system very large units, such as light-years, are necessary to measure the vast distances in space. The **speed of light** is the basis for this measurement (300,000 kilometers per second). Light could travel around the Earth's equator seven times in one second. A **light-year** is the distance light travels in one year, a distance of ten trillion kilometers.

It takes light from the Sun, our nearest star, a little more than 8 minutes to reach the Earth. Light reaching your eyes from other stars left those stars long ago. Excluding the Sun, light from the next closest star to Earth left that star over 4 years ago. Since the light that left distant stars long ago is just reaching the Earth, you are looking back in time (often hundreds of

What is the Earth's place in the universe?

Milky Way Galaxy
Diameter: 100,000 light years
Thickness: 10,000 light years

years) each time you view the stars.

The **solar system** is made of the Sun, its nine satellite planets, moons, and other celestial objects revolving around the Sun. The Sun is estimated to be about 5 billion years old and is expected to last about another 5 billion years. Average in size and temperature, the Sun is several light-years distant from the nearest other star.

Most stars exist in large systems of various shapes (spiral, elliptical, irregular) called **galaxies**. If you look in toward the center of our galaxy, the sky has a milky-looking appearance. This band of light across the dark night sky is what gives our galaxy its name, The **Milky Way Galaxy**.

Only one of billions of other stars in our spiral galaxy, the Sun is not near the center. The Sun orbits the galaxy in one of the outer spiral arms.

Groups of galaxies are called **clusters**. Astronomers estimate the universe to have a radius of 15 billion light-years, containing many clusters of galaxies all having billions of stars.

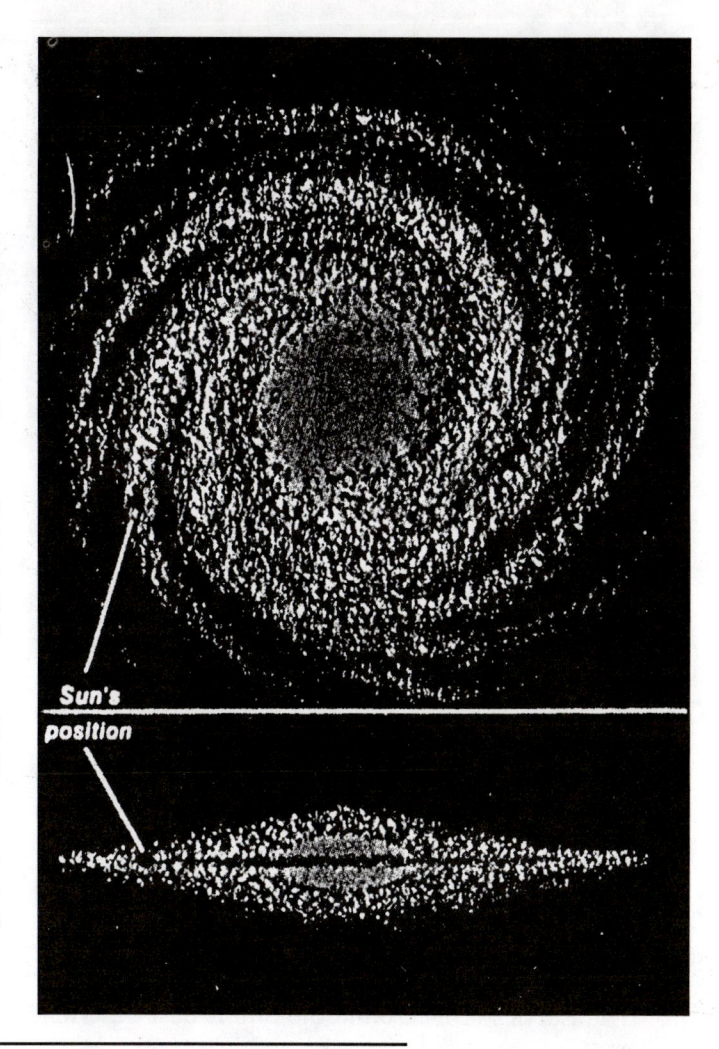

QUESTIONS FOR UNIT 8

1. A photograph showing circular star trails is evidence that the Earth
   1. rotates on its axis
   2. revolves around the Sun
   3. has a nearly circular orbit
   4. has a nearly spherical shape

2. The apparent angular diameter of the Sun was calculated by an observer in New York State once a month for four months. The diameters are shown in the data table.

Month	Angular Diameter
1	32'16"
2	32'30"
3	32'35"
4	32'31"

   Which statement is best supported by the data?
   1. The Earth rotates.
   2. The Sun rotates.
   3. The Earth is tilted 23½ degrees.
   4. The distance between the Earth and the Sun varies.

3. How would a three-hour time exposure photograph of stars in the northern sky appear if the Earth did not rotate?

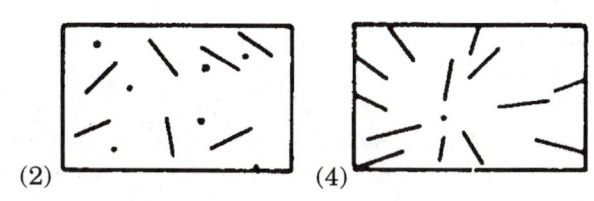

Base your answers to questions 4 through 8 on the *Earth Science Reference Tables*, the diagram below, and your knowledge of Earth science. The diagram shows the Earth's position in its orbit around the Sun at the beginning of each season. The Moon is shown at various positions as it revolves around the Earth.

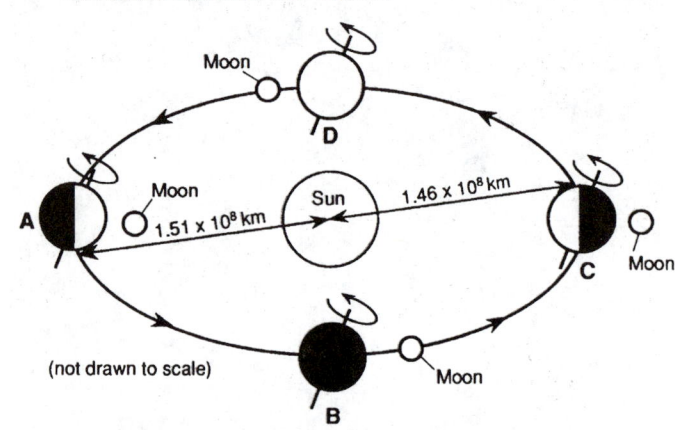

4 What is most likely represented by the symbol ↻ near the Earth's axis at each position?
 1 the direction of Earth's rotation
 2 the path of the Sun through the sky
 3 the changing tilt of Earth's axis
 4 convection currents in the atmosphere

5 The Earth's orbit around the Sun is best described as
 1 a perfect circle
 2 an oblate spheroid
 3 a very eccentric ellipse
 4 a slightly eccentric ellipse

6 Which position of the Earth represents the beginning of the winter season for New York State?
 (1) A   (3) C
 (2) B   (4) D

7 Which position of the Earth shows the Moon located where its shadow may sometimes reach the Earth?
 (1) A
 (2) B
 (3) C
 (4) D

8 As the Earth moves from position B to position C, what change will occur in the gravitational attraction between the Earth and the Sun?
 1 It will decrease, only.
 2 It will increase, only.
 3 It will decrease, then increase.
 4 It will remain the same.

Base your answers to questions 9 through 13 on the diagram below and your knowledge of Earth science. The diagram represents a plastic hemisphere upon which lines have been drawn to show the apparent paths of the Sun on four days at one location in the Northern Hemisphere. Two paths are dated. The protractor is placed over the north-south line. X represents the position of a vertical post.

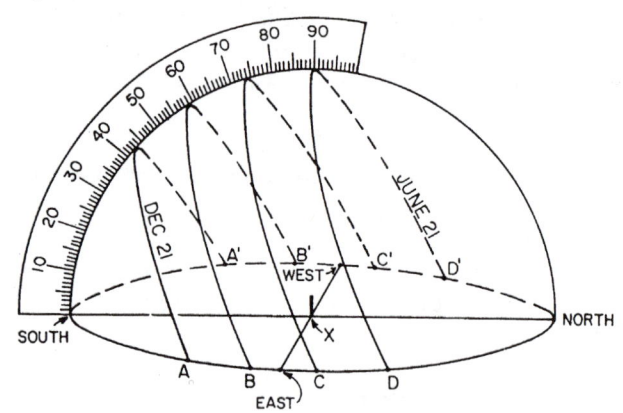

9 For which path is the altitude of the noon Sun 74°?
 (1) A-A'   (3) C-C'
 (2) B-B'   (4) D-D'

10 How many degrees does the altitude of the Sun change from December 21 to June 21?
 (1) 43°   (3) 66½°
 (2) 47°   (4) 74°

11 Which path of the Sun would result in the longest shadow of the vertical post at solar noon?
 (1) A-A'   (3) C-C'
 (2) B-B'   (4) D-D'

12 Which statement best explains the apparent daily motion of the Sun?
 1 The Earth's orbit is an ellipse.
 2 The Earth's shape is an oblate spheroid.
 3 The Earth is closest to the Sun in winter.
 4 The Earth rotates on its axis.

13 What is the latitude of this location?
 (1) 0°   (3) 66½° N
 (2) 23½° N   (4) 90° N

14 When observed from sunrise to sunset in New York State, the length of the shadow cast by a vertical pole will
 1 decrease, only
 2 increase, only
 3 first decrease, then increase
 4 first increase, then decrease

Base your answers to questions 15 through 18 on the diagrams below and your knowledge of Earth science. The diagrams represent four locations on the Earth's surface at the same time on March 21. Lines have been drawn to represent the apparent path of the Sun across the sky. The present position of the Sun, the position of Polaris, and the zenith (Z) are shown for an observer at each location.

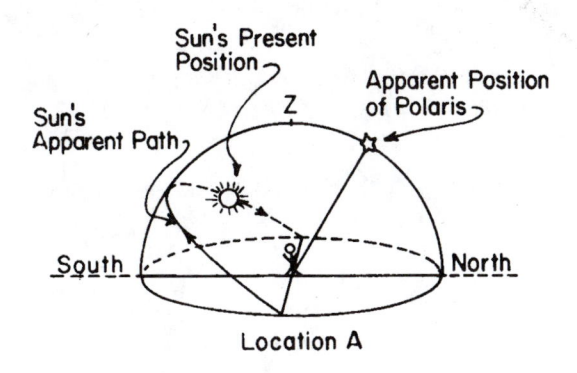

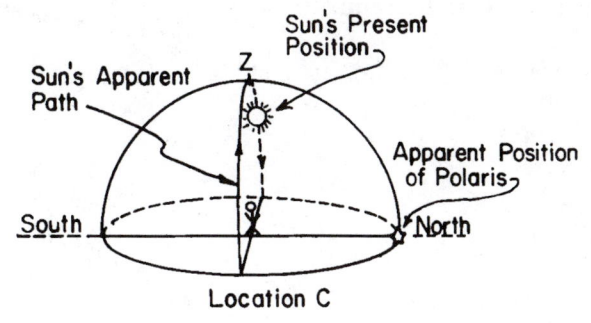

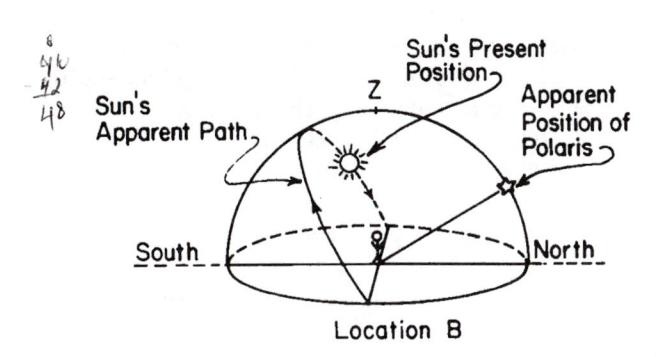

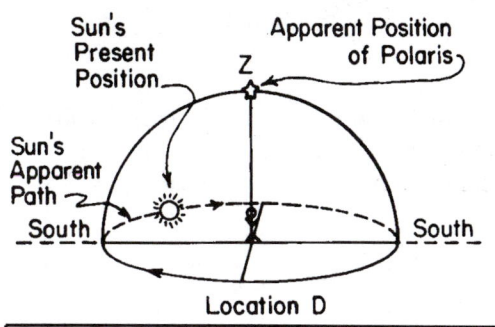

Location A

Location C

Location B

Location D

15 To an observer at location A, the Sun will appear to move from
 1 east to west at 15° per hour
 2 west to east at 15° per hour
 3 east to west at 1° per hour
 4 west to east at 1° per hour

16 The Sun's apparent path through the sky on this day is a direct result of the
 1 Sun's rotation
 2 Earth's rotation
 3 Sun's revolution around the Earth
 4 Earth's revolution around the Sun

17 What is the latitude of the observer at location D?
 (1) 90° N
 (2) 66½° N
 (3) 23½° N
 (4) 0°

18 What time of day is shown by the Sun's Present position at location A?
 1 morning
 2 noon
 3 afternoon
 4 midnight

19 The diagram below represents the elliptical orbit of the Earth around the Sun.

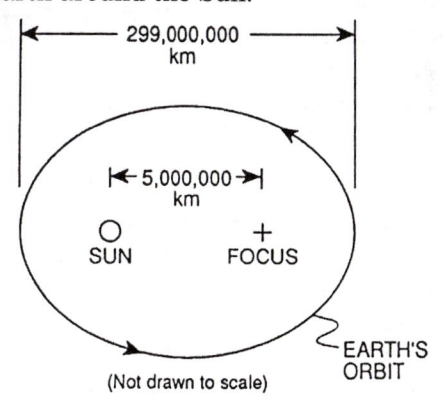

According to the distances shown in the diagram and the *Earth Science Reference Tables*, which equation should be used to find the eccentricity of the Earth's orbit?

1 eccentricity = $\dfrac{299{,}000{,}000 \text{ km}}{5{,}000{,}000 \text{ km}}$

2 eccentricity = $\dfrac{5{,}000{,}000 \text{ km}}{299{,}000{,}000 \text{ km}}$

3 eccentricity = 299,000,000 km − 5,000,000 km

4 eccentricity = $\dfrac{5{,}000{,}000 \text{ km}}{299{,}000{,}000 \text{ km} - 5{,}000{,}000 \text{ km}}$

UNIT EIGHT — EARTH IN SPACE — N&N© Page 169

20. A belt of asteroids is located an average distance of 503 million kilometers from the Sun. Between which two planets is this belt located?
1. Mars and Earth
2. Mars and Jupiter
3. Jupiter and Saturn
4. Saturn and Uranus

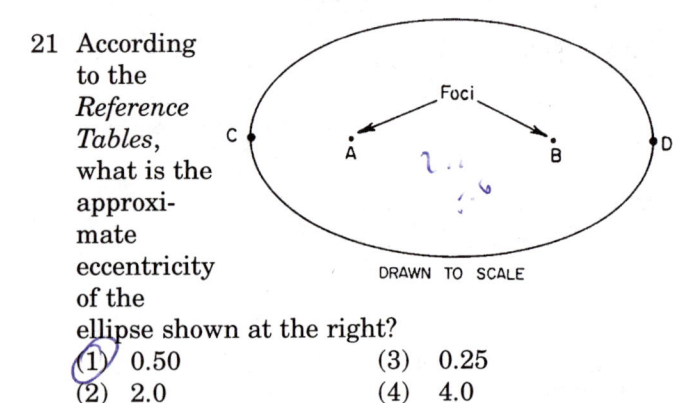

21. According to the *Reference Tables*, what is the approximate eccentricity of the ellipse shown at the right?
(1) 0.50
(2) 2.0
(3) 0.25
(4) 4.0

22. The diagram below represents the Earth's orbital path around the Sun. The Earth takes the same amount of time to move from A to B as from C to D.

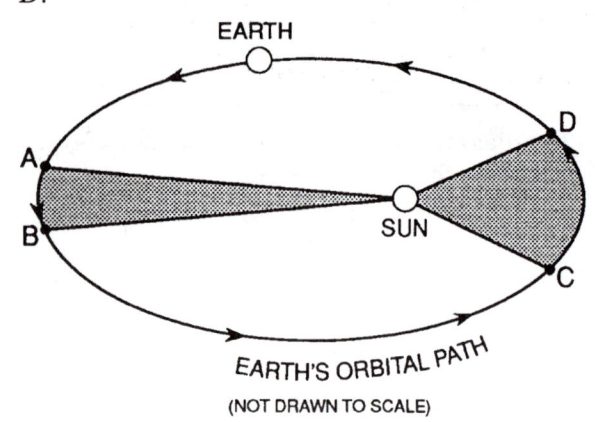

Which values are equal within the system?
1. The shaded sections of the diagram are equal in area.
2. The distance from the Sun to the Earth is the same at point A and at point D.
3. The orbital velocity of the Earth at point A equals its orbital velocity at point C.
4. The gravitational force between the Earth and the Sun at point B is the same as the gravitational force at point D.

23. Planet A has a greater mean distance from the Sun than planet B. On the basis of this fact, which further comparison can be correctly made between the two planets?
1. Planet A is larger.
2. Planet A's revolution period is longer.
3. Planet A's speed of rotation is greater.
4. Planet A's day is longer.

Questions 24 and 25 refer to the diagram which shows the Earth's orbit and the partial orbit of a comet on the same plane around the Sun.

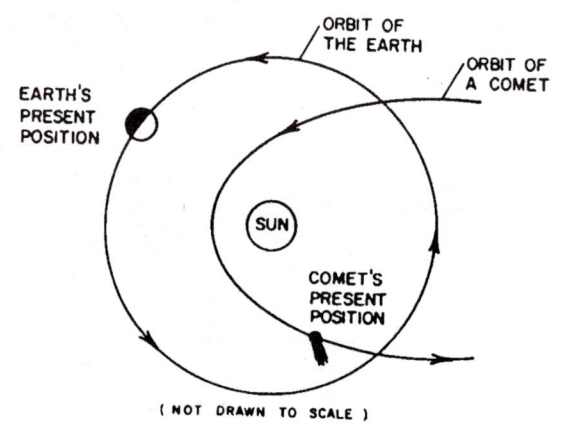

24. Which observation is true for an observer at the Earth's Equator at midnight on a clear night for the positions shown in the diagram?
1. The comet is directly overhead.
2. The comet is rising.
3. The comet is setting.
4. The comet is *not* visible.

25. Compared with the Earth's orbit, the comet's orbit has
1. less eccentricity
2. more eccentricity
3. the same eccentricity

Base your answers to questions 26 through 28 (found on the next page) on the *Earth Science Reference Tables*, the diagram below, and your knowledge of Earth science. The diagram represents the orbits of three planets, X, Y, and Z, around star A. Star A is located at one focus and point B is the other focus. Numbers 1 through 7 represent different positions of the three planets. The arrows show the direction of revolution.

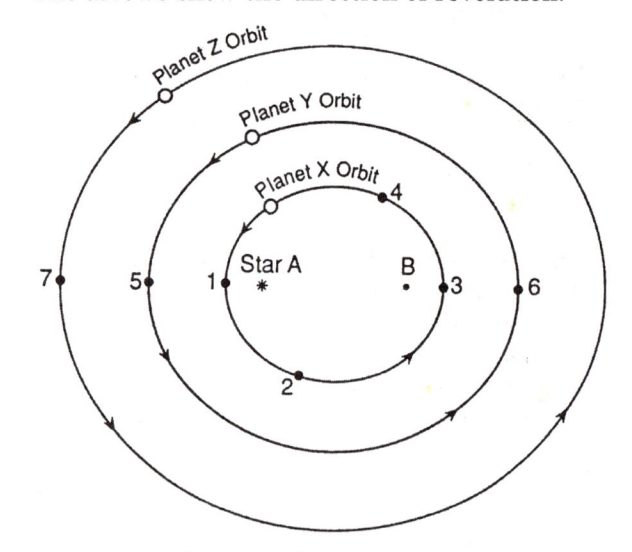

26 The orbital paths of these planets around star A can best be described as having
1   the same period of rotation
2   major axes of the same length
3   an elliptical shape, with star A at one focus
4   a circular shape, with star A at one focus

27 Which number indicates the position at which a planet would have the greatest gravitational attraction to star A? [Assume that all three planets have the same mass.]
(1) 7   (3) 3
(2) 6   (4) 5

28 At which position does planet X have the greatest orbital velocity?
(1) 1   (3) 3
(2) 2   (4) 4

29 The phases of the Moon are caused by the
1   Earth's revolution around the Sun
2   Moon's revolution around the Earth
3   Moon's varying distance from the Earth
4   Sun's varying distance from the Moon

30 The diagram shows the relative position of the Earth, Moon, and Sun for a one-month period.
Which diagram best represents the appearance of the Moon at position P when viewed from the Earth?
(1)   (2)   (3)   (4)

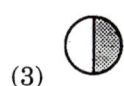

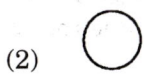

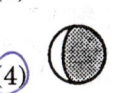

31 The diagram shows four different positions (W, X, Y, and Z) of the Moon in its orbit around the Earth.
In which position will the full moon phase be seen from the Earth?
1   W
2   X
3   Y
4   Z

32 A student drew the phase of the Moon observed from one location on the Earth on each of the dates shown below.

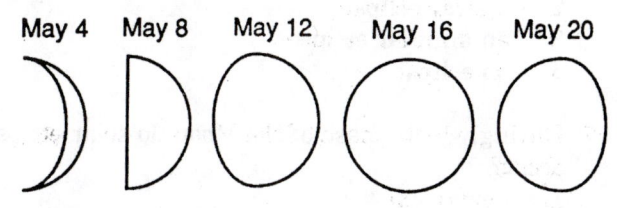

May 4   May 8   May 12   May 16   May 20

Which diagram best shows the Moon's phase on May 24?
(1)   (2)   (3)   (4)

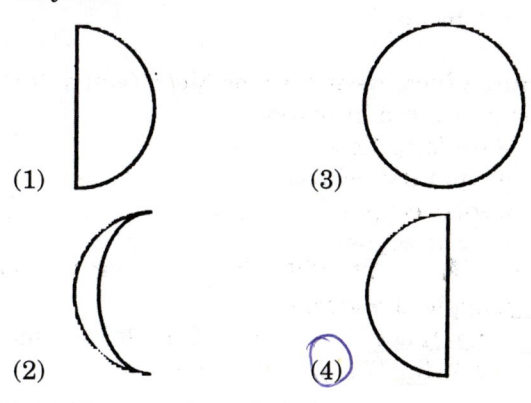

33 The graph below shows the changes in height of ocean water tides over the course of 2 days at one Earth location.

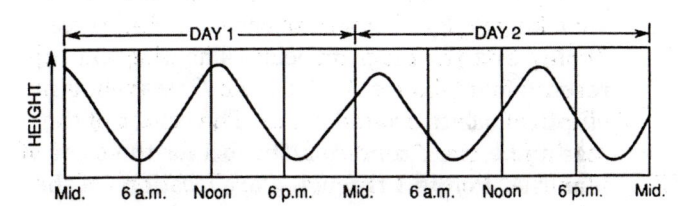

Which statement concerning these changes is best supported by the graph?
1   The changes are cyclic and occur at predictable time intervals.
2   The changes are cyclic and occur at the same time every day.
3   The changes are noncyclic and occur at sunrise and sunset.
4   The changes are noncyclic and may occur at any time.

34 The average time interval between one high tide and the next high tide is approximately
(1) 24 hours   (3) 3 hours
(2) 12 hours   (4) 6 hours

35 The high tides which occur at both the New Moon and the Full Moon are called
1   flood tides
2   neap tides
3   spring tides
4   ebb tides

36 When the Moon is completely covered within the Earth's umbra, which occurs?
1 a lunar eclipse
2 a solar eclipse
3 an annular eclipse
4 no eclipse

37 During which phase of the Moon do solar eclipses occur?
1 new moon
2 first quarter moon
3 last quarter moon
4 full moon

38 During which event does the Moon receive the least amount of insolation?
1 a stellar eclipse
2 an annular eclipse
3 a solar eclipse
4 a lunar eclipse

39 An example of a galaxy is
1 Ursa Major
2 the Milky Way
3 the solar system
4 Polaris

40 The diagram below represents the Milky Way Galaxy.

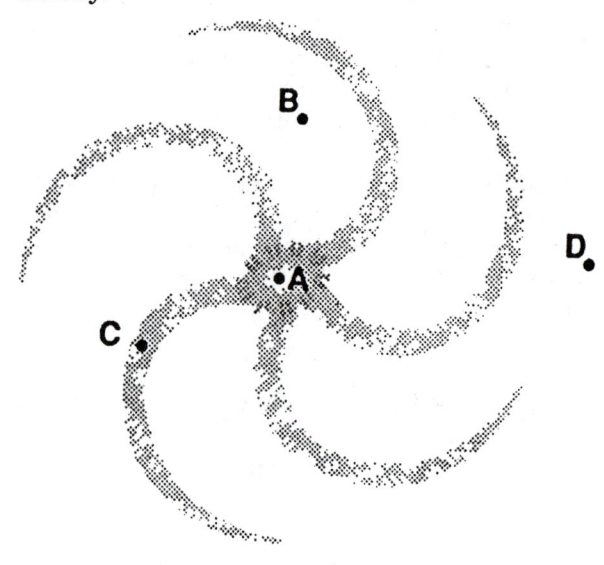

Which letter best represents the location of the Earth's solar system?
(1) A
(2) B
(3) C
(4) D

SKILL QUESTIONS

Base your answers to questions 1 through 5 on your knowledge of Earth science, the *Reference Tables*, and the diagram below. The diagram represents four planets A, B, C, and D, traveling in elliptical orbits around a star. The center of the star and letter *f* represent the foci for the orbit of planet A. Points 1 through 4 are locations of the orbit of planet A.

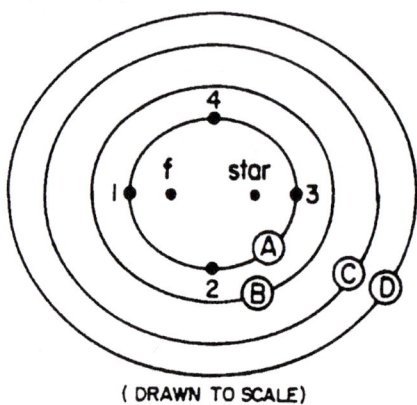

(DRAWN TO SCALE)

1 List the order of the planets from the shortest period of revolution to the longest.

2 If planets A, B, C, and D have the same mass and are located at the positions shown in the diagram, explain in a sentence which planet has the greatest gravitational attraction to the star and why.

3 Write the equation for determining the eccentricity of an ellipse.

4 Using the metric ruler in the *Reference Tables* and the equation for eccentricity, determine the eccentricity of planet A's orbit.

5 At which numbered position on the orbit of Planet A will Planet A have the greatest orbital velocity? Why?

Base your answers to questions 6 through 9 on your knowledge of Earth science, the *Reference Tables*, and the diagrams and table below. Diagram I represents the orbit of an Earth satellite, and diagram II shows how to construct an elliptical orbit using two pins and a loop of string. Table I shows the eccentricities of the orbits of the planets in the solar system.

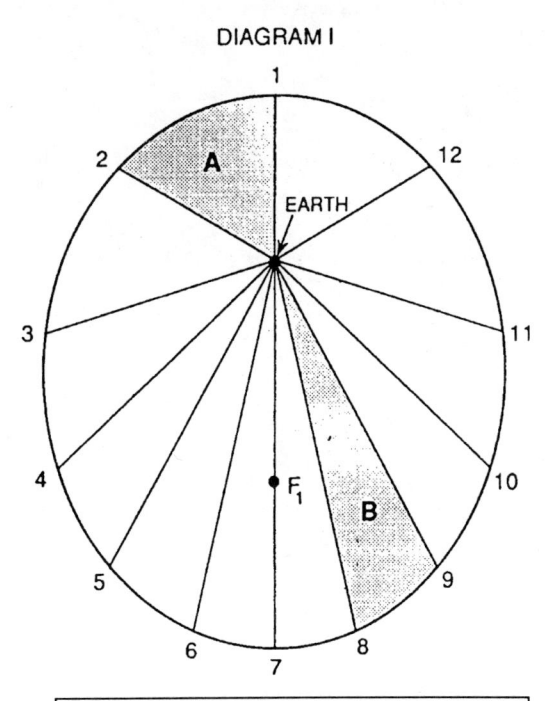

DIAGRAM I

DIAGRAM II

The satellite was at position 1 precisely at midnight on the first day. It arrived at position 2 the next midnight, 3 the next, and so on.

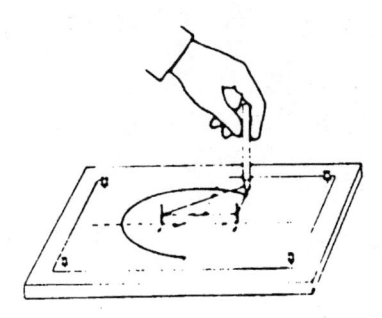

Table 1

Planet	Eccentricity of Orbit
Mercury	0.206
Venus	0.0007
Earth	0.017
Mars	0.093
Jupiter	0.048
Saturn	0.056
Uranus	0.047
Neptune	0.008
Pluto	0.250

6   Determine the eccentricity of the satellite's orbit.

7   The Earth satellite takes 24 hours to move between each numbered position on the orbit. In one sentence tell how area *A* (between positions 1 and 2) compares to area *B* (between positions 8 and 9)?

8   In one sentence, explain how moving the pins closer together in diagram II would effect the eccentricity of the ellipse being constructed.

9   According to Table I, which planet's orbit would most closely resemble a circle?

UNIT EIGHT – EARTH IN SPACE – N&N©

TOPIC F
ASTRONOMY

Vocabulary To Be Understood

"Big Bang"	Heliocentric Model	Spectroscope
Coriolis Effect	Jovian Planets	Spectrum
Doppler Shift	Red Shift	Terrestrial Planets
Epicycle	Revolution	Universe
Foucault Pendulum	Rotation	
Geocentric Model	Spectral Lines	

A. SOLAR SYSTEM MODELS

There are two models which can be used to explain the apparent motions of celestial objects: the **geocentric model** and the **heliocentric model**.

The **geocentric (Earth-centered) model** was an early attempt to illustrate the motions of the stars, Sun, Moon, and planets as seen from the Earth. The geocentric model shows a stationary Earth as the center with the Sun, Moon, planets, and stars revolving at different speeds in circular orbits. The Moon has the closest orbit to the Earth.

Although this model does provide an explanation of the daily motions of the Moon, Sun, and stars, it does not easily explain the complex motions of the planets. In order to explain the irregular motions of the planets against the uniform motion of the stars, the geocentric model shows the planets having two orbits. One orbit is around the Earth and a smaller secondary orbit, called an **epicycle**. The epicycles were an attempt to explain the irregular motions (speed and distance changes) of the planets.

The Sun and stars are seen with the largest circular orbits at great distances from the Earth.

Although the geocentric model cannot explain the terrestrial motions, **Coriolis Effect**, and the apparent change in the path of a **Foucault Pendulum**, it was still believed to be correct for thousands of years, because it seemed so obvious. From the Earth it looks like the Sun, Moon, planets and stars revolve around the Earth.

Since the 16th century the most commonly accepted model for celestial motions has been the **heliocentric (Sun-centered) model**. It is less complicated than the geocentric model and also explains the terrestrial motions.

How do the heliocentric and geocentric models explain apparent celestial motions?

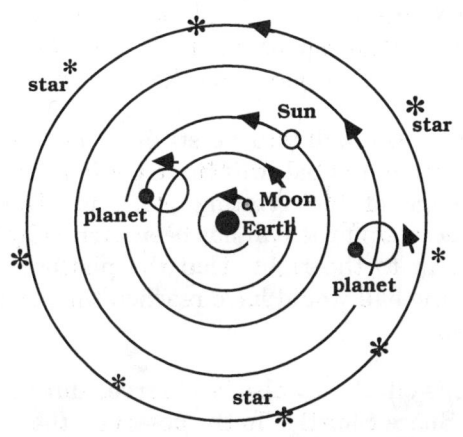

Celestial objects revolving about the Earth as depicted by the **Geocentric Model** which places the Earth as the center.

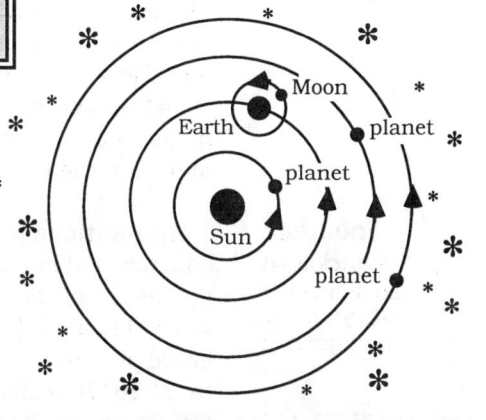

Earth and other celestial objects revolving about the Sun as depicted by the **Heliocentric Model** which places the Sun as the center.

TOPIC F – ASTRONOMY – N&N© Page 175

The Sun, rather than the Earth, is the "stationary" center of this model. The Moon orbits the Earth while the rotating Earth revolves around the Sun in an elliptical orbit. The planets also revolve around the Sun in elliptical orbits.

Outside of this solar system, the stars appear stationary due to great distances from the Earth. (Note: All the stars are actually moving.)

B. TERRESTRIAL OBSERVATIONS

By observing certain terrestrial evidence, Earth's motions can be determined. There are two main motions of the Earth:

- the Earth's **revolution**, a slightly elliptical orbit around the Sun (please see Section D of this Topic); and,
- the Earth's **rotation**, a spinning about the north – south axis.

MOTION AT THE EARTH'S SURFACE

The Earth rotates once in a 24 hour period, turning 360° or 15° per hour, on its axis. The surface rotational velocity is dependent on the observer's latitude. The closer to the equator – the faster the velocity, and the closer to the poles – the slower the velocity. This apparent speed change can be illustrated with the roller skating game "crack the whip," where skaters are strung out in a long line, turning around one skater in the center. The farther the skater is from the center of the spin, the faster the skater must go to maintain the turning chain.

How can the geocentric and heliocentric models be tested?

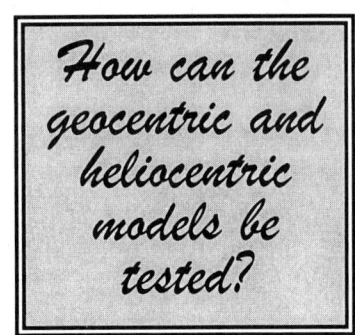

Foucault Pendulum
Back and forth swing is greatly exaggerated.

The rotational velocity at the equator is about 465 m/sec. As the observer moves north or south towards the poles, the speed proportionately decreases (at 20° N or S, 437 m/sec, at 60° N or S, 233 m/sec, and essentially 0 m/sec at the poles).

ROTATIONAL EVIDENCE

The two main evidences of the Earth's rotation are the apparent motion of the **Foucault pendulum** and the **Coriolis effect**.

The **Foucault pendulum** is a freely swinging pendulum, which, when allowed to swing without interference, appears to change direction in a manner that can be predicted.

The apparent direction change is due to the rotation of the Earth below the pendulum. The actual direction of the pendulum does not change.

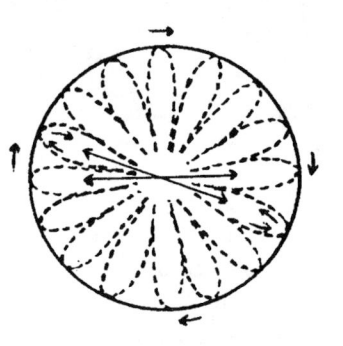

Foucault Pendulum
Pattern is greatly exaggerated.

All moving materials on the surface of the Earth (fluids in particular) tend to undergo horizontal deflection, known as the **Coriolis effect**. This predictable effect is caused by the Earth rotating. In the Northern Hemisphere currents and projectiles are deflected to their right, whereas in the Southern Hemisphere, they are deflected to their left. This effect accounts for the motions of winds and ocean currents near the surface of the Earth.

Watch the water drain from a sink or shower. If not interfered with, the water will always go down the drain in a clockwise direction (in the Northern Hemisphere). In the Southern Hemisphere, the water will drain counterclockwise.

The traditional model to explain the Coriolis effect is the spinning platform.

Two persons are standing on a rotating platform (baseball diamond), one at the center and one at the outside edge (perimeter). The platform is rotating counterclockwise. The person at the center throws a ball directly at the other person on the perimeter.

The flight of the ball will be straight, but to the observer the path of the ball will curve behind the person on the edge of the rotating platform. This is because the receiver of the ball has been carried by the moving platform to the right. Had the platform not been moving, the ball would have reached the receiver in a straight line.

The Coriolis effect may also be observed during the lift-off of the Space Shuttle. To the observer, the shuttle appears to be tilted as the Earth rotates below it.

Coriolis Effect

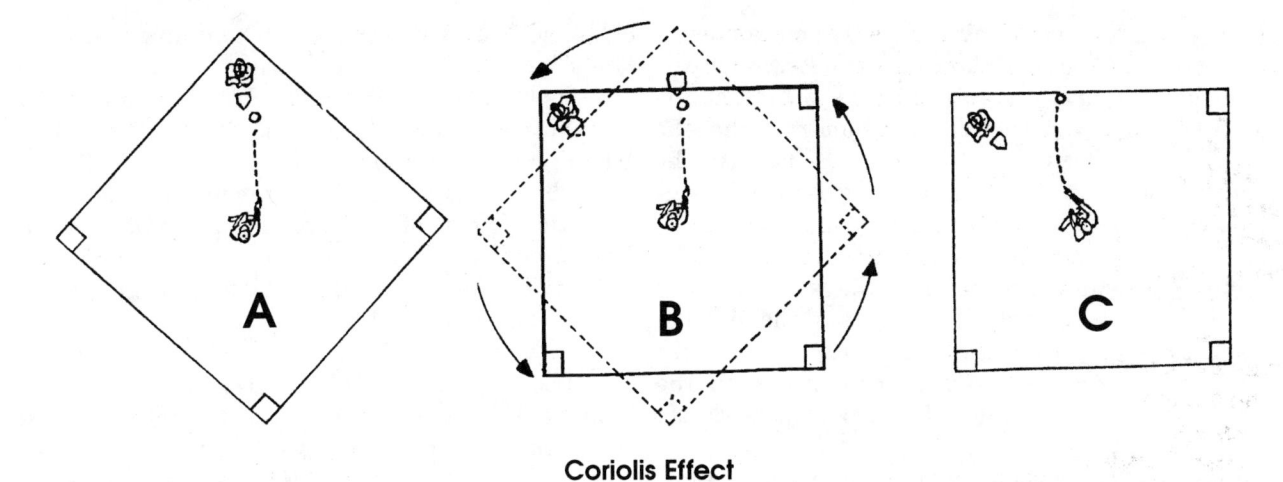

A ball being thrown between two persons riding on a rotating platform illustrates Coriolis effect.
A – On a stationary platform, the pitcher throws the baseball to the catcher in an apparent straight line.
B – As the platform rotates counterclockwise, the pitcher throws the baseball in the direction of the catcher, but the turning platform has carried the catcher counterclockwise away from the incoming ball.
C – From above the platform, the baseball appears to have curved to the right as the platform moved counterclockwise to the left, thus demonstrating the Coriolis Effect on the surface of the rotating Earth.

C. EARTH & OTHER PLANETS

The Earth is the only one of the nine planets in the solar system that has liquid water on its surface, because of the Earth's critical distance from the Sun. This water, along with the atmosphere, helps keep the temperature relatively constant. It is also this water that provided the environment necessary for the beginning, development, and sustenance of life.

How does Earth compare with nearby planets?

Since the average temperatures of the planets decrease as their distances from the Sun increase, and if the Earth were closer to the Sun, all the water would be in vapor form. A location farther from the Sun would result in all the water being solid ice.

The planets may be classified in different ways. One way is to divide them into those that are **Earth-like** (**terrestrial**) and those that are **Jupiter-like** (**Jovian**). Mercury, Venus, Earth, and Mars are terrestrial having rocky cores. Jupiter, Uranus, Saturn, and Neptune are Jovian, gas giants with low densities and thick atmospheres containing hydrogen, helium, methane, and ammonia. Pluto does not fit into either group and perhaps was not originally a planet.

Mercury, the planet closest to the Sun, has many similarities to the Moon. Like the Moon, it has no significant atmosphere. The lack of atmosphere means that incoming meteorites are not destroyed; hence, the surface of Mercury has impact craters much like the Moon. The tremendous temperature range from the lighted to the dark side of Mercury and lack of chemical weathering is also a result of the absence of an atmosphere.

Venus, similar in size, density, and mass to the Earth, is covered with thick clouds in an atmosphere made mostly of carbon dioxide. The greenhouse effect, resulting from so much carbon dioxide, results in blistering surface temperatures.

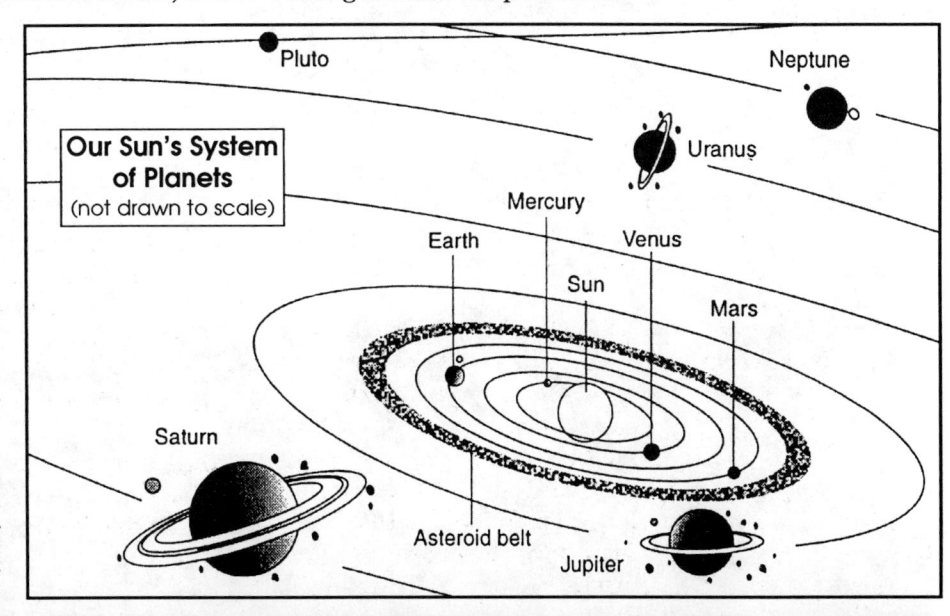

Our Sun's System of Planets (not drawn to scale)

TOPIC F – ASTRONOMY – N&N©   Page 177

An atmosphere containing free oxygen has developed only on the **Earth**. The small percentage of carbon dioxide in Earth's atmosphere helps to heat the atmosphere. However, scientists are concerned that the addition of more of carbon dioxide will increase the greenhouse effect on the Earth causing too much heating (global warming).

Like Venus, the atmosphere of **Mars** is composed primarily of carbon dioxide. Unlike Venus, the Martian atmosphere is very thin so there is little greenhouse effect. Temperatures are significantly colder on the crater marked surface of Mars than those on the Earth.

Although the environments of the other terrestrial planets are hostile to life, exploration, need, and technology may someday enable us to access the resources of these other worlds and perhaps make colonization of these planets possible.

D. THE UNIVERSE

When light is passed through a prism it is separated into its component colors of different wavelengths from red to violet. In this **spectrum of colors**, red has the longest wavelength and violet the shortest.

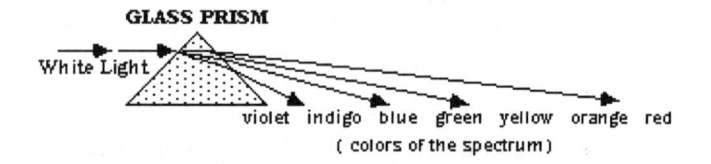

GLASS PRISM
White Light → [prism] → violet indigo blue green yellow orange red
(colors of the spectrum)

The **spectroscope**, which contains a prism, is an important astronomical instrument. By using a spectroscope to study the light from distant stars and galaxies, astronomers obtain information concerning the composition, temperature, and motion of the celestial objects. Each chemical element gives off a characteristic number of wavelengths; therefore, the combination of a star's elements which produce a pattern of **spectral lines** can be used to identify the star, rather like a fingerprint or bar code.

The **Doppler Shift** in spectral lines is an apparent change in wavelength caused by the relative motion of the source of light. If the spectral lines shift toward the red end of the spectrum, the light source is moving away from the observer; if they shift toward the violet end of the spectrum, the light source is moving toward the observer. The greater the degree in the shift of the lines, the faster the relative motion.

Edwin Hubble, (famed astronomer for whom the Hubble Space Telescope, HST was named) while studying the light from distant galaxies, noted a **red shift** in the spectra. He concluded that the galaxies are moving rapidly away from each other. The farther the galaxies are away from the Earth, the greater the red shift in their spectra. This has lead many astronomers to assume the universe is expanding.

What was the origin of the universe, and what will be its fate?

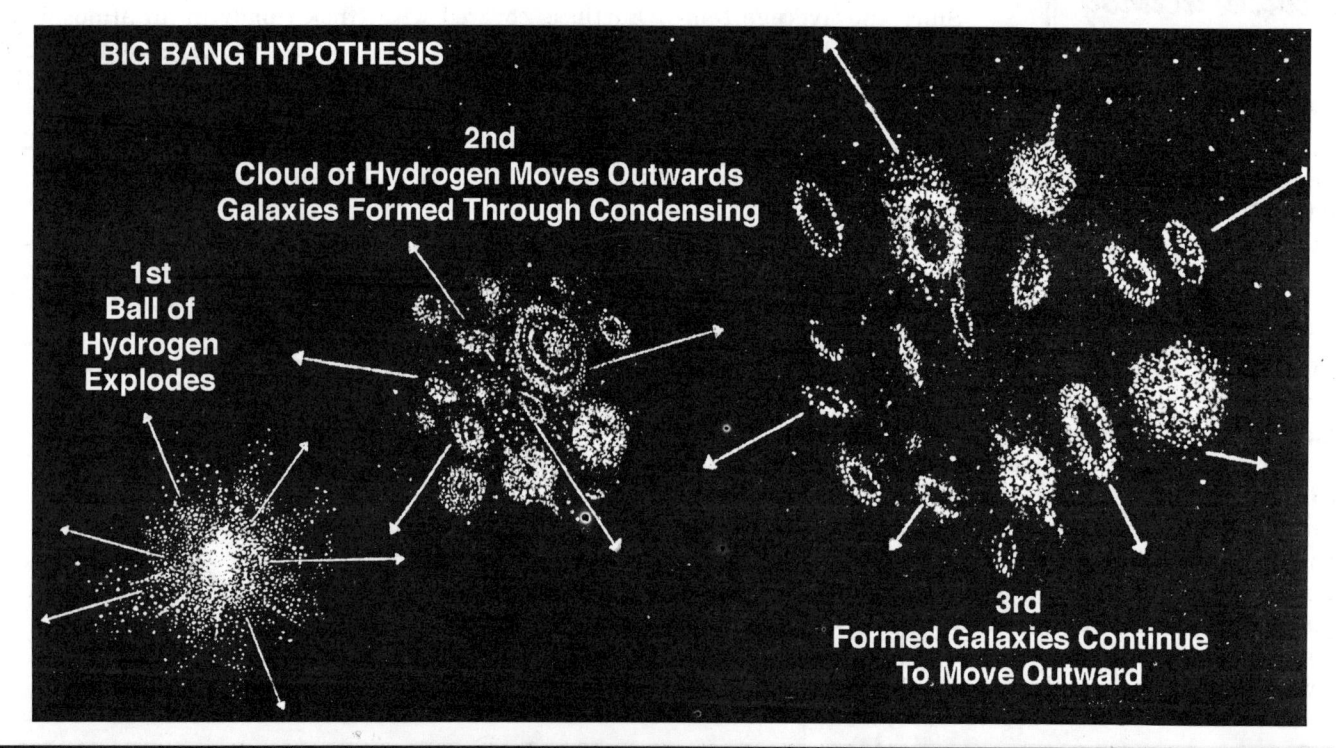

BIG BANG HYPOTHESIS

1st Ball of Hydrogen Explodes

2nd Cloud of Hydrogen Moves Outwards Galaxies Formed Through Condensing

3rd Formed Galaxies Continue To Move Outward

Standard Darkline Spectrum for an Element

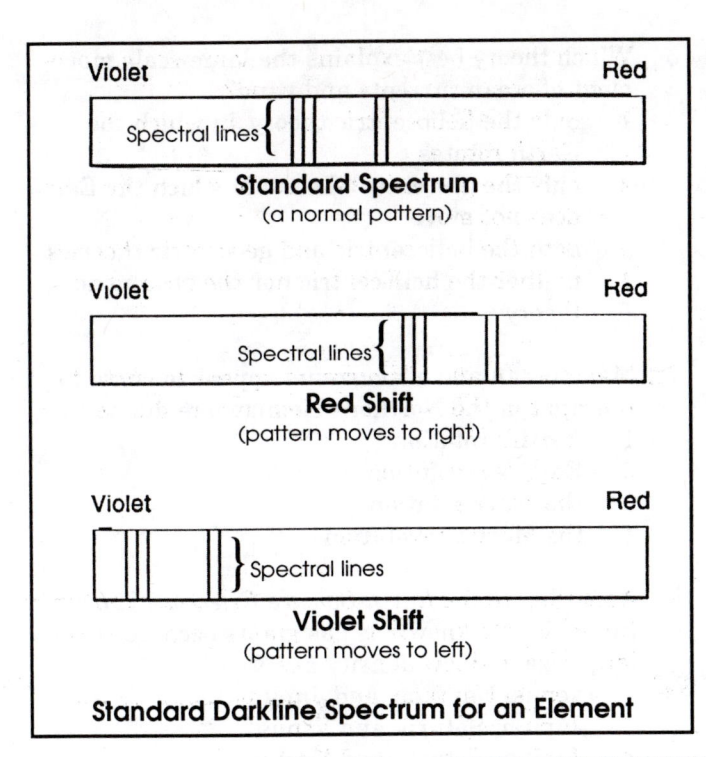

- Standard Spectrum (a normal pattern) — Violet to Red with Spectral lines
- Red Shift (pattern moves to right)
- Violet Shift (pattern moves to left)

Astronomers estimate that approximately 15 billion years ago the universe (entire celestial cosmos) began with a gigantic explosion called the **"Big Bang."** All the material of the universe is assumed to have been in one hot, supermassive, small volume. As this material exploded and expanded, it later condensed into galaxies of dust, gas, and stars that continue to fly apart due to the explosion. Radiation from this "Big Bang" can still be detected in space.

Will the universe stop expanding in 15 billion years or so and begin to contract back into that small supermassive volume, as some astronomers have suggested? The answer to that question depends upon the total mass of the universe which is a subject still under investigation.

QUESTIONS FOR TOPIC F

1. Which diagram best represents a heliocentric model of a portion of the solar system?
   [Key: E = Earth, P = Planet, S = Sun. Diagrams are not drawn to scale.]

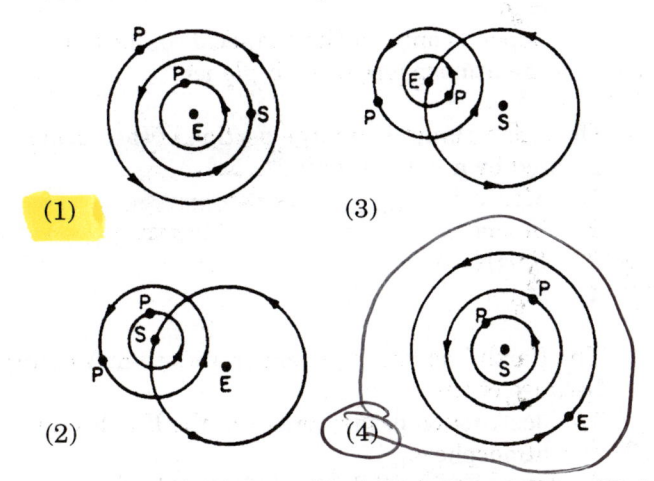

2. A Foucault pendulum is set in motion in New York State in a geographic north-south direction. Which observation will be made after a period of several hours?
   1. The pendulum appears to swing in a wide circle.
   2. The length of the pendulum's swing appears to increase gradually.
   3. The direction of the pendulum's swing appears to change in a predictable manner.
   4. The direction of the pendulum's swing appears to change in an *unpredictable* manner.

3. The diagram represents a Foucault pendulum in a building in New York State. Points A and A' are fixed points on the floor.

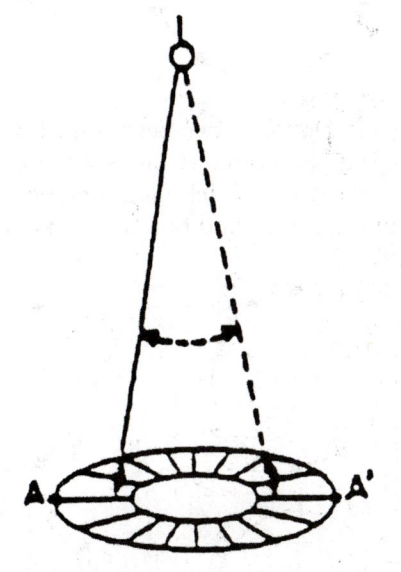

   As the pendulum swings for six hours, it will
   1. appear to change position due to Earth's rotation
   2. appear to change position due to Earth's revolution
   3. continue to swing between A and A' due to inertia
   4. continue to swing between A and A' due to air pressure

TOPIC F – ASTRONOMY – N&N©

Base your answers to questions 4 and 5 on the diagrams below, which represent two views of a swinging Foucault pendulum with a ring of 12 pegs at its base.

Diagram I (side view)

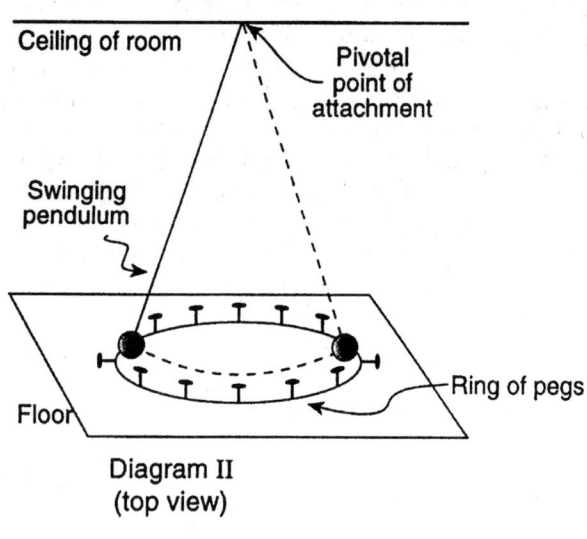

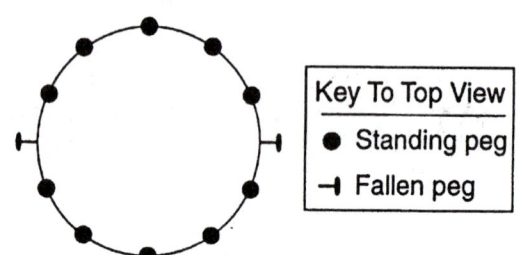

Diagram II (top view)

Key To Top View
● Standing peg
⊣ Fallen peg

4 Diagram II shows two pegs tipped over by the swinging pendulum at the beginning of the demonstration. Which diagram shows the pattern of standing pegs and fallen pegs after several hours?

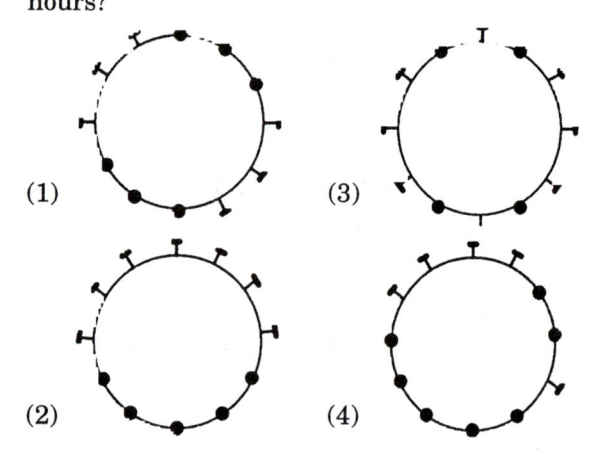

5 The predictable change in the direction of swing of a Foucault pendulum provides evidence that the
  1 Sun rotates on its axis
  2 Sun revolves around the Earth
  3 Earth rotates on its axis
  4 Earth revolves around the Sun

6 Which theory best explains the large-scale movement of ocean currents and wind?
  1 only the heliocentric theory, in which the Earth rotates
  2 only the geocentric theory, in which the Earth does not move
  3 both the heliocentric and geocentric theories
  4 neither the heliocentric nor the geocentric theory

7 Major ocean and air currents appear to curve to the right in the Northern Hemisphere due to
  1 Earth's rotation
  2 Earth's revolution
  3 the Sun's rotation
  4 the Moon's revolution

8 According to the *Earth Science Reference Tables*, three planets known as gas giants because of their large size and low density are
  1 Venus, Neptune, and Jupiter
  2 Jupiter, Saturn, and Venus
  3 Jupiter, Saturn, and Uranus
  4 Venus, Uranus, and Jupiter

9 The average temperature of the planets
  1 increases with greater distance from the Sun
  2 decreases with greater distance from the Sun
  3 has no relationship to the distance from the Sun
  4 depends only on the chemical composition of the atmosphere of each planet

10 On which planet is a large portion of the surface covered by a natural liquid?
  1 Mercury
  2 Venus
  3 Earth
  4 Mars

11 The Earth has fewer impact craters than Mercury because of the
  1 destruction of meteorites in the Earth's upper atmosphere
  2 more rapid subduction of crustal plates on Mercury
  3 slower weathering and erosion rates on the Earth
  4 faster rotational speed of Mercury

12 Earth and our solar system are
  1 older than the universe
  2 younger than the universe
  3 the same age as the universe

Base your answers to questions 13 through 17 on the diagram of the solar system below, the *Earth Science Reference Tables*, and your knowledge of Earth science.

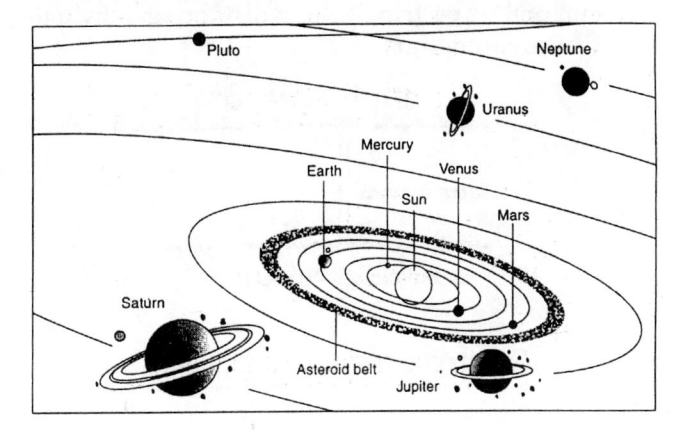

13 Which kind of model of the solar system is represented by the diagram?
 1 heliocentric model 3 sidereal model
 2 geocentric model 4 lunar model

14 If the Earth's distance from the Sun were doubled, the gravitational attraction between the Sun and Earth would be
 1 one-ninth as great
 2 nine times as great
 3 one-fourth as great
 4 four times as great

15 Which planet has the most eccentric orbit?
 1 Venus 3 Saturn
 2 Mars 4 Pluto

16 According to Kepler's Harmonic Law of Planetary Motion, the farther a planet is located from the Sun, the
 1 shorter its period of rotation
 2 shorter its period of revolution
 3 longer its period of rotation
 4 longer its period of revolution

17 On which planet would a measuring instrument placed at the planet's equator record the longest time from sunrise to sunset?
 1 Mercury 3 Earth
 2 Venus 4 Mars

18 The surface of Venus is much hotter than would be expected, considering its distance from the Sun. Which statement best explains this condition?
 1 Venus has many active volcanoes.
 2 Venus has a slow rate of rotation.
 3 The clouds of Venus are highly reflective.
 4 The atmosphere of Venus contains a high percentage of carbon dioxide.

19 The greatest difference in seasons would occur on a planet that has
 1 a circular orbit
 2 a slightly ellipitical orbit
 3 its axis of rotation perpendicular to the plane of its orbit around the Sun
 4 its axis of rotation inclined 45° to the plane of its orbit around the Sun

20 Impact craters are more obvious on the Moon and Mercury than on Earth because
 1 meteorites have not struck Earth
 2 weathering processes on Earth have removed most craters
 3 Earth is younger than Mercury of the Moon
 4 all meteorites burn up in Earth's atmosphere

21 Rock samples brought back from the Moon show absolutely no evidence of chemical weathering. This is most likely due to
 1 the lack of an atmosphere on the Moon
 2 extremely low surface temperatures on the Moon
 3 lack of biological activity on the Moon
 4 large quantities of water in the lunar "seas"

Base your answers to questions 22 and 23 on the diagram below which shows four planets orbiting the Sun.

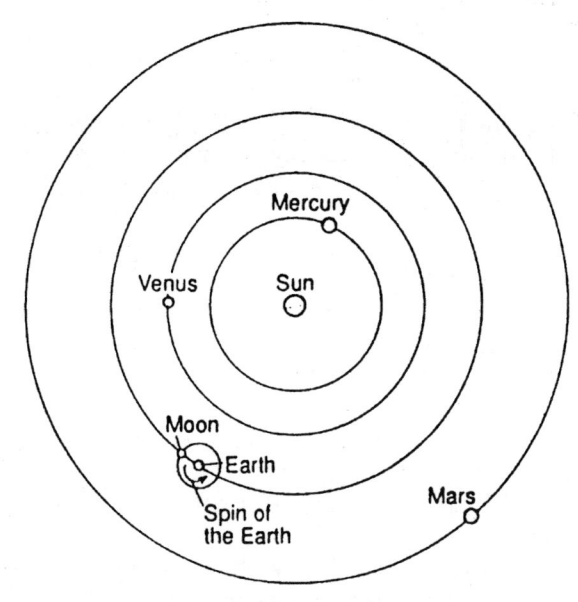

(not drawn to scale)

22 Which object will be visible from the Earth in the eastern sky after midnight?
 1 the Moon
 2 Mars
 3 Mercury
 4 Venus

TOPIC F – ASTRONOMY – N&N© Page 181

23 Which diagram best represents how Venus and the Moon would appear in the sky at sunset to an observer in New York State? [Diagrams are *not* drawn to scale.]

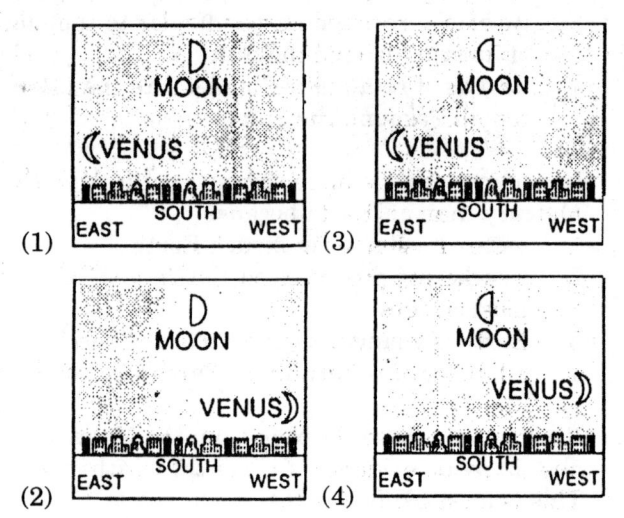

24 The presence of which atmospheric gas causes the high temperature of Venus?
 1 nitrogen (N_2)
 2 oxygen (O_2)
 3 hydrogen (H_2)
 4 carbon dioxide (CO_2)

25 Which planet's orbital shape would be most similar to Jupiter's orbital shape?
 1 Mercury   3 Pluto
 2 Venus   4 Uranus

26 A comparison of the age of the Earth obtained from radioactive dating and the age of the universe based on Galactic Doppler shifts suggest that
 1 the Earth is about the same age as the universe
 2 the Earth is immeasurably older than the universe
 3 the Earth was formed after the universe began
 4 the two dating methods contradict one another.

27 Which statement best describes how galaxies generally move?
 1 Galaxies move toward one another.
 2 Galaxies move away from one another.
 3 Galaxies move randomly.
 4 Galaxies do not move.

Base your answer to this question on the diagram below, the *Earth Science Reference Tables*, and your knowledge of Earth science. The diagram below shows reference lines on a standard spectrum and a spectrum from a distance galaxy moving away from Earth.

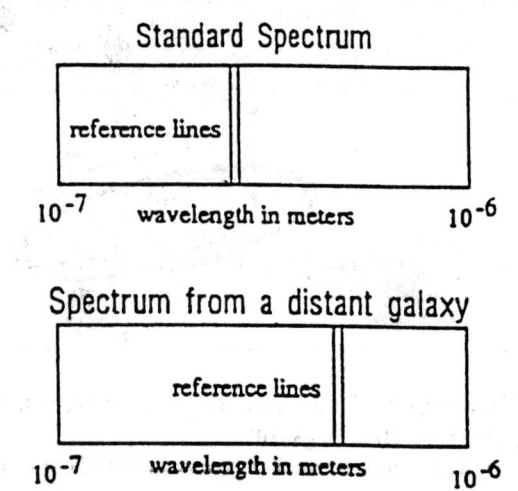

28 If the spectral lines produced by the light from a distant galaxy are
 1 shifted toward the violet, the galaxy is moving away from us
 2 shifted toward the violet, the galaxy is stationary
 3 are shifted toward the red, the galaxy is moving away from us
 4 shifted toward the red, the galaxy is moving towards us

29 In which group are the parts listed from oldest to youngest?
 1 universe, Milky Way, solar system
 2 solar system, Milky Way, universe
 3 Milky Way, solar system, universe
 4 universe, solar system, Milky Way

30 Background radiation detected in space is believed to be evidence that
 1 the Universe began with a primeval explosion
 2 the Universe is contracting
 3 all matter in the Universe is stationary
 4 galaxies are evenly spaced throughout the Universe

Page 182   N&N© SCIENCE SERIES – EARTH SCIENCE – MODIFIED PROGRAM

Unit Nine
Environmental Awareness

Vocabulary To Be Understood

Acid Rain	Chemical Pollutants	Pollution (solid, chemical, nuclear)
Aerobic Bacteria	Environment	Radioactive Wastes
Anaerobic Bacteria	Environmental Equilibrium	Technology
Biocide	Interrelationships	Toxic Wastes
Biological Pollutants	Natural Pollutants	Water Pollution

A. INTERRELATIONSHIPS

The Earth consists of a great variety of interrelated organisms and systems. Modern technology has given humans the capability to influence the **environmental equilibrium** rapidly and on a large scale. The influence on these organisms and systems may be constructive such as positive conservation or it may be negative and destructive.

A change in one part of the environment may cause a dramatic effect on other parts of the environment. For example, the building of dams, roadways, and communities have destroyed millions of acres of forest and natural wetlands.

On the other hand, **technology** has enabled humans to preserve many of nature's wonders. For example, major erosion of the American side of Niagara Falls during the first part of this century endangered the very survival of the falls. By changing the direction and flow of the water and by building unseen reinforcement in the cliffs, human engineering has been able to insure the continuance of the American Falls.

How can we describe the interdependence of Earth's living and nonliving systems?

B. ENVIRONMENTAL POLLUTION

Environmental pollution has reached serious levels in recent years, due to human neglect of the **environmental equilibrium**. The environment is considered polluted when the concentration of any substance or form of energy reaches a proportion that adversely affects humans, plants, and/or animals or the environment on which all living things depend.

Environmental pollutants are the result of both natural environmental disruptions and the technological oversights of humans (such as activities of individuals and industrial processes). The eruption of Mt. Saint Helens in the state of Washington, was an example of **natural pollution**. It produced millions of tons of ash and destroyed thousands of acres of plant and animal life. In addition, the airborne particles (such as dust) from the eruptions caused climatic and weather changes all across North America, affecting farm crops and air quality. Another example of a common natural pollutant is pollen.

People appear to have done much more damage than natural pollutants have done to the environmental equilibrium. This has been through the excessive addition of pollutants to the environment. Pollutants include such diverse materials as solids, liquids, gases, biologic organisms, and forms of energy such as heat, sound, and nuclear radiation and wastes.

Both natural processes and human pollution tend to vary with seasons, days of the week, and times of days. In nature, rivers tend to **purge** (churn up from the bottom) accumulated wastes each late summer or early fall. This is a "cleaning" process for the rivers, but leads to increased surface pollution, disruption of wildlife, and damaging river banks and water supplies. During weekdays and high traffic times, more air pollutants are added to the atmosphere, producing smog and industrial **toxins** (poisonous waste chemicals).

Water pollution is one of the most serious injuries to the landscape, since the Earth requires the purifying capabilities of water to clean and recycle many of Earth's natural resources. Major water pollutants include **heat** (caused by utility companies, industrial plants, and nuclear power facilities), **sewage** (as the overflow of improper waste management systems, home, industrial, and farm wastes), and **chemicals** such as phosphates (from fertilizers and detergents), heavy metals (such as mercury from several industrial processes), **PCBs** (wastes from manufacturing), and **oil spills** (from industrial accidents, well drilling rigs, and fuel tanker cleaning).

What pollutants affect the Earth's water?

Although technological advances have improved our quality of life in many ways, technological oversights have also brought about unplanned negative consequences.

SOURCES OF WATER POLLUTANTS

Pollutants are added to the hydrosphere through the activities of individuals, communities, and industrial processes. There is heightened public awareness of the harmful effects of pollution on the health, safely, and well being of humans, animals, plants, and the environment in general. To a large degree, the awareness is due to the recent discoveries of severely polluted streams, lakes, and ground water across all of North America, including the United States, Canada, and Mexico. For example, within the last two decades, increased health problems in the Love Canal area of western New York State led to the discovery of the landfill toxic-waste dumps.

For many years, industry has had the problem of what to do with the harmful wastes produced during the manufacturing of commercial products. Due to the ignorance of the effects produced by many of these toxic wastes and the economics of dumping these wastes, pollutants have been buried in the ground without the concern for, or knowledge of, ground water systems.

Over the years these poisons have seeped into ground water supplies and spread to surrounding areas, contaminating the water sources for communities, even miles away from the actual dump sites. Cleaning up these dumps has proven to be, at the least, very difficult and in many cases impossible.

Two decades ago, the lakes of the Adirondack region were known as some of the best fishing areas in New York State. However, many of the lakes (those over 2000 feet in elevation) now have little or no fish and plant life, due to the pollution of the lakes by **acid rain**.

What has killed some of the Adirondack lakes?

The wastes from industrial smoke and exhaust emissions (mostly sulfur dioxide and nitrogen oxide) pollute the air to such a great extent that precipitation has become acid. This acid rain falls into the lakes and surrounding water sheds, lowering the pH (acidity and alkalinity scale) of the water to such a degree as to make it impossible for animals and plants to survive. Aluminum concentrations toxic to fish are also released by ionized hydrogen exchange, as water percolates through the soil. Attempts to neutralize the lake water have met with only a limited success in small bodies of water. Acid rain may be one of the most serious pollution problems over the next few decades.

Other sources polluting the hydrosphere include, sewage waste from cities, towns, and communities, polluting lakes, streams, and rivers, such as the Hudson River, heated water waste from nuclear power plants and the ground water pollution around the industrial plants in the central and southern areas of New York State.

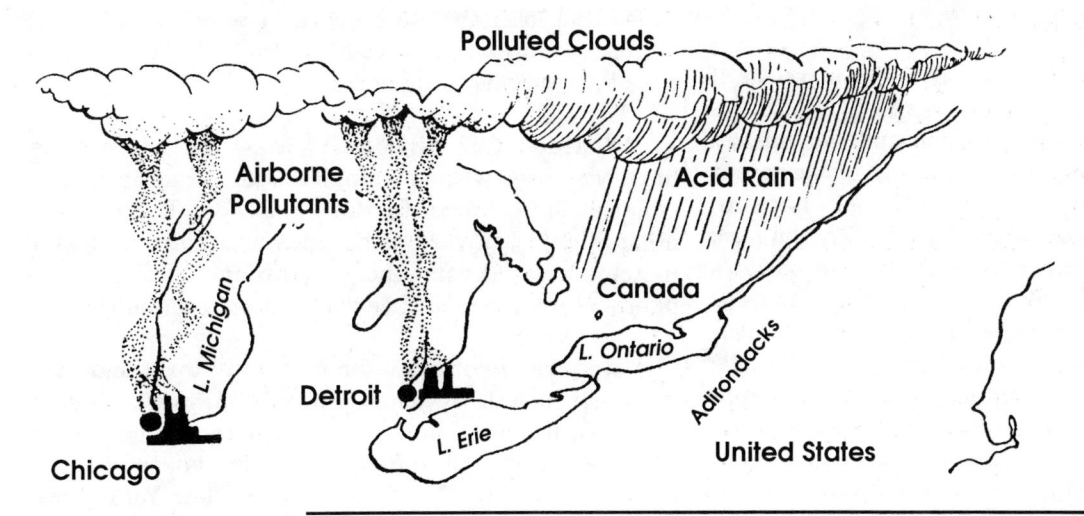

Acid Rain
The detrimental effects of acid rain know no boundaries. Airborne pollutants are transported by weather systems across international borders.

TYPES OF POLLUTANTS

Hydrospheric pollutants include dissolved and suspended materials such as organic and inorganic wastes, thermal energy effluent from industrial processes, radioactive substances, and the abnormal concentration of various organisms.

- **Organic wastes** include the sewage wastes from communities that contaminate lakes, rivers, and ground water supplies. The biggest danger from organic wastes is the spread of disease in wells and community drinking water supplies.

- **Inorganic wastes** include metal and plastic wastes, pesticide and herbicide often associated with landfills and community dumps, farming, and industry. Heavy metals, such as lead, mercury, and cadmium cause poisoning, health problems, cancer, and deformities in plants, animals, and humans.

- **Thermal wastes** often come from the cooling water of power plants and industrial complexes. The greatest effect of the heated water is the destruction of aquatic life caused by oxygen depletion in the warm water.

- **Radioactive wastes** come from a variety of sources, such as manufacturing, research, and nuclear power plants. Since many radioactive wastes remain dangerous for very long periods of times (sometimes millions of years), the major problem is how to keep them safe and store the radioactive wastes until they decay to harmless materials. The storage problem has led to other related nuclear material problems, such as safe transportation to storage areas and security in keeping the potentially dangerous materials out of the hands of terrorists.

- **Harmful organisms**, such as disease-causing microörganisms, tend to breed in toxic waste dumps and polluted ground water. Some microorganisms cause cancer, other serious diseases, and many other health problems.

The release of heated water from industrial and nuclear power plants and the increased activity of **aerobic bacteria** causes a loss of dissolved oxygen from the water. This form of pollution leads to an increase in the concentration of **anaerobic bacteria**, whose waste products are toxic. Biologic pollutants disrupt the normal living and nonliving cycles in the environment.

If the heated effluent pollutes the breeding grounds of fish, the fish may become extinct or move to other waters, thus destroying fishing and recreational water areas. This form of pollution can be observed in several areas of the Hudson River estuary.

CONCENTRATION OF POLLUTANTS

Most often the most severe pollution areas are in the vicinity of population centers. The highest concentrations of pollutants found in the Hudson River are generally located near cities, such as New York City, Poughkeepsie, and Albany. Generally, the same is true around lakes. Ground water pollution tends to be directly proportional to the size of the communities or industrial population. The larger the population, the greater the pollution problem.

LONG RANGE EFFECTS

Left uncontrolled and unchecked, the increase of the hydrospheric pollution could eventually cause the water sources of the Earth to become unfit for human use. Community groups, industry, and government have determined that this should not happen. Beginning with the *Clean Water Act of 1972*, the U.S. has spent billions of dollars to clean up the nation's water supply. In 1980, the United States Congress formed a "Superfund" for the clean up of toxic waste dumps. N.Y. State passed its own *Clean Water Act*, the *Freshwater* and *Saltwater Wetland Acts*, and the environmental quality regulations to protect the waters of New York State and the rights of citizen groups to know the impact on the environment of industrial and public projects.

C. NEGATIVE AFFECTS BY HUMANS

Both intentional and unintentional **technological oversights** have brought about unplanned consequences that have destroyed or reduced the quality of life in many landscape regions. In some cases, technological advances have produced waste products which humans do not know how to dispose of safely. In other cases, industrial wastes are disposed of carelessly and/or criminally.

In addition to **water pollution** (previously discussed), **air pollution** comes from cars, mass transportation vehicles, homes, industrial plants, and natural catastrophic events, such as forest fires and volcanic eruptions (e.g., Mt. St. Helens).

The major air pollutants include carbon monoxide (from burning), hydrocarbons (from fossil fuel burning cars, utilities, manufacturing), and particulates (dust, ash, smog). **Acid Rain** has now become a very serious result of air pollution. It develops from nitrogen oxides and sulfur dioxide, from industrial wastes and motor vehicles, combining with water droplets in the atmosphere.

One of the most serious problems with various air pollutants, including acid rain, is that weather fronts cause the pollutants to be carried to other landscape regions polluting the water and soil. The effects produced are not just localized.

Biocides, including **pesticides** and **herbicides**, have been used without complete knowledge of the possible harmful effects to the landscape. The environmental impact of various biocides has temporarily and in some cases permanently, contaminated the atmosphere, ground and surface water supplies, and the soil.

Disposal problems for the many toxic and even relatively harmless wastes of people's affluent life-style have become more and more serious in recent years. The discovery of hundreds of **toxic waste dumps**, such as the Love Canal in western New York State, have caused people and government to become concerned. Major disposal wastes often go into **landfills** and, when improperly disposed of, some wastes leak out of the dumps and into the ground water where they are transported to many different landscape regions.

Some of the disposal wastes are **solid** (such as cans, bottles, plastics, discarded appliances and cars), **chemical** (from chemical pesticides, industrial wastes and excesses, breakdown products from "thought to be" harmless wastes), and **nuclear** (radioactive particles, some remaining dangerous for millions of years).

The **human population growth** has risen rapidly over the years, in part due to the ability to control diseases and modify environments to produce greater amounts of food and provide adequate shelter.

Humans have been able to adjust the environment in ways that have made relatively uninhabitable land, habitable. However, this continued, unchecked growth in population (at an exponential rate) has far exceeded the food and shelter–producing capacities of many world ecosystems.

Agricultural lands, water sources, grazing grasslands for livestock, and forests have been so badly misused as to have resulted in the starvation and extinction of total populations of plants and animals.

High population density areas are most affected because of the concentration of the landscape pollution or the misuse of the landscape.

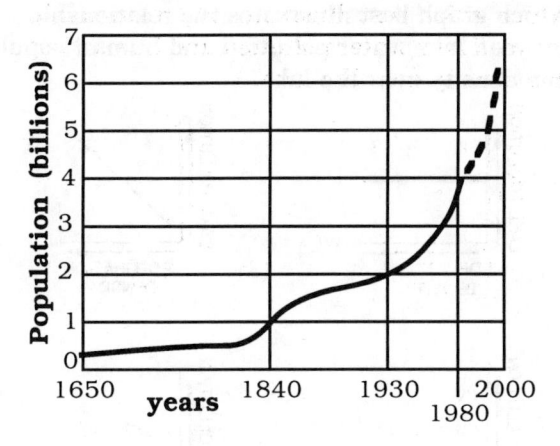

Environmental Population Growth
Overpopulation is considered by many scientists as the "ultimate threat" to the future stability of the environment.

D. POSITIVE AFFECTS BY HUMANS

Positive affects of human activities on the environment have come through an increased awareness of various ecological and landscape interactions. Humans have begun to intervene in the widespread destruction, through the efforts of individuals, community groups, and conservation clubs. There are attempts to clean up toxic waste dumps, reclaim wasted lands, and prevent continued disruption of the environment.

Population control methods have been developed to balance the rate of population growth with the environment's capabilities for food production and shelter. Most modern countries have developed laws, produced guidelines for family planning, and provided education for their populations. However, many "third world" and "developing" countries still have not been able to solve their increasing population growth rates.

The conservation of natural resources, such as reforestation efforts and cover-cropping techniques, have helped to reclaim lost ecosystems. Water conservation practices have led to the use of land that was previously unusable for agricultural needs. States, including New York State, have passed laws requiring the recycling of cans and bottles, and encouraging the conservation of materials such as paper, plastics, and fuels for energy. In addition to the obvious advantages for the environment, conservation measures have a great economic affect on individuals, employment, industry, and government.

Pollution controls put on industrial plants, public utilities, and automobiles have produced marked improvements in air and water quality. The use of unleaded gasoline in vehicles and special filters on industrial smoke stacks have reduced the incidence of smog and poor air quality alerts in cities, causing a decrease in respiratory problems for many Americans. Many towns and cities have made improvements in sewage disposal through the development and use of new sanitation techniques.

Environmental protection laws have been passed by state, local, and federal governments which regulate and guide the use of public lands and natural habitats. New York State designed a law (SEQR) which provides citizens with the opportunity to review and comment on the environmental impact of any proposed development, that may have a significant effect upon the landscape. Several state governments, including New York State, have passed freshwater and saltwater wetlands acts. These laws are designed to regulate and protect the large or unique wetland landscapes from development which would destroy them. These laws apply to both private and public owners and developers. These laws have helped to protect many species and maintain valuable landscapes for the future.

How can we live in balance with our natural environment?

The future may be better if people and technological advances continue to keep the survival of our environment as a primary consideration, when making decisions that affect the landscape. Careful planning, conservation, and education are necessary to insure a high standard of living for the future and protect the environment. As resources dwindle, alternatives must be found to meet our needs.

Research on complex issues such as global warming and ozone depletion is needed in order to understand and work to reduce the adverse effects of these problems.

Education must also be provided to instill:
- an awareness of the interdependence of all parts of the environment,
- a real concern for future generations, and
- a personal sense of responsibility and commitment for the survival of the environment.

QUESTIONS FOR UNIT 9

1. When the amounts of biologic organisms, sound, and radiation added to the environment reach a level that harms people, these factors are referred to as environmental
   1. interfaces
   2. pollutants
   3. phase changes
   4. equilibrium exchanges

2. Which is the least probable source of atmospheric pollution in heavily populated cities?
   1. human activities
   2. industrial plants
   3. natural processes
   4. automobile traffic

3. Pollution has become a serious problem chiefly because
   1. modern man's input of pollutants exceeds the rate at which nature can remove pollutants
   2. recent changes in landforms have disturbed the rate at which nature can remove pollutants
   3. there has been a gradual decrease in the amount of energy available for removal of pollutants
   4. there have been recent increases in natural environmental pollutants

4. The graph below shows the amount of ground water pollution at four different cities: A, B, C, and D. Ground water pollution tends to vary directly with population density.

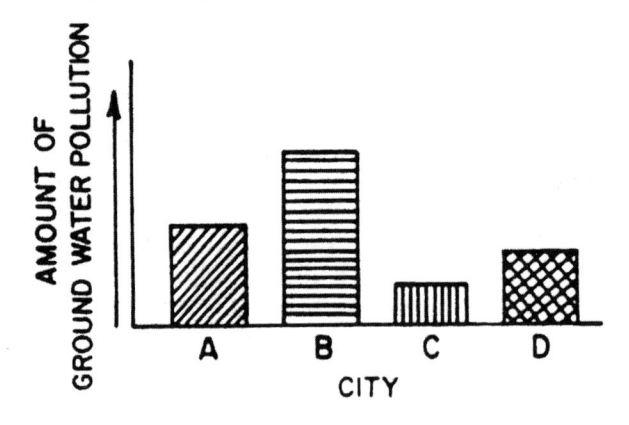

   In which order would these cities most likely be listed if ranked by population density *from highest* population *to lowest* population?

   (1) A, B, C, D
   (2) B, A, D, C
   (3) C, D, A, B
   (4) D, C, B, A

5. Which graph best illustrates the relationship between lake water pollution and human population density near the lake?

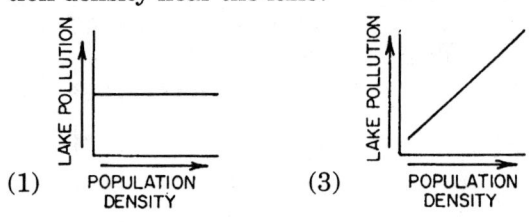

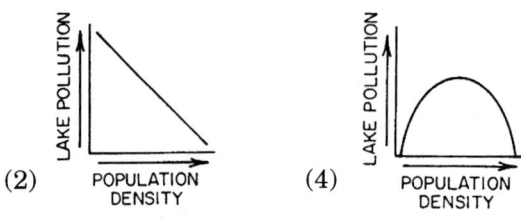

6. Air pollution from industrial processes may be most effectively reduced by
   1. growing more green plants to absorb the pollutants
   2. adding anti-irritants to the air
   3. removing waste materials before they enter the atmosphere
   4. precipitating the pollutants by heating the air

7. The map below illustrates the distribution of acid rain over the United States on a particular day. The isolines represent acidity measured in pH units.

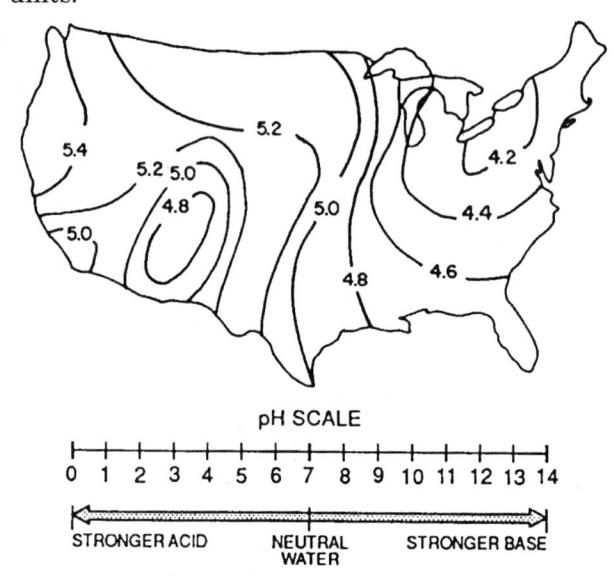

   According to the pH scale shown below the map, which region of the United States has the greatest acid rain problem?
   1. northeast
   2. northwest
   3. southeast
   4. southwest

8 Some scientists believe that high-flying airplanes and the discharge of fluorocarbons from coolants and making foam products are affecting the atmosphere. Which characteristic of the atmosphere do they believe is affected?
1   composition of the ozone layer of the stratosphere
2   wind velocity of the tropopause
3   location of continental polar highs
4   air movement in the doldrums

9 According to the diagram below, at which location would the water probably be most polluted?

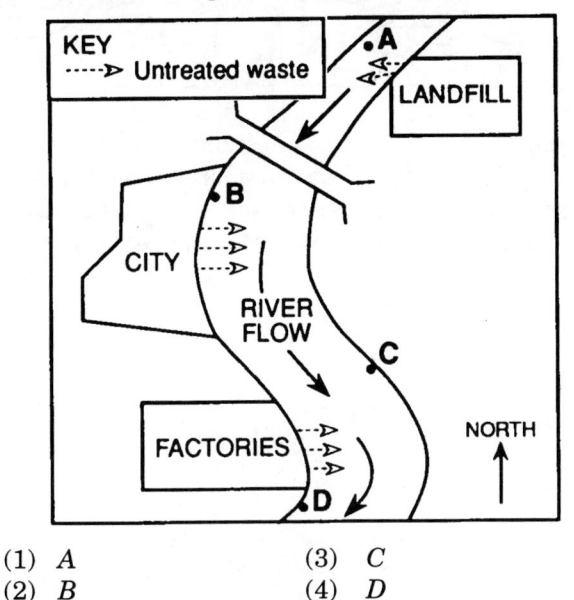

(1) A
(2) B
(3) C
(4) D

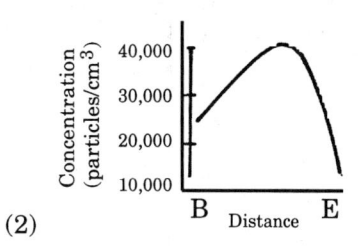

Base your answers to questions 10 through 14 on your knowledge of Earth science and the air pollution field map shown below. The isolines represent the concentration of pollutants measured in particles / cm³.

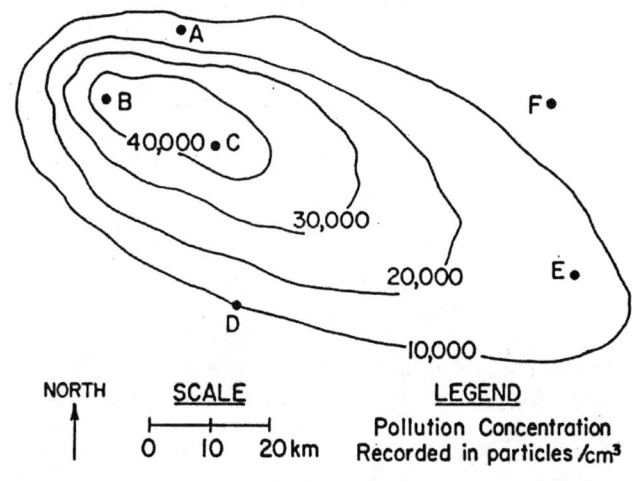

10 The major source of air pollution is most likely at point
(1) A
(2) B
(3) E
(4) D

11 The winds responsible for this air pollution pattern are most likely blowing from the
1   northeast
2   northwest
3   southeast
4   southwest

12 The most rapid increase in air pollution would be encountered when traveling between points
(1) A and B
(2) A and F
(3) C and D
(4) D and E

13 The air pollution field illustrated on the map is located in a heavily populated area. Which is the *least* probable source of the pollution?
1   human activities
2   industrial plants
3   automobile traffic
4   natural processes

14 Which graph best represents the relationship between the pollution concentration and distance from point B toward point E?

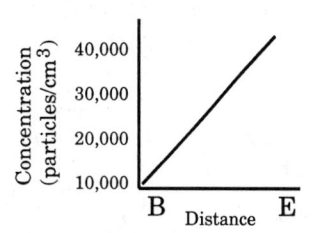

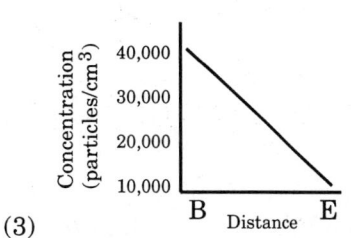

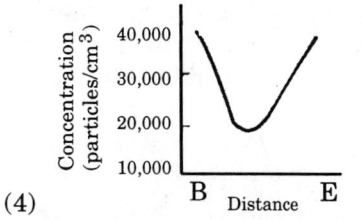

Base your answers to questions 15 through 18 on your knowledge of Earth science and on the map below which represents an area of New York State during April.

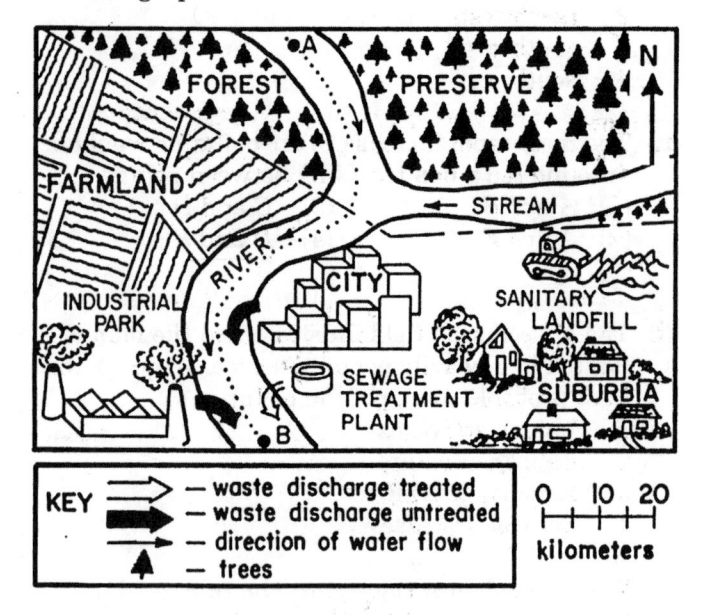

15 Which environmental region has probably been altered *least* by the activities of humans?
1 suburbia
2 sanitary landfill
3 farmland
4 forest preserve

16 Which graph best represents the probable number of bacteria (anaerobic) along line A-B in the river?

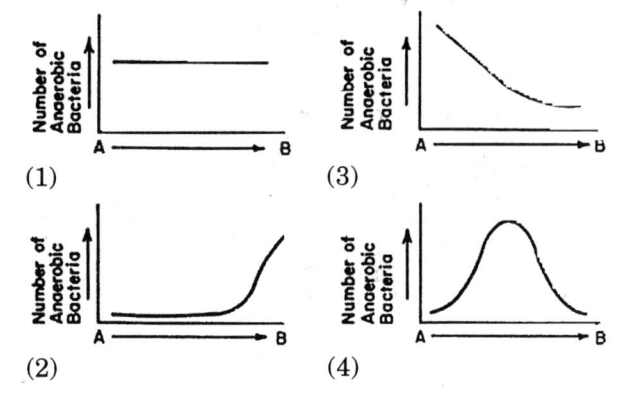

17 The people of this area defeated legislation that would have allowed the sale of a large section of the public owned forest preserve for the purpose of a second industrial park. They also passed a bond issue providing funds for an additional sewage treatment plant for the city. These actions are an indication that a majority of the voters
1 are opposed to higher taxes for any reason
2 feel technology can solve all the problems of the environment
3 are aware of the delicate balance in nature
4 feel that nature can take care of itself

18 Which diagram best illustrates the probable air pollution field of this area at an elevation of 100 meters on a windless spring afternoon?

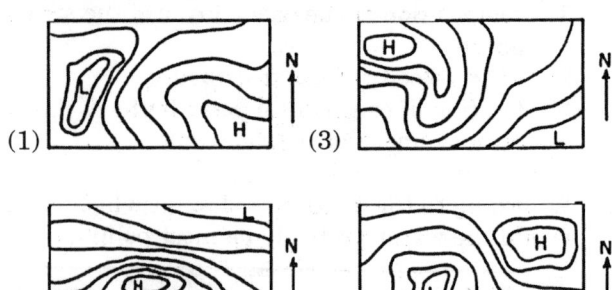

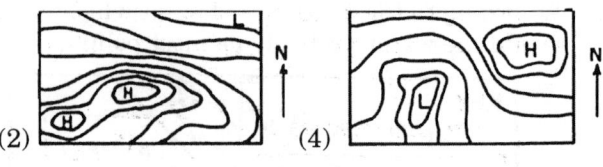

KEY:
H = High Pollution
L = Low Pollution

Base your answers to questions 19 through 23 on your knowledge of Earth science and on the diagram below which shows air, water, and noise pollution in a densely populated industrial area.

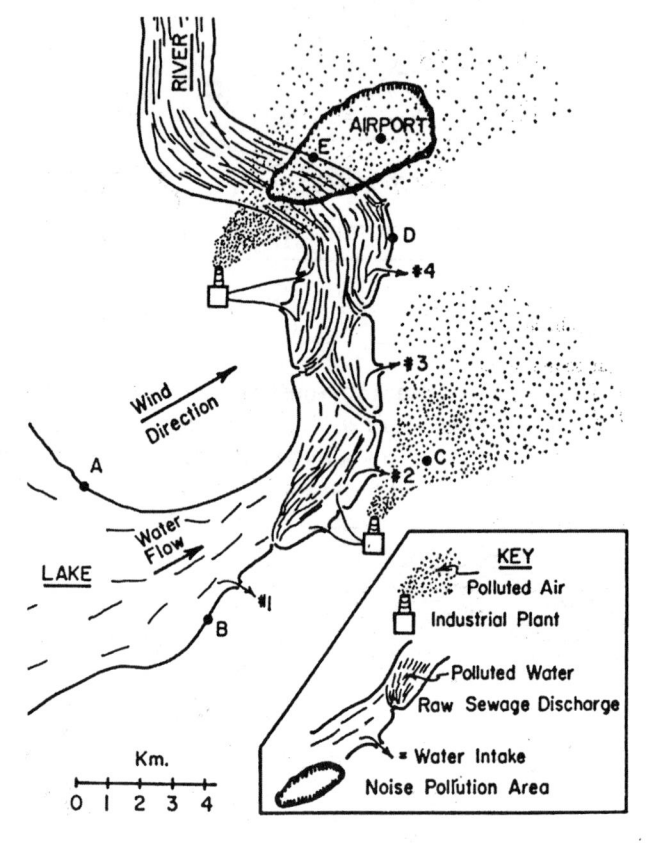

19 Air pollution would probably be greatest at which location?
(1) A
(2) B
(3) C
(4) D

20 Water pollution would probably be greatest at which location?
(1) A
(2) B
(3) C
(4) D

21 Noise pollution would be greatest at which location?
   (1) E   (3) C
   (2) B   (4) D

22 Which location is subjected to the greatest number of pollution factors?
   (1) A   (3) E
   (2) B   (4) D

23 If the water intakes supplied drinking water to the area, which intake would most likely require the most extensive purification procedures?
   (1) #1   (3) #3
   (2) #2   (4) #4

Base your answers to questions 24 and 25 on your knowledge of Earth science and on the graph below showing measurements of air pollutants as recorded in a city during a two-day period.

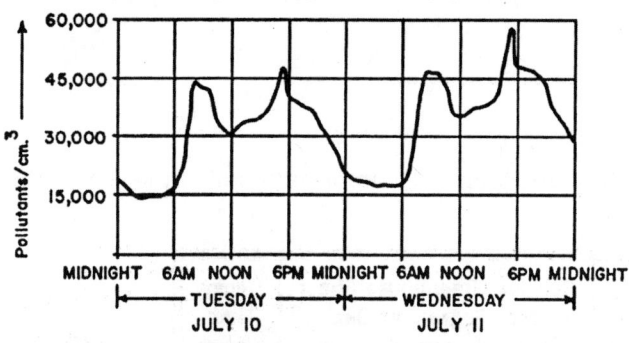

24 What is a probable cause for the increase in pollutants at 8 a.m. and 5 p.m. on the two days?
   1 change in insolation
   2 occurrence of precipitation
   3 high wind velocity
   4 heavy automobile traffic

25 On the basis of the trends indicated by the graph, at what time on Thursday, July 12, will the greatest amount of pollutants probably be observed?
   (1) 12 noon   (3) 3 a.m.
   (2) 5 p.m.   (4) 8 a.m.

Base your answers to questions 26 through 30 (found on the next page) on your knowledge of Earth science, the *Reference Tables*, and the diagrams and map. The graphs in diagram I show the sources of nitrogen and sulfur dioxide emissions in the U.S. Diagram II gives information about the acidity of Adirondack lakes. The map shows regions of the U.S. affected by acid rain.

DIAGRAM I

Sources of Nitrogen Emissions (24.5 million tons/yr.)
- Transportation 42%
- Electric Utilities 32%
- Other Combustion 26%

Sources of Sulfur Dioxide Emissions (29.7 million tons/yr.)
- Industrial 26%
- Electric Utilities 65%
- Other 9%

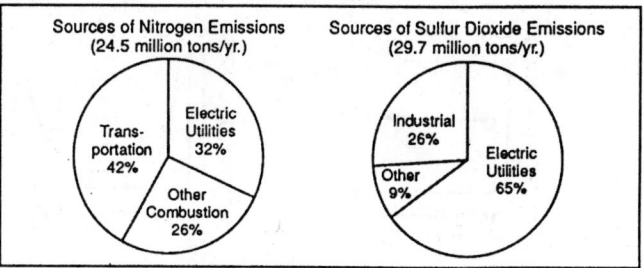

DIAGRAM II

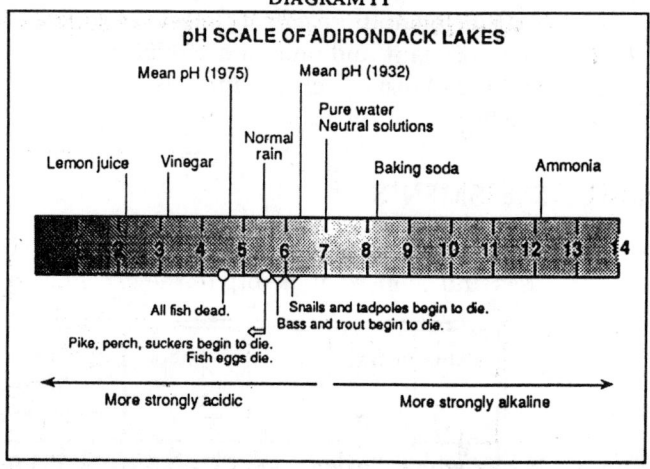

pH SCALE OF ADIRONDACK LAKES

REGIONS OF THE UNITED STATES SENSITIVE TO ACID RAIN

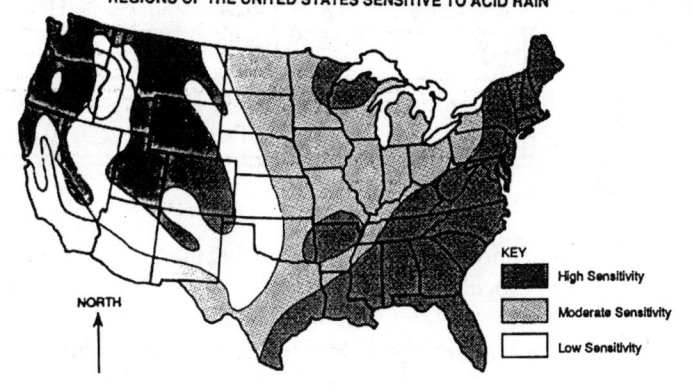

26 Which pH level of lake water would not support any fish life?
   (1) 7.0   (3) 5.0
   (2) 6.0   (4) 4.0

UNIT NINE – ENVIRONMENT – N&N©

27 Which graph best shows the acidity (pH) of Adirondack lakes since 1930?

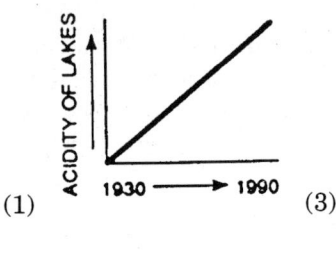

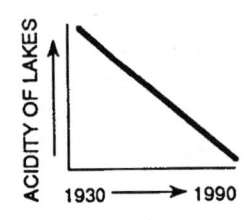

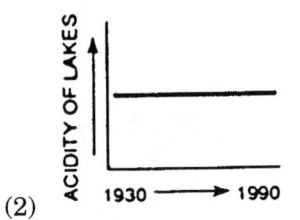

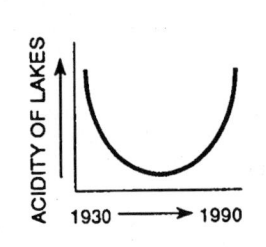

(1) graph showing acidity increasing from 1930 to 1990
(2) graph showing acidity constant from 1930 to 1990
(3) graph showing acidity decreasing from 1930 to 1990
(4) graph showing acidity decreasing then increasing from 1930 to 1990

28 The primary cause of acid rain is the
1   weathering and erosion of limestone rocks
2   decay of plant and animal organisms
3   burning of fossil fuels by humans
4   destruction of the ozone layer

29 Acid rain can best be reduced by
1   increasing the use of high-sulfur coal
2   controlling pollutants at the source
3   reducing the cost of petroleum
4   eliminating all use of nuclear energy

30 In addition to its effects on living organisms, acid rain may cause changes in the landscape by
1   decreasing chemical weathering due to an increase in destruction of vegetation
2   decreasing physical weathering due to less frost action
3   increasing the breakdown of rock material due to an increase in chemical weathering
4   increasing physical weathering of rock material due to an increase in the circulation of ground water

SKILL ASSESSMENTS

Base your answers to questions 1 and 2 on the table below which provides information about several different gases and their relationship to the greenhouse effect.

Name of Gas	Symbol	Concentration in Troposphere (parts per billion)	Relative Greenhouse Effect per Kilogram of Gas (in equal concentrations)	Decay Time (years)
Carbon dioxide	CO_2	353,000	1	120
Methane	CH_4	1,700	70	10
Ozone	O_3	10-50	1,800	0.1
Chlorofluorocarbon	CFC-11 CFC-12	0.28 0.48	4,000 6,000	6.5 120

1   In one or more complete sentences, state *one* reason that a scientist would recommend restricting the emission of chlorofluorocarbon CFC-12 into the atmosphere, based on the information in the table.

2   Which of the actions below would contribute more to the greenhouse effect? State a reason for your answer.

   a   releasing 1 kilogram of methane directly into the atmosphere

   or

   b   burning 1 kilogram of methane, resulting in the release of about 3 kilograms of carbon dioxide into the atmosphere

Base your answers to questions 3 through 4 on the diagram below, the following information and your knowledge of Earth science.

3   On an imaginary island with much plant-life and no predators, a doe and buck parent four fawns – two males (bucks) and two females (doe). The four offspring pair off into two new sets of parents, each having four offspring themselves. This population pattern continues through ten generations. Assuming no loss of

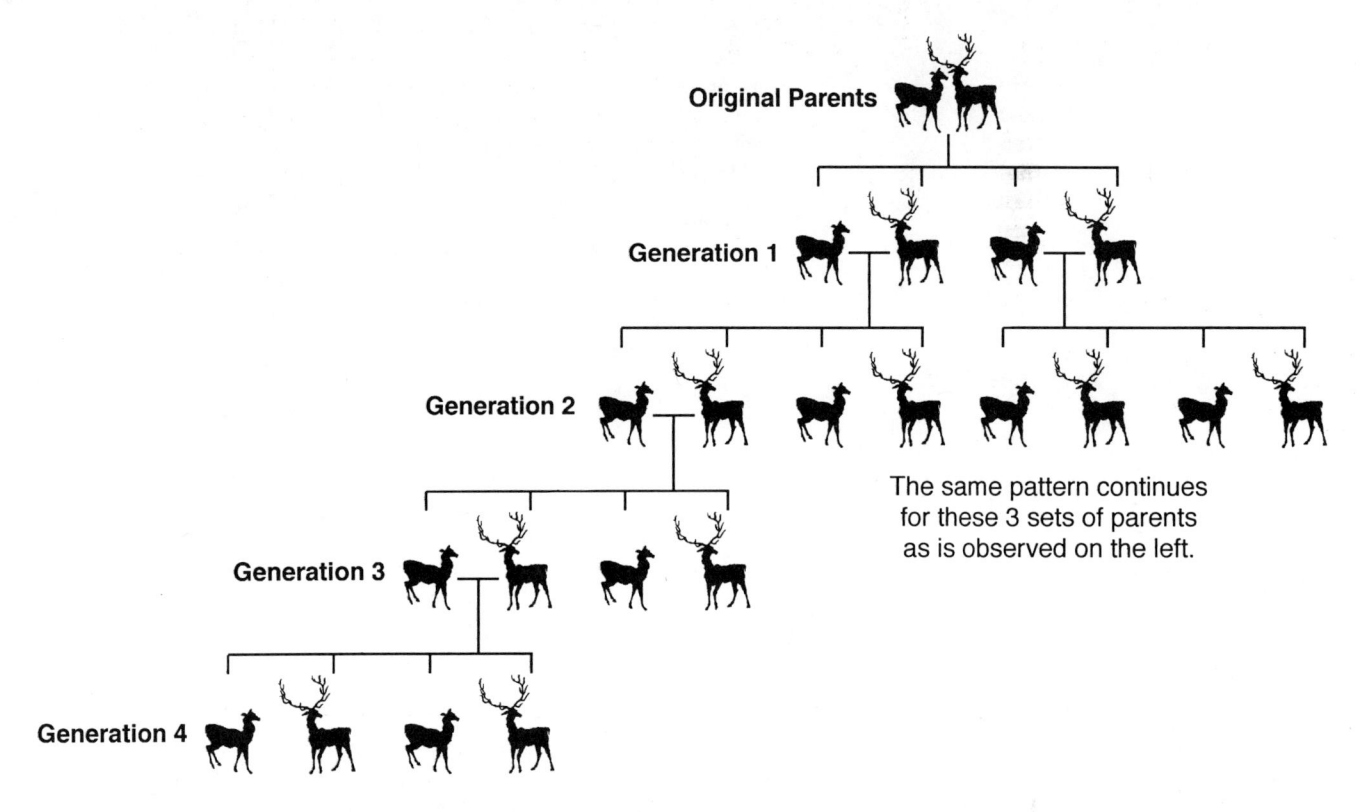

Original Parents

Generation 1

Generation 2

The same pattern continues for these 3 sets of parents as is observed on the left.

Generation 3

Generation 4

life or infertility in the deer population, what is the total number of fawns (the 10th generation) that will be born to the entire group of 9th generation parents?

4   In one or more complete sentences, describe at least *two* kinds of stress that this population growth could place on the island's environment.

Information for question 5: Like the deer population in the previous questions, the human population of the Earth also grows exponentially. Although it took two and a half million years for the Earth's human population to reach 5.4 billion, it will only take 40 years to double this number.

5   List three ways this human population growth will impact the environment of the Earth in the future. Be specific. Identify each impact as a "negative human effect" or a "positive human effect."

Index & Glossary
Modified Earth Science Program

Ablating (99) – the process of losing ice and snow from a glacier.

Absolute Age, or Date (107, 111) – dates in Earth's history arrived at by dating radioactive rocks and measured without reference to any other event.

Absorption (150) – opposite of radiation; the concentration of a substance through (permeating) a surface.

Absorption of Solar Energy (150) – concentration of the electromagnetic radiation of the Sun through the Earth's atmosphere, including ultraviolet, infrared, visible wavelengths.

Abyssal Plain (92) – flat ocean basin.

Acid Rain (184, 186) – weak acid precipitation from industrial waste gases.

Actual Evapotranspiration (152) – see water budget; due to evaporation and transpiration, the actual amount of water lost from a given area over a specific amount of time, expressed in millimeters of water.

Adiabatic Changes (140) – changes that occur without any loss or gain of energy, such as in heat changes.

Aerobic Bacteria (185) – bacteria that use oxygen in energy production.

Aerosols (151) – the mixtures of small particles suspended in a liquid or gas; such as fog, smog, muddy water, etc.

Air Mass (126) – large area of air within the lower atmosphere having generally the same temperature and humidity at any given level.

Air Masses (126) – arctic (A), polar (P), tropical (T), maritime (m), continental (c)

Air Pressure (122) – force exerted due to the weight of air; see weather.

Alpine Glacier (99) – glacier found in a mountain valley.

Altitude (19) – a celestial object's angular distance above the horizon; the vertical distance of one point above the Earth's surface at sea level.

Anaerobic Bacteria (185) – bacteria that do not use oxygen in their production of energy, but usually carry on a form of fermentation.

Anemometer (122) – instrument with rotating cups used to determine wind speed.

Angle of Insolation (148) – angle that the rays of the Sun hit the surface of the Earth; decreases with an increase in latitude; increases as the angle approaches perpendicular; varies with the time of day and day of the year.

Angular Diameter (165) – see Apparent Planetary Diameter

Annular Drainage (81) – ring-shaped drainage pattern often found on eroded domes.

Annular Eclipse (166) – eclipse that occurs when the Moon is too far from the Earth for the umbra of its shadow to reach the Earth.

Anticyclone (127) – HIGH, the opposite of a cyclone (LOW), having high pressure and clockwise winds in the northern hemisphere.

Aphelion (162, 163) – point on the Earth's orbit when the Earth is the farthest from the Sun (152 million kilometers); occurring on July 1st; point of Earth's slowest orbital speed.

Apparent Daily Motion (161) – perceived movement of celestial objects as seen from the Earth, circular, constant, daily, and cyclic.

Apparent Planetary Diameter (165) – not actual measured diameter, but the perceived diameter of a planet, changing according to the distance from observer.

Apparent Solar Day (161) – time (24+ hours) required for the Sun to cross a given meridian twice in succession.

Arc (148, 161) – portion of a circle through which the celestial objects rise in the east and set in west, from the observer's point of view on Earth's surface.

Arête (99) – sharp ridge formed between two Alpine or valley glaciers.

Arid Climate (152) – dry (deficit moisture) area of land, having greater potential evapotranspiration than precipitation for a majority of months in a year.

Asthenosphere (64) – the plastic-like part of the mantle beneath the lithosphere where convection currents are thought to occur.

Astronomical Unit (164) – mean distance of the Earth from the Sun; approximately 150,000,000 kilometers.

Atmosphere (21, 121, 140) – thin shell of gases surrounding the Earth, separated (stratified) into layers each having distinct characteristics; troposphere, stratosphere, mesosphere, and thermosphere.

Atmosphere, Selected Properties of Earth's (209) – Reference Table.

Atmospheric Transparency (151) – condition under which the Earth's atmosphere scatters, reflects, or absorbs the Sun's rays.

Atmospheric Variables (121) – weather changes in the atmosphere, including temperature, winds, moisture, air pressure, etc.

Axis, Earth's (19) – center of Earth between the North and South Poles, about which the Earth rotates.

Backwash (93) – water from breaking waves that gravity pulls back from the beach into the ocean.

Banding (45) – pattern of layers caused by differences in the crystal alignments of various minerals in many metamorphic rocks; type of foliating.

Barometer (122) – instrument used to measure air pressure.

Base Flow (147) – water moving from the local water table into streams.

Baymouth Bar (94) – sandbar that forms across the mouth of a bay.

Beach (94) – shoreline region of deposited particles; may contain sand, gravel, pebble, cobble and/or boulders.

Bedrock (79, 80, 108) – solid rock underneath soil or exposed rock at Earth's surface.

Bedrock, Generalized Geology in NYS (197) – Reference Table.

Bench Mark (53) – marker in the ground indicating the exact elevation of that location above sea level.

"Big Bang" (179) – theory that the universe began with a gigantic explosion.

Biocide (186) – substance harmful to plants and animals.

Biological Pollutant (185) – biologic organisms in concentrations that are harmful.

Breaker (93) – when the crest of a wave falls over onto the shore.

Calorie (139) – heat quantity unit; amount of energy required to raise the temperature of 1 gram of water through 1°C.; a large calorie (kilocalorie) is 1000 calories.

Capillarity (146) – upward movement of water against the force of gravity in a narrow space, such as a tube, plant vessel, or fine sand particles.

Capillary Water (147) – water that is found in the small spaces between fine grains of rock, sand, clay, or soil.

Carbon-14 Dating (111) – process for determining the absolute age of a fossil or other material containing the element Carbon-14 (a radioactive isotope of Carbon-12 with a half-life of 5,600 years).

Carbonate (43) a group made up of one or more metals combined with a carbon and three oxygen atoms (CO_3).

Carbonation (72) – the process of carbonic acid reaction with other materials.

Celestial Object (161) – "heavenly bodies," any object observed in the area above the Earth's atmosphere, including the Sun, Moon, stars, comets, planets, etc.

Celsius (149) – a temperature scale that registers the freezing point of water as 0° and the boiling point as 100° under normal atmospheric pressure.

Cementation (34) – process in some sedimentary rocks, in which various sized sediments are cemented (glued) together by the action of precipitated minerals, resembles man-made concrete.

Cenozoic Era (110) – most recent Earth history era representing 2 or 3% of the geologic time scale including modern plants, animals, and humans.

Center of Accumulation (99) – large area or center of accumulation for snow and ice of a continental glacier.

Centimeter Scale (195) – Reference Table.

Centrifugal Effect (20) – produced by Earth's rotation, tends to maintain movement away from the center or axis of the Earth.

Change of Phase (138) – see phase change.

Chemical Composition of Earth's Crust (205) – Reference Table.

Chemical Pollutants (184) – toxic materials such as phosphates and heavy metals that reach harmful levels in the environment.

Chemical Properties (31) – used for identification of minerals.

Chemical Sedimentary Rock (45) – sedimentary rock formed from dissolved mineral material which settles or precipitates out of water.

Chemical Weathering (71) – process that alters the chemical characteristics of rocks and minerals, such as oxidation and hydration.

Circumference of Earth (19, 21) – formula; see Earth Dimensions.

Cirque (99) – bowl-shaped depression gouged out of the side of a mountain by a glacier.

Cirrus Clouds (124) – high-altitude cloud composed of narrow bands or patches of thin, generally white, fleecy parts.

Classification System (7) – organized data, based on observable properties.

Clastic (44) – see fragmented sedimentary rocks.

Clean Air Act (186)

Clean Water Act (186)

Cleavage (32) – the tendency of a mineral to break along one or more smooth planes or surfaces.

Climate (72, 78, 152) – discounting local weather changes, the average or normal weather of a particular large Earth area.

Cloud (124) – type of aerosol in the atmosphere composed of suspended small water droplets and/or ice crystals.

Cloud Base (124) – the height of the bottom of the lowest clouds.

Cluster (167) – a group of galaxies.

Cold Front (127) – the interface, leading edge, of an air mass which has cooler temperatures than the preceding warmer air mass, usually associated with moisture and precipitation.

Colloid (75) – small particles, from 10^{-4} to 10^{-6} millimeters across, which tend to remain in solution for long periods of time.

Color (31) – property used for the identification of minerals.

Compression (34) – process involved in the production of some sedimentary and metamorphic rocks, in which the weight of deposited sediments, water, and/or Earth movements presses underlying sediments together.

Compressional Wave (Primary, P-wave) (61) – seismic wave (action is like the expansion and contraction of a spring) which travels at a speed of thousands of kilometers per hour, through solids, liquids, or gases.

Conclusion (7) – see inferences.

Condensation (11, 123, 139) – change in state from vapor (gas) to liquid, the loss of water vapor from warmer air onto a cooler surface, such as the condensation (fog) on an ice water glass.

Condensation Surface, or Nucleus (124) – solid surface onto which water vapor may condense to form liquid droplets, in the formation of clouds – water condenses on dust or other particles, in the formation of early morning dew – water condenses on any solid and cooler surface.

Conduction (137) – transfer of energy, usually heat, by contact from one atom to another atom within a liquid, gas, or solid.

Conglomerates (34) – rock consisting of unsorted pebbles and gravel embedded in cement.

Conservation of Natural Resources (187) – controlled use and systematic protection of natural resources, such as forests, soil, and water systems.

Constellation (161) – a group of stars that make up a recognizable pattern in the sky.

Constructional Forces (77) – see Uplifting (forces).

Contact Metamorphism (46, 107) – the process of rock changing due to contact with hot magma or lava.

Continental Climate (153) – average weather of a land mass, little affected by large bodies of water, characterized by extremes in temperature.

Continental Crust (62) – thick, low density upper part of the lithosphere that makes up the blocks (land mass) of a continent.

Continental Drift (53, 64) – theory, backed by continual evidences, that continents (Earth plates) are now, as well as in the past, shifting positions.

Continental Glacier (99) – a massive ice sheet covering a large area such as the Antarctic ice sheet.

Continental Margin (91) – the area along the edge of the continents made of the continental shelf, continental slope and continental rise.

Continental Polar Air Mass (126) – cP, usually a cold air mass originating in the land polar regions, such as in the northern most regions of Canada.

Continental Rise (92) – gently sloping surface of sediments at the base of the continental slope.

Continental Shelf (91) – the shallow submerged land sloping gently out from the shoreline.

Continental Slope (91) – the slope that leads from the continental shelf down to the deeper ocean

Continental Tropical Air Mass (126) – cT, warm air mass originating over warm land regions.

Contour Interval (24) – the difference in elevation between two consecutive contour lines.

Contour Line (24) – type of isoline on a topographic (contour) map, representing equal points of elevation.

Contour Map (24) – a topographic map, used as a model indicating elevations of the Earth's surface with the use of contour lines and symbols.

Convection (137) – transfer of energy, due to differences in substances' densities, in gases and liquids.

Convection Cell (current) (64, 138, 140) – circulatory motion in which heat energy is transferred from one place to another, due to density differences.

Convergence (140) – interfacing of air masses at the Earth's surface, in upper regions of the troposphere, making "air streams" or vertical currents.

Convergent Plate Boundary (54) – boundary between two colliding plates.

Coordinate System (22) – system or group of defined lines (may be imaginary lines) used for the determination or location of point(s) on a surface (such as graph, Longitude and Latitude).

Coriolis Effect (123, 140, 175-176) – observed path of an object (or fluid) at the surface of the Earth undergoing a predictable horizontal deflection; rightward deflection in the northern hemisphere and leftward in the southern hemisphere.

Correlation (108) – match up of rock ages and geologic events.

Counter Currents (93) – current that flows in an opposite direction to the flow of another current.

Crest (93) – the top of a wave

Crust (51, 62) – layer of granite or basalt rock forming the outer part of the Earth's lithosphere.

Crustal Plates (51) – large pieces of the lithosphere that move in relation to each other.

Crystal (32, 44) – Earth material having a repeating pattern of characteristic shapes, due to a material's internal atomic structure, such as a cube (halite) and a tetrahedron (silicate).

Crystalline Structure (32, 33, 43) – definite atomic pattern within a mineral (see Crystal).

Crystallization (31, 33) – formation of solid crystals into a rock, such as igneous rock, when the crystals separate from a magma solution.

Cumulus Clouds (124) – dense, white, fluffy, flat-based cloud with a multiple rounded top and a well-defined outline, usually formed by the ascent of thermally unstable air masses.

Cumuliform Clouds (124) – having the shape of cumulus clouds; see Cumulus Clouds.

Cumulonimbus Clouds (124) – extremely dense, vertically developed cumulus with a relatively hazy outline and a glaciated top extending to great heights, usually producing heavy rains, thunderstorms, or hailstorms.

Currents (92, 93) – movement or flow of "rivers" of water in a direction; see the specific type: warm, cold, counter-, density, surface, gyres, salinity, and turbidity currents.

Cycles (11) – usually an orderly manner in which events in time and space repeat.

Cyclic Change (10, 11) – a predictable change that occurs in a repeating pattern.

Cyclic Energy Transformation (164) – alternating changes of energy from kinetic to potential and potential to kinetic; as seen in the Earth's changes of energy and orbital speed around the Sun.

Cyclone (127) – also called LOW; low pressure air mass with counterclockwise winds in the northern hemisphere, including violent weather, such as tornadoes and hurricanes.

Daily Motion (175) – apparent motion in an arc path across the Earth's sky from east to west during each 24 hour period.

Daily Temperatures (149) – see Temperature and Insolation.

Decay, Radioactive (107, 111) – transformation or break down into a stable "daughter" element.

Deficit (152) – local condition when the actual evapotranspiration is not equal to the potential evapotranspiration, due to insufficient precipitation and water soil storage; see Drought.

Degree Metamorphism (46) – different metamorphic rocks formed from the same parent rock due to different pressure and temperature environments.

Dendritic Drainage (80) – random pattern of streams associated with plains and plateaus

Density (8, 9, 32, 76) – formula; mass of a material divided by the material's volume.

Density Current (93) – subsurface ocean current with a density greater than that of the surface water.

Density Variables (9) – characteristics such as temperature and pressure that causes changes in density and phase changes.

Deposition (76, 124) – the process of water vapor changing directly into ice crystals.

Deposition (75) – settling out of solution of sediments and minerals in an erosional system.

Destructional Forces (77) – see Leveling Forces.

Dew (124) – condensation occurring on the Earth's surface.

Dew Point Temperature (121-125) – temperature at which water vapor present in the air saturates air and begins to condense; dew forms.

Dew Point Temperatures (122, 206) – Reference Table.

Diastrophism (77) – process of deformation by which the major features of the Earth's crust, including continents, mountains, ocean beds, folds, and faults, are formed.

Dimensional Quantities (8) – time, length, or mass.

Direct (Vertical) Rays (148, 163) – rays of solar energy hitting the surface of the Earth at an angle of 90°, also called perpendicular rays.

Direct Relationship (10) – when both variables change in the same direction; see graphing.

Displaced Fossils (53) – example: marine fossils found in layers of sedimentary rock in mountains.

Displaced Sediments (73) – rock and mineral particles that are removed from their source and transported by water/wind to another place.

Distorted Structure (45) – resulting rock formations caused by Earth forces, such as heat and pressure, which bend, break, and fold rock layers.

Divergence (140) – following air mass convergence, ascending or descending, the spreading apart of air currents.

Divergent Plate Boundary (54) – boundary between two plates that are moving apart.

Doppler Shift (178) – apparent change in wavelength of light caused by relative motion of the source.

Drainage Patterns (80) – dendritic (random drainage over bedrock), trellised (parallel folds and faults), and radial (volcanic cones, young domes).

Drought (152) – a prolonged period of deficit weather conditions.

Drumlin (79, 100) – glacial hill shaped like the back of a spoon.

Dry-bulb Thermometer (121) – see Psychrometer.

Duration of Insolation (148) – length of time that the Sun's rays are received at a particular location on the Earth's surface; varies with latitude and season; Earth surface temperature is directly proportional to the duration of insolation.

Dynamic Equilibrium (12, 75) – a balance between two opposing processes going on at the same rate in a system, such as erosion and deposition and evaporation and condensation; refers to a landscape as well.

Earth (178) – third planet from the Sun, having a sidereal period of revolution about the sun of 365.26 days at a mean distance of approximately 149 million kilometers (92.96 million miles), an axial rotation period of 23 hours 56.07 minutes, an average radius of 6,374 kilometers (3,959 miles).

Earth Axis (148) – see Axis.

Earth Dimensions (19, 20) – includes the Earth's circumference, radius, diameter, volume, and surface area.

Earth's Interior, Inferred Properties of (63, 204) – Reference Table.

Earth Positions (22) see Coordinate System.

Earth Shape (19) – oblate spheroid with greater diameter at the equator than through the poles.

Earthquake (51, 61) – sudden trembling or shaking of the ground, usually caused by a shifting of rock layers along a fault or fissure under the Earth's surface.

Earthquake *P-wave* and *S-wave* Travel Time (11, 62, 205) – Reference Table.

Eccentricity of Ellipse (163) – formula; degree of the "out of roundness" of the ellipse, as determined by the distance between the two foci divided by the length of the major axis of the ellipse.

Eclipse (166) – shadows on Earth caused by the position of the Sun or Moon.

Electromagnetic Energy (137) – any energy radiated in transverse wave form, such as radio, sound, light, X-rays, etc.

Electromagnetic Spectrum (137, 147, 178) – wide range of wavelengths from lower frequencies such as radio waves short wave, AM, FM, TV, and radar to mid frequencies such as infrared rays, visible light, and ultraviolet, to high frequencies such as X-rays and gamma rays.

Electromagnetic Spectrum (208) – Reference Table.

Element (31) substance composed of atoms having an identical number of protons in each nucleus; cannot be reduced to simpler substances by normal chemical means.

Ellipse (163) – flattened circular path, having two foci (fixed radii); typical of the orbits of most all celestial objects and the Earth.

Energy (11, 46, 128, 137-140, 151, 164) – ability to do work; forms: kinetic and potential.

Energy Flow (11) – energy moving from one place in the environment to another.

Environment (10, 36, 77, 183) – the climatic and ecologic surroundings in which we live.

Environmental Equilibrium, or Balance (12, 183) – general stable and balanced state of the environment, changeable easily on a small scale.

Eon (109) – longest division of geologic time, containing two or more eras.

Epicenter (52, 62) – point on Earth's surface that is directly above the focus of an earthquake.

Epicycle (175) – smaller secondary orbit off a main orbit.

Equations and Proportions (210) – Reference Table.

Equator, Earth's (19) – center circumference of Earth, equally dividing Northern and Southern Hemispheres, 0° latitude.

Equilibrium, State of (12) – the tendency to remain unchanged.

Equinox (149, 163) – time at which the Sun's rays are directly perpendicular to the Earth's equator; equal day and night on the Earth; usually March 21st and September 23rd.

Era (109) – longest division of geologic time, made up of one or more periods.

Erosion (73-75, 78) – altering of the Earth's surface by the removing of rock, soil, and mineral pieces from one location to another by the action of water (liquid or solid) or wind.

Erosional-Depositional System (75) – the system involving the opposing processes of erosion and deposition, involving energy relationships and dynamic equilibrium.

Erratic (99) – large boulder left by a glacier.

Error (8) – difference between the actual and observed measurements.

Escarpment (80) – steep slope separating two gently sloping surfaces.

Esker (79) – long, narrow ridge of coarse gravel deposited by a stream flowing in or under a decaying glacial ice sheet.

Evaporation (11, 35, 138-139) – change of phase from liquid to vapor (gas) occurring at the surface of that liquid.

Evaporite (35) – form of sedimentary rock, caused by the precipitation of minerals from evaporating water, such as limestone, dolostone, gypsum, and salt.

Evapotranspiration (138, 152) – combination of both processes evaporation and transpiration.

Event (10) – the occurrence of a change in the environment.

Extinct (112) – no longer existing or living: an extinct species; no longer burning or active: an extinct volcano.

Extrusion (36, 44, 107) – mass of hardened lava at the Earth's surface, a type of igneous rock formation.

Extrusive Igneous Rock (44) – igneous rock that forms by the hardening of magma (hot liquid rock beneath the Earth's surface) after reaching the surface of the Earth.

Eye (128) – the central area of calm in a hurricane.

Fault (52-53, 77, 80, 108) – crack in the crust of the Earth along which rocks have moved.

Felsic (33) – Light-colored igneous rocks that are high in feldspar and quartz

Fetch (93) – expanse of open water over which the wind blows.

Field (23) – region of space which contains a measurable quantity at every point.

Field Values (125) – information shown on weather maps, such as temperature and pressure.

Finger Lake (100) – long narrow lake formed when glacial sediment dams up a river valley.

Flood (147) – overflowing of water onto land that is normally dry.

Focus (earthquake) (52) – point of origin of an earthquake.

Focus (geometric definition) (163) (plural is foci) – fixed point from which a radius of 360° is a circle or sphere; two fixed points (foci) are required to produce an ellipse or oblate spheroid.

Folded Strata (53, 77, 80, 108) – bend in the rock strata produced during the mountain-building process.

Foliation (35) – alignment of mineral flakes or bands in metamorphic rocks.

Fossil (35, 45, 53, 108-109, 112) – remains or traces of a once-living organism in sedimentary rock.

Fossil fuel (46) – fuels such as coal, oil, and natural gas which formed from organic matter in the ancient past.

Foucault Pendulum (175-176) – freely swinging pendulum, which when allowed to swing without interference appears to change direction in a predictable manner due to the Earth's rotation.

Fracture (32) – the way a mineral breaks if it does not have cleavage.

Fragmental Sedimentary Rock (45) – rocks formed by the compaction and cementation of sediments.

Friction (74, 93) – force found at the contact of two surfaces that offers resistance to motion, often producing heat or another form of energy.

Frictional Drag (92) – in the atmosphere, the slowing down of wind at the interface of the Earth surface caused by friction and the Coriolis effect.

Front (127) – the interface between two different air masses, such as the point of contact between warm and cold air masses.

Frost (124) – deposit of minute ice crystals formed when water vapor condenses at a temperature below freezing.

Galaxy (167) – a huge system of billions of stars.

Geocentric Model (175) – early attempt to explain the motions of celestial objects using the Earth as the stationary center for the orbiting celestial objects.

Geographic Poles (20) – actual axis points on the Earth, north and south, on which the Earth rotates.

Geologic History of NYS (112, 202-203) – Reference Table.
Geologic Time Scale (109) – geological periods, a scale of time that serves as a reference for correlating various events in the history of the Earth, divided into three main groups: eras, periods, and epochs, based on the study of rock history.
Geometry of a Sphere, Principle of (19) – as a sphere is rotated, the angle of a fixed point outside of the sphere as compared to the surface of that sphere, is equal to the angle of the sphere's rotation.
Glacier (75, 79, 99) – huge mass of ice slowly flowing over a land mass, formed from compacted snow in an area where snow accumulation exceeds melting and sublimation.
Glacial Features (79) – see individual features.
Glacial Ice (99) – usually glacial snow that has accumulated, compacted, and recrystallized.
Glacial Till (99) – unsorted sediment pushed, carried, or dragged by a glacier.
Glacial Valley (79) – rounded, often a river valley with characteristic U-shape.
Global Positioning System, or GPS (23) – modern navigation system capable of 3-D, precise latitude, longitude, and altitude with use of satellites.
Global Warming (152, 178) – the theory that the atmosphere of the Earth is becoming warmer caused primarily by increasing amounts of carbon dioxide and other greenhouse gases in the air.
Globe (19) – body with the shape of a sphere, especially a representation of the Earth in the form of a hollow ball.
Graded Bedding (76) – layering of sediment in a fashion where heavier and/or larger particles are on the bottom and lighter and/or smaller particles are on top, in decreasing size.
Gradient (24, 125) – formula; expression of the degree of change of a field quantity from place to place; may also be referred to as an average slope.
Grams (8) – metric unit of mass equal to one thousandth (10^{-3}) of a kilogram.
Graphing (10) – types of graphs illustrated.
Gravitation, Law of (20) – gravitational force is proportional to the square of the distance between the two centers of attracted objects.
Gravitational Force, or Gravity (20, 164, 165) – attraction between any two objects in the universe.
Gravity (20, 73, 74, 164) – natural phenomenon of attraction between massive bodies.
Greenhouse Effect (150-152, 178) – process which increases the atmospheric temperature of the Earth, due to the transmission of short wave radiation through the atmosphere, absorption and conversion to long wave radiation at the Earth's surface.
Greenwich Mean Time (23) – international reference time at Greenwich, England; the Prime Meridian with a longitude of 0°.
Groin (94) – walls built perpendicular to the shoreline to prevent longshore currents from removing sediment.
Ground Moraine (79) – unsorted glacial sediment covering the ground surface deposited directly by glacial ice.
Ground Water (145-6) – water that is found under the Earth's surface as the result of infiltration and storage.
Gyre (92) – large, circular pattern of surface currents found in the ocean.

Half-life (111) – time taken for half of a radioactive material to decay to its stable decay product; time for half of the atoms present to disintegrate.
Hardness (32) – the resistance of a mineral to being scratched.
Harmonic Law of Planetary Motion (163) – by Kepler states that a planet's period is the length of time required for the planet to orbit (revolve) once around the Sun.
Headland (94) – an area of high land jutting out into the ocean.
Heat of Fusion (139) – amount of latent heat involved in melting or freezing.
Heat of Vaporization (139) – amount of latent heat involved in evaporation or condensation.
Heliocentric Model (175-6) – a modern attempt to explain the motions of celestial objects using the Sun as the stationary center for the orbiting celestial objects and fixed star positions.
Herbicide (186) – chemical substance used to destroy or inhibit the growth of plants, especially weeds.
HIGH (127) – high pressure air mass, clockwise rotation in the northern hemisphere, more dense than a LOW; see Anticyclone.
High Noon (148) – 12:00 (noon) point in time when the Sun is at its highest altitude (zenith) on the observer's meridian.
Hook (94) – sandbar with a curved end.
Horizontal Displacement (53) – sideways shift of the Earth's surface along a transform fault or crack in the crust; see Faulting.
Horizontal Sorting (76) – sorting of sediments in a stream by decreasing velocity, larger particles first laid down, followed by smaller and smaller particles down stream.
Horn (99) – a pyramid-shaped mountain formed by glaciers; named after the Matterhorn in Switzerland.
Hot Spots (64) – high heat flows on Earth's surface.
Hubble, Edwin (178) – American astronomer who discovered (1929) that the velocities of nebulae increase with distance.
Hubble Space Telescope, or HST (178) – American satellite-based telescope for deep space study.
Humid Climate (152) – an area where the precipitation is greater than the potential evaporation for the majority of months in a year.
Hurricane (123, 128) – tropical cyclone with winds in excess of 119 km/hr (75mph).
Hydration (72) – the chemical reaction of water with other materials.
Hydrologic Cycle (145) – See Water Cycle
Hydrosphere (21) – thin layer of water which covers a majority (71%) of the Earth's surface.

Hypothesis (7) – tentative explanation that accounts for a set of facts and can be tested by further investigation.

Igneous Rock (33, 36, 44, 63, 107) – rock formed by the solidification and crystallization of magma or lava (hot molten rock).
Igneous Rock Identification (33, 200) – Reference Table.
Incident Insolation (151) – point and time of solar radiation hitting the Earth's surface; see Angle of Insolation and Insolation.
Index Contour Line (24) – usually every 5th line on a contour map which is printed darker and interrupted to give elevation.
Index Fossil (109) – fossil that is characteristic of a certain geologic time, sometimes referred to as a guide fossil.
Inertia (164) – the tendency of an object in motion to remain in motion in a straight path unless acted upon by an outside force.
Inferences (7) – interpretations (conclusions) based on observations.
Infiltration (146) – seeping and absorption of water into ground storage.
Infrared Rays (137, 150) – range of invisible radiation wavelengths from about 750 nanometers, just longer than red in the visible spectrum, to 1 millimeter, on the border of the microwave region.
Inner Core (62) – iron/nickel solid inner sphere (zone) of the Earth's interior.
Insolation (147-9) – Incoming solar radiation; Sun's energy that transmits through Earth's atmosphere and reaches Earth's surface.
Instruments (7) – tools used by the observer to improve on detail or extend the ability to obtain information and measurements.
Intensity of Insolation (148) – rate (amount and duration) of solar radiation reaching the Earth's surface.
Intensity Scale (52) – see Modified Mercalli Scale.
Interface (11) – boundary between materials at which a change in environmental equilibrium occurs involving a loss or gain in energy states.
Interglacial Period (100) – the time between periods of glaciation.
International Dateline (23) – imaginary line through the Pacific Ocean roughly corresponding to 180° longitude, to the east of which, by international agreement, the calendar date is one day earlier than to the west.
Interpretation (7) – see Inferences.
Interrelationships (183) – the mutual dependence between the Earth's living and nonliving systems.
Intrusion (36, 44, 107) – rock mass formed from liquid rock (magma) cooling below the Earth's surface, igneous rock.
Intrusive Igneous Rock (44, 107) – igneous rock formed below the Earth's surface by the hardening of magma (hot liquid rock).

Inverse Relationship (10) – when one variable increases as the other variable decreases; see Graphing.
Isobar (24, 122, 125) – type of isoline on a weather map used to indicate equal air pressure points.
Isoline (23, 125) – line representing equal values on a map or model (such as contours, isotherms, isobars) of field characteristics in two dimensions.
Isostasy (65) – condition of equilibrium in the Earth's crust in which masses of greatest density are lower than those of lesser density.
Isotherm (24, 125) – type of isoline on a weather map used to indicate equal temperature points.
Isotope (111) – a variety of an element that has the same atomic number, but a different atomic mass, due to a difference in the number of neutrons present in the nucleus, used for correlation studies when radioactive.
Iso-surface (24) – model representing field characteristics in 3 dimensions.

Jet Stream (128) – wavelike currents with high winds at upper levels which tend to control storm tracks.
Jetty (94) – rock barriers build on both sides of a harbor entrance to prevent the deposition of sediment from clogging the entrance.
Joint (108) – crack in a rock mass or rock where unlike a fault, no vertical or horizontal displacement has occurred.
Jovian Planets (177) – Jupiter-like planets; large gaseous giant planets; includes Jupiter, Uranus, Saturn, and Neptune.

Kame (79) – cone-shaped hill formed at the ice front of a glacier.
Kepler's Harmonic Law (163) – explanation of planetary motion; relates a planet's period to its distance from the Sun.
Kettle Hole (79) – depression in glacial sediment formed when a large block of buried ice melts.
Kettle Lake (100) – kettle hole filled with water.
Kinetic Energy (138, 164) – energy of action, motion or at work.

Lagoon (94) – shallow water behind a baymouth bar.
Land Breeze (123) – local wind blowing from the land toward the sea when the air pressure over the land is greater than the air pressure over the sea.
Land Derived Sediments (92) – particles eroded from solid land material carried to other location, such as the ocean.
Landfill (186) – method of solid waste disposal in which refuse is buried between layers of dirt so as to fill in or reclaim low-lying ground.
Landscape (76-79) – topography of the land, including the characteristics of the Earth's surface.
Landscape Regions, NYS (77, 196) – Reference Table.

Landslide (93) – downward sliding of a mass of earth and rock both under the ocean or on relatively dry hillsides or mountainsides.

Lapse Rate (207) – Reference Table.

Latent Heat (137, 138) – formula; energy released or absorbed during a phase change, such as a liquid to a gas, but with no temperature change involved.

Latitude (19, 22, 148, 153) – distance north or south of the equator measured in degrees (parallels) from 0° at the equator, to 90° at the geographic poles.

Latitudinal Climate Pattern (153) – e.g. zones, West to East belts, of long term weather, primarily due to factors of temperature, winds, ocean currents, moisture.

Lava (33) – magma that reaches the Earth's surface.

Length (8) – distance between the ends or sections of an object; usually measured in meters, centimeters, of millimeters.

Leveling Forces (Destructional) (77) – forces of weathering, erosion, transportation, deposition, and subsidence.

Light-year (166) – a unit of measurement equal to the distance light travels in one year; ten trillion kilometers.

Lithosphere (21, 22, 31, 53) – continuous outer solid rock shell of the Earth.

Local Water Budget (152) – system of accounting for an area's water yearly supply.

Long Waves, or *L-Waves* (61) – waves that travel along the Earth's surface.

Longitude (22-3) – distance east or west of the prime meridian measured in degrees from 0° at the prime meridian (runs through Greenwich, England – Greenwich Mean Time) to 180° east or west (in the Pacific Ocean – International Date Line).

Longshore Current (93) – current moving sediment parallel to the coastline.

LOW (127) – low pressure air mass, counterclockwise rotation in the northern hemisphere, less dense than a HIGH; see Cyclone.

Lunar Eclipse (166) – when the Moon passes into the shadow of the Earth.

Lunar Month (165) – average time between successive new or full moons, equal to 29 days 12 hours 44 minutes.

Luster (32) – the appearance of light reflected form a mineral's surface.

Mafic (33) – dark-colored igneous rocks high in iron and magnesium.

Magma (33) – molten rock material beneath the Earth's surface.

Magnetic Polarity (63-64) – see Mid-Oceanic Ridge.

Magnitude Scale (52) – see Richter Scale.

Major Axis (163) – the longest diameter of an ellipse

Mantle (62) – layer of the Earth between the crust and the core.

Mantle Convection Cells (64) – the movement of heat and matter caused by differences in density within the Earth's mantle.

Map Legend (24) – explanatory table or list of the symbols appearing on a map or chart.

Map Scale (24) – the ratio between the distance on a map and the distance on the Earth's surface.

Marine Climate (153) – long term weather characteristics of an area near water bodies, such as large oceans characterized by small seasonal temperature ranges and abundant precipitation.

Maritime Polar Air Mass (126) – mP – a cool, moist air mass originating over a cold water surface.

Marine Terrace (94) – series of flat areas which look like steps out of the ocean, cut by the action of waves or streams.

Maritime Tropical Air Mass (126) – mT – usually a warm air mass originating over tropical waters, such as the Caribbean region.

Mars (178) – fourth planet from the Sun, having a sidereal period of revolution about the Sun of 687 days at a mean distance of 227.8 million kilometers (141.6 million miles) and a mean diameter of approximately 6,726 kilometers (4,180 miles).

Mass (8) – amount (quantity) of matter which an object contains.

Mathematical Combinations (8) – in measurement, combinations of: e.g. basic dimensional quantities, density, pressure, volume, acceleration.

Meander (74-75) – curving pattern of a river due to erosion and deposition.

Measurement (8) – use of time, length, or mass as a basic dimensional quantity (numerical).

Melting (139) – changing from a solid to a liquid state by application of heat or pressure or both.

Meridians (22) – grid "lines" (imaginary great circles) running between the North Pole and the South Pole; used to measure longitude.

Mesosphere (21) – portion of the atmosphere from about 30 to 80 kilometers (20 to 50 miles) above the Earth's surface, characterized by temperatures that decrease from 10°C to -90°C (50°F to -130°F) with increasing altitude.

Mesozoic Era (109) – third era of geologic time, including the Triassic Period, the Jurassic Period, and the Cretaceous Period and characterized by the development of flying reptiles, birds, and flowering plants and the appearance and extinction of dinosaurs.

Metamorphic Rock (35-36, 45-46, 107) – rocks formed by the effect of heat pressure and/or chemical action on other rocks, a recrystallization of pre-existing rocks.

Metamorphic Rock Identification (35, 201) – Reference Table.

Meteorologist (12, 128) – studies, reports, and forecasts weather conditions, as on television.

Meters (8) – international standard unit of length, approximately equivalent to 39.37 inches.

Metric System (8) – decimal system of units based on the meter as a unit length, the kilogram as a unit mass, and the second as a unit time.

Mid-Ocean Ridge (54, 63-64) – mountain ridge in mid-ocean, such as the Mid-Atlantic Ridge, which extends for about 64,000 kilometers roughly parallel to continental margins.
Milky Way Galaxy (167) – the name of our galaxy.
Millibar (122) – unit used to measure air pressure.
Mineral (31, 43, 44) – inorganic (nonliving) crystalline, solid substance with a definite chemical (atomic) shape and composition.
Model (24) – description or representation of an idea or concept which helps to illustrate actions or information (for example, models can be used to illustrate the Earth's shape and size).
Modified Mercalli Scale (52) – earthquake intensity scale based upon damage.
Moho (63) – shortened form of Mohorovicic discontinuity which is the crust-mantle boundary.
Mohs' Scale (32) – list of ten minerals from softest (#1) to hardest (#10) used to determine relative hardness of other minerals.
Moisture (72, 121, 140, 152) water in any form in the atmosphere and other places.
Moisture Capacity (121) – amount of water that can be held by an air mass, cold air generally hold less water than warm air; amount of moisture that can be held by soil, see Absolute Humidity.
Moisture Deficit (152) – see Drought.
Monomineralic Rock (31) – rock composed of just one mineral type.
Moon (165-166) – Earth's satellite.
Moraines (79, 100) – features made of glacial till.
Mountain (77, 153) – elevated landscape with distorted rock structure.

National Hurricane Center (128) – in Florida, tracks hurricanes and issues appropriate warnings.
National Severe Storm Forecast Center (129) – in Kansas City predicts and tracks tornadoes.
National Weather Service (128) – uses satellites to track weather systems and issue appropriate warnings.
Natural Pollutants (183) – materials produced by nature having a negative effect on people, plants, animals, or property; i.e. pollen or volcanic ash.
Neap Tide (166) – tides of the smallest tidal range occurring at the quarter phases of the Moon.
Nonrenewable Resources (46) – resources that are used much faster and in greater amounts than they can form.
Noon (163) – point in time when the Sun is directly on observer's meridian.
North Pole (19, 22) – 90° north latitude.
North Star (19) – see Polaris.
Nuclear Waste (185) – long-term harmful by-products of nuclear reactions.

Oblate Spheroid (19) – slightly flattened sphere; shape of the Earth, flattened at the geographic poles and bulging at the equator.

Observation (7) – use of senses in measuring and collecting data concerning environment.
Occluded Front (127) – interface formed when a cold front overtakes a warm front.
Ocean (91, 92) – entire body of salt water that covers more than 70 percent of the Earth's surface.
Ocean Basin (92) – see Abyssal Plain.
Ocean Currents (92) – see the specific current.
Ocean Currents, Surface (198) – Reference Table.
Ocean Trenches (92) – deep, sharp valleys under the ocean associated with colliding plates.
Oceanic Crust (54, 62) – thinner, more dense part of the Earth's crust composed of basaltic material.
Ocean-Floor Spreading (63-64) – theory supported by past and present evidence that the ocean floor is moving outwards from the mid-ocean ridge.
Orbit (163) – path of a celestial object, satellite, and/or Earth about a center, usually an ellipse.
Orbital Velocity (Speed) (163) – the speed of an object at any given time in its orbit; usually changing due to distances from its gravitational center.
Organic Evolution (112) – theory of change, an explanation of how new species develop by punctuated (rapid) or gradual (slow) changes.
Organic Sedimentary Rock (45) – rock formed from the remains of plant and/or animal material.
Organic Sediments (92) – sediments formed from skeletal remains and shells of microscopic marine organisms.
Organic Substance (45, 185) – material containing the element Carbon, usually associated with living or once living things.
Orogeny (46, 110) – mountain building processes.
Orographic Effect (124, 153) – effect that mountains have on weather and climate; blockage of precipitation from the leeward side of mountains.
Outcrop (108) – exposed bedrock.
Outer Core (62) – liquid Earth zone between the inner core of the Earth and the mantle, like the inner core composed of iron and nickel.
Outwash Plain (79, 100) – horizontal layers of sorted glacial material deposited in front of the glacier by the meltwaters of the glacier.
Oxidation (71) – the chemical reaction of oxygen with other materials.
Oxide (43) – binary compound of an element or a radical with oxygen (O_2).
Ozone (150) – a form of oxygen containing three atoms in the molecule (O_3).

P-waves (61) see Primary Wave.
Paleozoic Era (109) – geologic time that includes the Cambrian, Ordovician, Silurian, Devonian, Mississippian, Pennsylvanian, and Permian periods and is characterized by the appearance of marine invertebrates, primitive fishes, land plants, and primitive reptiles.
Parallelism of the Axis (148) – the Earth's axis at any place in its orbit is parallel to the axis in any other place in the Earth's orbit.

Parallels (22) – grid lines on a map or globe; another term used to determine latitude.

Parent Rock (45) – the original rock from which a metamorphic rock forms.

Particle Size (Transported) to Water Velocity Relationship (75, 206) – Reference Table.

Penumbra (166) – partial shadow cast by the Earth, Moon, or Sun during an eclipse.

Percent Error (8, 9) – deviation formula, mathematical expression of a calculated error in percent (%).

Perihelion (162, 163) – point on the Earth's orbit when it is closest to the Sun; a distance of 147 million kilometers, usually occurring January 1st.

Period (109) – unit of geological time, longer than an epoch and shorter than an era.

Period of Revolution (163) – amount of time an object takes to make one complete orbit around its center, in the case of the Earth, 365+ days to orbit the Sun.

Permeability (146) – rate at which moisture passes through a material.

Permeability Rate (146) – speed at which water pass through a porous material; see Permeability and Porosity.

Perpendicular Insolation (148, 163) – vertical rays of the Sun; 90° radiation at Earth's surface.

Pesticides (186) – chemical used to kill pests, especially insects.

Phase Change (138, 139) – change of a material through states of solid, liquid, and gas.

Phases (States) of Matter (9) – solid, liquid, and gas.

Phases of the Moon (165) – changes in the amount of the illuminated surface of the Moon as seen from the Earth, cyclic over a $29^{1}/_{2}$ day period.

Physical Constants (210) – Reference Table.

Physical Properties (31) – used for identification of minerals.

Physical Weathering (71) – the process that alters the physical characteristics of rocks/minerals, generally leading to breaking into smaller pieces.

Plain (77) – low elevation landscape, gentle slopes and relatively stable, often composed of horizontal layers of sedimentary rocks.

Planetary Motions (175) – the non uniform movement of planets.

Planetary Period (163) – the time it takes a planet to make one revolution arount the Sun.

Planetary Wind and Moisture Belts in the Troposphere (209) – Reference Table.

Planetary Wind Belts (140, 154) – zones on the Earth where winds generally blow in one direction only, such as the prevailing southwest winds of the U.S.

Plate Tectonic Theory (53, 63) – idea that there are six large crustal plates, and many smaller ones, moving on the surface of Earth in a way that can be calculated and predicted.

Plates (77) – solid pieces of the Earth's lithosphere.

Plateau (77) – high elevation landscape, relatively stable with little or no distortion of the rock layers.

Pleistocene Epoch (100) – time beginning two or three million years ago and ending about 10,000 years ago; called the Great Ice Age.

Pointer Stars (19, 161) – outer two stars of Big Dipper which align with Polaris, the end star of the Little Dipper.

Polaris (19, 23, 161) – North Star, used in navigation since it is almost directly over the north geographic pole of the Earth.

Pollutants (183-6) – solids, liquids, gases, biological organisms, forms of energy such as heat, sound and nuclear radiation (see Pollution).

Pollution (183-6) – when the concentration of any substance or form of energy reaches a proportion that adversely affects man, his property, or the plant and animal life.

Pollution of Water (184) – examples: PCBs, oil spills.

Polymineralic Rock (31) – rock composed of more than one (several) mineral types.

Population Density (81) – areas affected by landscape pollution and the misuse of the landscape due to concentrated population.

Population Growth in Humans (186, 187) – exponential increase in Earth's human population.

Porosity (146) – percent of open space in a volume of a certain material.

Potential Energy (138, 151, 164) – stored energy or energy at rest as the result of its state or position.

Potential Evapotranspiration (152) – estimated amount of water loss that can occur due to heat energy available.

Precambrian Era (109) – oldest and largest division of geologic time, preceding the Cambrian Period, often subdivided into the Archeozoic and Proterozoic eras, and characterized by the appearance of primitive forms of life.

Precipitation (rock formation definition) (34, 35, 75) – type of sedimentation (deposition) involved in the production of evaporites, in which dissolved solids come out of solution.

Precipitation (weather definition) (124, 145, 152) – generally from clouds, the falling of water as liquid (rain) or solid (ice, hail, and snow).

Prediction (10) – the use of natural evidence to predetermine the scope and direction of a future environmental change.

Predominant Agent (74) – main or primary agent (cause affect) of a specific action, such as in weathering and erosion on Earth the main agent is running water.

Pressure (8) – combination of gravitational force compared to surface area.

Pressure Belt (140-141, 154) – band of high or low pressure in the atmosphere caused by regions of rising or settling air.

Pressure Gradient (125) – degree of difference over a specified distance of high and low pressures, the greater the pressure gradient difference, the greater the wind speeds.

Pressure Scale (208) – Reference Table.

Primary Wave, or *P-wave* (61) – compressional wave generated by an earthquake which travels through solids, liquids, and gases.

Prime Meridian (23) – 0° longitude on the Earth, passing through Greenwich, England.

Principle of Geometry of a Sphere (19) – as a sphere is rotated, the angle of a fixed point outside of the sphere as compared to the surface of that sphere, is equal to the angle of the sphere's rotation.

Principle of Superposition (107, 108) – idea that the oldest bed in a sequence of horizontal sedimentary rock layers is the one on the bottom.

Principle of Uniformitarianism (111) – all geologic phenomena may be explained as the result of existing forces having operated uniformly from the origin of the Earth to the present time.

Probability of Occurrence (128) – chance that a specific event will occur, often used in the prediction of weather systems.

Properties of Earth's Atmosphere (209) – Reference Tables.

Psychrometer (121) – instrument consisting of a dry bulb and wet bulb thermometer used to determine the dew point temperature and the relative humidity.

Purge (184) – in bodies of water, materials churn up from the bottom.

Quiet Medium (76) – still water, ice, or air in which settling occurs, generally in a graded bedding, a process which is more complex than in a moving medium.

Radial Drainage (80) – streams radiate out like the spokes of a wheel from a central high elevation such as on a volcano.

Radiation (137) – object's electromagnetic wave transmission.

Radiational Cooling (140) – decrease in temperature at the Earth's surface caused by energy radiating out into space.

Radiative Balance (152) – average energy levels on the Earth remain constant due to the Earth giving off as much energy as it receives.

Radiative Balance (152) – stable Earth temperature; gaining as much energy as the Earth gives off.

Radioactive Dating (107, 111) – process of determining the age of rock by measuring the half-life of radioactive materials in the rock.

Radioactive Isotope (111) – materials that decay from more unstable forms into more stable forms; see Half-life.

Radioactive Waste (185-186) – see Nuclear Waste.

Radioactivity (Decay) (107, 111) – secondary source of energy for the Earth; spontaneous and natural nuclear breakdown from unstable to stable atomic forms, energy is released, the process has a constant and predictable rate, not affected by environmental changes.

Random Reflection (151) – reflection of insolation due to aerosols in the atmosphere; dust and water droplets increase the amount of random reflection, causing a decrease in the amount of insolation reaching the Earth's surface; see Scattering.

Rate of Decay (107) see Half-life, Radioactive Dating.

Recharge (152) – replacement of water by infiltration into the soil storage area.

Recrystallization (35, 45) – formation of new crystalline materials by the enlargement of preexisting crystals by the action of thermal metamorphism.

Red Shift (178-9) – spectral lines move toward the red end of the spectrum from a celestial object moving away from the Earth; see Doppler Shift.

Reflection (151) – change in electromagnetic wave direction due to the non penetration of a wave into a surface; a smooth surface will reflect a wave at the same angle at which it strikes the surface.

Refraction (52, 62) – change in direction when an electromagnetic wave goes from one material to another material with a different density.

Regional Metamorphism (46) – recrystallization of rocks over a large area caused by extreme pressure associated with mountain building.

Relative Age (Date) (107, 108) – dates in the Earth's history determined with reference to other events helpful in determining relationships in a time line.

Relative Humidity (121-2) – ratio of the mass of water vapor per unit volume of the air to the mass of water vapor per unit volume of saturated air at the same temperature.

Relative Humidity (212) – Reference Table.

Residual Sediment (74) – sediment that remains at the site of weathering.

Resource Conservation (46) – see Conservation of Natural Resources.

Reversal of Magnetic Polarity (64) – reference to changing polarity as observed in rocks because of the Earth's magnetic poles reversing.

Revolution (149, 163, 176) – orbiting of one body around another body.

Richter Scale (52) – earthquake magnitude scale based upon energy released.

Rock (31) – relatively hard, naturally formed mineral or petrified matter; stone.

Rock Cycle (36) – model to explain the changes in rocks and the formation of sedimentary and nonsedimentary rocks, igneous and metamorphic rocks.

Rock Cycle in Earth's Crust (36, 200) – Reference Table.

Rock Formation (109) – body or mass of rock with similar features and characteristics.

Rock Forming Minerals (31) – minerals, mostly silicates, that form 90% of the Earth's crust.

Rock Properties (33-36, 44-46) – characteristics of various rocks.

Rock Resistance (79) – characteristic of rock types to resist the forces of change, including weathering and erosion.

Rotation (149, 161, 176) – spinning of an object about its own axis.

Runoff (147, 152) – water that does not infiltrate the soil storage area and flows over the land surface to lakes, streams, and oceans.

S-wave (61) – see Shear Wave and Secondary Wave.

Salinity (91) – a measure of dissolved solids in sea water; saltiness.

Salinity Current (93) – density current caused by water that has a greater than average salinity.

Satellite (164) – any object which is held by another object's force of gravitation, around which it revolves (orbits); the Moon is the Earth's satellite; a man-made object which orbits the Earth.

Saturation Point (121) – point at which the air is completely filled with water vapor to the air's maximum capacity, after which water will condense.

Scattering (151) – wave movement in different directions due to reflection.

Sea Arch (94) – feature formed when waves cut through a headland.

Sea Breeze (123) – local wind blowing from the sea toward the land when the air pressure is higher over the sea than over the land.

Sea Cliff (94) – feature formed where waves attack steeply sloping land along a coast.

Sea Water, Composition of (91) – see chart.

Sea-floor spreading (63) – the process by which ocean plates move apart allowing new ocean crust to form between them.

Seamount (92) – submerged cone-shaped mountain peak.

Seasonal Lag (150) – the maximum surface temperature occurs after the time of maximum insolation and the minimum surface temperature occurs after the time of minimum insolation.

Seasons (148, 162) – divisions of the year caused by climatic changes, angle of insolation, Earth tilt; generally, spring, summer, fall, and winter.

Secondary Wave, or S-wave (61) – transverse earthquake waves which only travel through solids and travel more slowly than P-waves.

Sediment (34, 71, 75-76, 92) – rock particles that are produced and/or transported by erosion and weathering.

Sediment Laden Flow (93) – movement of an erosional transport agent containing some forms of sediment, such as a glacier, turbidity current.

Sedimentary Rock (34, 45) – rock formed from compaction and cementation of sediment.

Sedimentary Rock Identification (34, 201) – Reference Table.

Sedimentation (75, 94) – settling out of solution of sediments, including minerals, in an erosional system; deposition; see Deposition.

Seismic Wave (52, 61) – wave that radiates from the point of origin of an earthquake, moving in all directions through solid rock.

Seismograph (52, 61) – very delicate instrument that detects and records passing earthquake waves.

Sensory Preception (7) use of senses of sight, hearing, touch, smell, taste.

Settling Rate (75-76) – time required for a certain sediment to settle out of water or air.

Shadow Zone (62) – a band from 102° to 143° away from an earthquake in which neither P-waves nor S-waves are recorded on seismographs.

Shear Wave (Secondary, S-wave) (61) – wave that causes individual rock particles to vibrate at right angles to the direction that the wave is traveling; cannot pass through liquids.

Sidereal Month (165) – one complete revolution of the Moon around the Earth – $27\frac{1}{3}$ days.

Silicates (43, 44) – minerals containing silicon-oxygen tetrahedra.

Silicon-Oxygen Tetrahedron (43) – structural model of a silicate mineral.

Sink (11) – portion of an energy system with lower energy concentrations, into which energy usually flows.

Snowfield (99) – area where snow accumulates and recrystallizes into the ice that feeds glaciers.

Soil Association (73) – unit of soil classification, including the characteristics of the soil, composition, porosity, permeability, structure, and the ability to support life.

Soil Conservation (73) – using the soil in ways to preserve and protect it.

Soil Formation (72) – production of soil, particles of rocks and minerals, and organic matter.

Soil Horizons (72) – layers of soil produced as a result of the weathering processes and biologic activity; examples: horizontal topsoil, horizontal subsoil.

Soil Storage (152) – amount of water held below Earth's surface in the soil.

Soil Storage Change (152) – amount of water either removed or added to a soil storage area; usage and recharge, respectively.

Solar Eclipse (166) – eclipse that occurs when the Moon's umbra reaches the Earth's surface.

Solar Electromagnetic Spectrum (137) – full range of wavelengths emitted from the Sun with the maximum intensity occurring in the visible region.

Solar Energy (140) – any energy forms radiated from the Sun.

Solar Radiation (137, 147) – energy from the Sun.

Solar System (167, 177) – orbiting system of the Earth, planets, and moons with the Sun as the center of revolution.

Solar System Data (208) – Reference Table.

Solidification (33) – see Crystallization.

Solstice (149, 163) – times when the Sun's rays are perpendicular (at zenith) to the $23\frac{1}{2}°$ north latitude (about June 21st), summer solstice and $23\frac{1}{2}°$ south latitude (about December 21st), winter solstice.

Solution (75) – dissolved minerals carried in water.

Solvent (91) – substance in which another substance is dissolved, forming a solution; substance, usually a liquid, capable of dissolving another substance.

Solvent, Universal (91) – water.

Sorted and Unsorted Particles (146) – selection of various materials, based on size, the more similar the particle sizes, the greater the sorting, the greater the difference in the particle sizes, the less the sorting.

Sorting of Sediments (76) – manner in which materials in suspension settle out of a transport medium in a definite pattern.

Source (11) – portion of an energy system with the highest energy concentrations, from which energy usually flows.

Source Region (126) – place on the Earth where an air mass forms.

South Pole (22) – 90° South latitude.

Species (112) – most specific part of the classification system, or two organisms of the same species are able to mate and produce fertile offspring.

Specific Gravity (32) – the ratio of the weight of a mineral to the weight of an equal volume of water.

Specific Heat (139) – amount of heat necessary to raise the temperature of 1 gram of any substance 1°C, measured in calories; specific heat of water is 1; most other substances have specific heats of less than 1.

Spectral Lines (178) – lines corresponding to various wavelengths seen in an elements spectrum.

Spectroscope (178) – astronomical instrument used to study the light from celestial objects.

Spectrum (178, 208) – see Electromagnetic Spectrum.

Speed of Light (166) – 300,000 kilometers per second.

Spit (94) – narrow strips of sand deposited by longshore currents.

Spring Tides (166) – tides of maximum tidal range occurring at New and Full Moon phases.

Station Model (125) – on a weather map, describes weather conditions at a reporting station.

Station Model, Weather Maps (207) – Reference Table.

Stationary Front (127) – interface of two air masses that do not move.

Storm Track (128) – path taken by the center of a storm or low pressure system.

Strata (53) – layers of rock material, usually sedimentary rock.

Stratiform Clouds (124) see Stratus Clouds.

Stratosphere (21) – region of the atmosphere above the troposphere and below the mesosphere.

Stratus Clouds (124) – low-altitude cloud formation consisting of a horizontal layer of gray clouds.

Streak (32) – the color of the powder of a mineral.

Stream Bed (75) – interface of the water and bottom of a stream, including the rock particles and bottom materials.

Stream Discharge (74, 147) – measurement of amount of water passing a certain point in a stream in a certain amount of time; rate of flow in volume.

Stream Drainage Pattern (80) – pattern that forms due to the way water drains across the land in a stream or river system.

Stream Velocity (74) – graphs comparing with discharge, slope, and particle size carried.

Striations (99) – scratches in rock caused by glaciers dragging rocks over the surface.

Subduction Zones (54) – form from the collision of plates with the denser ocean plate diving down (subducting) into the mantle.

Sublimation (124) – the process by which a solid changes directly to a gas.

Sublimation (9) – phase change from a solid to a gas without a liquid phase, as in the sublimation of ice to water vapor without melting to a liquid form.

Subsidence (53) – act of sinking or settling of the Earth's surface.

Summer Solstice (149, 163) – in Northern Hemisphere occurs about June 21, when the Sun is in the zenith at the tropic of Cancer.

Sun (161, 163, 167) – star, center of solar system.

Sun's Path (162) – see Apparent Motions

Superposition, Principle of (107) – "youngest" rock layers appear on top of "oldest" rock layers.

Surf (93) – foaming water formed by breaking waves.

Surface Ocean Currents (198) – Reference Tables.

Surplus Water (147) – surface water that neither evaporates nor infiltrates, but is runoff.

Suspension (75) – particles transported in all levels of water in a stream held up by the motion of the stream.

Swash (93) – the water that surges forward onto the beach from the breaking waves.

Synodic Month (165) – see Lunar Month.

Synoptic Weather Map (126) – broad "bird's eye view" of the weather.

Technology (183) – application of modern science, especially to industrial or commercial objectives.

Technological Oversights (81, 186) – types; human errors involving the environment.

Temperature (120, 149) – specific degree of hotness or coldness as indicated on or referred to a standard scale; see Celsius.

Terminal Moraine (79, 100) – large ridge of glacial till marking the farthest advance of glacial ice.

Terrestrial (Earth-like) Planets (177) – having rocky cores; includes Mercury, Venus, Earth, and Mars.

Terrestrial Radiation (152) – outgoing Earth radiation.

Tetrahedral Unit (43) – as seen in the bonds between oxygen and silicon; the basic building block of the silicates.

Texture (44) – appearance and feel of a surface; used in rock identification.

Thermal Metamorphism (46) – see contact metamorphism by heat.

Thermosphere (21) – outermost shell of the atmosphere, between the mesosphere and outer space, where temperatures increase steadily with altitude.

Tidal Range (166) – difference between high and low tide water levels.

Tidal Wave (93) – see Tsunami.

Tides (165-166) – periodic variation in the surface level of the oceans and of bays, gulfs, inlets, and estuaries, caused by gravitational attraction of the Moon and Sun.

Till (99) – unsorted glacial sediment.

Tilting (53) – Earth movement resulting in a change in the position of rock layers.

Time (8) – number, as of years, days, or minutes, representing such an interval separating two points on this continuum; a duration.

Topographic Map (24) – see Contour Map.

Tornado (128) – small, violent storm with a characteristic funnel-shaped cloud.
Toxic Waste Dump (186) – location of disposal of environmentally harmful materials.
Toxins (184) – poisonous substance, especially a protein, that is produced by living cells or organisms and is capable of causing disease when introduced into the body tissues but is often also capable of inducing neutralizing antibodies or antitoxins.
Transform Plate Boundary (54-55) – boundary between plates that are grinding past each other.
Transition Zones (46) – boundary between rock types.
Transported Sediments (74) – erosional product moved from the source of weathering to a different location.
Transporting Agents (74) – actions that affect erosion and move sediments from one place to another.
Transporting Systems (73) – all agents involved in erosion and movement; e.g. erosion, the transporting agent, energy, and the material moved.
Trellis Drainage (80) – drainage pattern with parallel mainstreams and right-angled tributaries found in regions with parallel folds or faults.
Tropic of Cancer (163) – 23½° North latitude.
Tropic of Capricorn (163) – 23½° South latitude.
Trough (93) – bottom part of a wave.
Tsunami (52, 93) – seismic sea wave commonly referred to as a tidal wave although it is not caused by tides.
Turbidity Current (91, 92, 93 – sediment-laden density current.

U-shaped Valley (99) – round bottomed valleys carved as a result of glaciers.
Ultraviolet Rays (137) – range of invisible radiation wavelengths from about 4 nanometers, on the border of the x-ray region, to about 380 nanometers, just beyond the violet in the visible spectrum.
Unbra (166) – completely dark portion of the shadow cast by the Earth, Moon, or the Sun during an eclipse.
Unconformity (108) – surface of erosion between rock layers of different ages indicating that deposition was not continuous.
Universe (166) – all matter and energy, including Earth, the galaxies and all therein, and the contents of intergalactic space, regarded as a whole.
Uplifting (53, 77) – vertical displacement of the Earth's surface; faulting.

Vapor Pressure (138) – pressure exerted by water within the atmosphere.
Variable (10) – factors involved in change.
Vein (108) – mineral deposits that have filled a rock crack or permeable zone; regularly shaped and lengthy occurrence of an ore; a lode.
Venus (177) – second planet from the Sun, having an average radius of 6,052 kilometers (3,760 miles), a mass 0.815 times that of Earth, and a sidereal period of revolution about the Sun of 224.7 days at a mean distance of approximately 108.1 million kilometers (67.2 million miles).
Vertical Displacement (53) – faulting in which a portion of the Earth's surface is either uplifted or subsides.
Vertical Sorting (76) – graded bedding; layering of sediment so that the largest, densest particles are on the bottom of the layer and the smallest, least dense particles are on the top.
Violet Shift (179) – on a darkline spectrum for an element, the pattern moves to the left indicating that the light source is moving toward the Earth; see Doppler Shift.
Visible Light (137, 150) – wavelengths of the electromagnetic spectrum perceptible to the eye.
Volcanic Time Markers (109) – layers of ash from a volcano used in dating geological events.
Volume (8) – amount of space occupied by a three-dimensional object or region of space, expressed in cubic units.
Vortex (129) – the center of a tornadoes funnel-shaped cloud.

"Walking the Outcrop" (108) – observing the landscape through actual local observation.
Warm Front (127) – the interface, leading edge, of an air mass which has warmer temperatures than the preceding cooler air mass.
Wastes (185) – types of, including organic, inorganic, thermal and radioactive.
Water (74) – hydrogen and oxygen (H_2O); universal solvent.
Water Budget (152) – system of accounting for an area's water yearly supply.
Water Cycle (145) – also, hydrologic cycle; Earth system in which water is continually moving from the atmosphere to the Earth and from the Earth to the atmosphere.
Water Pollution (184) – types.
Water Table (146) – within the ground surface of the Earth, top of the zone of saturation.
Water Vapor (121) – gaseous state of water in air.
Wave Height (93) – vertical distance between the crest and trough of a wave.
Wave Refraction (94)– the bending of a line of waves as it approaches the shore.
Wave Velocity (61)
Wavelength (93, 137) – distance between two successive wave crests.
Weather (121, 141) – state of the atmosphere at a given time and place, with respect to variables such as temperature, moisture, wind velocity, and barometric pressure.
Weather Forecasting (141)
Weather Map Information (125)
Weather Map Symbols (125)
Weathering (71) – physical and chemical processes that change the surface of the Earth.
Wegener, Alfred (53) – see Continental Drift.
Weight (8) – unit measure of gravitational force.

Wet-bulb Thermometer (121) – see Psychrometer.
Wind (122) – movement of air caused by Earth systems.
Wind Vane (122) – weather instrument used to determine the direction of the wind.
Winter Solstice (149, 163) – occurs about December 21, when the Sun is over the tropic of Capricorn.

X-ray (137) – relatively high-energy photon with wavelength in the approximate range from 0.01 to 10 nanometers.

Yearly Temperatures (150)

Zone of Aeration (146) – portion of ground through which water passes until the water reaches the zone of saturation.
Zone of Convergence (140) – region of low pressure.
Zone of Divergence (140) – region of high pressure.
Zone of Saturation (146) – portion of ground with an upper boundary called the water table.
Zone, Stratified (21) – group of layered materials or regions, each having distinct characteristics.

PRACTICE EXAM
MODIFIED EARTH SCIENCE PROGRAM
JUNE 1996

Part I

Answer all 40 questions in this part.

Directions (1–40): For *each* statement or question, select the word or expression that, of those given, best completes the statement or answers the question. Record your answer on the separate answer paper in accordance with the directions on the front cover of this booklet. [40]

1 Measurements of the Sun's altitude at the same time from two different Earth locations a known distance apart are often used to determine the

   1 circumference of the Earth
   2 period of the Earth's revolution
   3 length of the major axis of the Earth's orbit
   4 eccentricity of the Earth's orbit

2 The data table below gives information on mineral hardness.

MINERAL HARDNESS

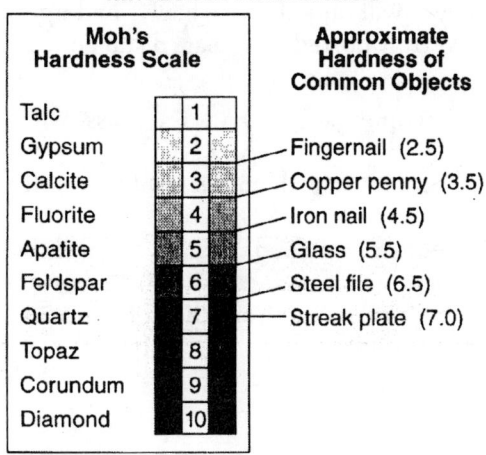

Moh's Hardness Scale		Approximate Hardness of Common Objects
Talc	1	
Gypsum	2	Fingernail (2.5)
Calcite	3	Copper penny (3.5)
Fluorite	4	Iron nail (4.5)
Apatite	5	Glass (5.5)
Feldspar	6	Steel file (6.5)
Quartz	7	Streak plate (7.0)
Topaz	8	
Corundum	9	
Diamond	10	

Moh's scale would be most useful for

   1 identifying a mineral sample
   2 finding the mass of a mineral sample
   3 finding the density of a mineral sample
   4 counting the number of cleavage surfaces of a mineral sample

3 Which diagram best represents a sample of the metamorphic rock gneiss? [Diagrams show actual size.]

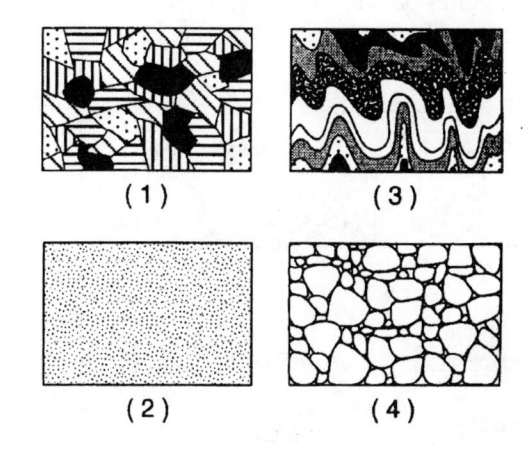

4 Which two igneous rocks could have the same mineral composition?

   1 rhyolite and diorite
   2 pumice and scoria
   3 peridotite and andesite
   4 gabbro and basalt

5 According to the *Earth Science Reference Tables*, much of the surface bedrock of the Adirondack Mountains consists of

   1 slate and dolostone
   2 gneiss and quartzite
   3 limestone and sandstone
   4 conglomerate and red shale

6 The theory of plate tectonics suggests that
   1 the continents moved due to changes in the Earth's orbital velocity
   2 the continents' movements were caused by the Earth's rotation
   3 the present-day continents of South America and Africa are moving toward each other
   4 the present-day continents of South America and Africa once fit together like puzzle parts

7 Contact zones between tectonic plates may produce trenches. According to the *Earth Science Reference Tables*, one of these trenches is located at the boundary between which plates?
   1 Australian and Pacific
   2 South American and African
   3 Australian and Antarctic
   4 North American and Eurasian

8 Which cross section best represents the general bedrock structure of New York State's Allegheny Plateau?

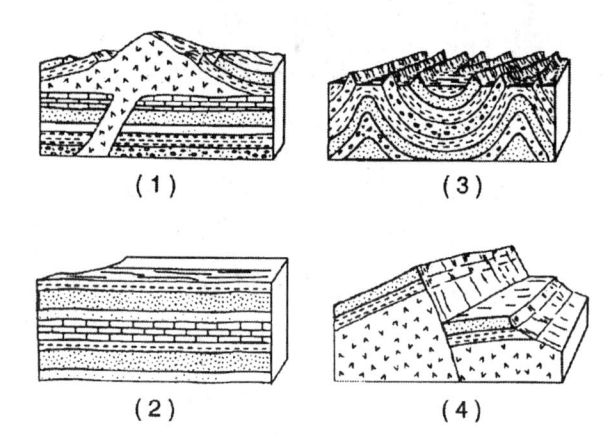

(1)   (2)   (3)   (4)

9 Which statement best describes a stream with a steep gradient?
   1 It flows slowly, producing a V-shaped valley.
   2 It flows slowly, producing a U-shaped valley.
   3 It flows rapidly, producing a V-shaped valley.
   4 It flows rapidly, producing a U-shaped valley.

10 The diagram below represents a geologic cross section.

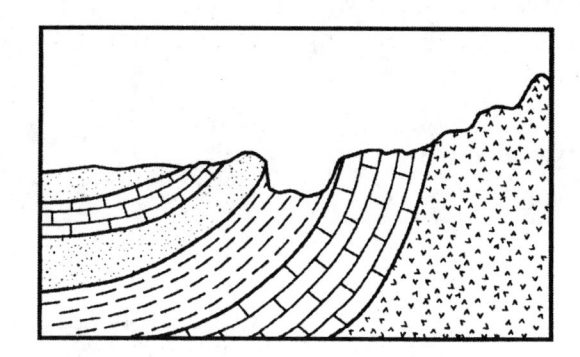

Which rock type appears to have weathered and eroded the most?

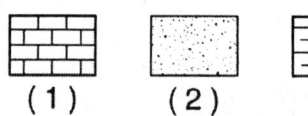

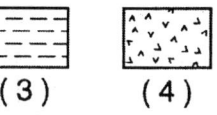

(1)   (2)   (3)   (4)

11 In which type of landscape are meandering streams most likely found?
   (1) gently sloping plains
   (2) regions of waterfalls
   (3) steeply sloping hills
   (4) V-shaped valleys

12 What change will a pebble usually undergo when it is transported a great distance by streams?
   1 It will become jagged and its mass will decrease.
   2 It will become jagged and its volume will increase.
   3 It will become rounded and its mass will increase.
   4 It will become rounded and its volume will decrease.

13 A large, scratched boulder is found in a mixture of unsorted, smaller sediments forming a hill in central New York State. Which agent of erosion most likely transported and then deposited this boulder?

1 wind
2 ocean waves
3 a glacier
4 running water

14 The map below represents a river as it enters a lake.

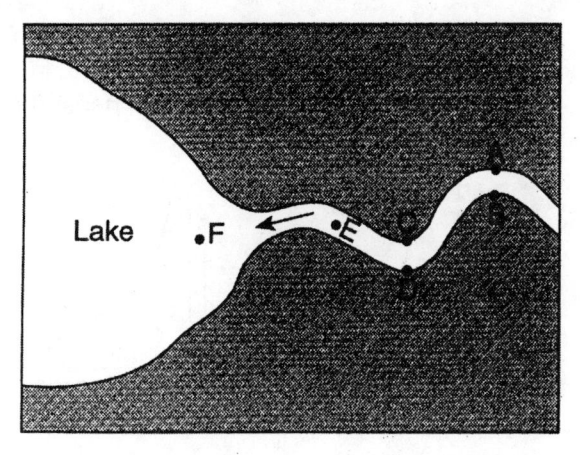

At which locations is the amount of deposition greater than the amount of erosion?

(1) A, C, and E
(2) B, C, and F
(3) B, D, and F
(4) A, D, and E

15 The velocity of a stream is decreasing. As the velocity approaches zero, which size particle will most likely remain in suspension?

1 pebble
2 sand
3 clay
4 boulder

16 What is the relative age of a fault that cuts across many rock layers?

1 The fault is younger than all the layers it cuts across.
2 The fault is older than all the layers it cuts across.
3 The fault is the same age as the top layer it cuts across.
4 The fault is the same age as the bottom layer it cuts across.

17 In order for an organism to be used as an index fossil, the organism must have been geographically widespread and must have

1 lived on land
2 lived in shallow water
3 been preserved by volcanic ash
4 existed for a geologically short time

18 The cartoon below is a humorous look at geologic history.

Early Pleistocene mermaids

If Early Pleistocene mermaids had existed, their fossil remains would be the same age as fossils of

1 armored fish
2 mastodonts
3 trilobites
4 dinosaurs

19 According to the *Earth Science Reference Tables*, when did the Jurassic Period end?

(1) 66 million years ago
(2) 144 million years ago
(3) 163 million years ago
(4) 190 million years ago

20 The table below gives information about the radioactive decay of carbon-14. [Part of the table has been left blank for student use.]

Half-Life	Mass of Original C-14 Remaining (grams)	Number of Years
0	1	0
1	$\frac{1}{2}$	5,700
2	$\frac{1}{4}$	11,400
3	$\frac{1}{8}$	17,100
4		
5		
6		

What is the amount of the original carbon-14 remaining after 34,200 years?

(1) $\frac{1}{8}$ g (3) $\frac{1}{32}$ g
(2) $\frac{1}{16}$ g (4) $\frac{1}{64}$ g

Note that question 21 has only three choices.

21 If a sample of a radioactive substance is crushed, the half-life of the substance will
1 decrease
2 increase
3 remain the same

22 According to the *Earth Science Reference Tables*, which geologic event is associated with the Grenville Orogeny?
1 the formation of the ancestral Adirondack Mountains
2 the advance and retreat of the last continental ice sheet
3 the separation of South America from Africa
4 the initial opening of the Atlantic Ocean

23 In which atmospheric layer is most water vapor found?
1 troposphere   3 thermosphere
2 stratosphere 4 mesosphere

24 In order for clouds to form, cooling air must be
1 saturated and have no condensation nuclei
2 saturated and have condensation nuclei
3 unsaturated and have no condensation nuclei
4 unsaturated and have condensation nuclei

25 Which graph best represents the relationship between air temperature and air density in the atmosphere?

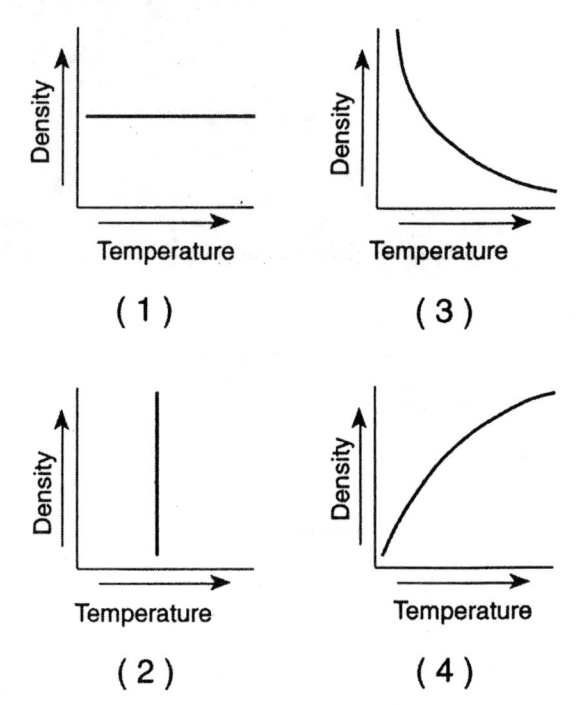

(1) (2) (3) (4)

26 The greatest source of moisture entering the atmosphere is evaporation from the surface of
1 the land
2 the oceans
3 lakes and streams
4 ice sheets and glaciers

27 Which angle of the Sun above the horizon produces the greatest intensity of sunlight?
(1) 70°   (3) 40°
(2) 60°   (4) 25°

28 The diagram below represents a cross section of air masses and frontal surfaces along line AB. The dashed lines represent precipitation.

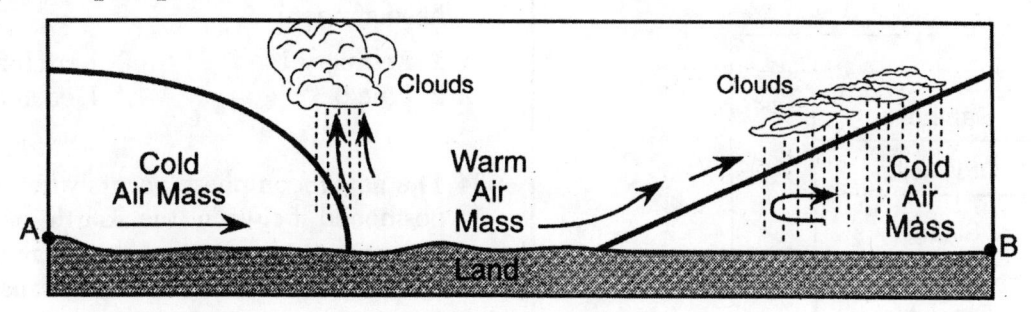

Which weather map best represents this frontal system?

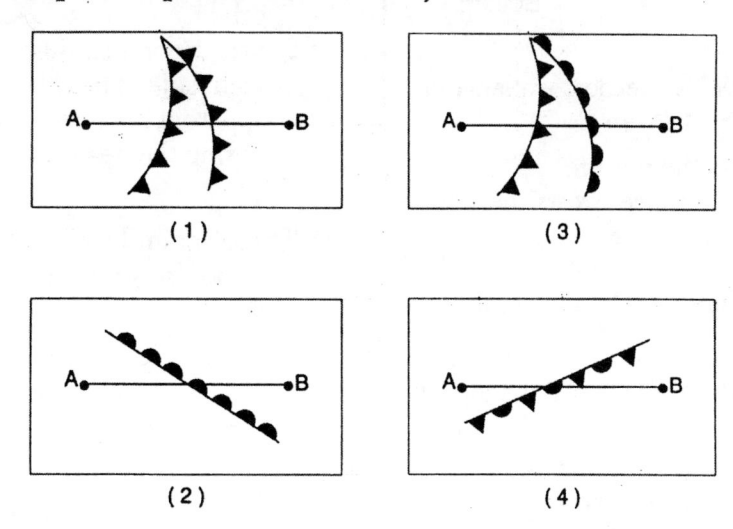

29 The cross section below shows several locations in the State of Washington and the annual precipitation at each location. The arrows represent the prevailing wind direction.

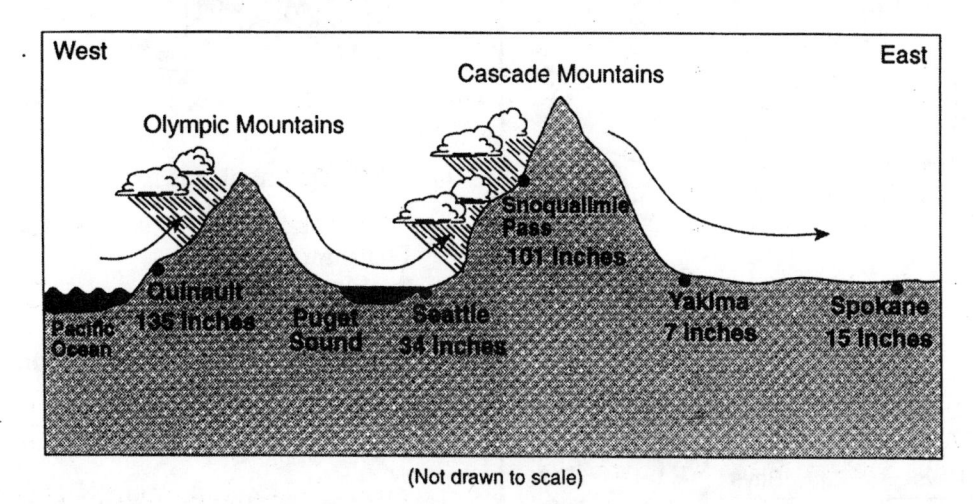

Why do the windward sides of these mountain ranges receive more precipitation than the leeward sides?

1 Sinking air compresses and cools.
2 Sinking air expands and cools.
3 Rising air compresses and cools.
4 Rising air expands and cools.

30 The diagram below represents a cross section of a series of rock layers of different geologic ages.

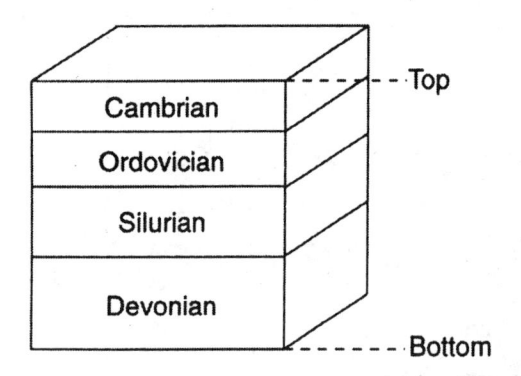

```
Cambrian      -- Top
Ordovician
Silurian
Devonian
              -- Bottom
```

Which statement provides the best explanation for the order of these rock layers?

1 The oldest layer is on the bottom.
2 A buried erosional surface exists between layers.
3 The layers have been overturned.
4 The Permian layer has been totally eroded.

31 In New York State, which day has the shortest period of daylight?

1 March 21   3 September 21
2 June 21   4 December 21

32 Which diagram shows the position of the Earth relative to the Sun's rays during a winter day in the Northern Hemisphere?

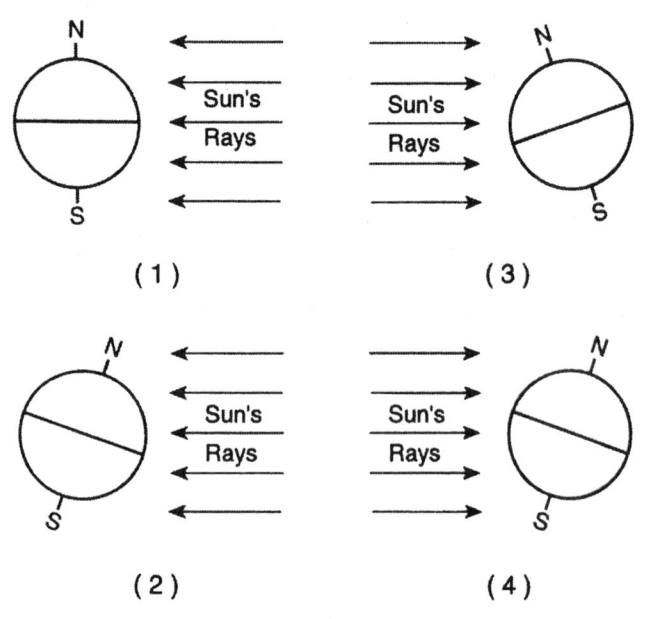

33 To an observer located at the Equator, on which date would the Sun appear to be directly overhead at noon?

1 February 1   3 March 21
2 June 6   4 December 21

34 The new-moon phase occurs when the Moon is positioned between the Earth and the Sun. However, these positions do not always cause an eclipse (blocking) of the Sun because the

1 Moon's orbit is tilted relative to the Earth's orbit
2 new-moon phase is visible only at night
3 night side of the Moon faces toward the Earth
4 apparent diameter of the Moon is greatest during the new-moon phase

35 The diagram below shows an instrument made from a drinking straw, protractor, string, and rock.

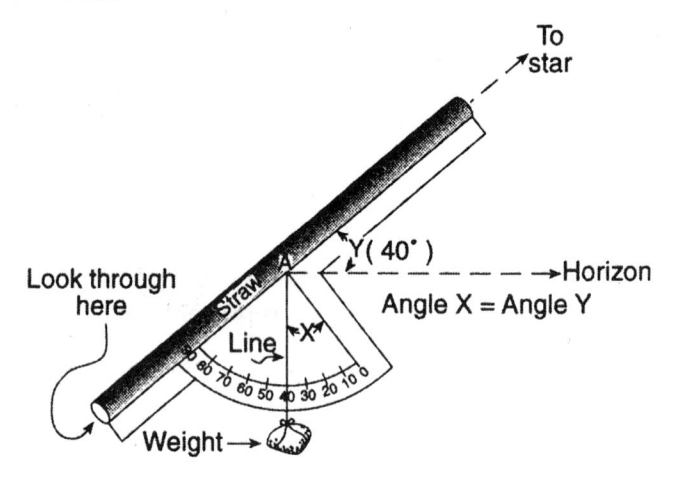

This instrument was most likely used to measure the

1 distance to a star
2 altitude of a star
3 mass of the Earth
4 mass of the suspended weight

Base your answers to questions 36 through 40 on the *Earth Science Reference Tables*, the weather map below, and your knowledge of Earth science. The weather map shows a hurricane that was located over southern Florida. The isobars show air pressure in inches of mercury. Letters A through D represent four widely separated locations.

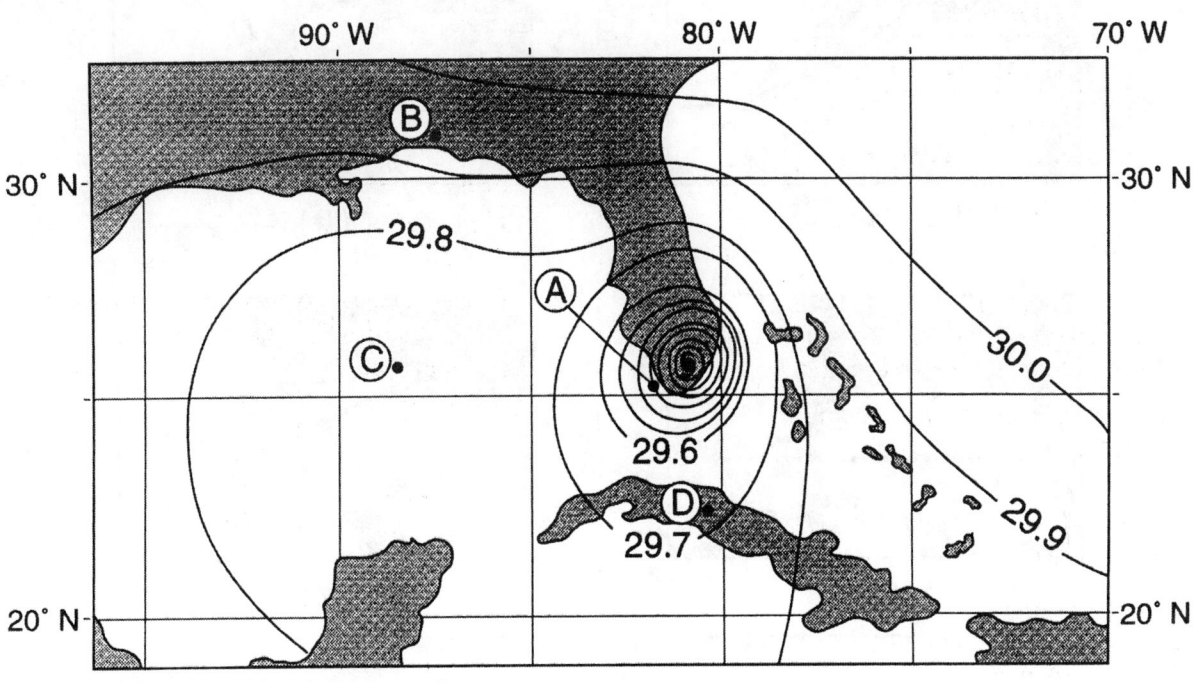

KEY - ᕒ = Hurricane center

36 What is the latitude and longitude at the center of the hurricane?

(1) 26° N 81° W (3) 34° N 81° W
(2) 26° N 89° W (4) 34° N 89° W

37 At which location were the winds of this hurricane the strongest?

(1) A (3) C
(2) B (4) D

38 What was the direction of movement of surface winds associated with this hurricane?

1 counterclockwise and away from the center
2 counterclockwise and toward the center
3 clockwise and away from the center
4 clockwise and toward the center

39 Which station model best represents some of the atmospheric conditions at location A?

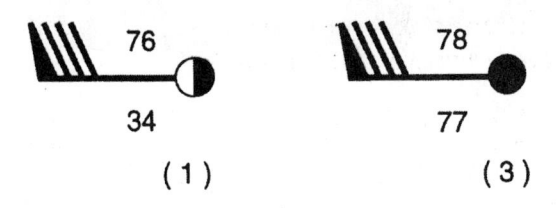

(1) 76 / 34
(2) 71 / 82
(3) 78 / 77
(4) 63 / 45

40 Which map best shows the most likely track of this hurricane?

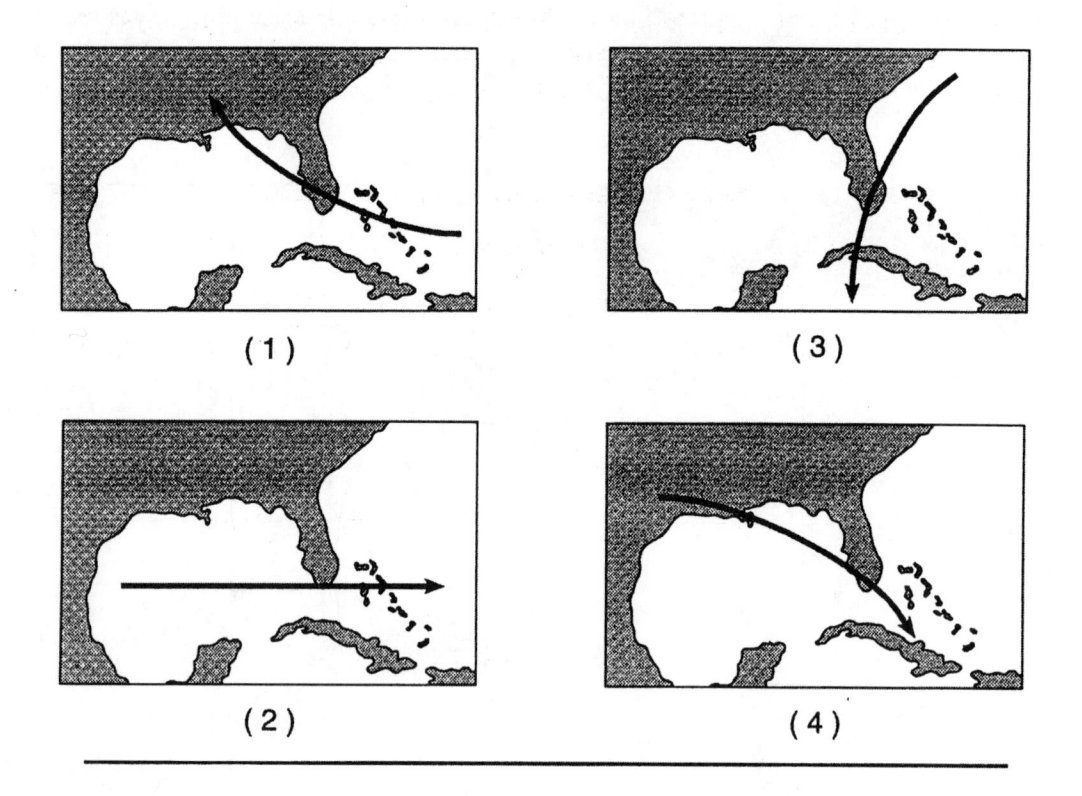

(1)

(2)

(3)

(4)

Part II

This part consists of six groups, each containing ten questions. Choose any one of these six groups. Be sure that you answer all ten questions in the single group chosen. Record the answers to these questions on the separate answer paper in accordance with the directions on the front cover of this booklet. [10]

Group A — Rocks and Minerals

If you choose this group, be sure to answer questions 41–50. Some questions may require the use of the *Earth Science Reference Tables*.

41 The most abundant element in the Earth's crust is

  1 nitrogen
  2 oxygen
  3 silicon
  4 hydrogen

42 Which sedimentary rock is formed by compaction and cementation of land-derived sediments?

  1 siltstone
  2 dolostone
  3 rock salt
  4 rock gypsum

43 The physical properties of minerals result from their

  1 density and color
  2 texture and color of streak
  3 type of cleavage and hardness
  4 internal arrangement of atoms

44 The bedrock of the flat areas on the Moon is mostly basalt. This fine-grained igneous rock was most likely formed by the

  1 cementing and compacting of sediments
  2 changes caused by heat and pressure on pre-existing rocks
  3 slow cooling of magma deep under the surface
  4 rapid cooling of molten rock in lava flows

45 Heat and pressure due to magma intrusions may result in

  1 vertical sorting
  2 graded bedding
  3 contact metamorphism
  4 chemical evaporites

46 In the diagram below, each angle of the triangle represents a 100 percent composition of the mineral named at that angle. The percentage of the mineral decreases toward 0 percent as either of the other angles of the triangle is approached. Letter A represents the mineral composition of an igneous rock.

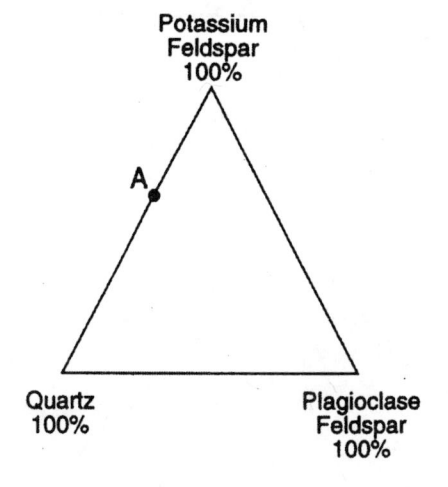

Rock A is a coarse-grained igneous rock that can best be identified as

  1 rhyolite
  2 pumice
  3 granite
  4 gabbro

47 What is one difference between the metamorphic rocks quartzite and hornfels?

  1 Hornfels is foliated; quartzite is nonfoliated.
  2 Hornfels contains plagioclase; quartzite does *not* contain plagioclase.
  3 Hornfels is produced by regional metamorphism; quartzite is produced by contact metamorphism.
  4 Hornfels is medium grained; quartzite is fine grained.

48 In which part of the Earth are felsic rocks most likely to be found?

   1 continental crust   3 plastic mantle
   2 oceanic crust   4 rigid mantle

49 Which symbol represents the sedimentary rock with the smallest grain size?

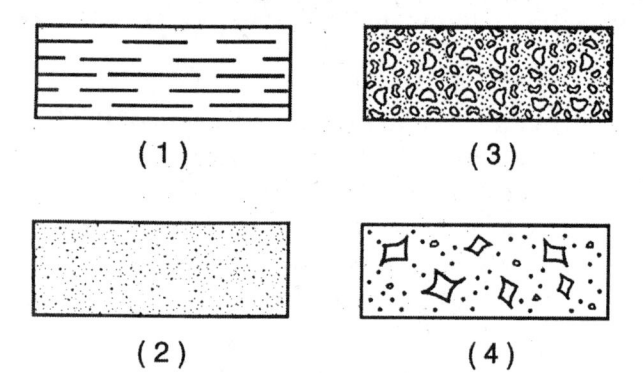

50 Which diagram best represents the silicon-oxygen tetrahedron of which talc, feldspar, and quartz are composed?

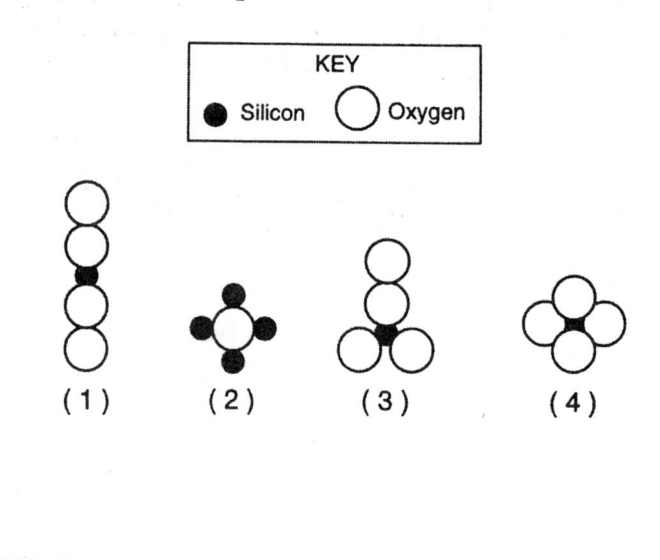

Group B — Plate Tectonics

If you choose this group, be sure to answer questions 51–60.

Base your answers to questions 51 through 55 on the *Earth Science Reference Tables*, the map below, and your knowledge of Earth science. The map shows mid-ocean ridges and trenches in the Pacific Ocean. Specific areas *A*, *B*, *C*, and *D* are indicated by shaded rectangles.

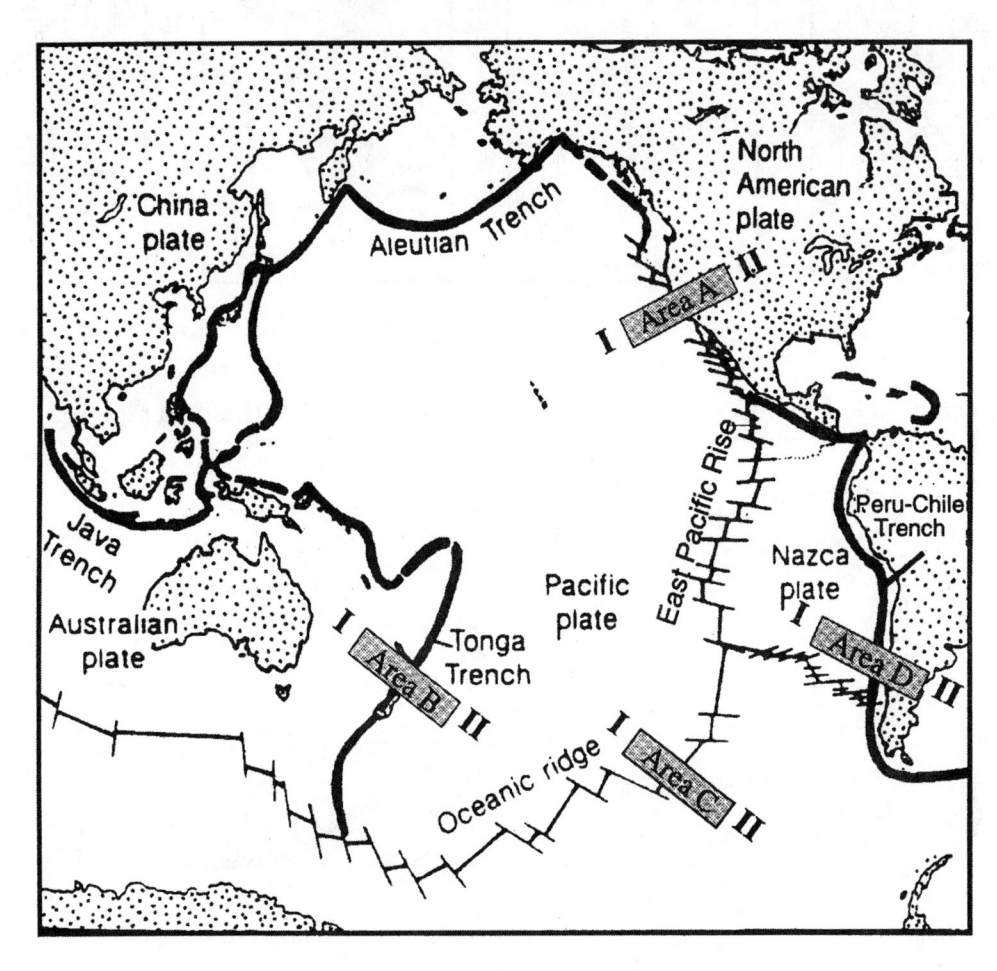

51 Movement of the crustal plates shown in the diagram is most likely caused by
  1 the revolution of the Earth
  2 the erosion of the Earth's crust
  3 shifting of the Earth's magnetic poles
  4 convection currents in the Earth's mantle

52 The crust at the mid-ocean ridges is composed mainly of
  1 shale   3 granite
  2 limestone   4 basalt

53 Mid-ocean ridges such as the East Pacific Rise and the Oceanic Ridge are best described as
  1 mountains containing folded sedimentary rocks
  2 mountains containing fossils of present-day marine life
  3 sections of the ocean floor that contain the youngest oceanic crust
  4 sections of the ocean floor that are the remains of a submerged continent

54 Which map best shows the direction of movement of the oceanic crustal plates in the vicinity of the East Pacific Rise (ridge)?

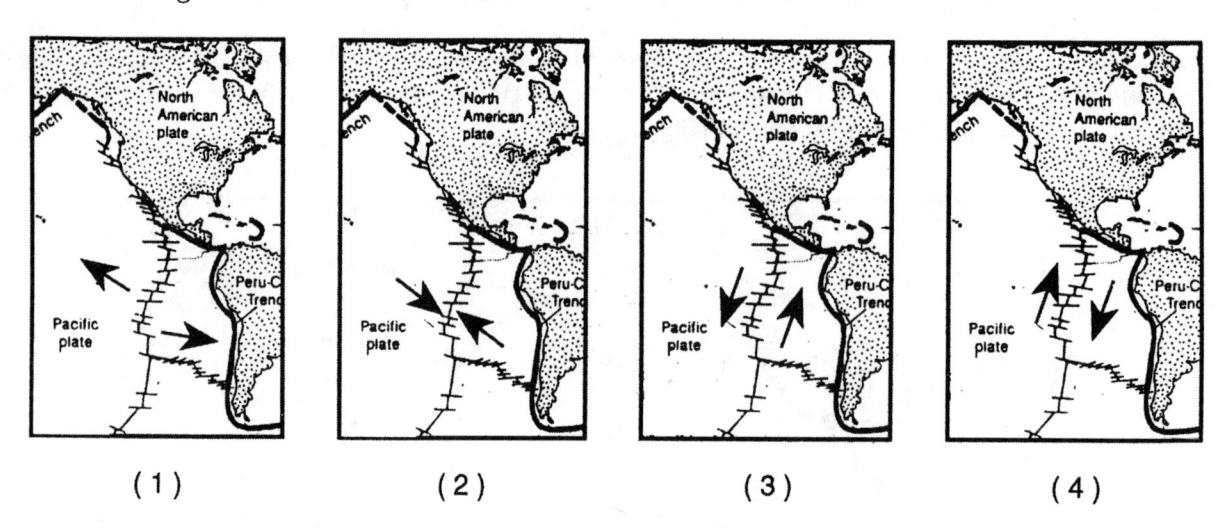

(1)   (2)   (3)   (4)

55 The cross section below represents an area of the Earth's crust within the map region.

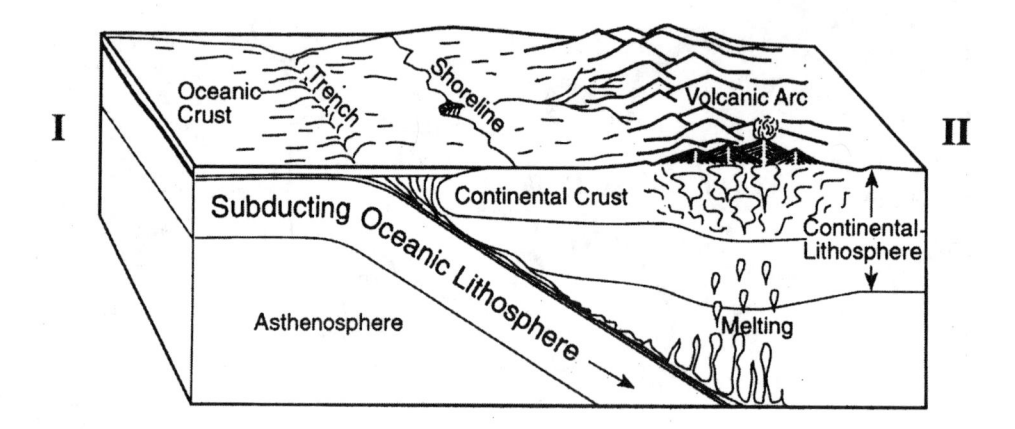

Which shaded rectangular area on the map does this cross section represent?
1 Area A
2 Area B
3 Area C
4 Area D

56 A seismogram recorded at a seismic station is shown below.

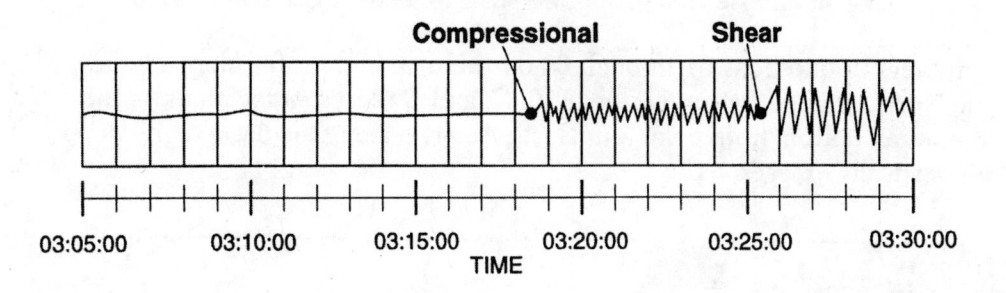

Which information can be determined by using this seismogram?

1. depth of the earthquake's focus
2. direction to the earthquake's focus
3. location of the earthquake's epicenter
4. distance to the earthquake's epicenter

57 Which statement best describes the relationship between the travel rates and travel times of earthquake P-waves and S-waves from the focus of an earthquake to a seismograph station?

(1) P-waves travel at a slower rate and take less time.
(2) P-waves travel at a faster rate and take less time.
(3) S-waves travel at a slower rate and take less time.
(4) S-waves travel at a faster rate and take less time.

58 An earthquake's P-wave traveled 4,800 kilometers and arrived at a seismic station at 5:10 p.m. At approximately what time did the earthquake occur?

(1) 5:02 p.m.
(2) 5:08 p.m.
(3) 5:10 p.m.
(4) 5:18 p.m.

59 The rock between 2,900 kilometers and 5,200 kilometers below the Earth's surface is inferred to be

1. an iron-rich solid
2. an iron-rich liquid
3. a silicate-rich solid
4. a silicate-rich liquid

60 Where is the thickest part of the Earth's crust?

1. at mid-ocean ridges
2. at transform faults
3. under continental mountain ranges
4. under volcanic islands

Group C — Oceanography
If you choose this group, be sure to answer questions 61–70.

Base your answers to questions 61 through 63 on the map and profile shown below. The map shows the major areas of the North Atlantic Ocean. Letters A, B, C, and D represent locations on the ocean floor. The profile represents the ocean bottom from point X in North America along the dashed line to point Y in Africa. Note that the profile is vertically exaggerated.

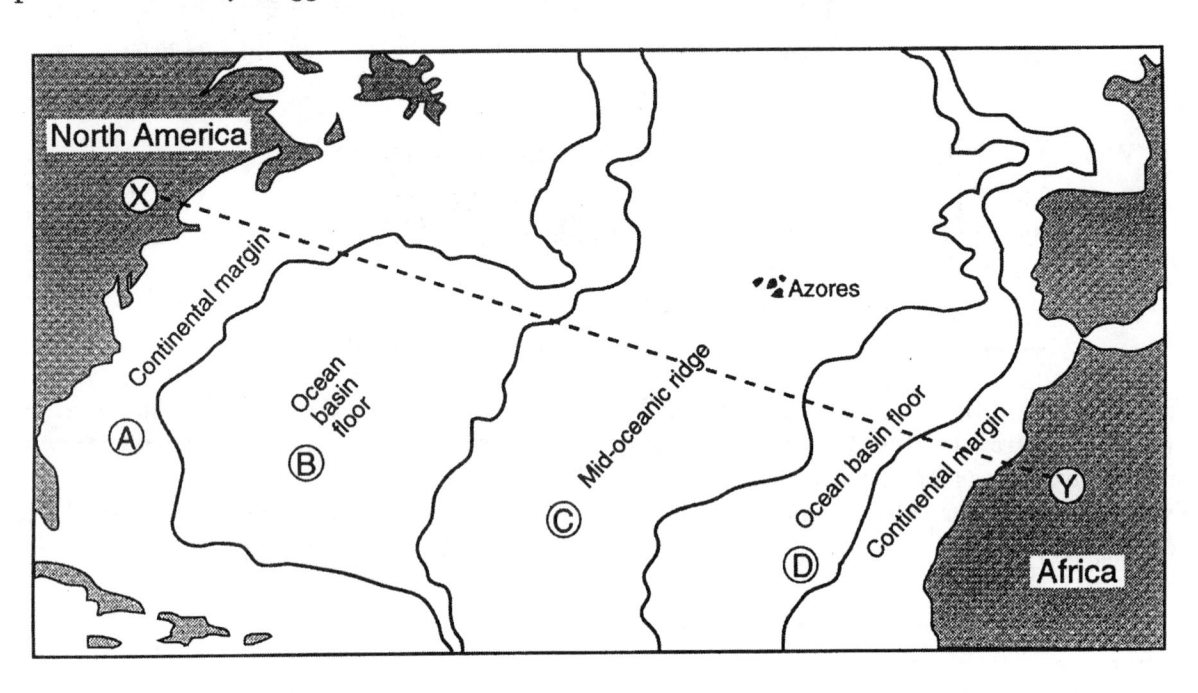

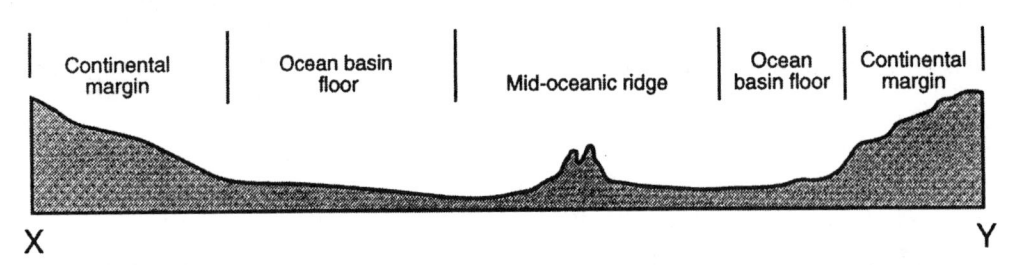

61 Classification of the ocean bottom into the areas shown is based on the
1 distance from continental landmasses
2 topography of the ocean floor
3 age of ocean-bottom rocks
4 type of ocean-bottom sediments

62 At which location would land-derived sediments most likely be accumulating on the ocean bottom?
(1) A
(2) B
(3) C
(4) D

63 Which statement about the age of ocean-floor rocks is correct?
1 All ocean-floor rocks are generally the same age.
2 Rocks at location C are generally older than rocks at locations A and B.
3 Rocks at location C are generally younger than rocks at locations A and B.
4 Igneous rocks at location D are generally younger than rocks at location C.

Base your answers to questions 64 through 67 on the diagram below, which shows ocean waves approaching a shoreline. A groin (a short wall of rocks perpendicular to the shoreline) and a breakwater (an offshore structure) have been constructed along the beach. Letters *A*, *B*, *C*, *D*, and *E* represent locations in the area.

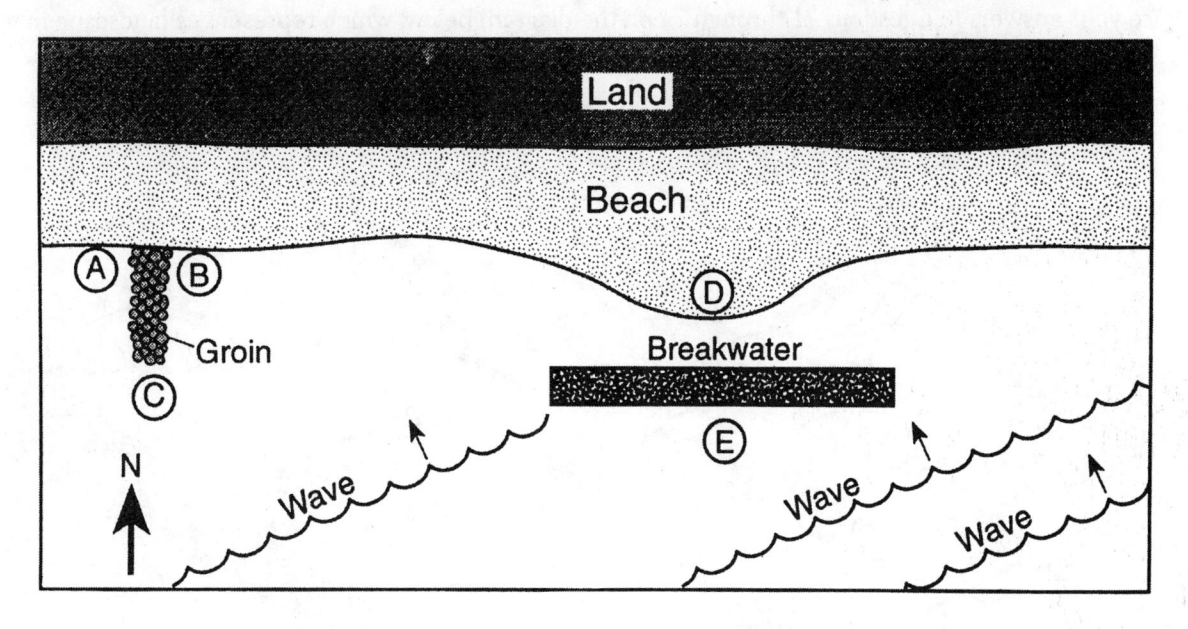

64 What is the most common cause of the approaching waves?

 1 underwater earthquakes
 2 variations in ocean-water density
 3 the gravitational effect of the Moon
 4 winds at the ocean surface

65 This shoreline is located along the east coast of North America. Which ocean current would most likely modify the climate of this shoreline?

 1 Florida Current   3 Brazil Current
 2 Canaries Current   4 California Current

66 At which location will the beach first begin to widen due to sand deposition?

 (1) *A*   (3) *C*
 (2) *B*   (4) *E*

Note that question 67 has only three choices.

67 The size of the bulge in the beach at position *D* will

 1 decrease
 2 increase
 3 remain the same

68 The curved pattern of surface currents in the North Atlantic is most affected by

 1 the Earth's rotation
 2 convection in the Earth's mantle
 3 density differences of ocean water
 4 gravitational attraction of the Moon

69 What is the source of most dissolved minerals in seawater?

 1 weathering of seafloor rocks
 2 weathering and erosion of continental rocks
 3 deep-ocean organic-matter sediments
 4 gases from underwater volcanic eruptions

70 When the seafloor moves as a result of an underwater earthquake and a large tsunami develops, what will most likely occur?

 1 Deep-ocean sediments will be transported over great distances.
 2 No destruction will occur near the origin of the earthquake.
 3 The direction of the tsunami will be determined by the magnitude of the earthquake.
 4 Severe destruction will occur in coastal areas.

Group D — Glacial Processes

If you choose this group, be sure to answer questions 71–80.

Base your answers to questions 71 through 74 on the diagram below, which represents a landscape in which sediments were deposited by a continental glacier. Letters *A*, *B*, *C*, *D*, and *E* represent locations in this area.

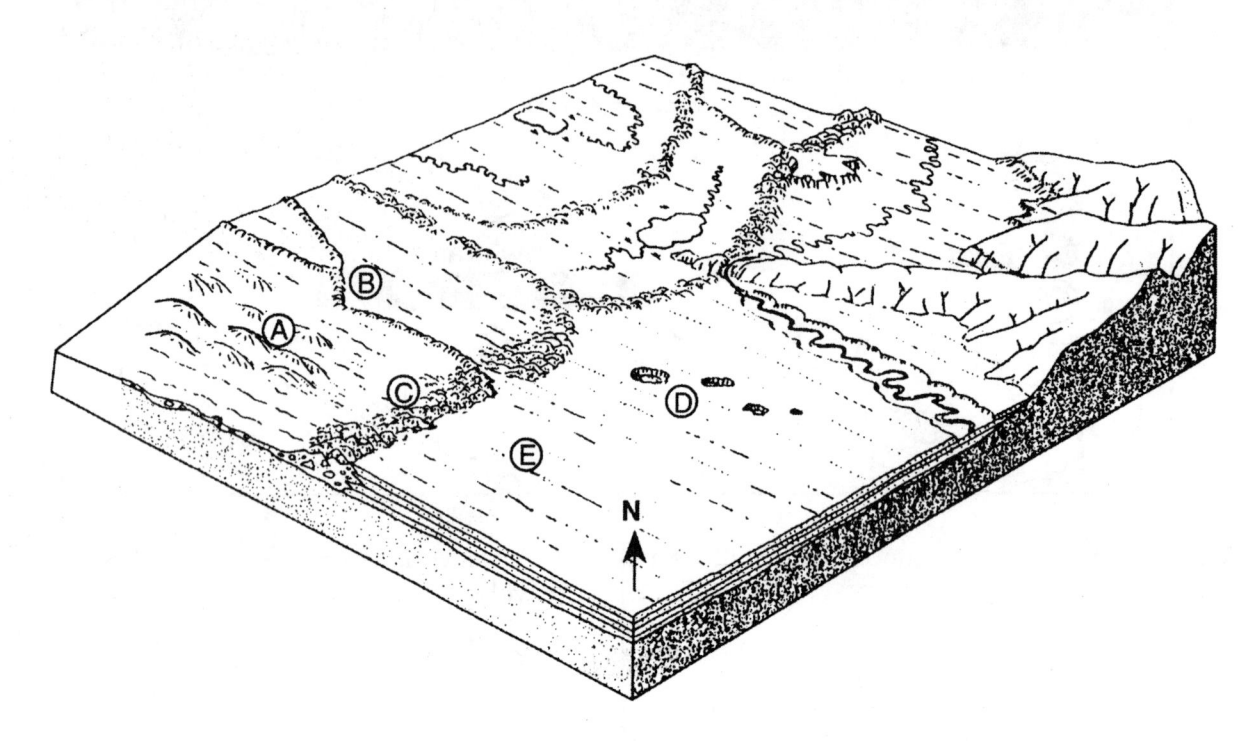

71 The sediments deposited at location *E* are best described as

1 sorted and unlayered
2 sorted and layered
3 unsorted and unlayered
4 unsorted and layered

72 A terminal moraine that marks the farthest advance of the glacier is found at location

(1) *A*   (3) *C*
(2) *B*   (4) *D*

73 Features formed by the melting of isolated ice blocks surrounded by sediment are found near which location?

(1) *A*   (3) *C*
(2) *B*   (4) *D*

74 Which arrow best represents the direction of ice movement that formed the deposits shown in the diagram?

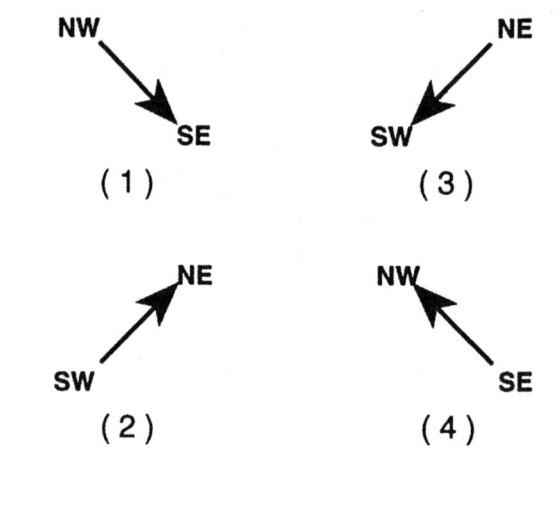

(1) NW → SE
(2) SW → NE
(3) NE → SW
(4) SE → NW

75 Which statement identifies a result of glaciation that has had a positive effect on the economy of New York State?
 1 Large amounts of oil and natural gas were formed.
 2 The number of usable water reservoirs was reduced.
 3 Many deposits of sand and gravel were formed.
 4 Deposits of fertile soil were removed.

76 The record of glaciation in New York State provides a source of information about
 1 changes in bedrock composition
 2 changes in the global climate
 3 movements of crustal plates
 4 plants and animals from the Paleozoic Era

77 Because of glaciation, New York State presently has soils that are best described as
 1 deep and residual
 2 rich in gemstone minerals
 3 unchanged by glaciation
 4 thin and rocky

78 At the present time, glaciers occur mostly in areas of
 1 high latitude or high altitude
 2 low latitude or low altitude
 3 middle latitude and high altitude
 4 middle latitude and low altitude

79 Wooden stakes were placed on a glacier in a straight line as represented by A–A' in the diagram below. The same stakes were observed later in the positions represented by B–B'.

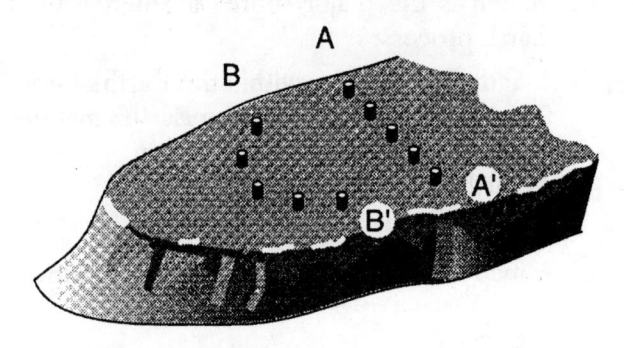

The pattern of movement of the stakes provides evidence that
 1 glacial ice does not move
 2 glacial ice is melting faster than it accumulates
 3 the glacier is moving faster in the center than on the sides
 4 friction is less along the sides of the glacier than in the center

80 Which landscape region is a flat plain consisting mainly of unsorted clays, gravels, sands, scratched pebbles, boulders, and cobbles?
 1 the Adirondacks   3 the Catskills
 2 Long Island   4 the Taconics

Group E — Atmospheric Energy

If you choose this group, be sure to answer questions 81–90. Some questions may require the use of the *Earth Science Reference Tables*.

81 Which is the major source of energy for most Earth processes?
1 radioactive decay within the Earth's interior
2 convection currents in the Earth's mantle
3 radiation received from the Sun
4 earthquakes along fault zones

82 A map of the United States is shown below.

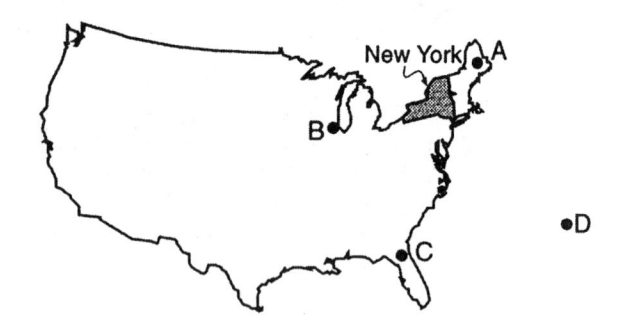

Weather conditions in which location would be of most interest to a person predicting the next day's weather for New York State?

(1) A (3) C
(2) B (4) D

83 Which statement about electromagnetic energy is correct?

(1) Violet light has a longer wavelength than red light.
(2) X rays have a longer wavelength than infrared waves.
(3) Radar waves have a shorter wavelength than ultraviolet rays.
(4) Gamma rays have a shorter wavelength than visible light.

84 The graph below shows the air temperature and air pressure recorded over a 30-day period at one location.

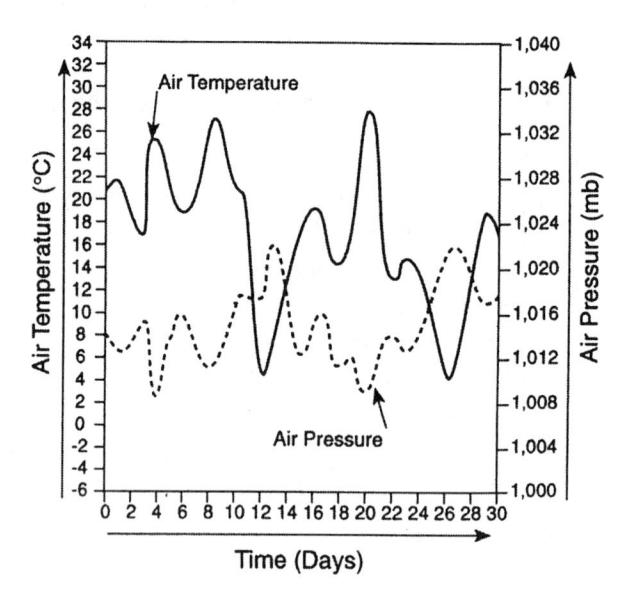

What were the approximate air temperature and air pressure readings on day 6?

(1) 10°C and 1,024 mb
(2) 19°C and 1,016 mb
(3) 10°C and 1,016 mb
(4) 19°C and 1,024 mb

85 Which planetary wind pattern is present in many areas of little rainfall?
1 Winds converge and air sinks.
2 Winds converge and air rises.
3 Winds diverge and air sinks.
4 Winds diverge and air rises.

86 Which weather conditions are most probable when the moisture content of the air increases, resulting in a lower atmospheric pressure?
1. sunny and fair
2. cold and windy
3. partly cloudy, with skies becoming clear
4. cloudy, with a chance of precipitation

87 On a sunny day at the beach, the dark-colored sand gets hot while the water stays cool because the sand
1. reflects less energy and has a lower specific heat than the water
2. reflects less energy and has a higher specific heat than the water
3. reflects more energy and has a lower specific heat than the water
4. reflects more energy and has a higher specific heat than the water

88 During which phase change does water absorb the most heat?
1. freezing
2. melting
3. condensation
4. evaporation

89 What is the wet-bulb temperature when the air temperature is 16°C and the relative humidity is 71%?
(1) 11°C
(2) 13°C
(3) 3°C
(4) 19°C

90 The diagram below represents the percentage of total incoming solar radiation that is affected by clouds.

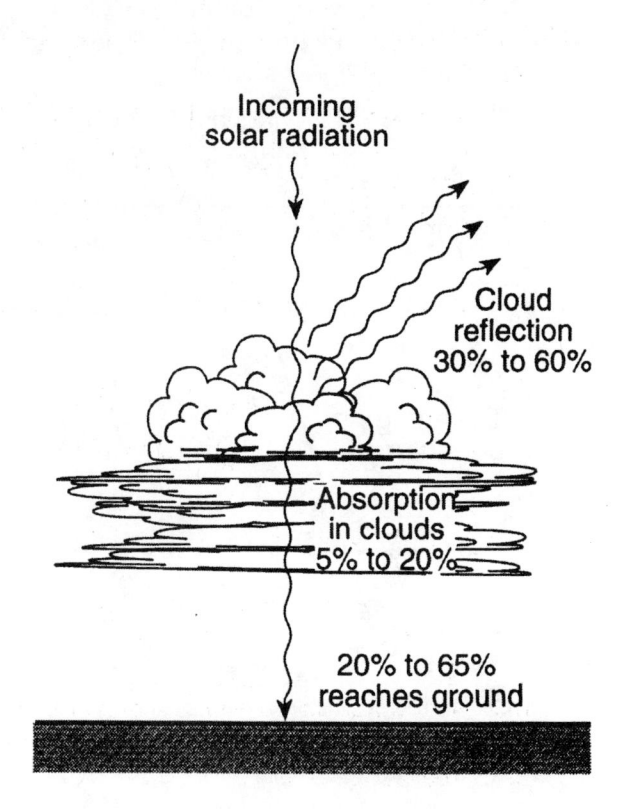

What percentage of incoming solar radiation is reflected or absorbed on cloudy days?
(1) 100%
(2) 35% to 80%
(3) 5% to 30%
(4) 0%

Group F — Astronomy

If you choose this group, be sure to answer questions 91–100.

Base your answers to questions 91 and 92 on the diagrams below, which represent two views of a swinging Foucault pendulum with a ring of 12 pegs at its base.

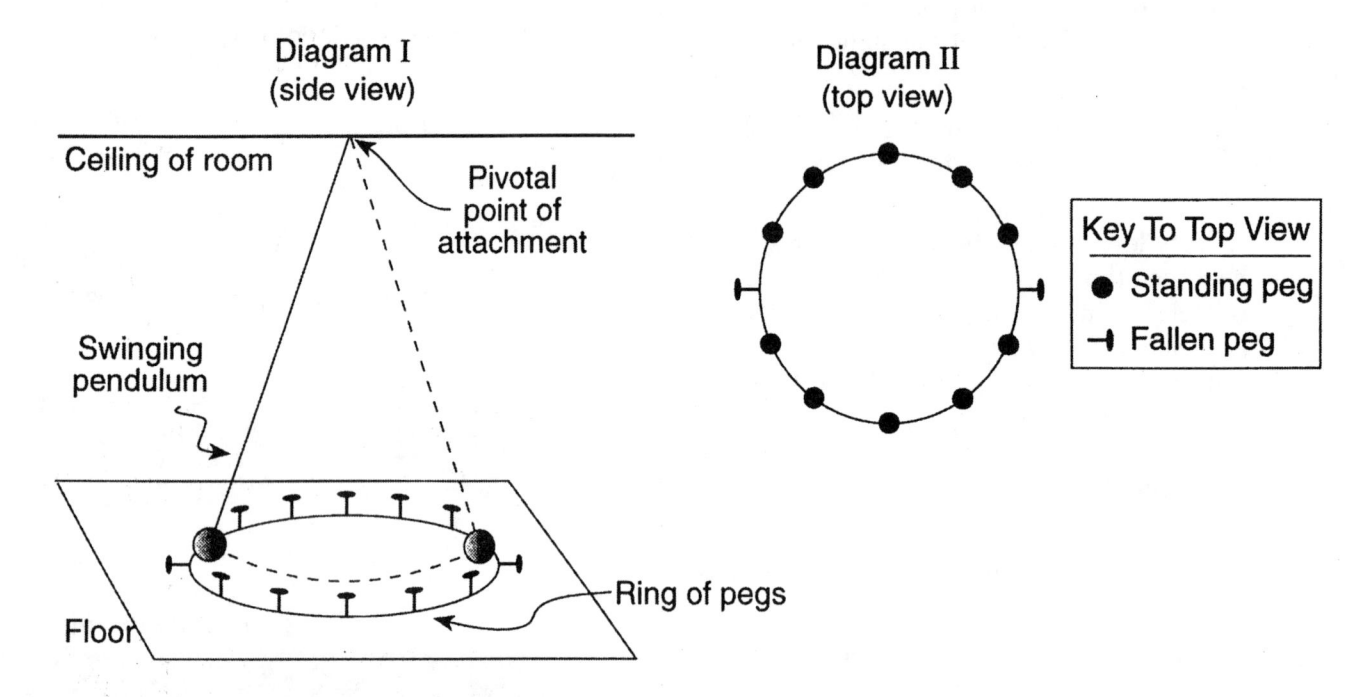

Diagram I
(side view)

Diagram II
(top view)

Key To Top View
● Standing peg
⊣ Fallen peg

91 Diagram II shows two pegs tipped over by the swinging pendulum at the beginning of the demonstration. Which diagram shows the pattern of standing pegs and fallen pegs after several hours?

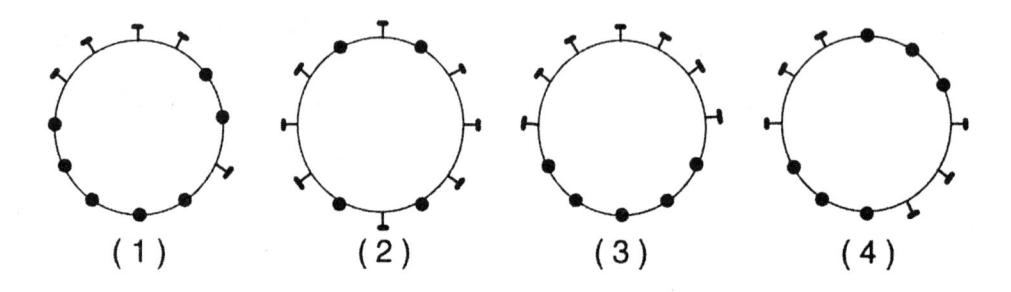

(1)   (2)   (3)   (4)

92 The predictable change in the direction of swing of a Foucault pendulum provides evidence that the

1 Sun rotates on its axis
2 Sun revolves around the Earth
3 Earth rotates on its axis
4 Earth revolves around the Sun

93 Which planet's orbital shape would be most similar to Jupiter's orbital shape?
1 Uranus   3 Venus
2 Pluto   4 Mercury

94 The diagram below shows several planets at various positions in their orbits at a particular time.

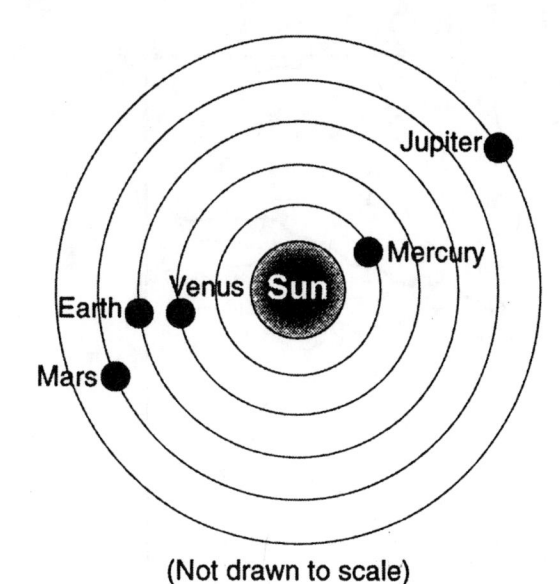

(Not drawn to scale)

Which planet would be visible from the Earth at night for the longest period of time when the planets are in these positions?
1 Mercury   3 Mars
2 Venus   4 Jupiter

95 According to current data, the Earth is apparently the only planet in our solar system that has
1 an orbiting moon
2 an axis of rotation
3 atmospheric gases
4 liquid water on its surface

96 Which statement best describes the geocentric model of our solar system?
1 The Earth is located at the center of the model.
2 All planets revolve around the Sun.
3 The Sun is located at the center of the model.
4 All planets *except* the Earth revolve around the Sun.

97 Which planet's day is longer than its year?
1 Mercury   3 Mars
2 Venus   4 Jupiter

Note that question 98 has only three choices.

98 Compared to the distances between the planets of our solar system, the distances between stars are usually
1 much less
2 much greater
3 about the same

99 In what way are the planets Mars, Mercury, and Earth similar?
1 They have the same period of revolution.
2 They are perfect spheres.
3 They exert the same gravitational force on each other.
4 They have elliptical orbits with the Sun at one focus.

100 The symbols below represent the Milky Way galaxy, the solar system, the Sun, and the universe.

◯ = Milky Way Galaxy
⬭ = Solar System
• = Sun
☐ = Universe

Which arrangement of symbols is most accurate?

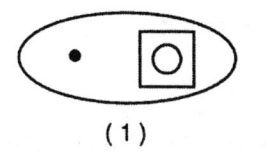

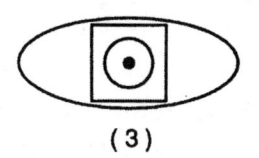

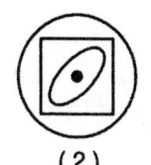

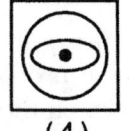

(1)   (2)   (3)   (4)

Part III

This part consists of questions 101 through 115. Be sure that you answer all questions in this part. Record your answers in the spaces provided on the separate answer paper. You may use pen or pencil. Some questions may require the use of the *Earth Science Reference Tables*. [25]

Base your answers to questions 101 through 104 on the topographic map of Cottonwood, Colorado, below. Points *A*, *B*, *X*, and *Y* are marked for reference.

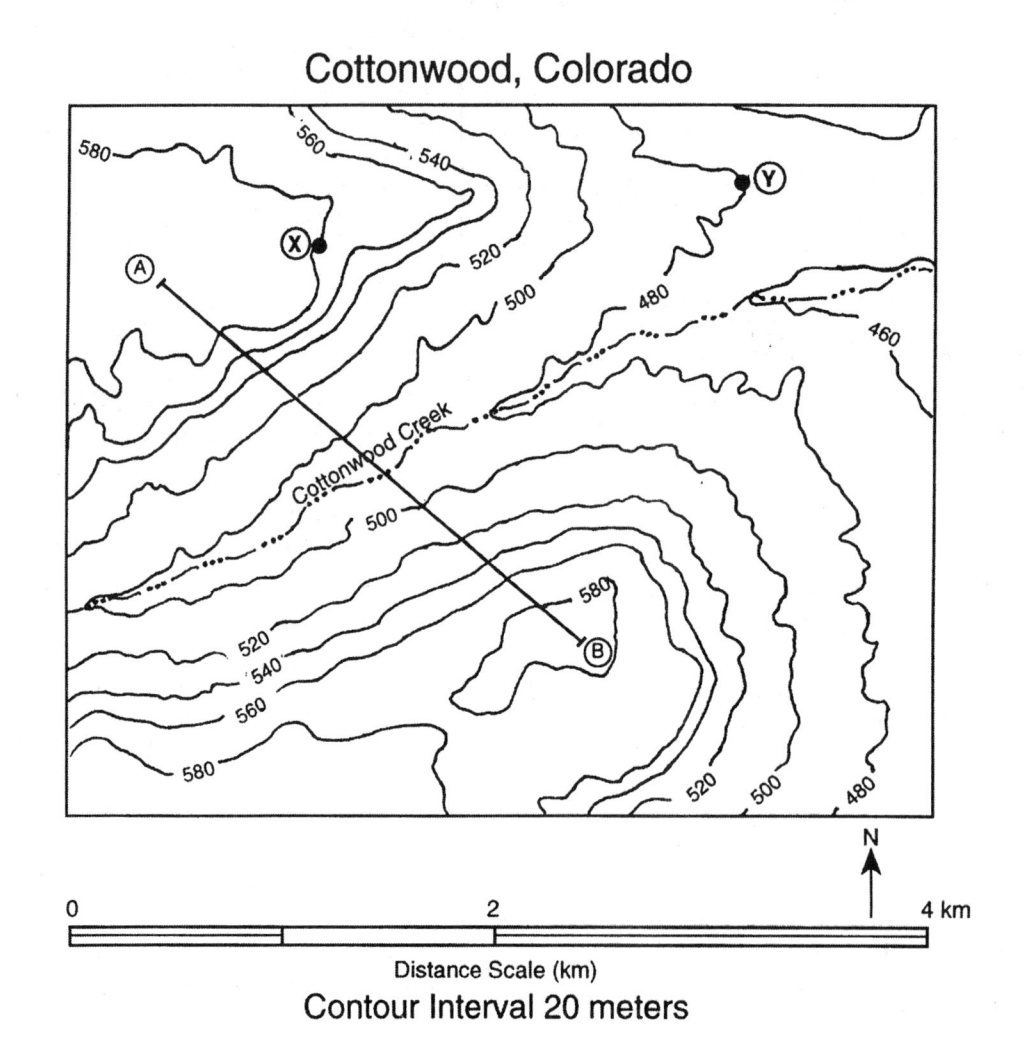

Cottonwood, Colorado

Distance Scale (km)
Contour Interval 20 meters

101 State the general direction in which Cottonwood Creek is flowing. [1]

102 State the highest possible elevation, to the *nearest meter*, for point *B* on the topographic map. [1]

103 On the grid provided *on your answer paper*, draw a profile of the topography along line *AB* shown on the map. [3]

104 In the space provided *on your answer paper*, calculate the gradient of the slope between points *X* and *Y* on the topographic map, following the directions below.

 a Write the equation for gradient. [1]
 b Substitute data from the map into the equation. [1]
 c Calculate the gradient and label it with the proper units. [2]

Base your answers to questions 105 through 108 on the information below and on your knowledge of Earth science.

The climate of an area is affected by many variables such as elevation, latitude, and distance to a large body of water. The effect of these variables on average surface temperature and temperature range can be represented by graphs on grids that have axes labeled as shown below. [The same grids appear on your answer paper.]

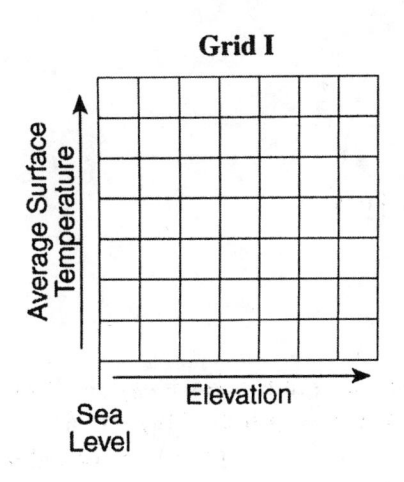

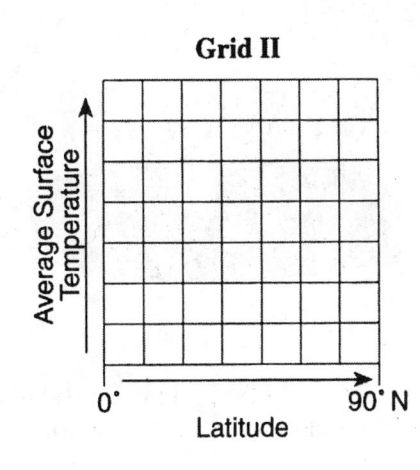

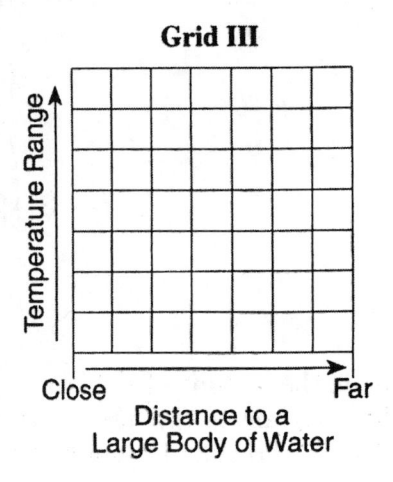

Grid I — Average Surface Temperature vs. Elevation (Sea Level to higher)

Grid II — Average Surface Temperature vs. Latitude (0° to 90° N)

Grid III — Temperature Range vs. Distance to a Large Body of Water (Close to Far)

105 On Grid I *on your answer paper*, draw a line to show the relationship between elevation and average surface temperature. [1]

106 On Grid II *on your answer paper*, draw a line to show the relationship between latitude and average surface temperature. [1]

107 On Grid III *on your answer paper*, draw a line to show the relationship between distance to a large body of water and temperature range. [1]

108 The climate of most locations near the Equator is warm and moist.
  a Explain why the climate is warm. [1]
  b Explain why the climate is moist. [1]

Base your answers to questions 109 through 111 on the diagram below and on your knowledge of Earth science. The diagram represents the apparent path of the Sun on the dates indicated for an observer in New York State. The diagram also shows the angle of Polaris above the horizon. [The same diagram appears on your answer paper.]

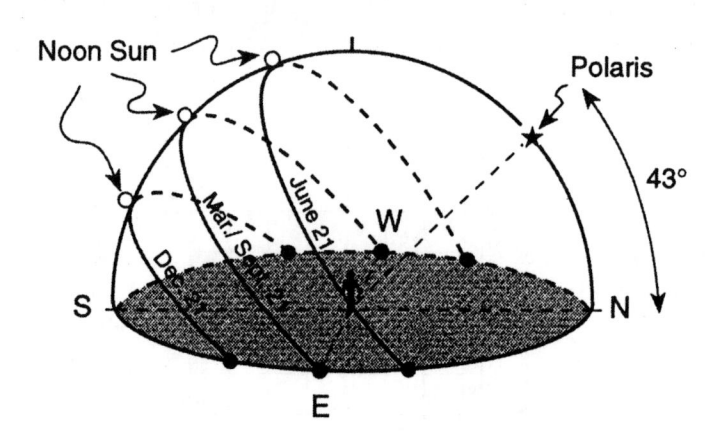

109 State the latitude of the location represented by the diagram to the *nearest degree*. Include the latitude direction in your answer. [2]

110 On the diagram *on your answer paper*, label the zenith. [1]

111 On the diagram *on your answer paper*, draw the apparent path of the Sun on May 21. Mark the position of sunrise on May 21 and label it *Sunrise*. [2]

Base your answers to questions 112 through 114 on the weather station data shown in the table below.

Air Temperature	21°C
Barometric Pressure	993.1 mb
Wind Direction	From the east
Windspeed	25 knots

112 State the air temperature in degrees Fahrenheit. [1]

113 State the barometric pressure in its proper form, as used on a station model. [1]

114 On the station model *on your answer paper*, draw a line with feathers to indicate the wind direction and speed. [2]

Base your answer to question 115 on the newspaper article below and your knowledge of Earth science.

Legislation Protects Ozone

The governor of New York signed environmental legislation that restricted the use of ozone-depleting chemicals employed in refrigeration systems, air-conditioners, and fire extinguishers.

The law restricts, and in some cases bans, the sale of chlorofluorocarbons and halons. Both have been found to contribute to the destruction of the Earth's ozone layer, which protects the Earth from dangerous ultraviolet rays of the Sun.

115 Using one or more complete sentences, state one reason that ultraviolet rays are dangerous. [2]

EARTH SCIENCE
PROGRAM MODIFICATION EDITION

JUNE 1996

ANSWER PAPER

Part I Credits	....	....
Part II Credits	....	....
Part III Credits	....	....
Performance Test Credits	.....	....
Local Project Credits	....	_____
Total (Official Regents) Examination Mark	....	....
Reviewer's Initials:		_____

Student ....

Teacher .... School ....

Grade (circle one)   8   9   10   11   12

Record all of your answers on this answer paper in accordance with the instructions on the front cover of the test booklet.

Part I (40 credits)

1	1	2	3	4	**15**	1	2	3	4	**29**	1	2	3	4
2	1	2	3	4	**16**	1	2	3	4	**30**	1	2	3	4
3	1	2	3	4	**17**	1	2	3	4	**31**	1	2	3	4
4	1	2	3	4	**18**	1	2	3	4	**32**	1	2	3	4
5	1	2	3	4	**19**	1	2	3	4	**33**	1	2	3	4
6	1	2	3	4	**20**	1	2	3	4	**34**	1	2	3	4
7	1	2	3	4	**21**	1	2	3		**35**	1	2	3	4
8	1	2	3	4	**22**	1	2	3	4	**36**	1	2	3	4
9	1	2	3	4	**23**	1	2	3	4	**37**	1	2	3	4
10	1	2	3	4	**24**	1	2	3	4	**38**	1	2	3	4
11	1	2	3	4	**25**	1	2	3	4	**39**	1	2	3	4
12	1	2	3	4	**26**	1	2	3	4	**40**	1	2	3	4
13	1	2	3	4	**27**	1	2	3	4					
14	1	2	3	4	**28**	1	2	3	4					

Part II (10 credits)

Answer the questions in only one of the six groups in this part. Be sure to mark the answers to the group you choose in accordance with the instructions on the front cover of the test booklet. Leave blank the spaces for the five groups of questions you do not choose to answer.

Group A
Rocks and Minerals

41	1	2	3	4
42	1	2	3	4
43	1	2	3	4
44	1	2	3	4
45	1	2	3	4
46	1	2	3	4
47	1	2	3	4
48	1	2	3	4
49	1	2	3	4
50	1	2	3	4

Group B
Plate Tectonics

51	1	2	3	4
52	1	2	3	4
53	1	2	3	4
54	1	2	3	4
55	1	2	3	4
56	1	2	3	4
57	1	2	3	4
58	1	2	3	4
59	1	2	3	4
60	1	2	3	4

Group C
Oceanography

61	1	2	3	4
62	1	2	3	4
63	1	2	3	4
64	1	2	3	4
65	1	2	3	4
66	1	2	3	4
67	1	2	3	
68	1	2	3	4
69	1	2	3	4
70	1	2	3	4

Group D
Glacial Processes

71	1	2	3	4
72	1	2	3	4
73	1	2	3	4
74	1	2	3	4
75	1	2	3	4
76	1	2	3	4
77	1	2	3	4
78	1	2	3	4
79	1	2	3	4
80	1	2	3	4

Group E
Atmospheric Energy

81	1	2	3	4
82	1	2	3	4
83	1	2	3	4
84	1	2	3	4
85	1	2	3	4
86	1	2	3	4
87	1	2	3	4
88	1	2	3	4
89	1	2	3	4
90	1	2	3	4

Group F
Astronomy

91	1	2	3	4
92	1	2	3	4
93	1	2	3	4
94	1	2	3	4
95	1	2	3	4
96	1	2	3	4
97	1	2	3	4
98	1	2	3	
99	1	2	3	4
100	1	2	3	4

Part III (25 credits)
Answer all questions in this part.

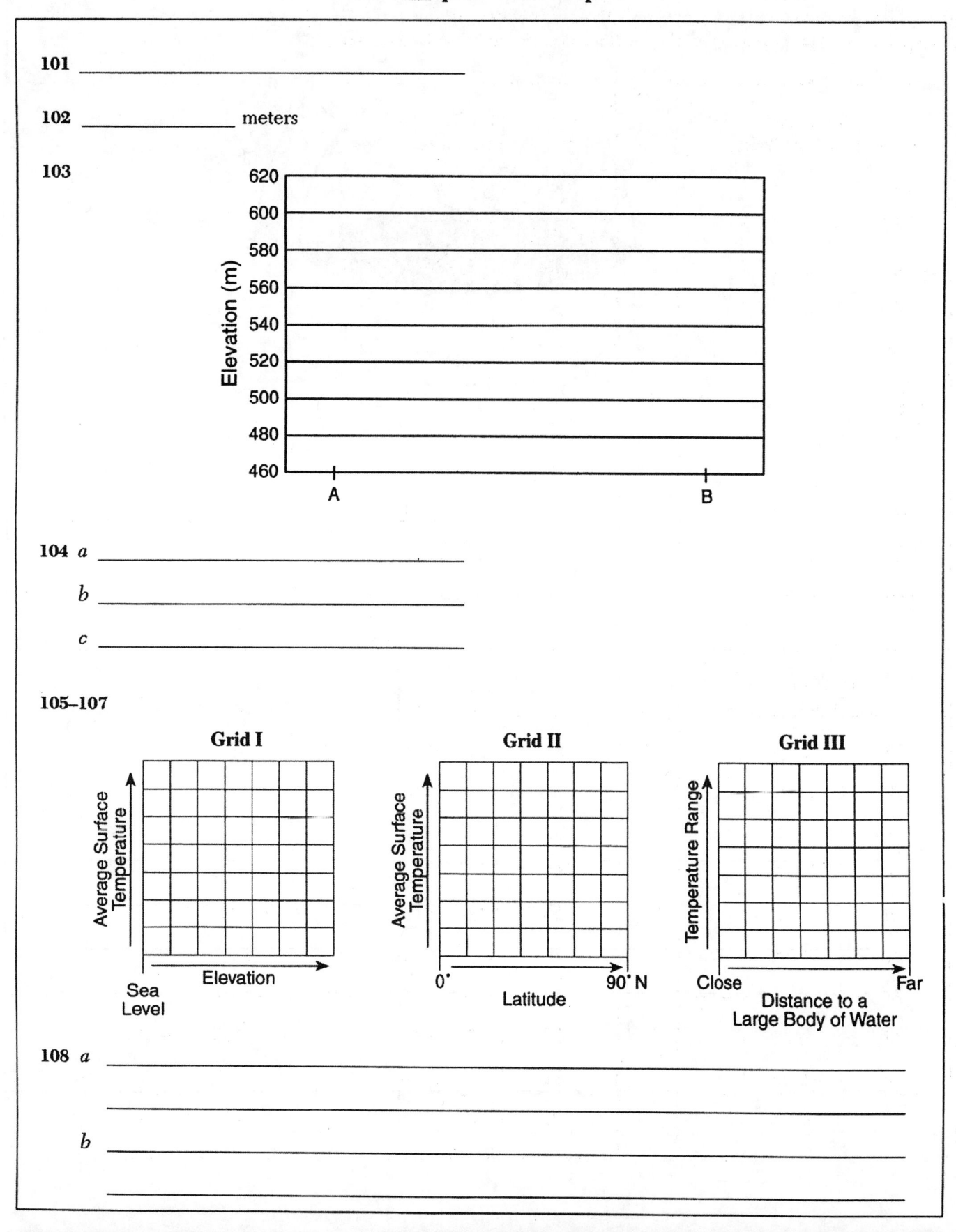

101 _____

102 _____ meters

103

Elevation (m)
620
600
580
560
540
520
500
480
460

A   B

104 a _____

b _____

c _____

105–107

Grid I
Average Surface Temperature vs Elevation (Sea Level →)

Grid II
Average Surface Temperature vs Latitude (0° → 90° N)

Grid III
Temperature Range vs Distance to a Large Body of Water (Close → Far)

108 a _____
_____

b _____
_____

109 _____

110–111

Noon Sun — Polaris — 43°
Mar./S... June 21
Dec...
S W N
E

112 _____ °F

113 _____

114

○

N ↑

115 _____
_____
_____
_____

PRACTICE EXAM
MODIFIED EARTH SCIENCE PROGRAM
JUNE 1997

Part I
Answer all 40 questions in this part.

Directions (1–40): For *each* statement or question, select the word or expression that, of those given, best completes the statement or answers the question. Record your answer on the separate answer paper in accordance with the directions on the front page of this booklet. Some questions may require the use of the *Earth Science Reference Tables*. [40]

1 A large earthquake occurred at 45° N 75° W on September 5, 1994. Which location in New York State was closest to the epicenter of the earthquake?

  1 Buffalo   3 Albany
  2 Massena   4 New York City

2 Isolines on the map below show elevations above sea level, measured in meters.

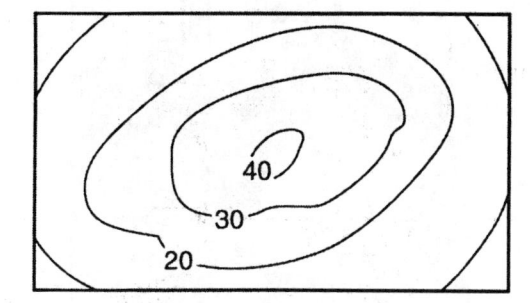

What is the highest possible elevation represented on this map?

  (1) 39 m   (3) 49 m
  (2) 41 m   (4) 51 m

3 A conglomerate contains pebbles of limestone, sandstone, and granite. Based on this information, which inference about the pebbles in the conglomerate is most accurate?

  1 They had various origins.
  2 They came from other conglomerates.
  3 They are all the same age.
  4 They were eroded quickly.

5 Which statement about the formation of a rock is best supported by the rock cycle?

  1 Magma must be weathered before it can change to metamorphic rock.
  2 Sediment must be compacted and cemented before it can change to sedimentary rock.
  3 Sedimentary rock must melt before it can change to metamorphic rock.
  4 Metamorphic rock must melt before it can change to sedimentary rock.

6 Which granite sample most likely formed from magma that cooled and solidified at the slowest rate?

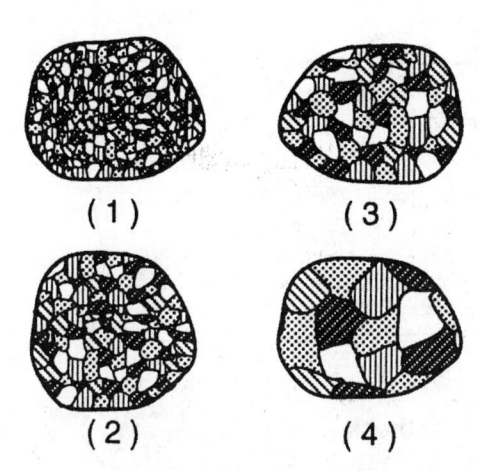

7 The interpretation that the Earth's outer core is liquid was made primarily from

  1 deep-sea drilling data   3 seismic data
  2 magnetic data   4 satellite data

9 The cartoon below presents a humorous view of Earth science.

The cartoon character on the right realizes that the sand castle will eventually be

1 compacted into solid bedrock
2 removed by agents of erosion
3 preserved as fossil evidence
4 deformed during metamorphic change

10 A deposit of rock particles that are scratched and unsorted has most likely been transported and deposited by

1 wind
2 glacial ice
3 running water
4 ocean waves

11 The diagram below shows a process called frost wedging.

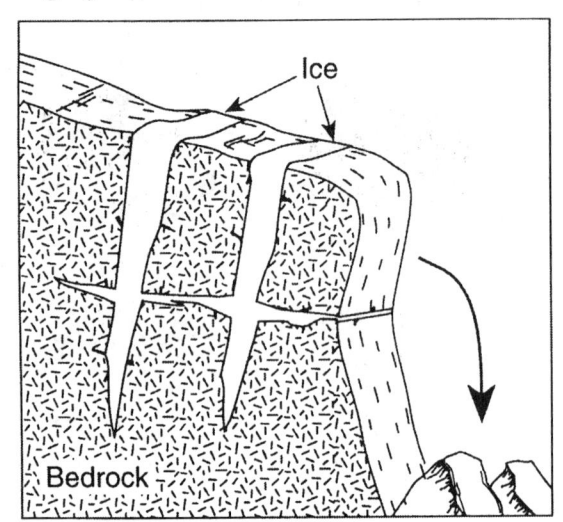

Frost wedging is an example of

1 weathering
2 cementing
3 metamorphism
4 deposition

12 The diagram below shows a soil profile formed in an area of granite bedrock. Four different soil horizons, A, B, C, and D, are shown.

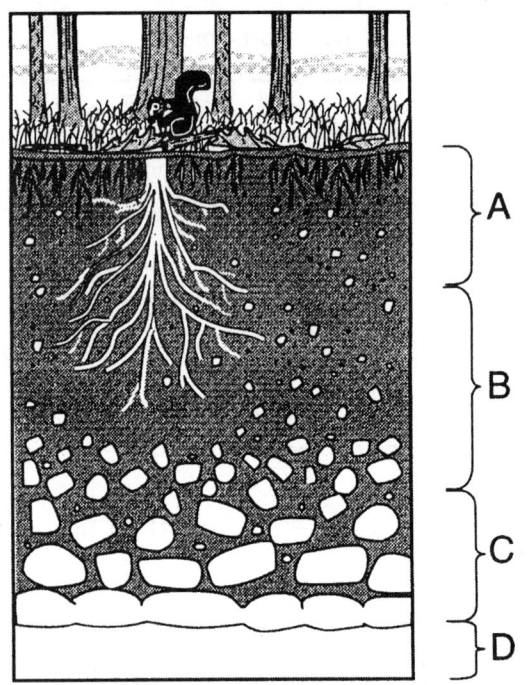

Which soil horizon contains the greatest amount of material formed by biological activity?

(1) A
(2) B
(3) C
(4) D

13 The map below shows the ancient location of evaporating seawater, which formed the Silurian-age deposits of rock salt and rock gypsum now found in some New York State crustal bedrock.

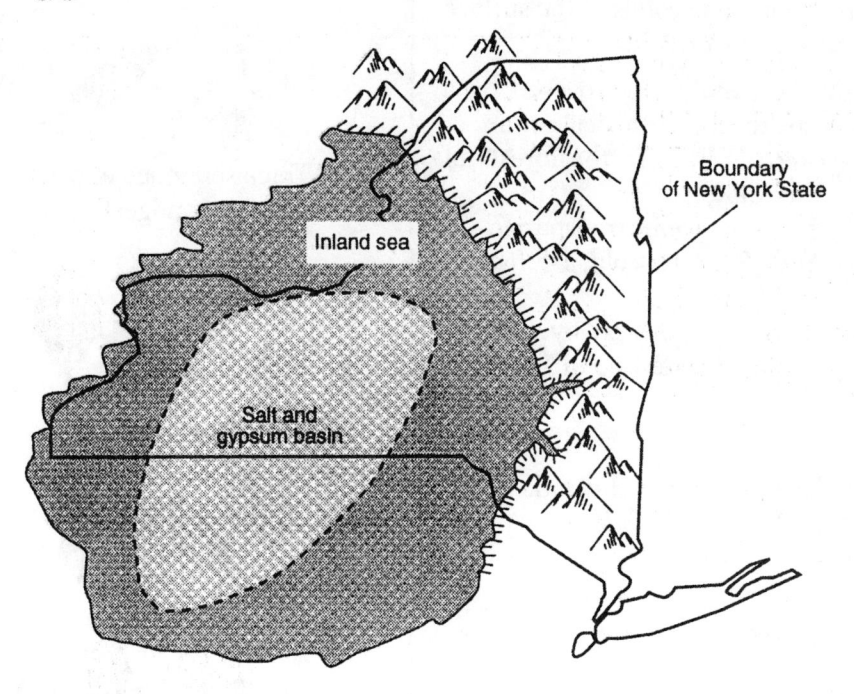

Within which two landscape regions are these large rock salt and rock gypsum deposits found?

1 Hudson Highlands and Taconic Mountains
2 Tug Hill Plateau and Adirondack Mountains
3 Erie-Ontario Lowlands and Allegheny Plateau
4 the Catskills and Hudson-Mohawk Lowlands

14 The diagrams below represent geologic cross sections from two widely separated regions.

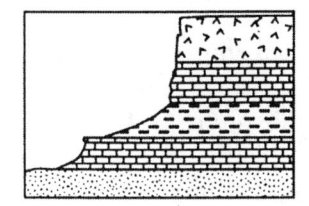

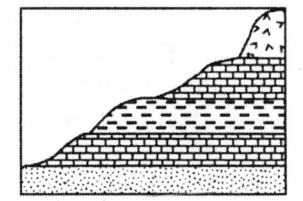

The layers of rock appear very similar, but the hillslopes and shapes are different. These differences are most likely the result of

1 volcanic eruptions
2 earthquake activity
3 soil formation
4 climate variations

15 Which statement correctly describes an age relationship in the geologic cross section below?

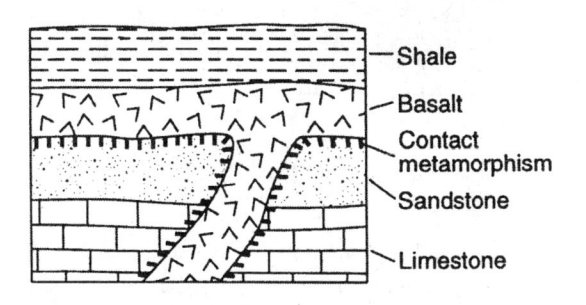

1 The sandstone is younger than the basalt.
2 The shale is younger than the basalt.
3 The limestone is younger than the shale.
4 The limestone is younger than the basalt.

16 Present-day corals live in warm, tropical ocean water. Which inference is best supported by the discovery of Ordovician-age corals in the surface bedrock of western New York State?

1 Western New York State was covered by a warm, shallow sea during Ordovician time.
2 Ordovician-age corals lived in the forests of western New York State.
3 Ordovician-age corals were transported to western New York State by cold, freshwater streams.
4 Western New York was covered by a continental ice sheet that created coral fossils of Ordovician time.

17 The diagram below represents the radioactive decay of uranium-238.

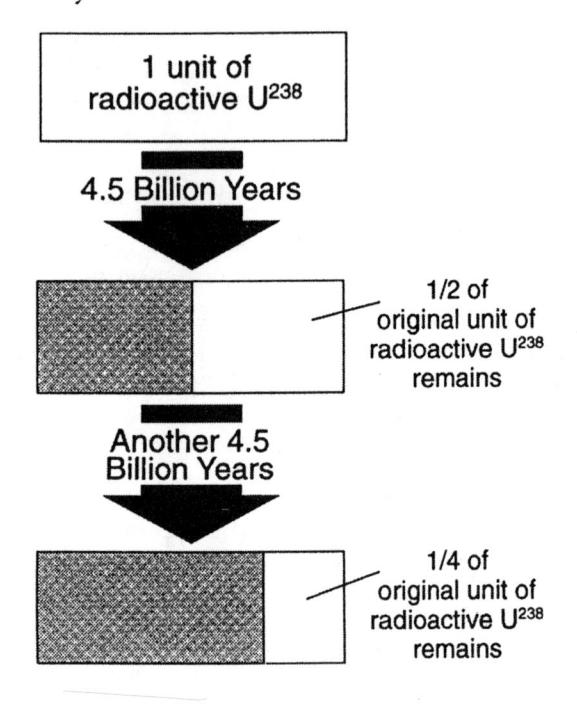

Shaded areas on the diagram represent the amount of

1 undecayed radioactive uranium-238 (U^{238})
2 undecayed radioactive rubidium-87 (Rb^{87})
3 stable carbon-14 (C^{14})
4 stable lead-206 (Pb^{206})

18 By which process does water vapor change into clouds?

1 condensation
2 evaporation
3 convection
4 precipitation

19 A geologist collected the fossils shown below from locations in New York State.

Which sequence correctly shows the fossils from oldest to youngest?

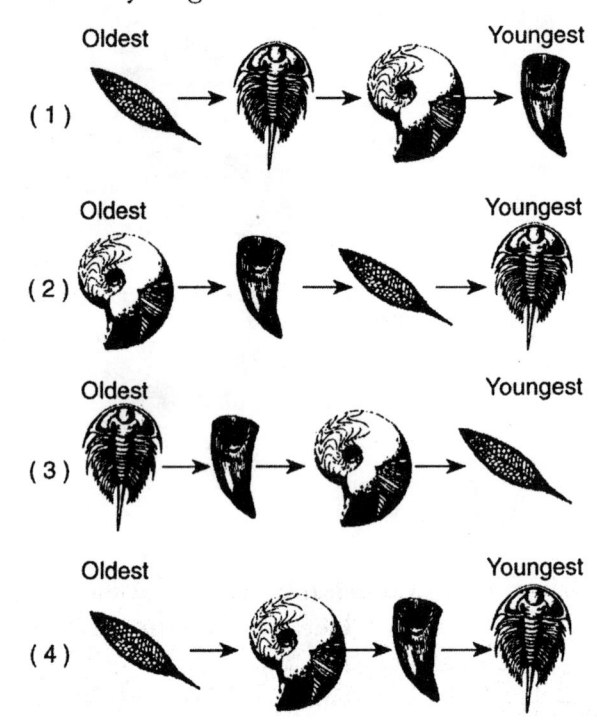

20 Which list shows atmospheric layers in the correct order upward from the Earth's surface?

1 thermosphere, mesosphere, stratosphere, troposphere
2 troposphere, stratosphere, mesosphere, thermosphere
3 stratosphere, mesosphere, troposphere, thermosphere
4 thermosphere, troposphere, mesosphere, stratosphere

21 Which statement best explains why precipitation occurs at frontal boundaries?

1 Cold fronts move slower than warm fronts.
2 Cold fronts move faster than warm fronts.
3 Warm, moist air sinks when it meets cold, dry air.
4 Warm, moist air rises when it meets cold, dry air.

Base your answers to questions 22 and 23 on the weather instrument shown in the diagram below.

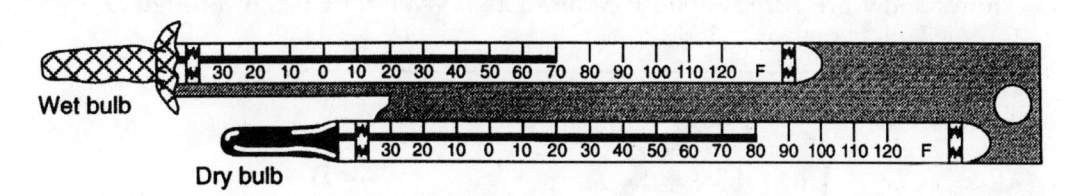

22 What are the equivalent Celsius temperature readings for the Fahrenheit readings shown?

1 wet 21°C, dry 27°C
2 wet 26°C, dry 37°C
3 wet 70°C, dry 80°C
4 wet 158°C, dry 176°C

23 Which weather variables are most easily determined by using this weather instrument?

1 air temperature and windspeed
2 visibility and wind direction
3 relative humidity and dewpoint
4 air pressure and cloud type

24 How do clouds affect the temperature at the Earth's surface?

1 Clouds block sunlight during the day and prevent heat from escaping at night.
2 Clouds block sunlight during the day and allow heat to escape at night.
3 Clouds allow sunlight to reach the Earth during the day and prevent heat from escaping at night.
4 Clouds allow sunlight to reach the Earth during the day and allow heat to escape at night.

25 A low-pressure system is shown on the weather map below.

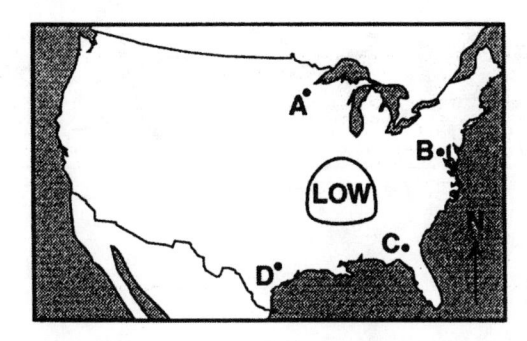

Toward which point will the low-pressure system move if it follows a normal storm track?

(1) A
(2) B
(3) C
(4) D

26 Compared to an inland location of the same elevation and latitude, a coastal location is likely to have

1 warmer summers and cooler winters
2 warmer summers and warmer winters
3 cooler summers and cooler winters
4 cooler summers and warmer winters

27 Which graph best represents the relationship between average yearly temperature and latitude?

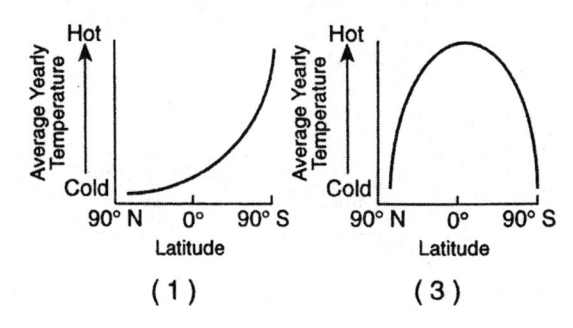

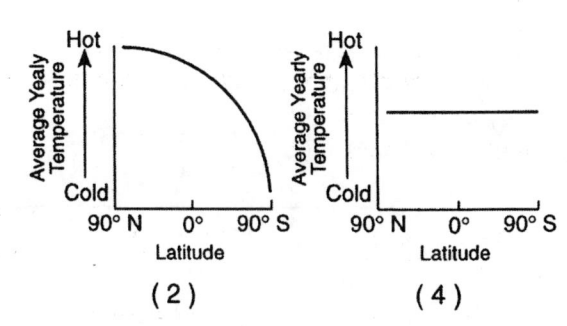

Base your answers to questions 28 through 30 on the *Earth Science Reference Tables* and the weather map below. The map shows a low-pressure system. Weather data is given for cities A through D.

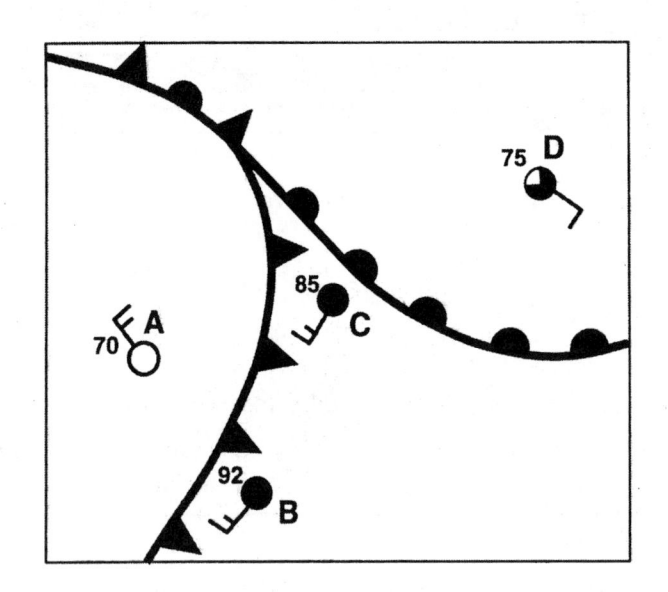

28 Which map correctly shows the locations of the continental polar (cP) and maritime tropical (mT) air masses?

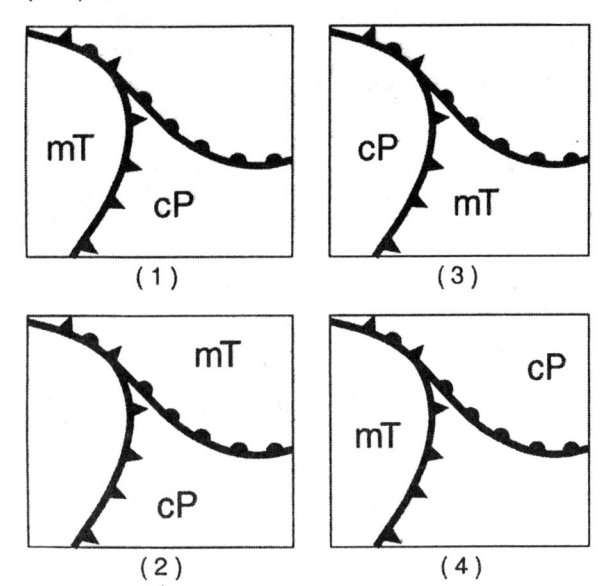

29 Which city is *least* likely to have precipitation starting in the next few hours?
(1) A
(2) B
(3) C
(4) D

30 Which map correctly shows arrows indicating the probable surface wind pattern?

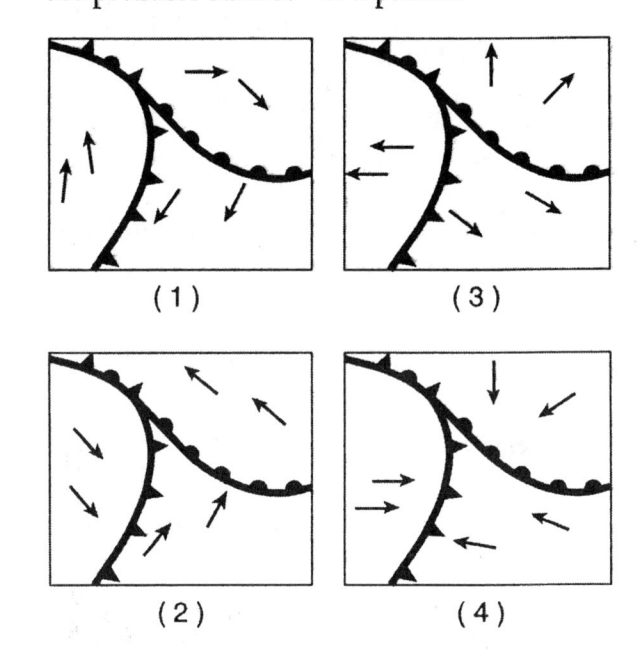

Base your answers to questions 31 through 33 on the *Earth Science Reference Tables* and the map below. The map represents a view of the Earth looking down from above the North Pole, showing the Earth's 24 standard time zones. The Sun's rays are striking the Earth from the right. Points A, B, C, and D are locations on the Earth's surface.

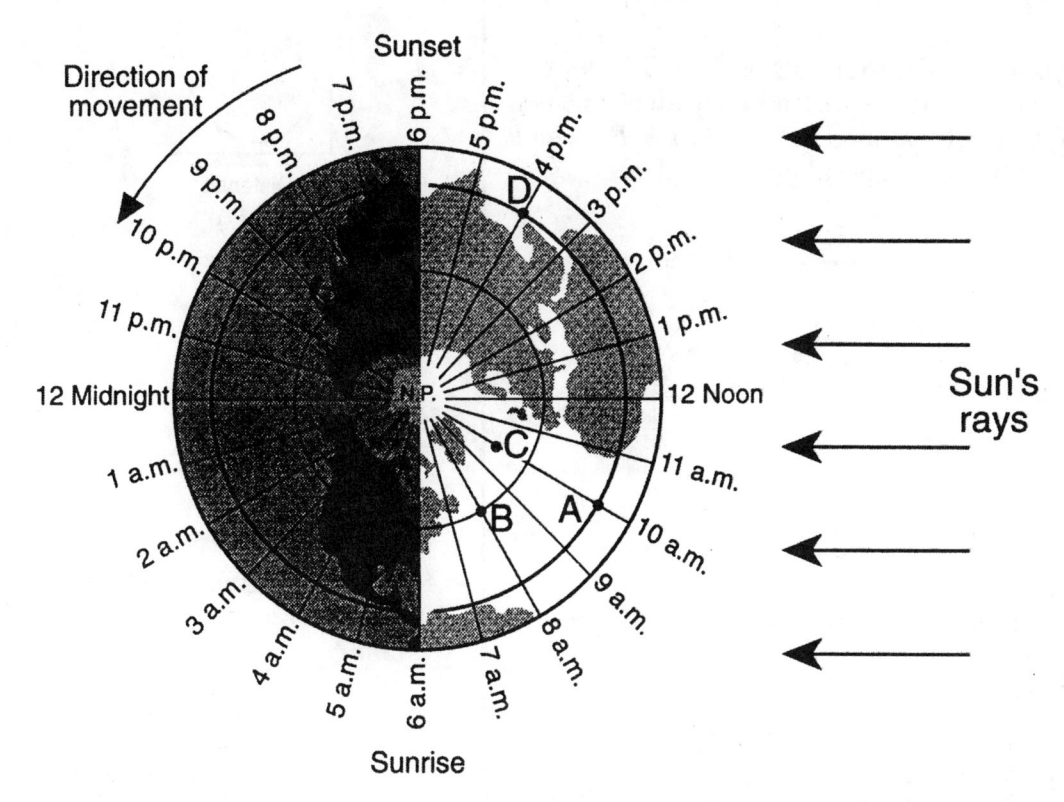

31 At which position would the altitude of the North Star (Polaris) be greatest?

(1) A   (3) C
(2) B   (4) D

32 Which date could this diagram represent?

1 January 21   3 June 21
2 March 21   4 August 21

33 Areas within a time zone generally keep the same standard clock time. In degrees of longitude, approximately how wide is one standard time zone?

(1) $7\frac{1}{2}°$ (3) $23\frac{1}{2}°$
(2) 15°   (4) 30°

34 Why do the locations of sunrise and sunset vary in a cyclical pattern throughout the year?

1 The Earth rotates on a tilted axis while revolving around the Sun.
2 The Sun rotates on a tilted axis while revolving around the Earth.
3 The Earth's orbit around the Sun is an ellipse.
4 The Sun's orbit around the Earth is an ellipse.

35 Compared to Jupiter and Saturn, Venus and Mars have greater

1 periods of revolution
2 orbital velocities
3 mean distances from the Sun
4 equatorial diameters

36 Billions of stars in the same region of the universe are called

1 solar systems
2 asteroid belts
3 constellations
4 galaxies

Base your answers to questions 37 and 38 on the diagram below, which represents the path of a planet in an elliptical orbit around a star. Points A, B, C, and D indicate four orbital positions of the planet.

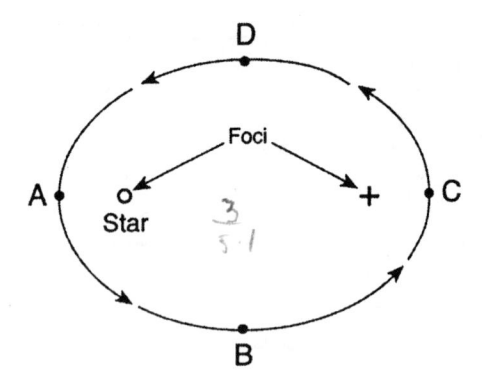

37 The eccentricity of the planet's orbit is approximately

(1) 0.18
(2) 0.65
(3) 1.55
(4) 5.64

38 Which graph best represents the gravitational attraction between the star and the planet?

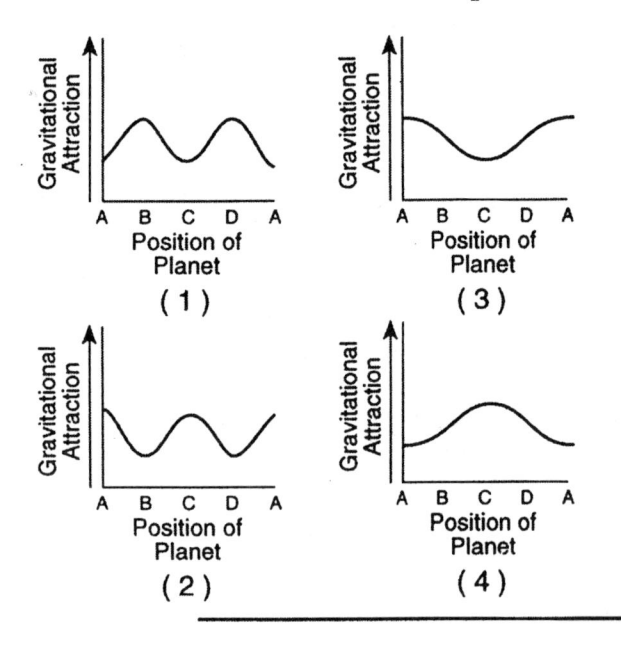

39 Which graph best represents the most common relationship between the amount of air pollution and the distance from an industrial city?

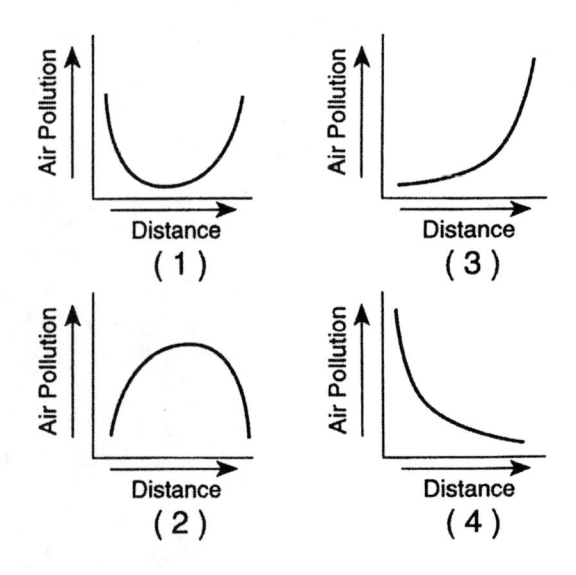

40 Which diagram best shows how air inside a greenhouse warms as a result of energy from the Sun?

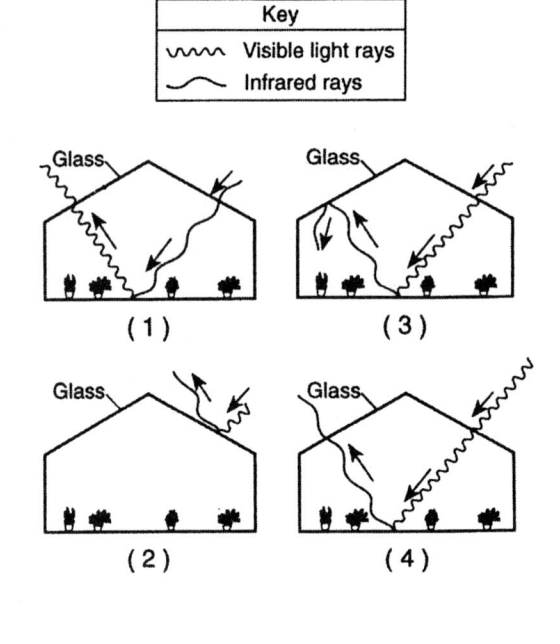

Part II

This part consists of six groups, each containing ten questions. Choose any one of these six groups. Be sure that you answer all ten questions in the single group chosen. Record the answers to these questions on the separate answer paper in accordance with the directions on the front cover of this booklet. Some questions may require the use of the *Earth Science Reference Tables*. [10]

Group A — Rocks and Minerals

If you choose this group, be sure to answer questions 41–50.

Base your answers to questions 41 through 43 on the cross section below. The cross section shows the surface and subsurface rock formations near New York City.

Geologic Section Across the Hudson River

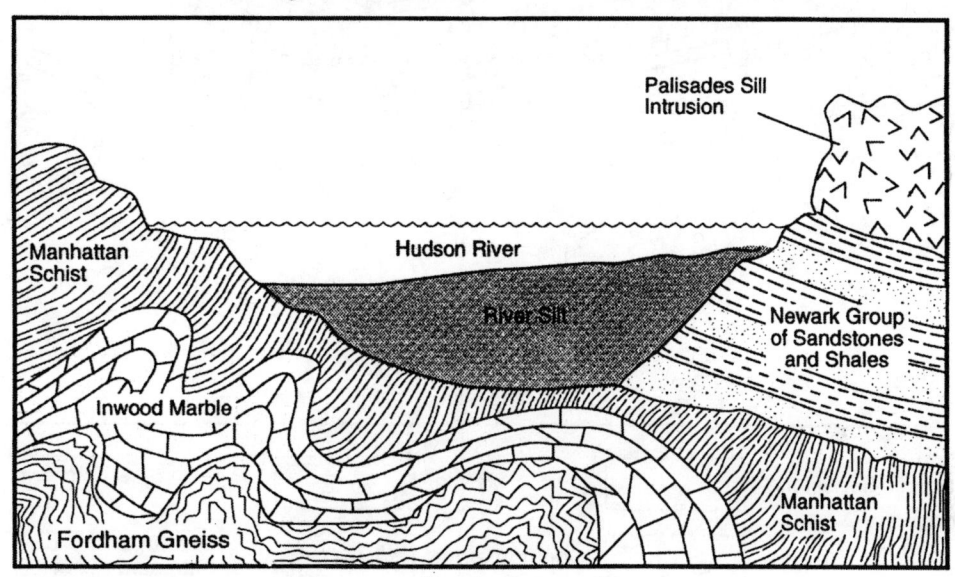

(Not drawn to scale)

41 The portion of the Palisades sill that contains large crystals of plagioclase feldspar and pyroxene is considered to be similar in texture and composition to

  1 obsidian   3 basalt glass
  2 granite   4 gabbro

42 Which rock formation was originally limestone?

  1 Palisades sill   3 Inwood marble
  2 Fordham gneiss   4 Manhattan schist

43 The rock types shown on the left side of this geologic cross section were mainly the result of

  1 heat and pressure exerted on previously existing rock
  2 melting and solidification of crustal rocks at great depths
  3 tectonic plate boundaries diverging at the mid-ocean ridge
  4 compaction and cementation of sediments under ocean waters

Base your answers to questions 44 and 45 on the diagram below, which represents a cross section of an area of the Earth's crust. Letters A through F represent rock units.

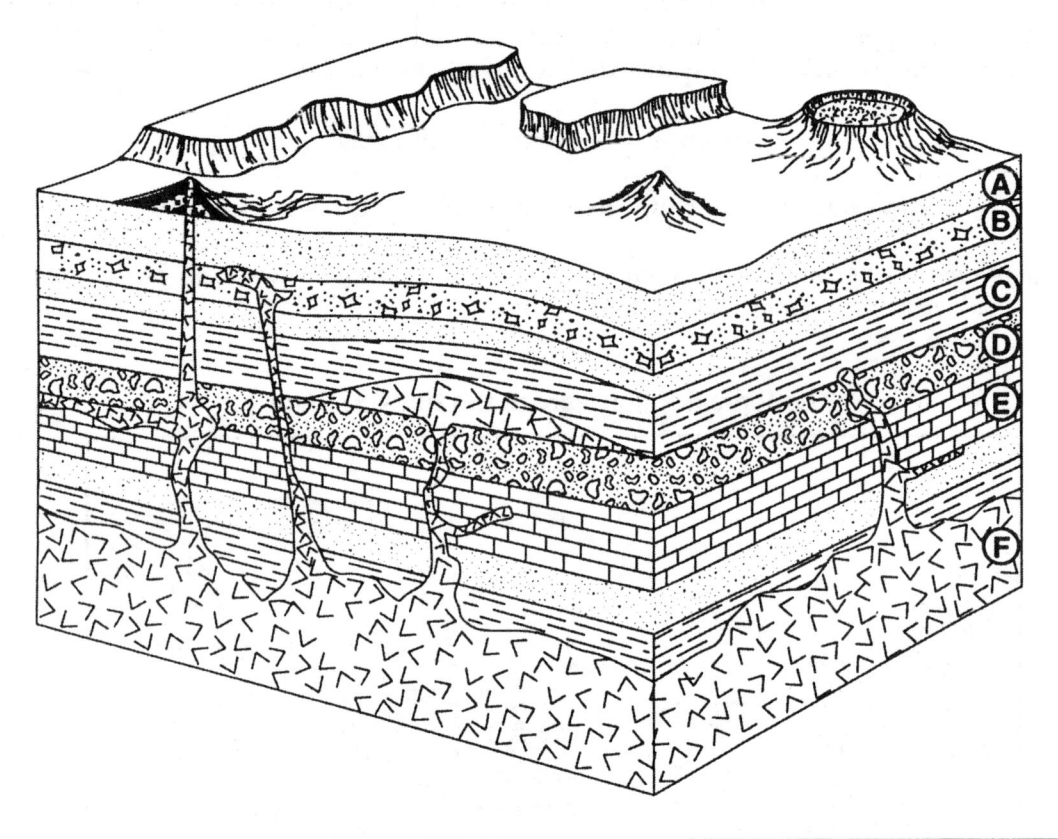

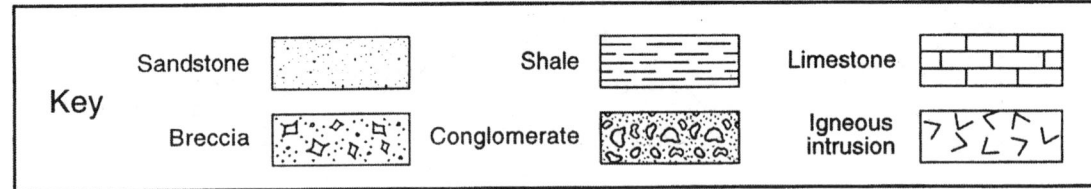

44 Which rock most likely had a chemical origin?
 (1) A
 (2) E
 (3) C
 (4) F

45 Which statement best describes how rock layers B and D are different from each other?
 1 One is intrusive and the other is extrusive.
 2 One is clastic and the other is nonclastic.
 3 One is foliated and the other is nonfoliated.
 4 One has angular fragments and the other has rounded fragments.

Base your answers to questions 46 through 48 on the table below, which gives the properties of four varieties of the mineral garnet.

Garnet				
Variety	Composition	Density (g/cm³)	Hardness	Typical Color
Pyrope	$Mg_3Al_2Si_3O_{12}$	3.6	7 to 7.5	Deep red to nearly black
Almandine	$Fe_3Al_2Si_3O_{12}$	4.3	7 to 7.5	Brownish red
Spessartine	$Mn_3Al_2Si_3O_{12}$	4.2	7 to 7.5	Orange red
Grossular	$Ca_3Al_2Si_3O_{12}$	3.6	6.5 to 7	Yellowish green

Chemical Symbols

Al — aluminum
Ca — calcium
Fe — iron
Mg — magnesium
Mn — manganese
O — oxygen
Si — silicon

46 Which variety is correctly described as a calcium garnet?
 1 pyrope
 2 almandine
 3 spessartine
 4 grossular

47 Garnets such as almandine are generally found in metamorphic rocks. In which metamorphic rocks are garnets most likely to be found?
 1 schist and gneiss
 2 gneiss and quartzite
 3 quartzite and marble
 4 marble and schist

48 In which New York State landscape region are garnets most likely to be found in surface bedrock?
 1 Newark Lowlands
 2 Adirondack Mountains
 3 Erie-Ontario Lowlands
 4 Allegheny Plateau

49 The cleavage or fracture of a mineral is normally determined by the mineral's
 1 density
 2 oxygen content
 3 internal arrangement of atoms
 4 position among surrounding minerals

50 Oxygen is the most abundant element by volume in the Earth's
 1 inner core
 2 troposphere
 3 hydrosphere
 4 crust

Group B — Plate Tectonics

If you choose this group, be sure to answer questions 51–60.

Base your answers to questions 51 and 52 on the diagrams below of geologic cross sections of the upper mantle and crust at four different Earth locations, A, B, C, and D. Movement of the crustal sections (plates) is indicated by arrows, and the locations of frequent earthquakes are indicated by ✱. Diagrams are not drawn to scale.

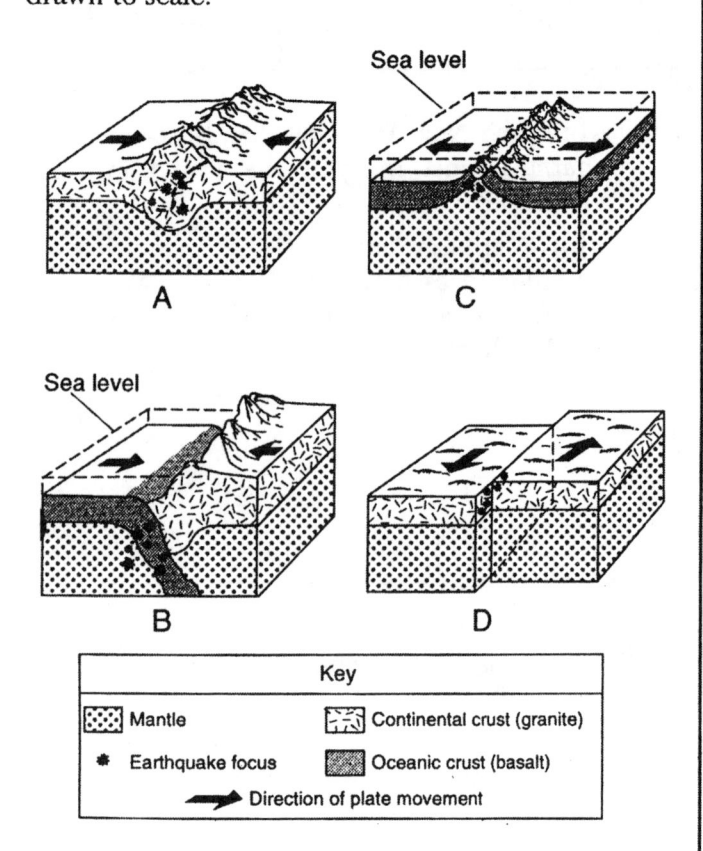

Key
- Mantle
- Earthquake focus
- Continental crust (granite)
- Oceanic crust (basalt)
- Direction of plate movement

51 Which location best represents the boundary between the African plate and the South American plate?
(1) A
(2) B
(3) C
(4) D

52 Which diagram represents plate movement associated with transform faults such as those causing California earthquakes?
(1) A
(2) B
(3) C
(4) D

53 The diagram below represents a cross section of a portion of the Earth's crust.

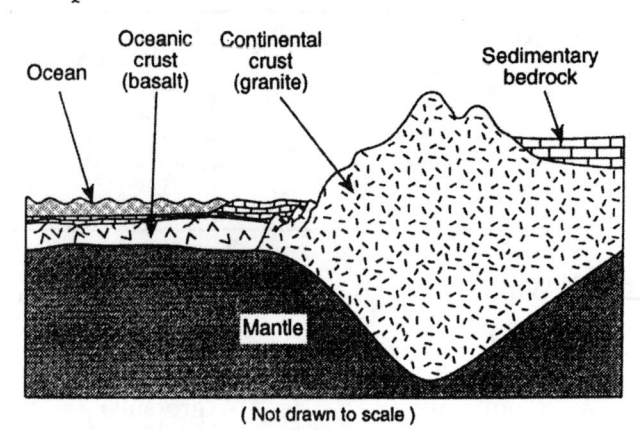

Which statement about the Earth's crust is best supported by the diagram?
1 The oceanic crust is thicker than the mantle.
2 The continental crust is thicker than the oceanic crust.
3 The continental crust is composed primarily of sedimentary rock.
4 The crust is composed of denser rock than the mantle is.

54 At a depth of 2,000 kilometers, the temperature of the stiffer mantle is inferred to be
(1) 6,500°C
(2) 4,200°C
(3) 3,500°C
(4) 1,500°C

55 Hot springs on the ocean floor near the mid-ocean ridges provide evidence that
1 convection currents exist in the asthenosphere
2 meteor craters are found beneath the oceans
3 climate change has melted huge glaciers
4 marine fossils have been uplifted to high elevations

56 Which geologic event occurred most recently?
1 initial opening of the Atlantic Ocean
2 formation of the Hudson Highlands
3 formation of the Catskill delta
4 collision of North America and Africa

Base your answers to questions 57 through 60 on the four seismograms below. The seismograms show the arrival of *P*-waves and *S*-waves from the same earthquake at four different seismograph stations.

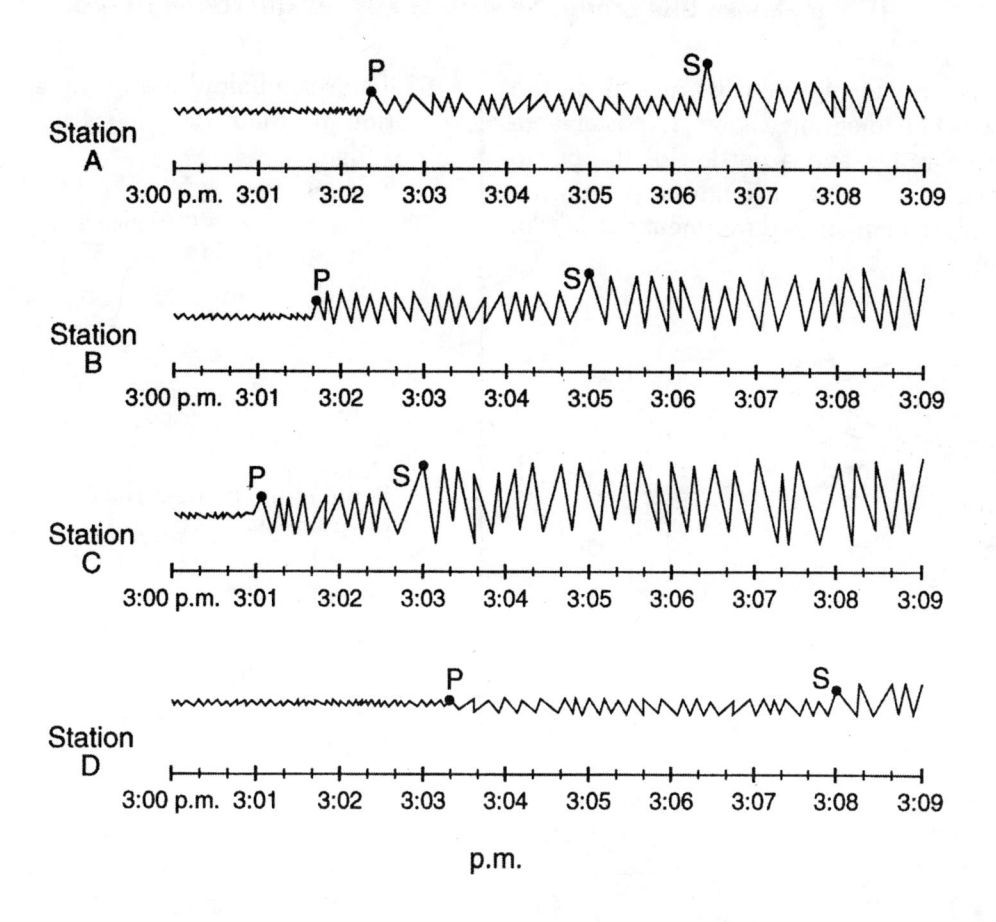

57 Which station is farthest from the epicenter of the earthquake?
(1) A (3) C
(2) B (4) D

58 How many seismograms are needed to locate the epicenter of this earthquake?
1 Any one of the seismograms may be used.
2 Any two of the seismograms may be used.
3 Any three of the seismograms may be used.
4 All four seismograms must be used.

59 What is the distance between station *A* and the epicenter of the earthquake?
(1) 1,000 km (3) 2,600 km
(2) 2,000 km (4) 3,200 km

60 A fifth seismic station, located 5,600 kilometers from the earthquake epicenter, recorded the arrival of the *P*-wave at 3:06 p.m. What time did the earthquake occur?
(1) 2:05 p.m. (3) 2:34 p.m.
(2) 2:13 p.m. (4) 2:57 p.m.

Group C — Oceanography

If you choose this group, be sure to answer questions 61–70.

Base your answers to questions 61 and 62 on the diagram below. The diagram shows a coastal area with a mountain range and a portion of the ocean floor. A turbidity current through a submarine canyon has formed a fan-shaped sediment deposit on the ocean floor.

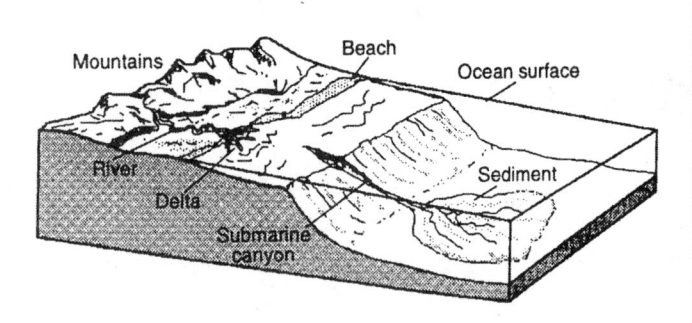

61 Which diagram best represents a cross section of the sediment in the fan-shaped deposit?

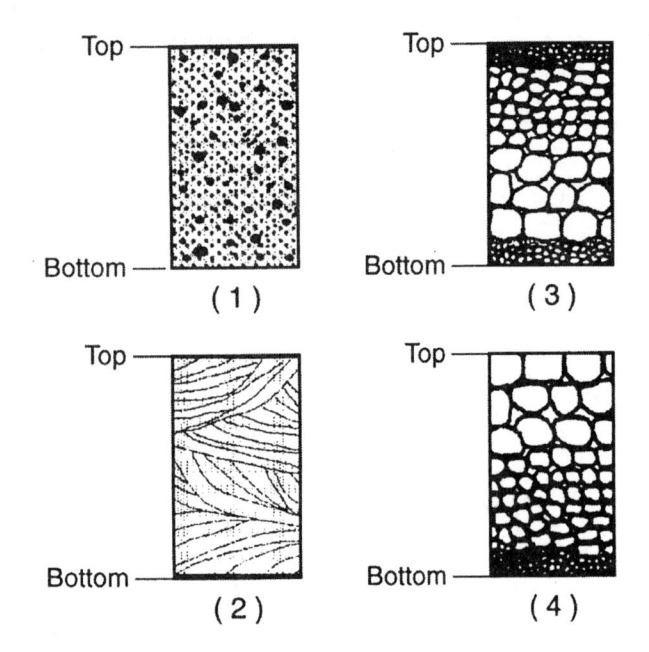

62 In which region of the ocean is the submarine canyon located?

   1 tidal zone   3 deep ocean basin
   2 continental margin   4 mid-ocean ridge

63 The scale below shows the age of rocks in relation to their distance from the Mid-Atlantic Ridge.

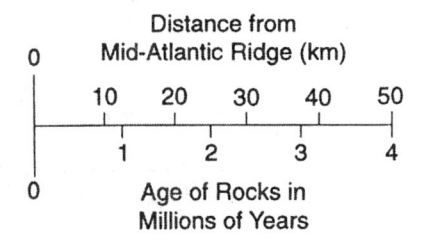

Some igneous rocks that originally formed at the Mid-Atlantic Ridge are now 37 kilometers from the ridge. Approximately how long ago did these rocks form?

(1) 1.8 million years ago
(2) 2.0 million years ago
(3) 3.0 million years ago
(4) 45.0 million years ago

64 The shaded areas of the map below indicate concentrations of pollutants along the coastlines of North America.

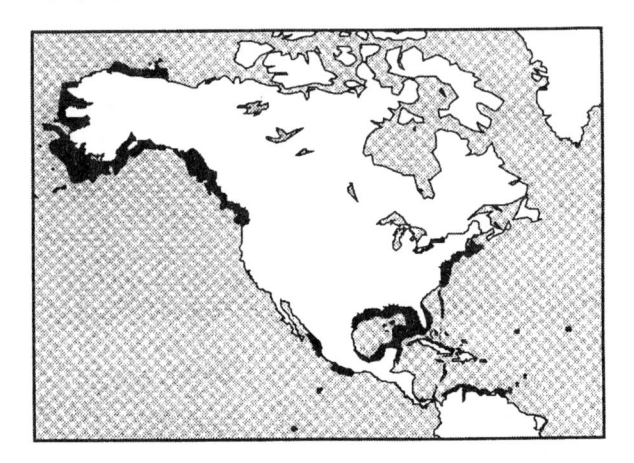

Polluting material may have been carried to the Alaska area by the

1 California Current
2 North Pacific Current
3 Florida Current
4 Labrador Current

Base your answers to questions 65 and 66 on the diagram below. The diagram represents a shoreline with waves approaching at an angle. The exposed bedrock of the wavecut cliff is granite. Arrow A shows the direction of the longshore current and arrow B shows the general path of wave travel.

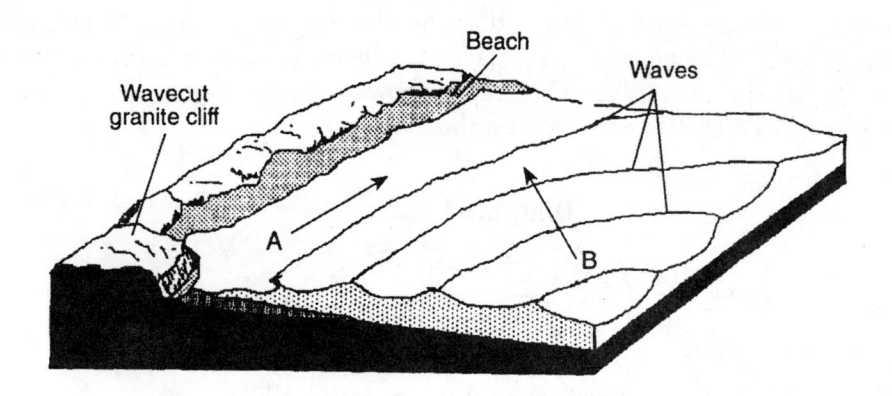

65 Which minerals are most likely to be found in the beach sand?

1 olivine and hornblende
2 pyroxene and plagioclase feldspar
3 plagioclase feldspar and olivine
4 quartz and potassium feldspar

66 A large storm with high winds that develops out at sea is most likely to result in

1 decreased erosion along the shoreline
2 increased deposition along the shoreline
3 increased wave height near the shore
4 unchanged shoreline features

Base your answers to questions 67 through 70 on the *Earth Science Reference Tables* and the graphs below. Graph I shows the yearly precipitation and evaporation at different latitudes, and graph II shows the salinity of the ocean at different latitudes. Salinity is a measure of the total amount of dissolved minerals in seawater expressed as parts per thousand.

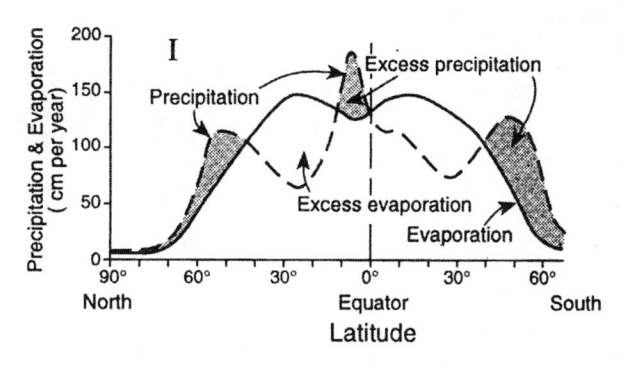

67 Compared to the amount of precipitation at the North Pole, the amount of precipitation at the Equator is

1 half as much
2 twice as much
3 about the same
4 more than ten times greater

68 At which latitude is ocean salinity *least*?

(1) 90° N
(2) 30° N
(3) 0°
(4) 60° S

69 The concentration of dissolved minerals in seawater would be increased by

1 rapid evaporation
2 heavy rainfall
3 melting glaciers
4 tsunamis

70 What is the source of most of the dissolved minerals that cause surface salinity?

1 ice sheets
2 industrial pollutants
3 continental erosion
4 tropical storms

Group D — Glacial Processes

If you choose this group, be sure to answer questions 71–80.

Base your answers to questions 71 through 74 on the diagrams below. Diagram I shows an imaginary present-day continent covered by an advancing glacial ice sheet. Isolines called isopachs are drawn, representing the thickness of the ice sheet in meters. Diagram II shows a cross section of the glacier with the land beneath it along reference line XY. Point A is a location on the glacier.

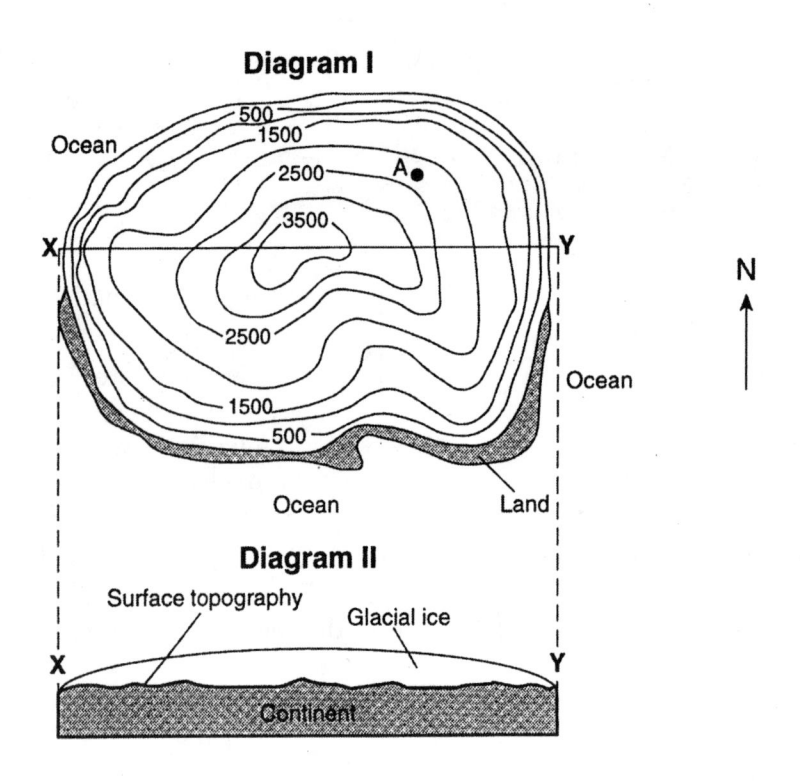

Diagram I

Diagram II

71 Chemical analysis of an ice sample taken from the central core of the glacier would most likely be used to study

1 subsurface bedrock
2 past atmospheric conditions
3 the glacier's rate of movement
4 the exact location of the ice sheet

72 Which statement best describes the movement of this continental glacier?

1 The glacier is advancing from north to south, only.
2 The glacier is advancing from south to north, only.
3 The glacier is moving outward in all directions from the central zone of accumulation.
4 The glacier is moving inward from all directions toward the center of the continent.

73 What is the approximate thickness of the ice at location A?

(1) 1800 m
(2) 2250 m
(3) 2800 m
(4) 3400 m

74 Which statement best explains why the glacier originally formed on this continent?

1 The accumulation of yearly rainwater froze every winter.
2 The accumulation of snow during the cold season exceeded the melting of ice during the warm season.
3 Icebergs from the surrounding sea accumulated.
4 The continent has a low latitude and a low elevation.

Base your answers to questions 75 through 78 on the diagrams below. Diagram I shows melting ice lobes of a continental glacier during the Pleistocene Epoch. Diagram II represents the landscape features of the same region at present, after the retreat of the continental ice sheet. Letters A through F indicate surface features in this region.

Diagram I

Diagram II

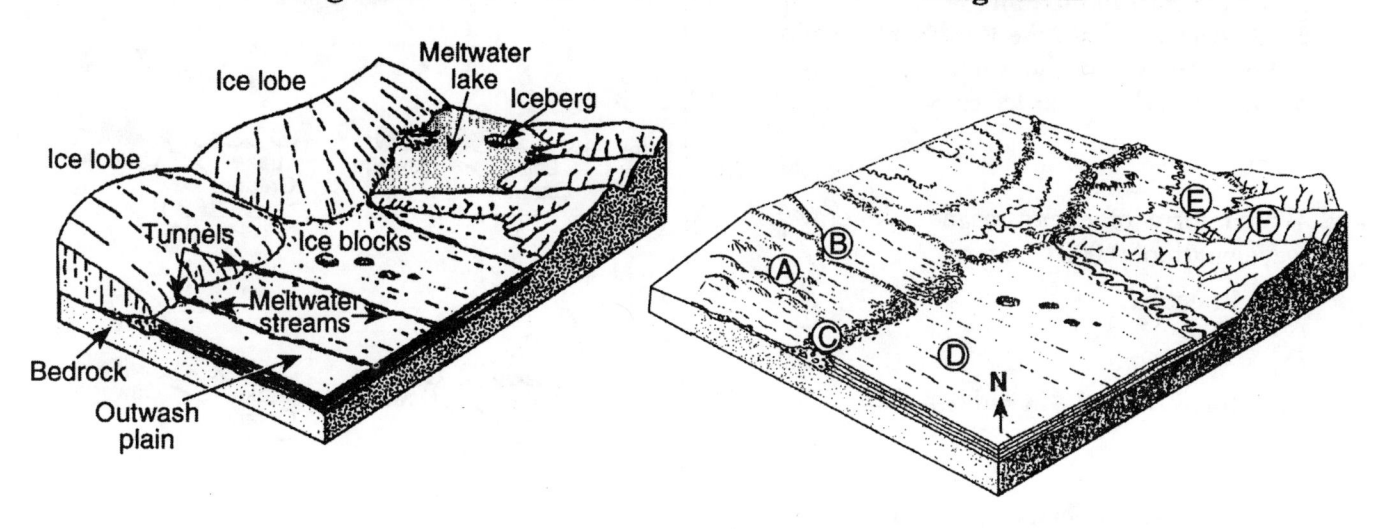

75 Which erosional feature most likely formed on the surface of the bedrock under the glacial ice?

1 sorted sands
2 sand dunes
3 parallel grooves
4 a V-shaped valley

76 Which features in diagram II are composed of till directly deposited by the glacial ice?

(1) A and C
(2) B and D
(3) C and E
(4) E and F

77 Which fossil has been found in glacial areas like those represented by D in diagram II?

1 placoderm fish
2 ammonoid
3 stromatolite
4 mastodont

Note that question 78 has only three choices.

78 In the interval between the time represented by diagram I and the time represented by diagram II, sea level most likely had

1 decreased
2 increased
3 remained the same

79 Which New York State landscape feature was formed primarily as a result of glacial deposition?

1 Adirondack Mountains
2 Hudson-Mohawk Lowlands
3 Tug Hill Plateau
4 Long Island

80 Glacial movement is caused primarily by

1 gravity
2 erosion
3 Earth's rotation
4 global winds

Group E — Atmospheric Energy

If you choose this group, be sure to answer questions 81–90.

81 Which source provides the most energy for atmospheric weather changes?
 1 radiation from the Sun
 2 radioactivity from the Earth's interior
 3 heat stored in ocean water
 4 heat stored in polar ice caps

82 Which form of electromagnetic energy has a wavelength of 0.0001 meter?
 (1) ultraviolet
 (2) infrared
 (3) FM and TV
 (4) shortwave and AM radio

83 Daily weather forecasts are based primarily on
 1 ocean currents
 2 seismic data
 3 phases of the Moon
 4 air-mass movements

84 The diagram below shows air rising from the Earth's surface to form a thunderstorm cloud.

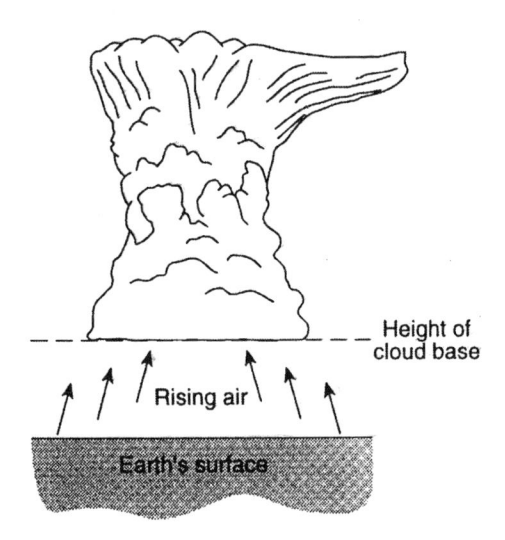

According to the Lapse Rate chart, what is the height of the base of the thunderstorm cloud when the air at the Earth's surface has a temperature of 20°C and a dewpoint of 12°C?
 (1) 1.0 km   (3) 3.0 km
 (2) 1.5 km   (4) 0.7 km

85 Which cross section best shows the normal movement of the air over Oswego, New York, on a very hot summer afternoon?

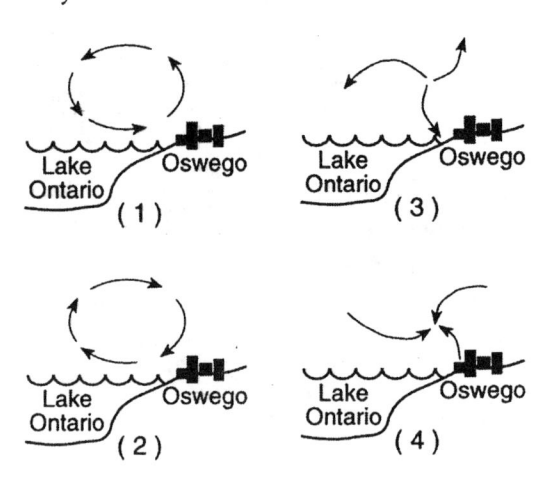

86 The diagram below represents an activity in which an eye dropper was used to place a drop of water on a spinning globe. Instead of flowing due south toward the target point, the drop appeared to follow a curved path and missed the target.

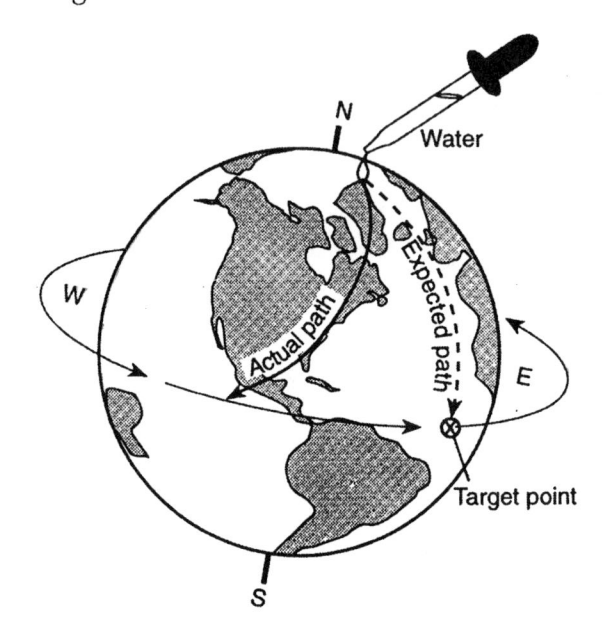

This curved-path phenomenon most directly affects the Earth's
 1 tilt   3 wind belts
 2 Moon phases   4 tectonic plates

Base your answers to questions 87 through 90 on the graph below. The graph shows the results of a laboratory activity in which a sample of ice at –50°C was heated at a uniform rate for 80 minutes. The ice has a mass of 200 grams.

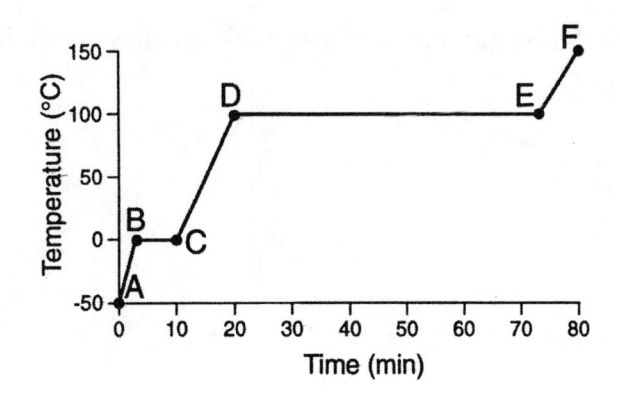

87 What was the temperature of the water 20 minutes after heating began?
(1) 70°C
(2) 100°C
(3) 110°C
(4) 150°C

88 Which change could shorten the time needed to melt the ice completely?
1 using colder ice
2 stirring the sample more slowly
3 reducing the initial sample to 100 grams of ice
4 reducing the number of temperature readings taken

89 What was the total amount of energy absorbed by the sample during the time between points *B* and *C* on the graph?
(1) 200 calories
(2) 800 calories
(3) 10,800 calories
(4) 16,000 calories

90 During which interval of the graph is a phase change occurring?
(1) *A* to *B*
(2) *E* to *F*
(3) *C* to *D*
(4) *D* to *E*

Group F — Astronomy

If you choose this group, be sure to answer questions 91–100.

91 Which graph best illustrates the average temperatures of the planets in the solar system?

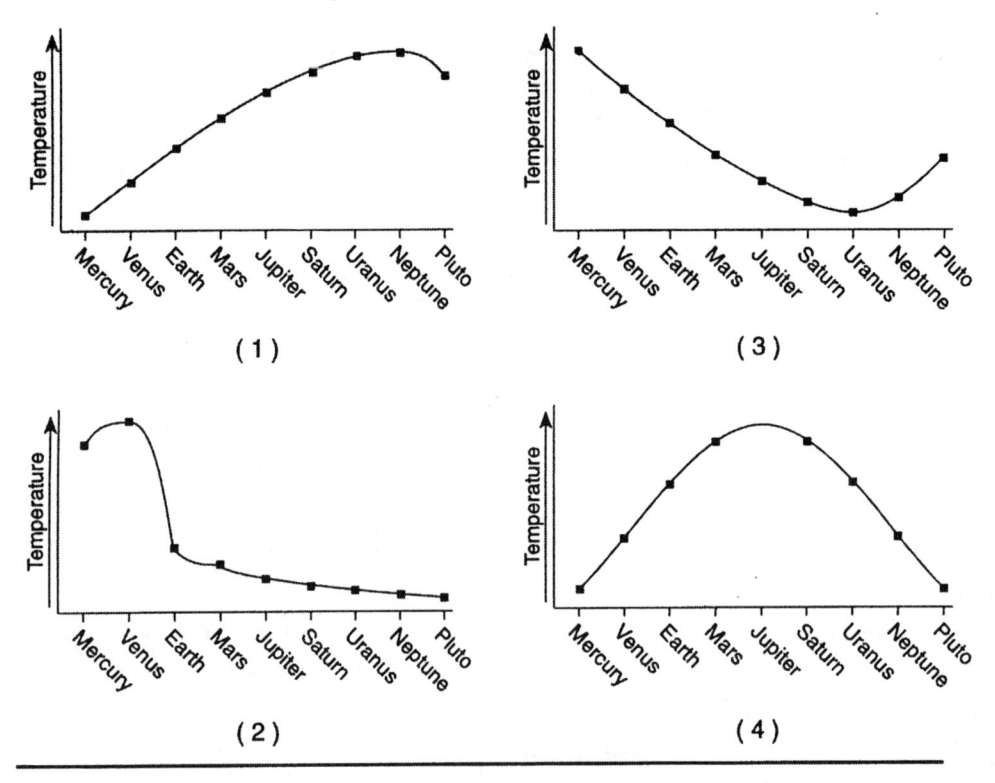

(1) (2) (3) (4)

92 A belt of asteroids is located an average distance of 503 million kilometers from the Sun. Between which two planets is this belt located?

1 Mars and Jupiter
2 Mars and Earth
3 Jupiter and Saturn
4 Saturn and Uranus

93 Why are impact structures (craters) more common on the surface of Mars than on the surfaces of Venus, Earth, and Jupiter?

1 Mars has the greatest surface area and receives more impacts.
2 The tiny moons of Mars are breaking into pieces and showering its surface with rock fragments.
3 Mars has a strong magnetic field that attracts iron-containing rock fragments from space.
4 The thin atmosphere of Mars offers little protection against falling rock fragments from space.

94 The diagram below represents part of the night sky including the constellation Leo. The black circles represent stars. The open circles represent the changing positions of one celestial object over a period of a few weeks.

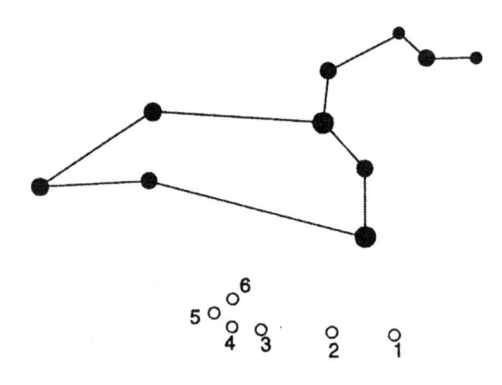

The celestial object represented by the open circles most likely is

1 a galaxy
2 a planet
3 Earth's Moon
4 another star

Base your answers to questions 95 and 96 on the graphs below. The graphs show the composition of the atmospheres of Venus, Earth, Mars, and Jupiter.

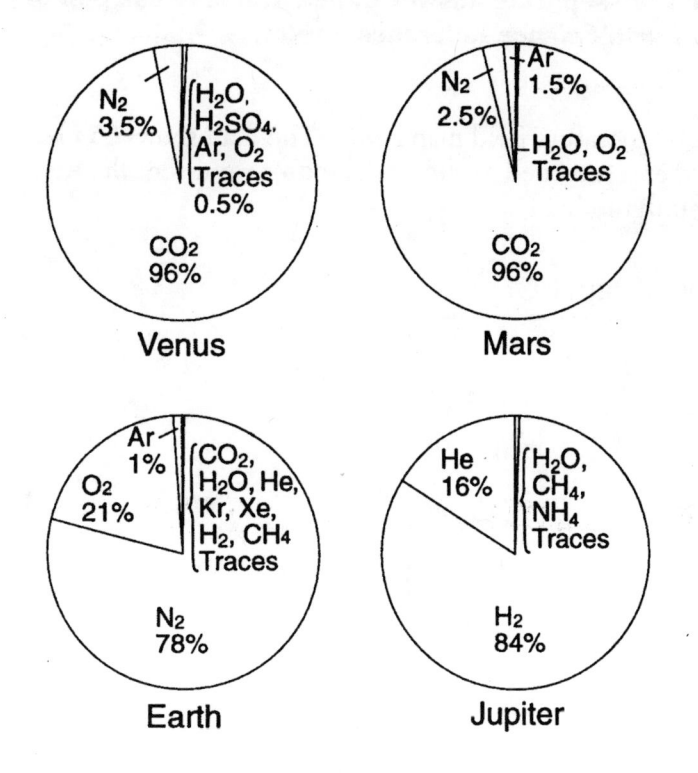

95 Which gas is present in the atmospheres of Venus, Earth, and Mars but is *not* present in the atmosphere of Jupiter?

1 argon (Ar)
2 methane (CH_4)
3 hydrogen (H_2)
4 water vapor (H_2O)

96 Which planet has an atmosphere composed primarily of CO_2 and a period of rotation greater than its period of revolution?

1 Venus
2 Mercury
3 Earth
4 Mars

97 In which type of model are the Sun, other stars, and the Moon in orbit around the Earth?

1 heliocentric model
2 tetrahedral model
3 concentric model
4 geocentric model

98 In 1851, the French physicist Jean Foucault constructed a large pendulum that always changed its direction of swing at the same rate in a clockwise direction. According to Foucault, this change in direction of swing was caused by the

1 Moon's rotation on its axis
2 Moon's revolution around the Earth
3 Earth's rotation on its axis
4 Earth's revolution around the Sun

99 Which planet has vast amounts of liquid water at its surface?

1 Venus
2 Mars
3 Jupiter
4 Earth

100 The diagram below represents a standard dark-line spectrum for an element.

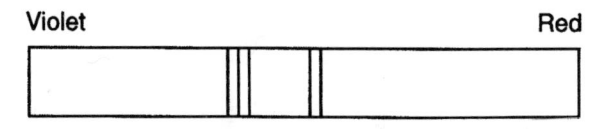

The spectral lines of this element are observed in light from a distant galaxy. Which diagram represents these spectral lines?

Part III

This part consists of questions 101 through 120. Be sure that you answer all questions in this part. Record your answers in the spaces provided on the separate answer paper. You may use pen or pencil. Some questions may require the use of the *Earth Science Reference Tables*. [25]

Base your answers to questions 101 and 102 on the temperature field map below. The map shows 25 measurements (in °C) that were made in a temperature field and recorded as shown. The dots represent the exact locations of the measurements. *A* and *B* are locations within the field.

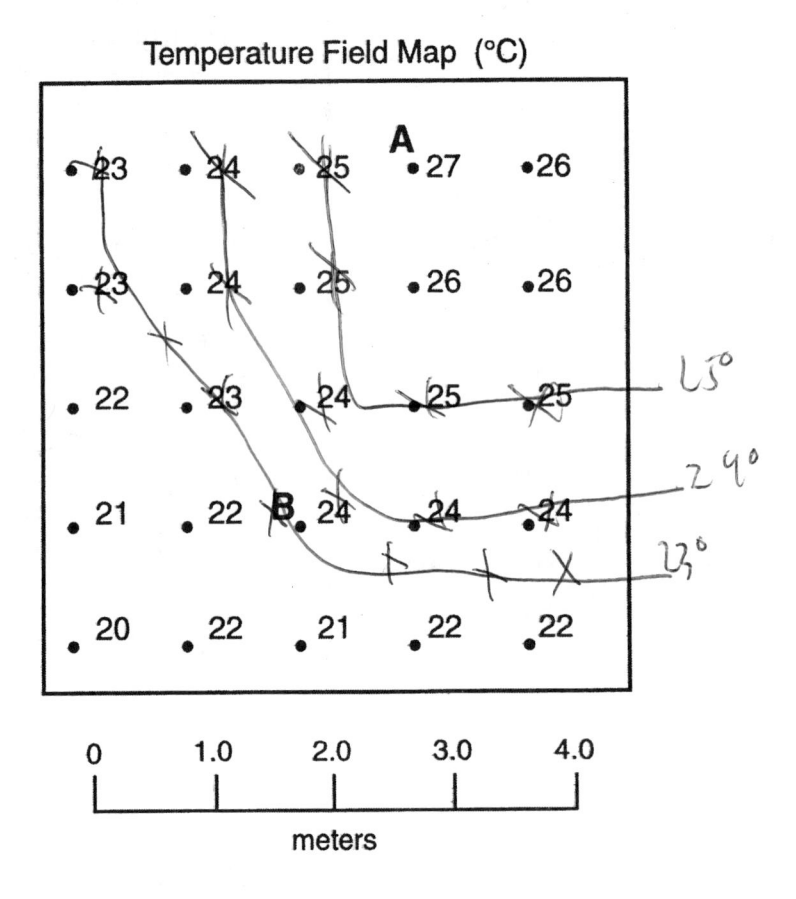

Temperature Field Map (°C)

101 On the temperature field map provided *on your answer paper,* draw three isotherms: the 23°C isotherm, the 24°C isotherm, and the 25°C isotherm. [2]

102 In the space provided *on your answer paper*, calculate the temperature gradient between locations *A* and *B* on the temperature field map, following the directions below.
   a Write the equation for gradient.   [1]
   b Substitute data from the map into the equation.   [1]
   c Calculate the gradient and label it with the proper units.   [1]

Base your answers to questions 103 through 105 on the diagram below. The diagram represents the supercontinent Pangaea, which began to break up approximately 220 million years ago.

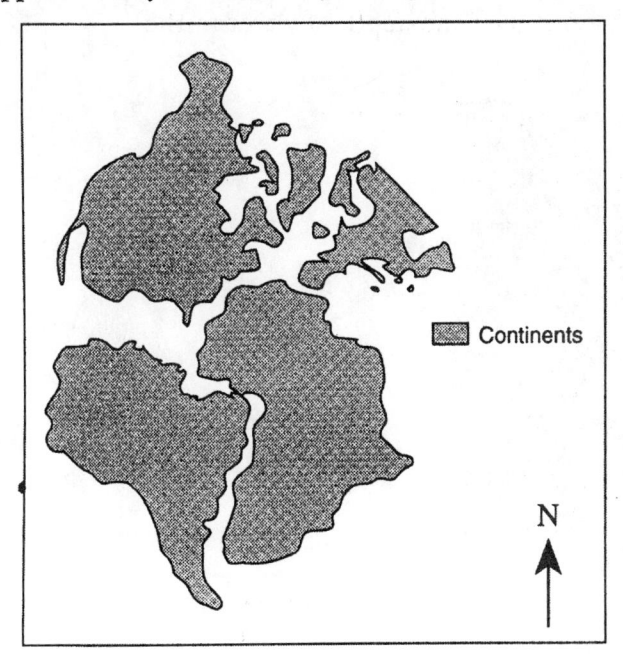

103 During which geologic period within the Mesozoic Era did the supercontinent Pangaea begin to break apart? [1]   *Triassic*

104 State one form of evidence that supports the inference that Pangaea existed. [1]   *They look like pieces of a puzzle.*

105 State the compass direction toward which North America has moved since Pangaea began to break apart. [1]   *Southwest*

Base your answers to questions 106 through 108 on the information below.

A mountain is a landform with steeply sloping sides whose peak is usually thousands of feet higher than its base. Mountains often contain a great deal of nonsedimentary rock and have distorted rock structures caused by faulting and folding of the crust.

A plateau is a broad, level area at a high elevation. It usually has an undistorted, horizontal rock structure. A plateau may have steep slopes as a result of erosion.

106 State why marine fossils are *not* usually found in the bedrock of the Adirondack Mountains. [1]   *no water is near the Adirondack mts.*

107 State the agent of erosion that is most likely responsible for shaping the Catskill Plateau so that it physically resembles a mountainous region. [1]   *Erosion*

108 State the approximate age of the surface bedrock of the Catskills. [1]

302 million yrs old.

Base your answers to questions 109 through 112 on the diagram and the stream data table below.

The diagram represents a stream flowing into a lake. Arrows show the direction of flow. Point P is a location in the stream. Line XY is a reference line across the stream. Points X and Y are locations on the banks. The data table gives the depth of water in the stream along line XY.

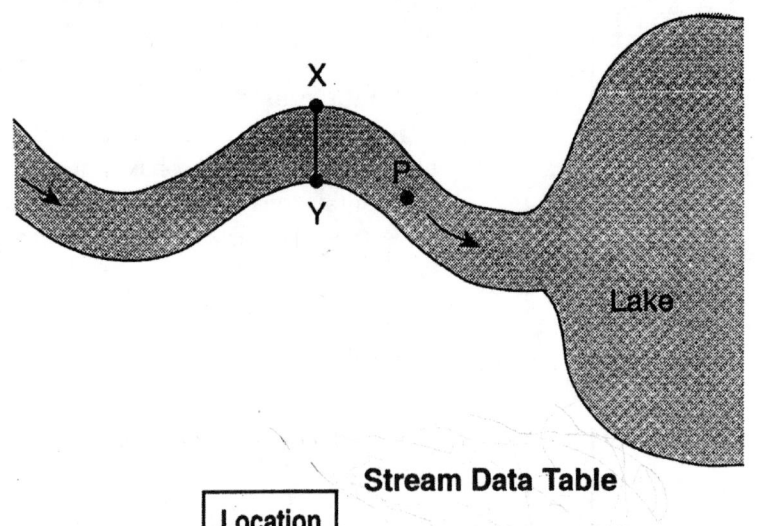

Stream Data Table

	Location X							Location Y
Distance from X (meters)	0	5	10	15	20	25	30	35
Depth of Water (meters)	0	5.0	5.5	4.5	3.5	2.0	0.5	0

Directions (109–110): Use the information in the data table to construct a profile of the depth of water. Use the grid provided *on your answer paper*, following the directions below.

109 On the vertical axis, mark an appropriate scale for the depth of water. Note that the zero (0) at the top of the axis represents the water surface. [1]

21

110 Plot the data for the depth of water in the stream along line XY and connect the points. (Distance is measured from point X.) [2]

Example:

56

111 State why the depth of water near the bank at point X is different from the depth of water near the bank at point Y. [1]

because of Erosion

112 At point P, the water velocity is 100 centimeters per second. State the name of the largest sediment that can be transported by the stream at point P. [1]

cobbles/pebbles

Base your answers to questions 113 through 115 on the weather maps below. The weather maps show the positions of a tropical storm at 10 a.m. on July 2 and on July 3.

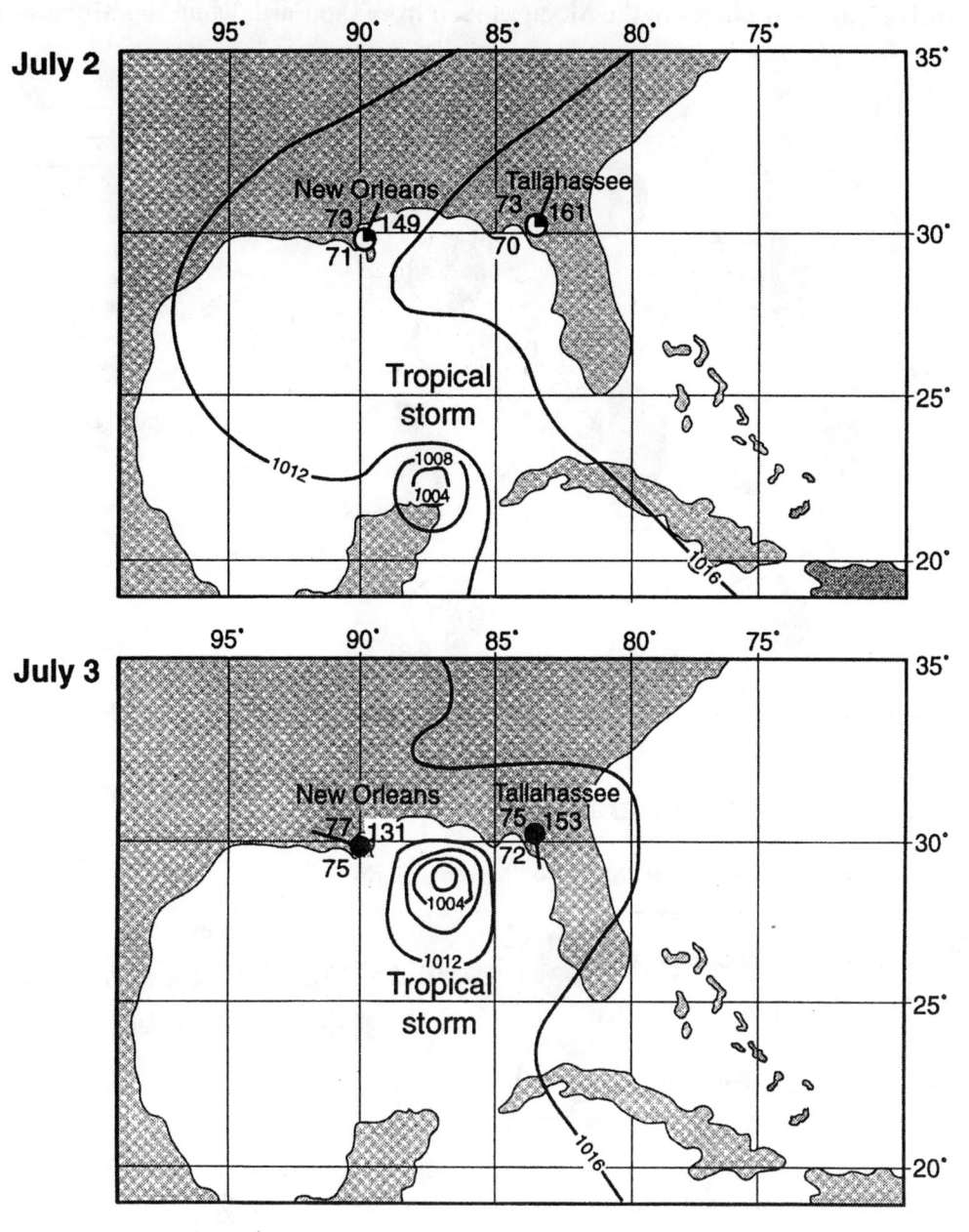

113 State the dewpoint temperature in Tallahassee on July 2. [1]

73°

114 Windspeed has been omitted from the station models. In one or more sentences, state how an increase in the storm's windspeed from July 2 to July 3 could be inferred from the maps. [2] *The windspeed increased*

115 The storm formed over warm tropical water. State what will most likely happen to the windspeed when the storm moves over land. [1] *it decreased*

Base your answers to questions 116 through 118 on the diagrams below. Diagram I represents the Moon orbiting the Earth as viewed from space above the North Pole. The Moon is shown at 8 different positions in its orbit. Diagram II represents phases of the Moon as seen from the Earth when the Moon is at position 2 and at position 4.

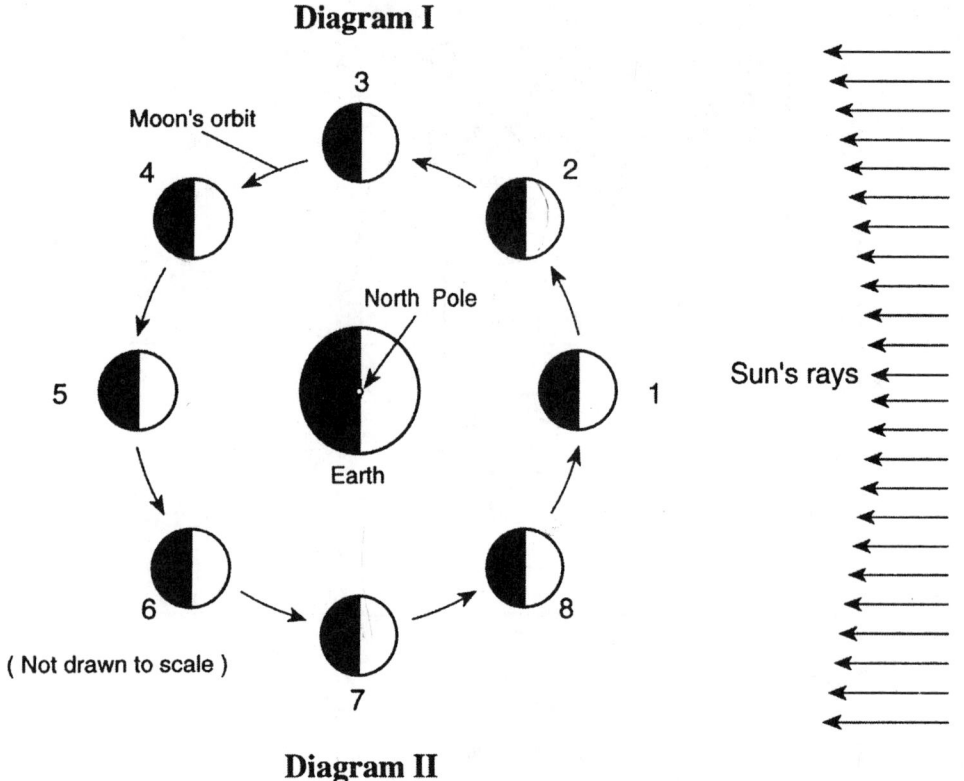

Diagram I

(Not drawn to scale)

Sun's rays →

Diagram II

Phase of the Moon as seen from the Earth

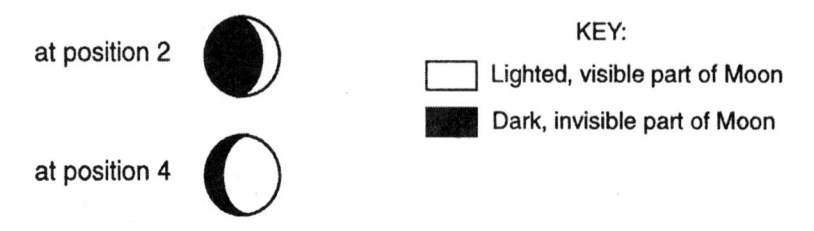

at position 2
at position 4

KEY:
☐ Lighted, visible part of Moon
■ Dark, invisible part of Moon

116 Shade the circle provided *on your answer paper* to illustrate the Moon's phase as seen from the Earth when the Moon is at position 7. [1]

117 State the two positions of the Moon at which an eclipse could occur. [1]
1, 5

118 State the approximate length of time required for one complete revolution of the Moon around the Earth. [1] 12

Page 280   N&N© SCIENCE SERIES – EARTH SCIENCE – MODIFIED PROGRAM

Base your answers to questions 119 and 120 on the graph below. The graph shows the average water temperature and the dissolved oxygen levels of water in a stream over a 12-month period. The level of dissolved oxygen is measured in parts per million (ppm).

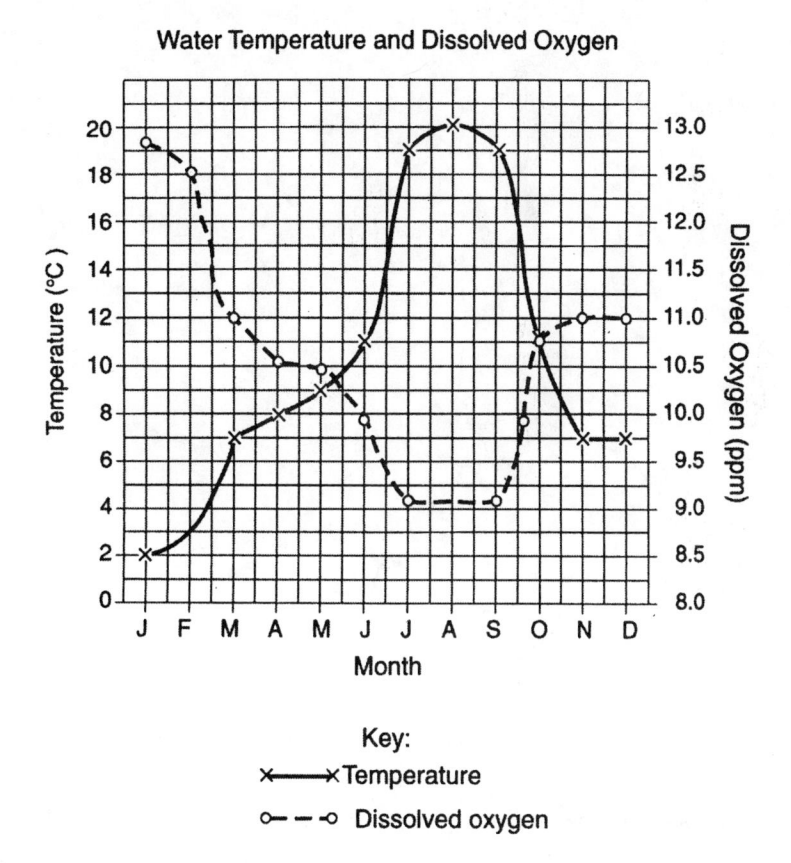

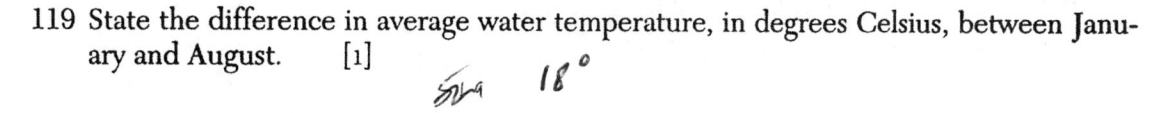

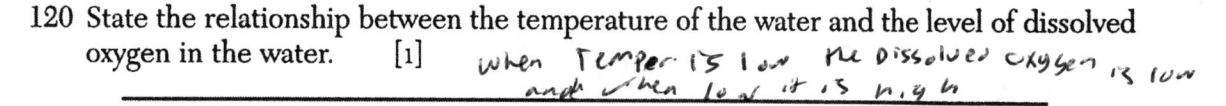

Water Temperature and Dissolved Oxygen

Key:
×——× Temperature
○− −○ Dissolved oxygen

119 State the difference in average water temperature, in degrees Celsius, between January and August. [1]   *18°*

120 State the relationship between the temperature of the water and the level of dissolved oxygen in the water. [1]   *when Temper is low the Dissolved oxygen is low and when low it is high*

EARTH SCIENCE
PROGRAM MODIFICATION EDITION
JUNE 1997

ANSWER PAPER

Part I Credits	....
Part II Credits	....
Part III Credits	....
Performance Test Credits	....
Local Project Credits	....
Total (Official Regents) Examination Mark	....
Reviewer's Initials: _____

Student ....

Teacher .... School ....

Grade (circle one)   8   9   10   11   12

Record all of your answers on this answer paper in accordance with the instructions on the front cover of the test booklet.

Part I (40 credits)

1	1	2	3	4	15	1	2	3	4	29	1	2	3	4
2	1	2	3	4	16	1	2	3	4	30	1	2	3	4
3	1	2	3	4	17	1	2	3	4	31	1	2	3	4
4	1	2	3	4	18	1	2	3	4	32	1	2	3	4
5	1	2	3	4	19	1	2	3	4	33	1	2	3	4
6	1	2	3	4	20	1	2	3	4	34	1	2	3	4
7	1	2	3	4	21	1	2	3		35	1	2	3	4
8	1	2	3	4	22	1	2	3	4	36	1	2	3	4
9	1	2	3	4	23	1	2	3	4	37	1	2	3	4
10	1	2	3	4	24	1	2	3	4	38	1	2	3	4
11	1	2	3	4	25	1	2	3	4	39	1	2	3	4
12	1	2	3	4	26	1	2	3	4	40	1	2	3	4
13	1	2	3	4	27	1	2	3	4					
14	1	2	3	4	28	1	2	3	4					

Part II (10 credits)

Answer the questions in only one of the six groups in this part. Be sure to mark the answers to the group you choose in accordance with the instructions on the front cover of the test booklet. Leave blank the spaces for the five groups of questions you do not choose to answer.

Group A
Rocks and Minerals

41	1	2	3	4
42	1	2	3	4
43	1	2	3	4
44	1	2	3	4
45	1	2	3	4
46	1	2	3	4
47	1	2	3	4
48	1	2	3	4
49	1	2	3	4
50	1	2	3	4

Group B
Plate Tectonics

51	1	2	3	4
52	1	2	3	4
53	1	2	3	4
54	1	2	3	4
55	1	2	3	4
56	1	2	3	4
57	1	2	3	4
58	1	2	3	4
59	1	2	3	4
60	1	2	3	4

Group C
Oceanography

61	1	2	3	4
62	1	2	3	4
63	1	2	3	4
64	1	2	3	4
65	1	2	3	4
66	1	2	3	4
67	1	2	3	4
68	1	2	3	4
69	1	2	3	4
70	1	2	3	4

Group D
Glacial Processes

71	1	2	3	4
72	1	2	3	4
73	1	2	3	4
74	1	2	3	4
75	1	2	3	4
76	1	2	3	4
77	1	2	3	4
78	1	2	3	
79	1	2	3	4
80	1	2	3	4

Group E
Atmospheric Energy

81	1	2	3	4
82	1	2	3	4
83	1	2	3	4
84	1	2	3	4
85	1	2	3	4
86	1	2	3	4
87	1	2	3	4
88	1	2	3	4
89	1	2	3	4
90	1	2	3	4

Group F
Astronomy

91	1	2	3	4
92	1	2	3	4
93	1	2	3	4
94	1	2	3	4
95	1	2	3	4
96	1	2	3	4
97	1	2	3	4
98	1	2	3	4
99	1	2	3	4
100	1	2	3	4

Practice Exam
Modified Earth Science Program
June 1998

Part I

Answer all 40 questions in this part.

Directions (1–40): For *each* statement or question, select the word or expression that, of those given, best completes the statement or answers the question. Record your answer on the separate answer paper in accordance with the directions on the front page of this booklet. Some questions may require the use of the *Earth Science Reference Tables*. [40]

Base your answers to questions 1 through 3 on the contour map below. Elevations are expressed in feet. The ▲ indicates the exact elevation of the top of Basket Dome.

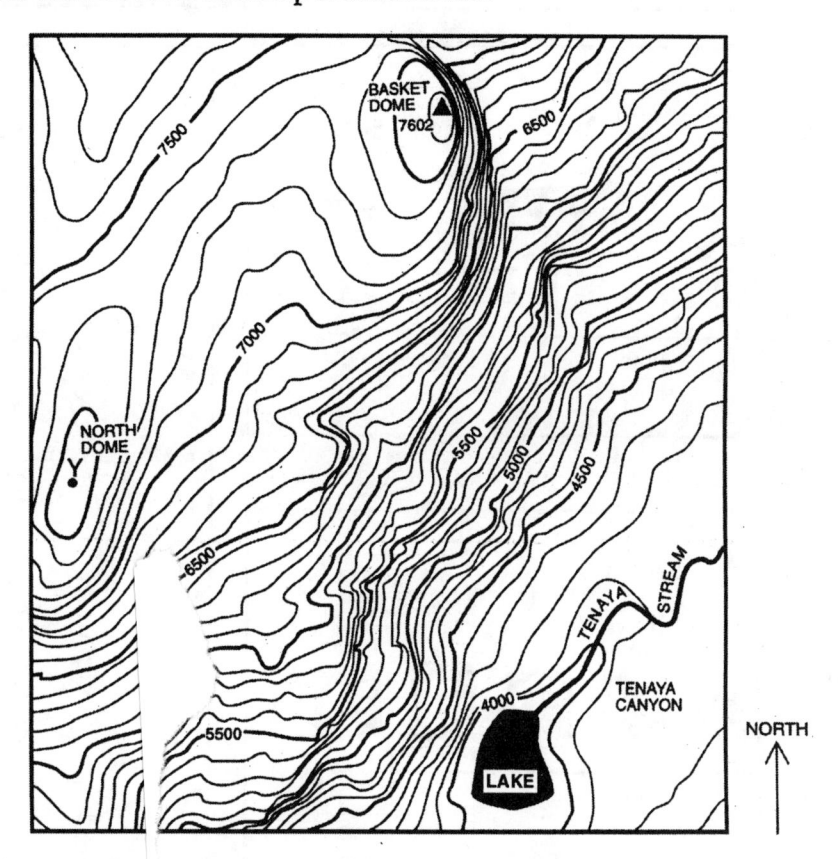

1 What is the highest possible elevation of point Y on North Dome?
  (1) 7,500 ft
  (2) 7,590 ft
  (3) 7,599 ft
  (4) 7,601 ft

2 Forty years ago, the highest elevation of Basket Dome was 7,600 feet. What is the rate of crustal uplift for Basket Dome?
  (1) 0.05 ft/yr
  (2) 2 ft/yr
  (3) 5 ft/yr
  (4) ft/yr

3 In which general direction does Tenaya Stream flow?
  1 southeast to northwest
  2 northwest to southeast
  3 southwest to northeast
  4 northeast to southwest

4 Each series below, labeled A through D, represents a sequence of events over the passage of time.

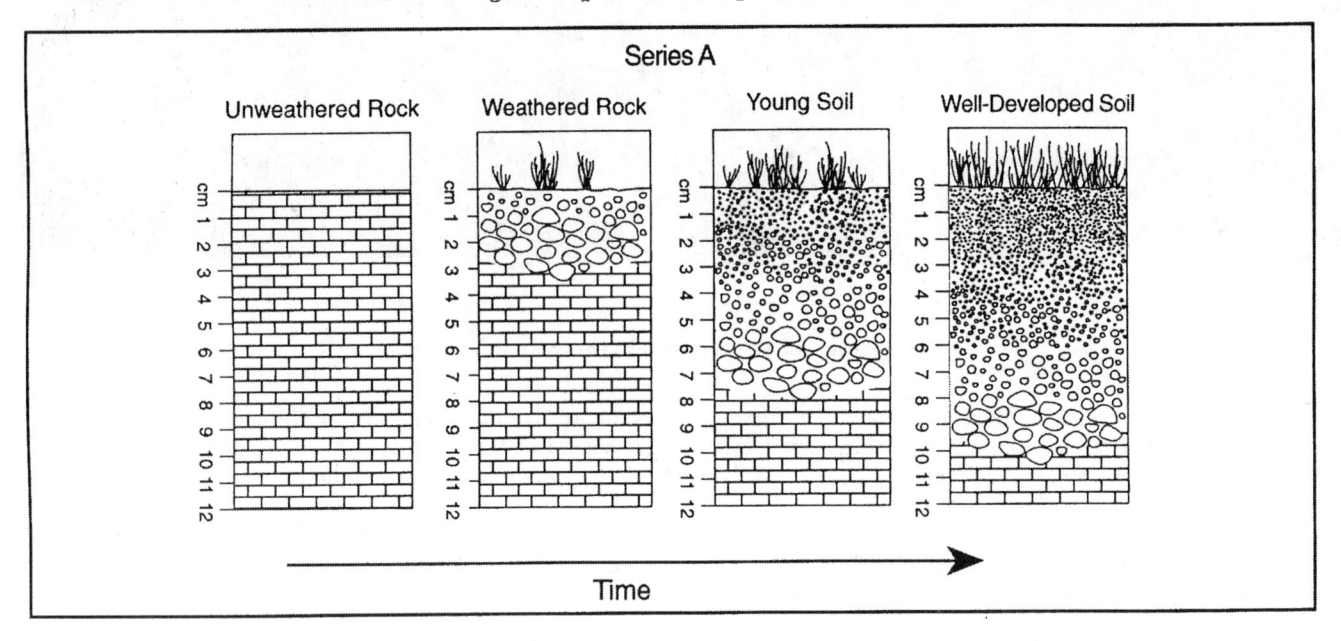

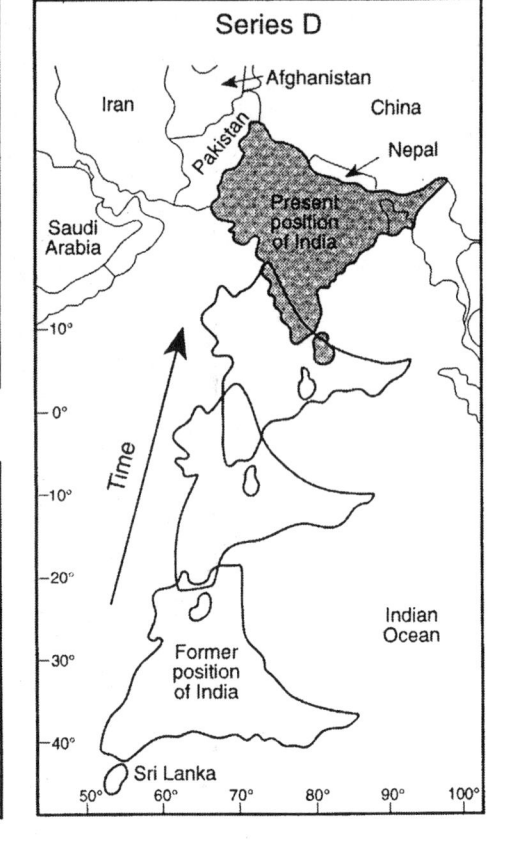

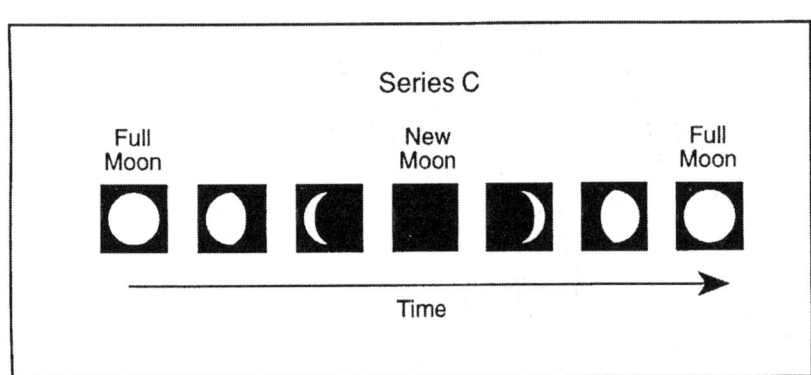

Series A
Unweathered Rock → Weathered Rock → Young Soil → Well-Developed Soil
Time →

Series B
Ancestral Elephant → ... → Modern Elephant
Time →

Series C
Full Moon → ... → New Moon → ... → Full Moon
Time →

Series D
Former position of India → Present position of India
Time ↑

Which series would take the *least* amount of time to complete?

(1) A
(2) B
(3) C
(4) D

5 Which New York State landscape region is located at 42° N 75° W?
  1 Erie-Ontario Lowlands
  2 the Catskills
  3 Hudson-Mohawk Lowlands
  4 Tug Hill Plateau

6 The diagrams below represent photographs of a large sailboat taken through telescopes over time as the boat sailed away from shore out to sea. The number above each diagram shows the magnification of the telescope lens.

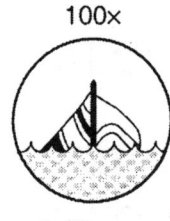

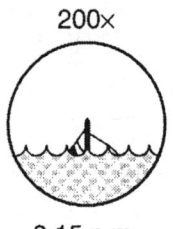

   50×   100×   200×

   12:15 p.m. 1:45 p.m.   3:15 p.m.

Which statement best explains the apparent sinking of this sailboat?
  1 The sailboat is moving around the curved surface of Earth.
  2 The sailboat appears smaller as it moves farther away.
  3 The change in density of the atmosphere is causing refraction of light rays.
  4 The tide is causing an increase in the depth of the ocean.

7 Which rocks usually have the mineral quartz as part of their composition?
  1 conglomerate, gabbro, rock salt, and schist
  2 breccia, fossil limestone, bituminous coal, and siltstone
  3 shale, scoria, gneiss, and marble
  4 granite, rhyolite, sandstone, and hornfels

8 Which sedimentary rock may have both a chemical origin and an organic origin?
  1 limestone   3 rock salt
  2 rock gypsum   4 shale

9 The diagrams below represent four different mineral samples.

Which mineral property is best represented by the samples?
  1 density   3 hardness
  2 cleavage   4 streak

10 The photograph below represents a mountainous area in the Pacific Northwest.

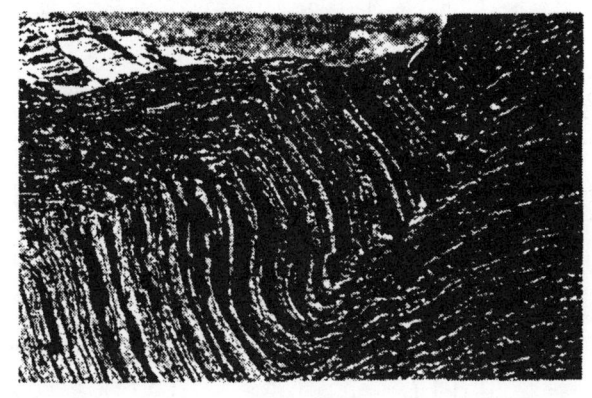

Scientists believe that sedimentary rocks like these represent evidence of crustal change because these rocks were
  1 formed by igneous intrusion
  2 faulted during deposition
  3 originally deposited in horizontal layers
  4 changed from metamorphic rocks

11 Which feature is commonly formed at a plate boundary where oceanic crust converges with continental crust?
  1 a mid-ocean ridge   3 a transform fault
  2 an ocean trench   4 new oceanic crust

Base your answers to questions 12 and 13 on the map below, which shows a portion of California along the San Andreas Fault zone. The map gives the probability (percentage chance) that an earthquake strong enough to damage buildings and other structures will occur between the present time and the year 2024.

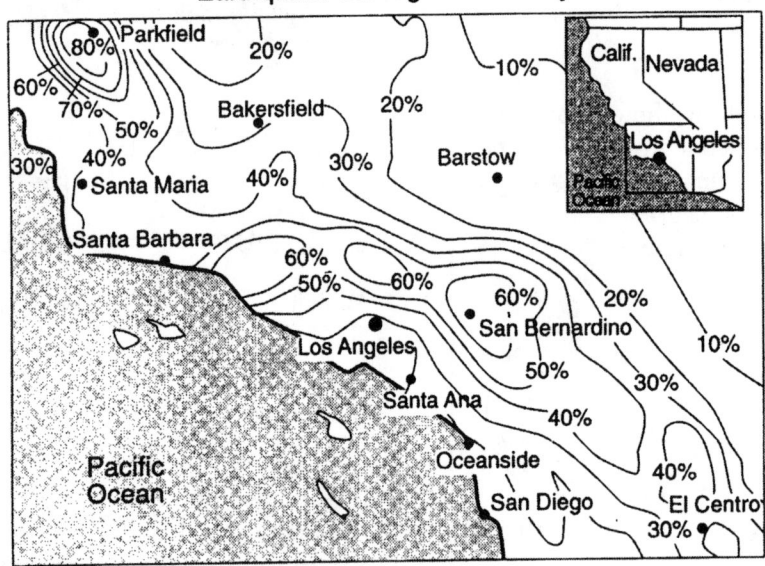

12 Which map represents the most likely location of the San Andreas Fault line?

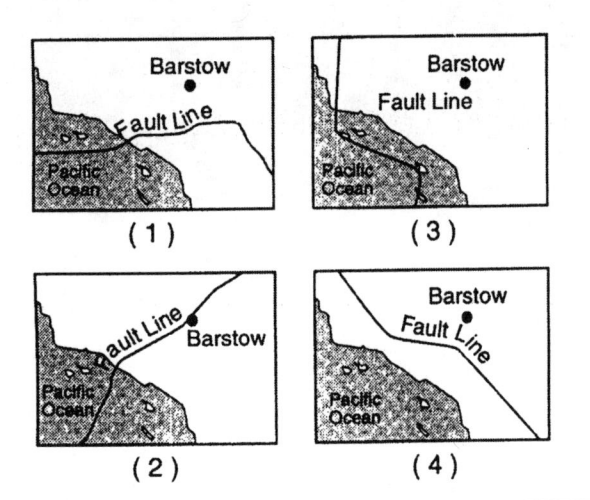

13 Which city has the greatest danger of damage from an earthquake?

1 Parkfield
2 San Diego
3 Santa Barbara
4 San Bernardino

14 The diagram below is a map view of a stream flowing through an area of loose sediments. Arrows show the location of the strongest current.

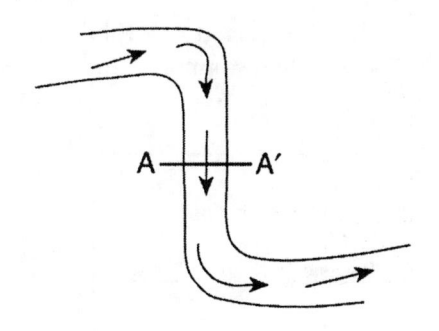

Which stream profile best represents the cross section from A to A'?

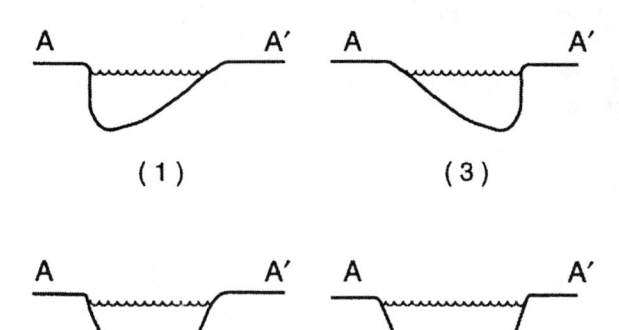

(1) (2) (3) (4)

15 The diagram below represents a landscape area.

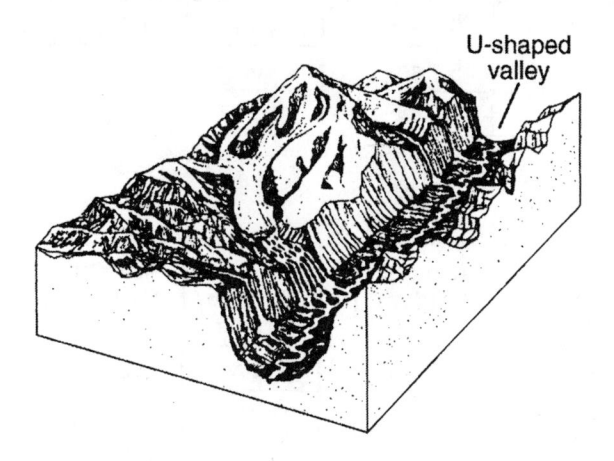

The main valley in this landscape area resulted mostly from
1 chemical weathering
2 volcanic activity
3 glacial erosion
4 stream erosion

16 Which New York State landscape region is composed mostly of intensely metamorphosed surface bedrock?
1 Hudson Highlands
2 Allegheny Plateau
3 Atlantic Coastal Plain
4 Erie-Ontario Lowlands

17 A geologic cross section is shown below.

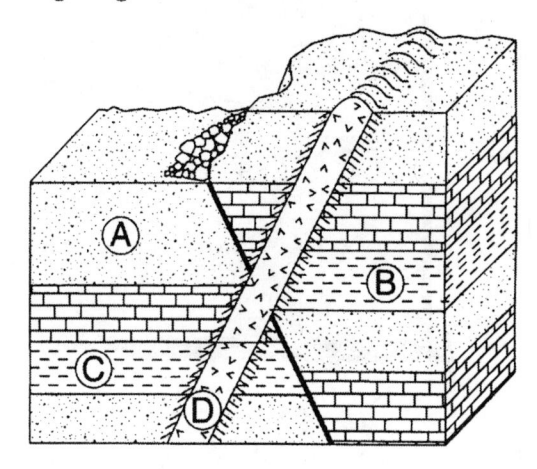

Key
Sandstone
Limestone
Contact metamorphism
Shale
Igneous rock

The most recently formed rock unit is at location
(1) A
(2) B
(3) C
(4) D

18 Based on studies of fossils found in subsurface rocks near Buffalo, New York, scientists have inferred that the climate of this area during the Ordovician Period was much warmer than the present climate. Which statement best explains this change in climate?
1 The Sun emitted less sunlight during the Ordovician Period.
2 Earth was farther from the Sun during the Ordovician Period.
3 The North American Continent was nearer to the Equator during the Ordovician Period.
4 Many huge volcanic eruptions occurred during the Ordovician Period.

Base your answers to questions 19 and 20 on the diagram and data table for the laboratory activity described below. Diagrams are not drawn to scale.

Laboratory Activity

Different combinations of the particles shown in the data table were placed in a tube filled with a thick liquid and allowed to fall to the bottom. The tube was then stoppered and quickly turned upside down, allowing the particles to settle.

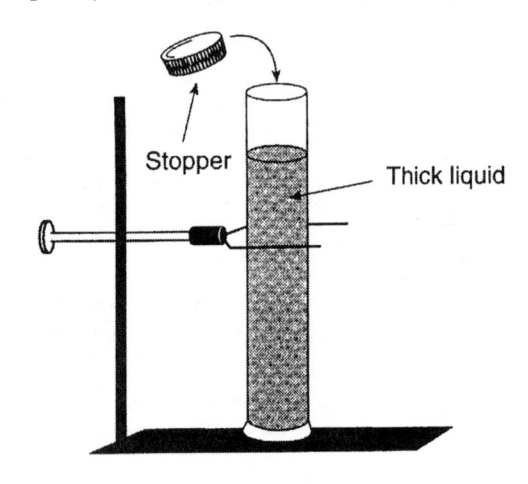

Data Table — Particles Used in Activities

Particle	Diameter	Density
	15 mm Al (aluminum)	2.7 g/cm^3
	15 mm Fe (iron)	7.9 g/cm^3
	15 mm Pb (lead)	11.4 g/cm^3

19 Which diagram represents the sorting that most likely occurred when the tube was turned upside down and the particles of the three different metals were allowed to settle?

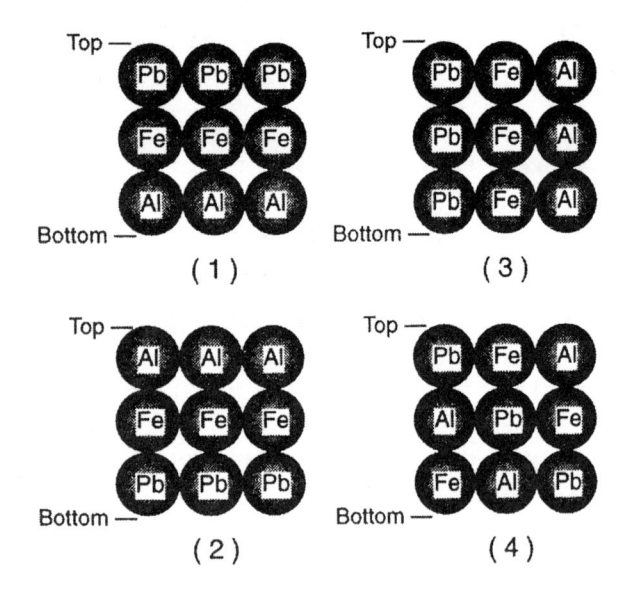

(1) Top: Pb Pb Pb / Fe Fe Fe / Al Al Al : Bottom
(2) Top: Al Al Al / Fe Fe Fe / Pb Pb Pb : Bottom
(3) Top: Pb Fe Al / Pb Fe Al / Pb Fe Al : Bottom
(4) Top: Pb Fe Al / Al Pb Fe / Fe Al Pb : Bottom

20 In another activity, round, oval, and flat aluminum particles with identical masses were dropped individually into the tube. Which table shows the most likely average settling times of the different-shaped particles?

(1)
Particle Shape	Average Settling Time
Round	5.1 sec
Oval	5.1 sec
Flat	5.1 sec

(2)
Particle Shape	Average Settling Time
Round	5.1 sec
Oval	3.2 sec
Flat	6.7 sec

(3)
Particle Shape	Average Settling Time
Round	6.7 sec
Oval	5.1 sec
Flat	3.2 sec

(4)
Particle Shape	Average Settling Time
Round	3.2 sec
Oval	5.1 sec
Flat	6.7 sec

21 The diagrams below represent the rock layers and fossils found at four widely separated rock outcrops.

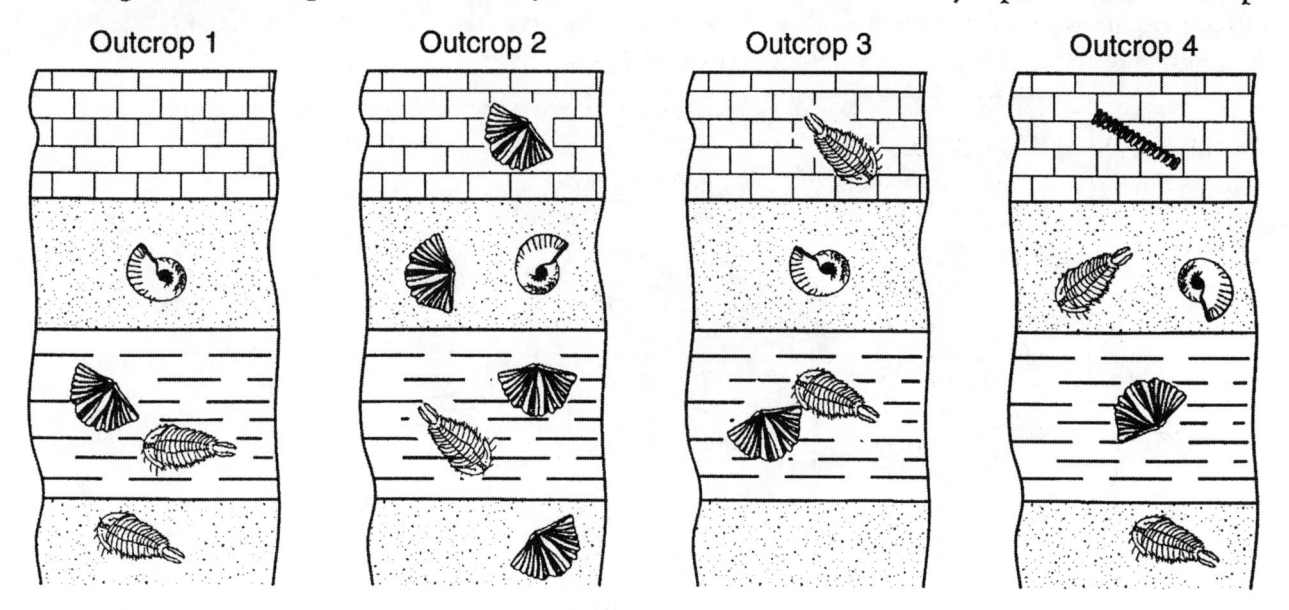

Outcrop 1   Outcrop 2   Outcrop 3   Outcrop 4

Which fossil appears to be the best index fossil?

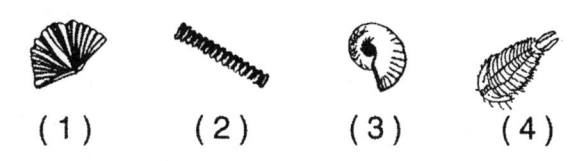

(1)   (2)   (3)   (4)

22 Which column best represents the relative lengths of time of the major intervals of geologic history?

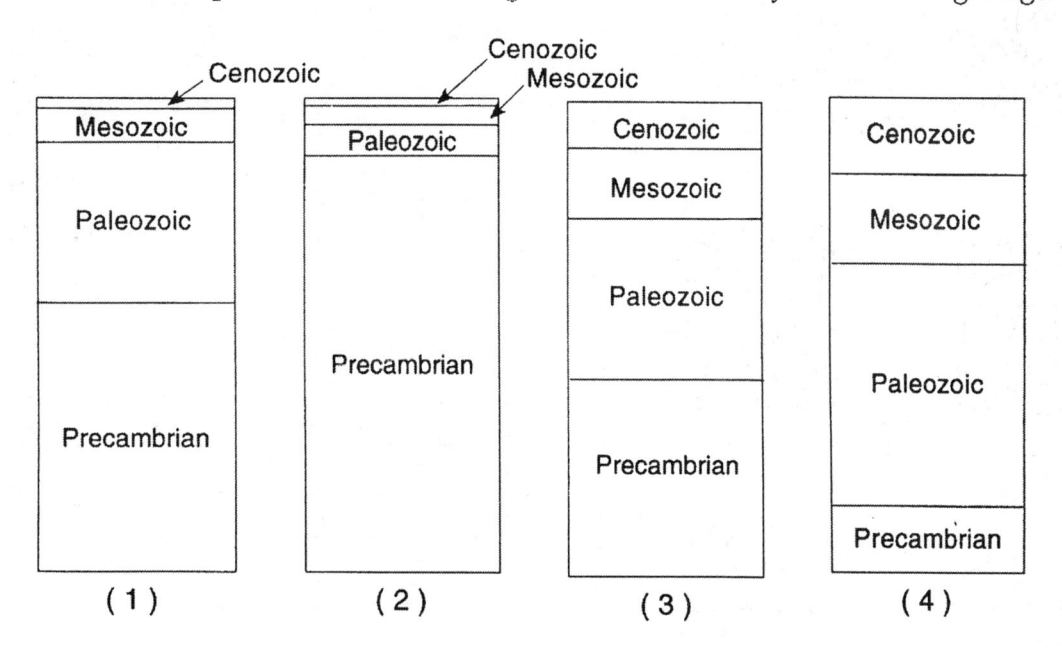

(1)   (2)   (3)   (4)

Note that question 23 has only three choices.

23 The map below represents a satellite image of Hurricane Gilbert in the Gulf of Mexico. Each **X** represents the position of the center of the storm on the date indicated.

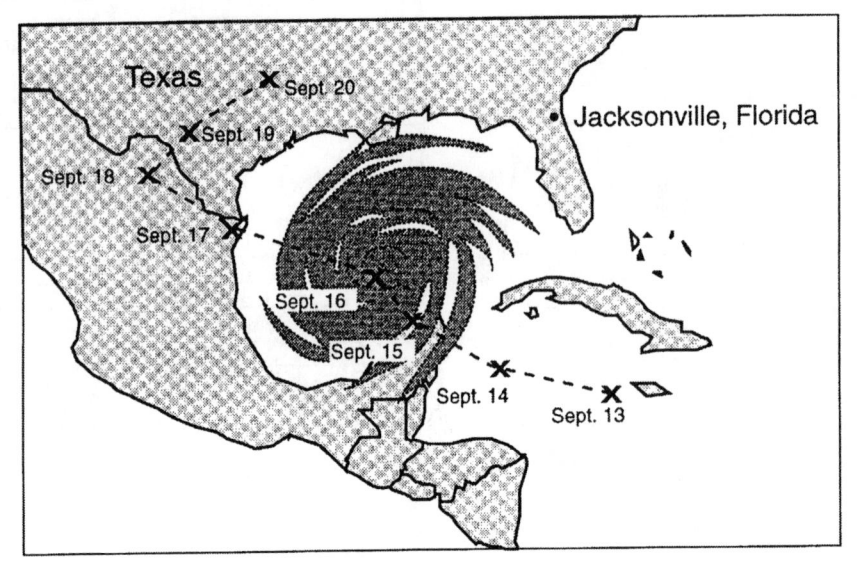

Compared to its strength on September 16, the strength of Hurricane Gilbert on September 18 was

1 less
2 greater
3 the same

---

24 Which graph best represents the radioactive decay of uranium-238 into lead-206?

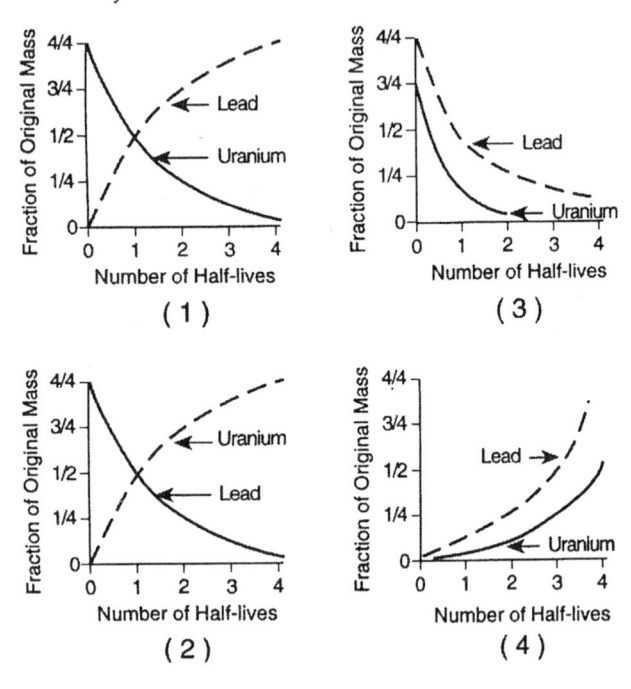

25 Winds are blowing from high-pressure to low-pressure systems over identical ocean surfaces. Which diagram represents the area of greatest windspeed? [Arrows represent wind direction.]

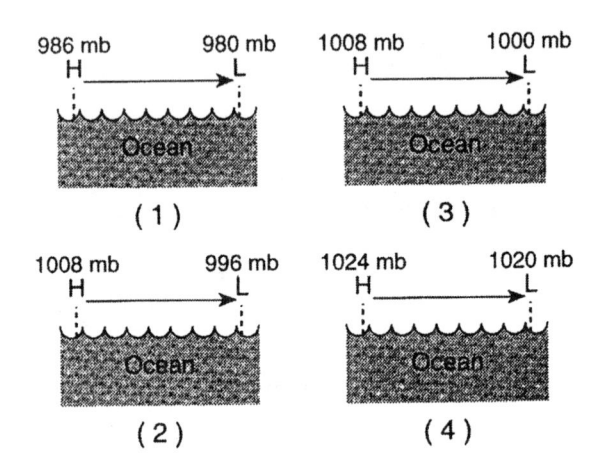

Base your answers to questions 26 through 29 on the weather map below. The map shows a weather system that is affecting part of the United States.

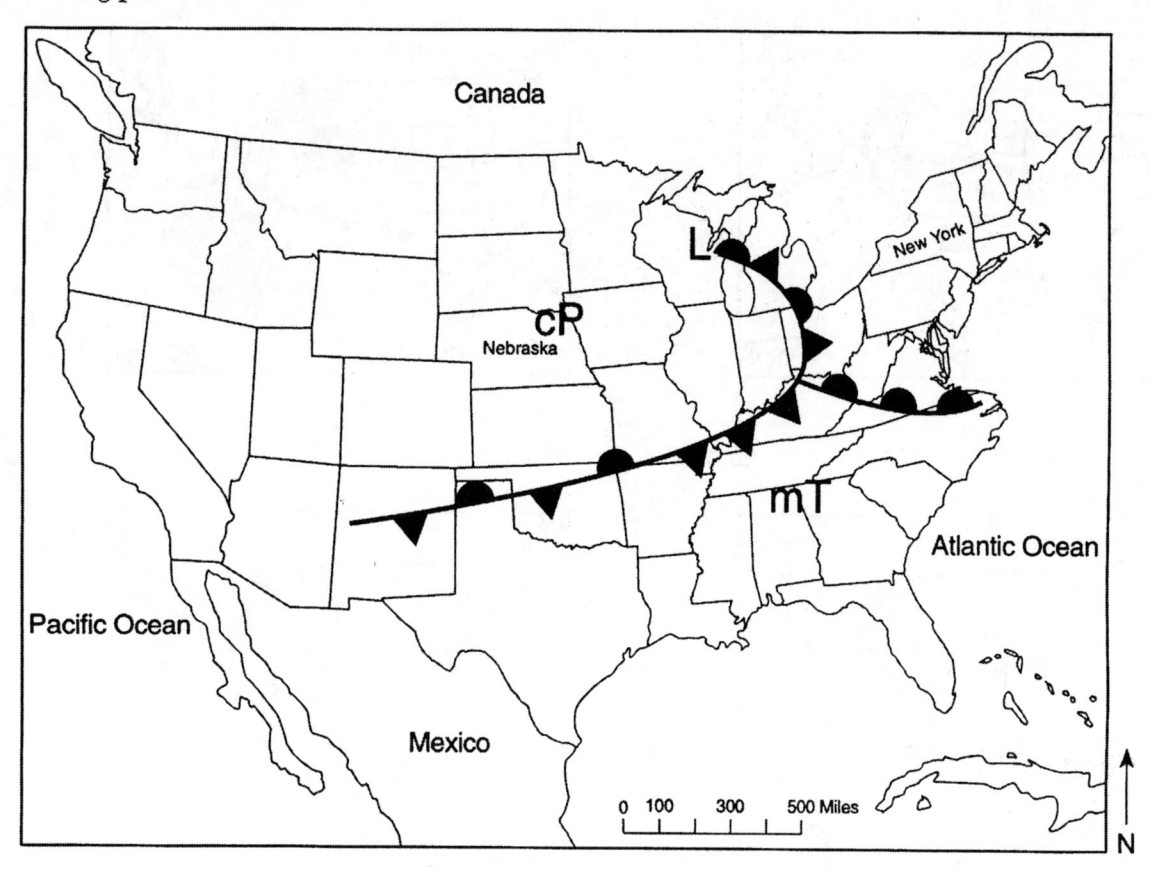

26 Which diagram shows the surface air movements most likely associated with the low-pressure system?

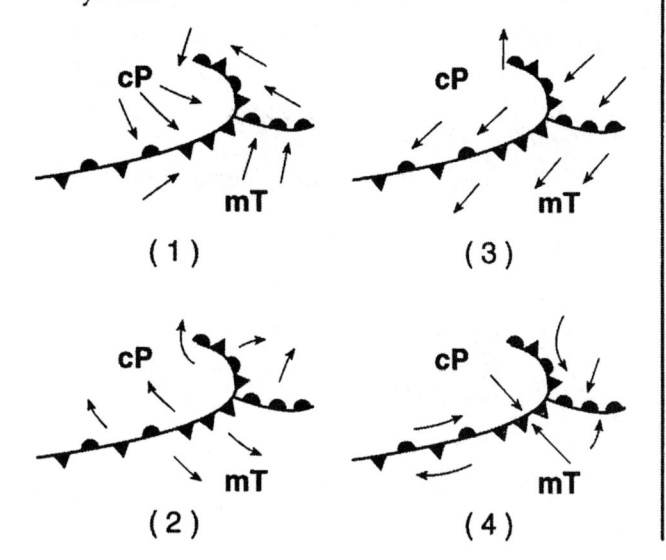

27 What is the total number of different kinds of weather fronts shown on this weather map?
(1) 1 (3) 3
(2) 2 (4) 4

28 The air mass influencing the weather of Nebraska most likely originated in
1 the northern Pacific Ocean
2 the northern Atlantic Ocean
3 central Canada
4 central Mexico

PRACTICE EXAM – JUNE 1998 – MODIFIED EARTH SCIENCE PROGRAM – N&N©

29 Which map shows the area where precipitation is most likely occurring? [Shaded areas represent precipitation.]

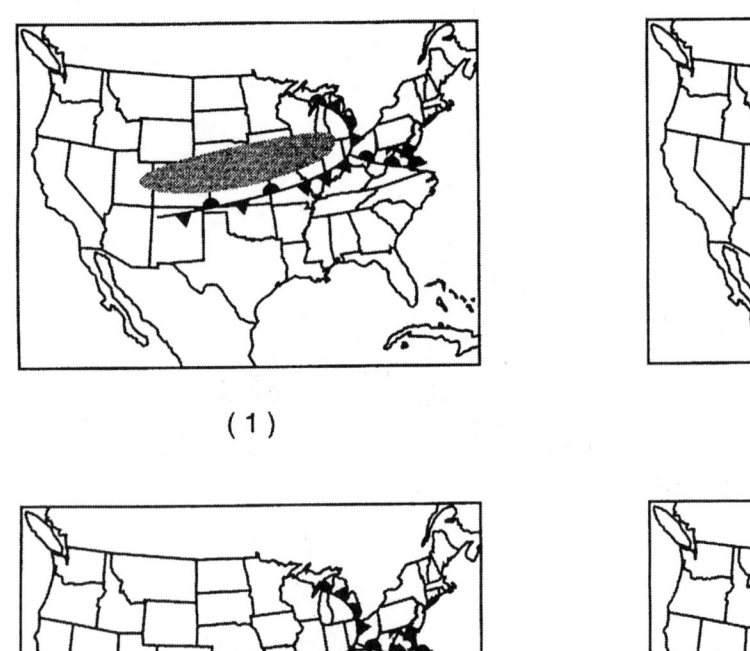

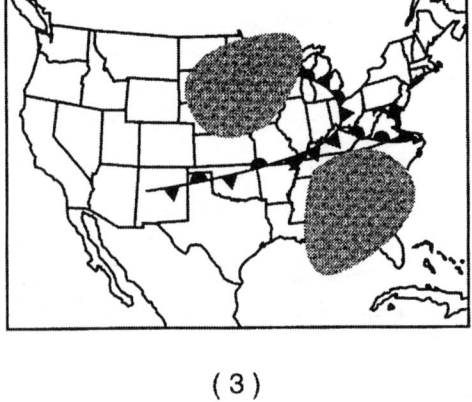

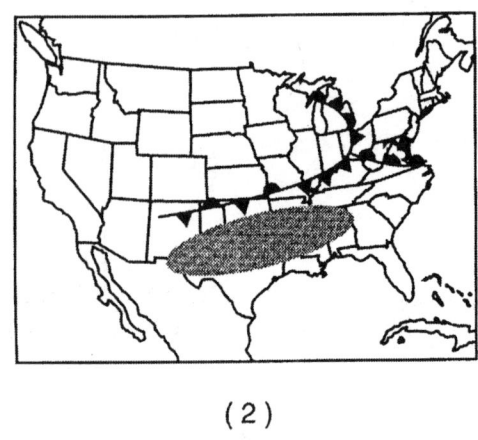

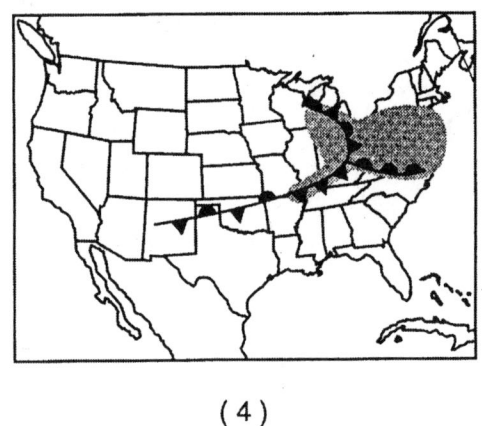

(1)

(2)

(3)

(4)

30 Locations in New York State are warmer in summer than in winter because in summer

1 the solar radiation reaching Earth's surface is more intense, and the number of daylight hours is fewer
2 the solar radiation reaching Earth's surface is more intense, and the number of daylight hours is greater
3 the solar radiation reaching Earth's surface is less intense, and the number of daylight hours is fewer
4 the solar radiation reaching Earth's surface is less intense, and the number of daylight hours is greater

31 A parcel of air has a dry-bulb temperature of 16°C and a wet-bulb temperature of 10°C. What are the dewpoint and relative humidity readings of the air?

(1) –10°C dewpoint and 14% relative humidity
(2) –10°C dewpoint and 45% relative humidity
(3) 4°C dewpoint and 14% relative humidity
(4) 4°C dewpoint and 45% relative humidity

Base your answers to questions 32 and 33 on the diagrams below. Diagram I shows a house located in New York State. Diagram II shows a solar collector that the homeowner is using to help heat the house.

Diagram I

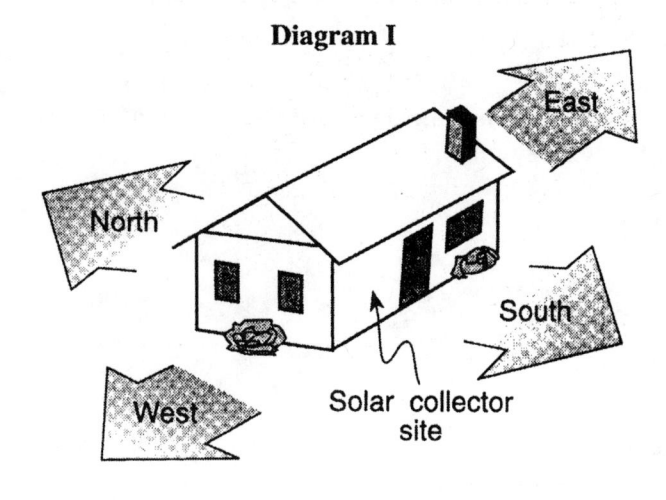

Diagram II
Solar Collector

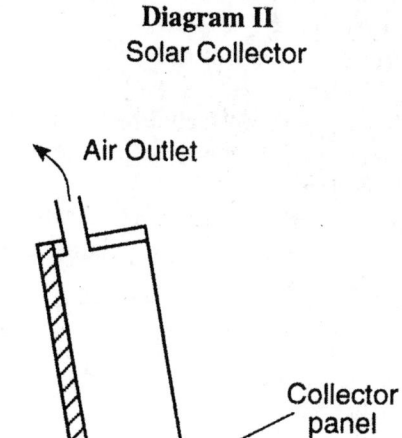

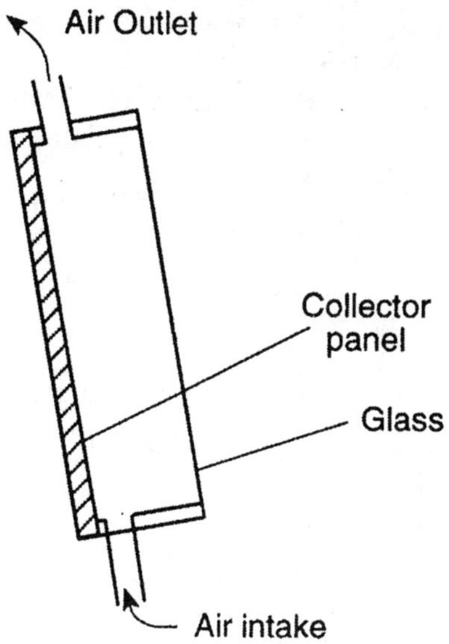

32 Which side view shows the correct placement of the solar collector on the side of this house to collect the maximum amount of sunlight?

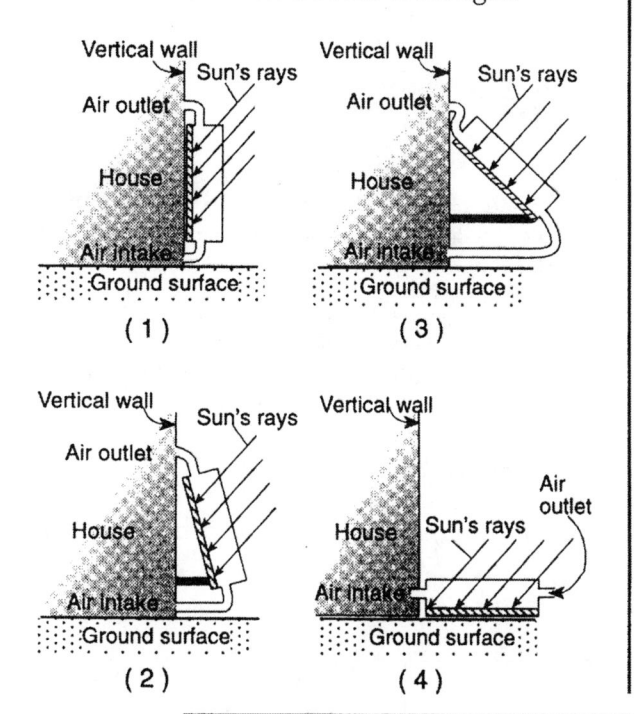

33 Which diagram best represents both the wavelength of visible light entering this house through a window and the wavelength of infrared rays being given off by a chair?

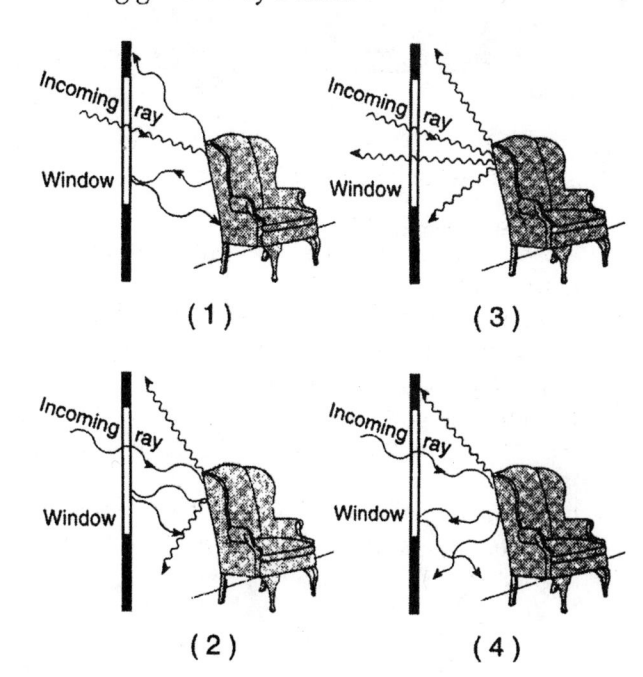

PRACTICE EXAM – JUNE 1998 – MODIFIED EARTH SCIENCE PROGRAM – N&N©

34 Which fact provides the best evidence that Earth's axis is tilted?
1 Locations on Earth's Equator receive 12 hours of daylight every day.
2 The apparent diameter of the Sun shows predictable changes in size.
3 Planetary winds are deflected to the right in the Northern Hemisphere and to the left in the Southern Hemisphere.
4 Winter occurs in the Southern Hemisphere at the same time that summer occurs in the Northern Hemisphere.

35 The diagram below represents a planet revolving in an elliptical orbit around a star.

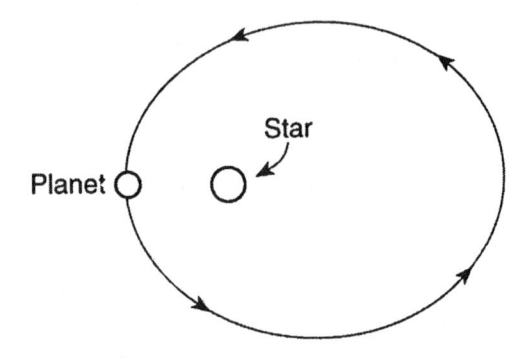

As the planet makes one complete revolution around the star, starting at the position shown, the gravitational attraction between the star and the planet will
1 decrease, then increase
2 increase, then decrease
3 continually decrease
4 remain the same

36 A student in New York State observed that the altitude of the Sun at noon is decreasing each day. During which month could the student have made these observations?
1 January
2 March
3 May
4 October

37 Which member of the solar system has a diameter of 3.48×10^3 kilometers?
1 Pluto
2 Earth
3 Earth's Moon
4 the Sun

38 The diagram below represents a portion of the solar system.

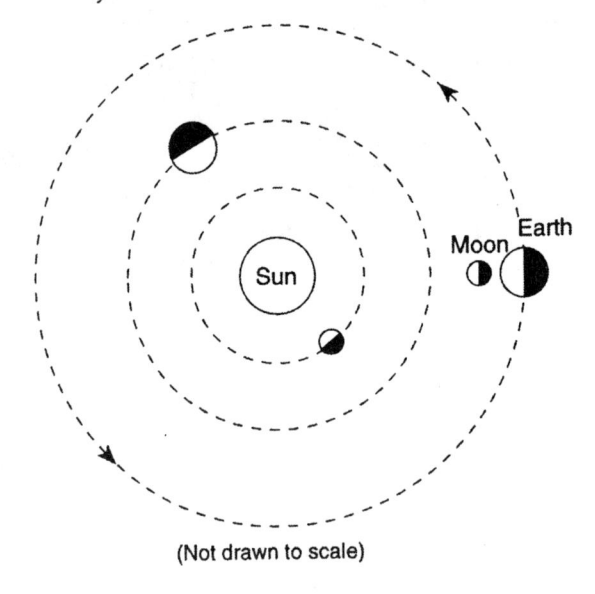

(Not drawn to scale)

In addition to Earth, which planets are represented by the diagram?
1 Saturn and Pluto
2 Mercury and Venus
3 Uranus and Neptune
4 Jupiter and Mars

39 Tropical rain forests remove carbon dioxide gas from Earth's atmosphere. The destruction of the rain forests could affect Earth's overall average temperature because
1 more of Earth's reradiation would be absorbed by the atmosphere
2 more sunlight would be reflected back to space by Earth's atmosphere
3 more visible light would be absorbed by Earth's atmosphere
4 more ultraviolet light would be transmitted through Earth's atmosphere

40 The maps below show changes occurring around a small New York State lake over a 30-year period.

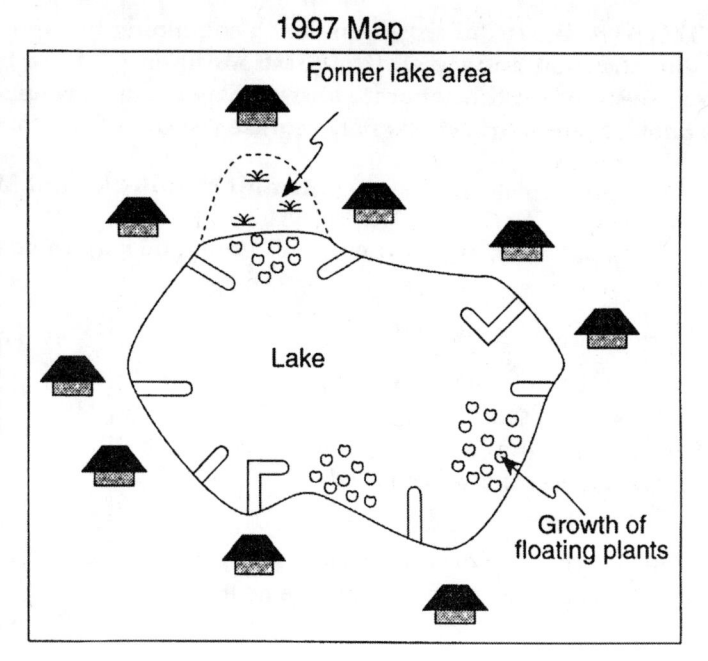

1967 Map

1997 Map

Which graph shows the probable changes in the quality of ground water and lake water in this region from 1967 to 1997? [Ground water is water that has infiltrated beneath Earth's surface.]

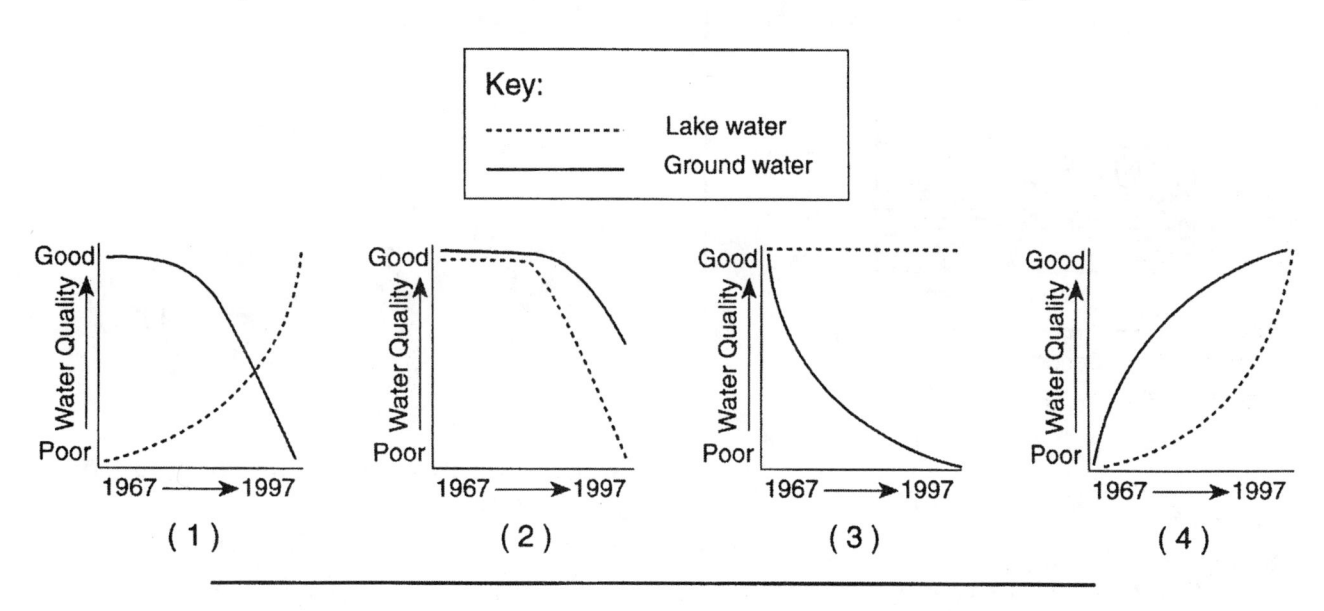

Key:
- - - - - - Lake water
———— Ground water

(1) (2) (3) (4)

Part II

This part consists of six groups, each containing five questions. Choose any *two* of these six groups. Be sure that you answer all five questions in each of the two groups chosen. Record the answers to these questions on the separate answer paper in accordance with the directions on the front cover of this booklet. Some questions may require the use of the *Earth Science Reference Tables*. [10]

Group A — Rocks and Minerals

If you choose this group, be sure to answer questions 41–45.

41 Slate is formed by the
1 deposition of chlorite and mica
2 foliation of schist
3 metamorphism of shale
4 folding and faulting of gneiss

42 Which property of a mineral most directly results from the internal arrangement of its atoms?

1 volume   3 crystal shape
2 color   4 streak

43 The diagram below shows a cross section through a portion of Earth's crust.

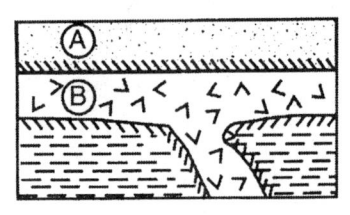

Key
- Sandstone
- Shale
- Granite
- Contact metamorphism

Rock found in the zone between *A* and *B* is nonfoliated and fine grained. This rock is most likely

1 metaconglomerate   3 marble
2 gneiss   4 quartzite

44 The graph below represents the percentage of each mineral found in a sample of igneous rock. Which mineral is represented by the letter *X* in the graph?

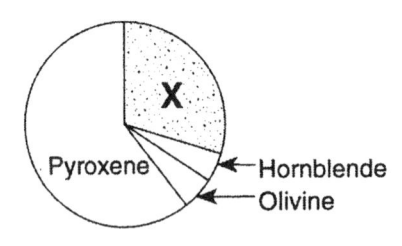

1 potassium feldspar   3 quartz
2 plagioclase feldspar   4 biotite

45 The diagram below represents top and side views of models of the silicate tetrahedron.

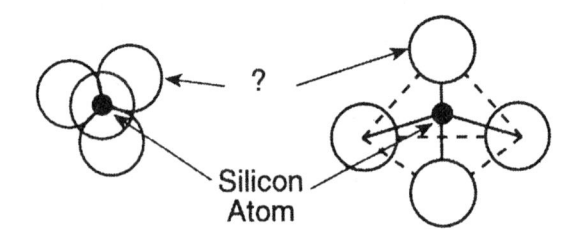

Which element combines with silicon to form the tetrahedron?

1 oxygen   3 potassium
2 nitrogen   4 hydrogen

Group B — Plate Tectonics

If you choose this group, be sure to answer questions 46–50.

46 Which cross-sectional diagram of Earth correctly shows the paths of seismic waves from an earthquake traveling through Earth's interior?

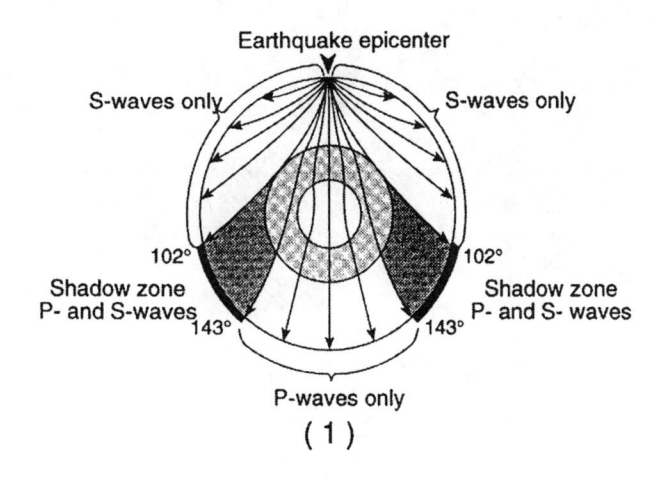

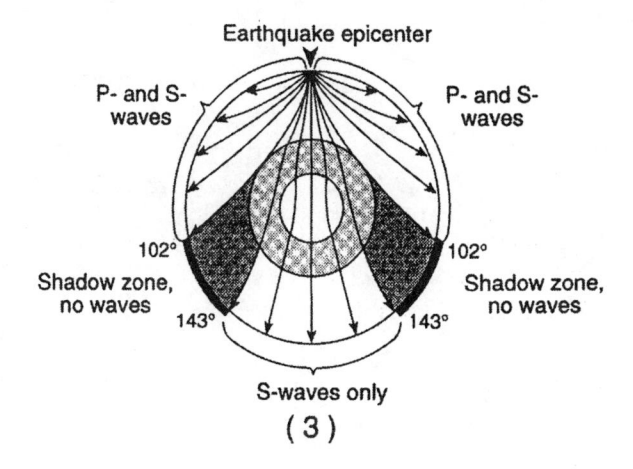

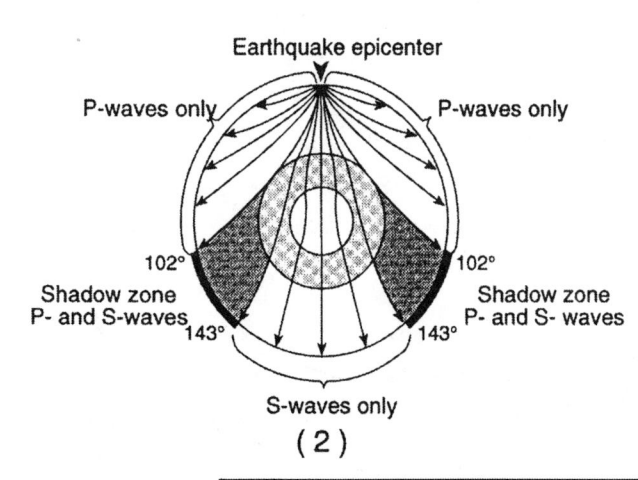

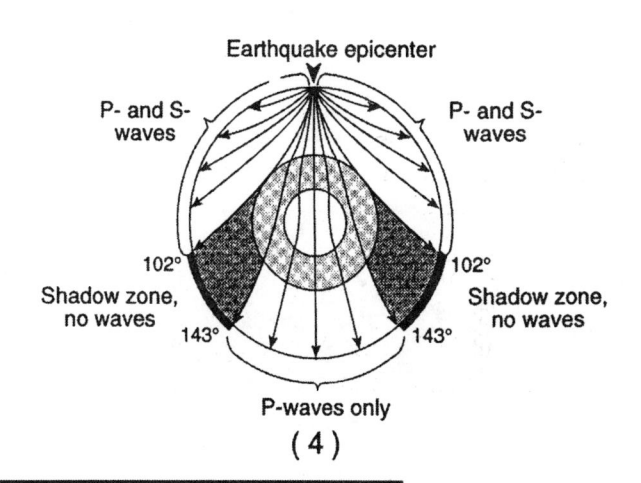

47 The actual temperature at the boundary between the stiffer mantle and the outer core is estimated to be approximately

(1) 1.5°C (3) 3000°C
(2) 250°C (4) 5000°C

48 How far from an earthquake epicenter is a city where the difference between the P-wave and S-wave arrival times is 6 minutes and 20 seconds?

(1) 1.7×10^3 km (3) 3.5×10^3 km
(2) 9.9×10^3 km (4) 4.7×10^3 km

49 Compared to oceanic crust, continental crust is generally

1 older and thinner
2 older and thicker
3 younger and thinner
4 younger and thicker

50 Which map best represents the general pattern of magnetism in the oceanic bedrock near the mid-Atlantic Ridge?

Group C — Oceanography

If you choose this group, be sure to answer questions 51–55.

51 The cartoon below presents a humorous look at ocean wave action.

"Here comes another big one, Roy, and here — we — goooooowheeeeeeeoool!"

The ocean waves that are providing enjoyment for Roy's companion are the result of the

1 interaction of the hydrosphere with the moving atmosphere
2 interaction of the lithosphere with the moving troposphere
3 absorption of short-wave radiation in the stratosphere
4 absorption of energy in the asthenosphere

52 What is the main source of dissolved salts in the ocean?

1 human activities
2 minerals carried from the land by rivers
3 precipitation from storm fronts
4 weathered basalts at mid-ocean ridges

53 Diagrams A, B, C, and D below represent a sequence of events that occurred in the deep ocean and that resulted in a tsunami. Diagram A represents the first event of the sequence, and diagram D represents the fourth event.

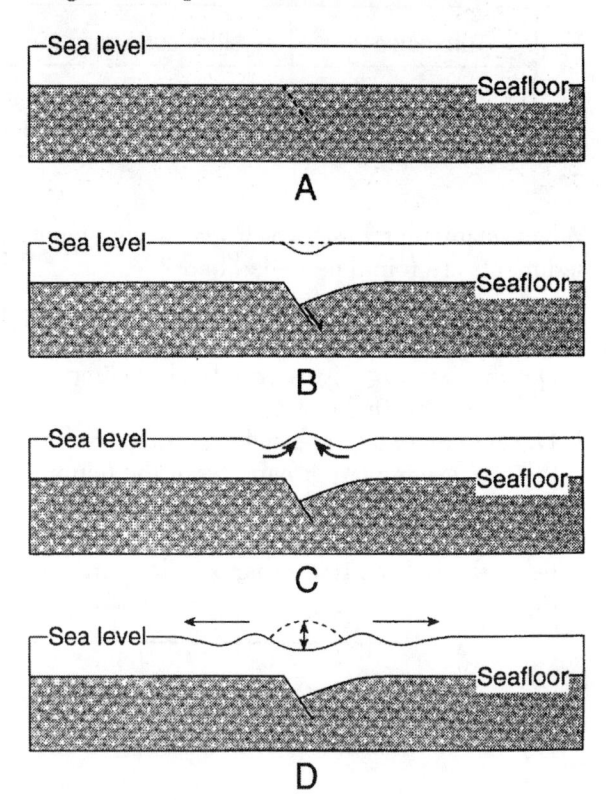

A

B

C

D

The surface waves shown in diagram D were most likely caused by

1 a submarine landslide
2 folding of the ocean floor
3 displacement by a fault
4 strong winds from a hurricane

Base your answers to questions 54 and 55 on the table below, which lists the four main classes of ocean-floor sediments and shows the origin and an example of each sediment.

Ocean-Floor Sediments

Classification	Origin	Examples
Lithogenic	Land-derived	Muds and clays
Biogenic	Shells of microscopic organisms	Oozes
Turbigenic	Turbidity currents	Graded beds in the deep ocean
Authigenic	Ocean water by chemical precipitation directly on the ocean floor	Manganese nodules

54 Which statement best explains why turbigenic sediments are found in graded beds?
  1 The Moon's gravitational force causes the cyclic pattern of sediment deposition.
  2 Ocean-floor organisms sort fresh sediments into layers of similar sizes.
  3 During cementation, smaller particles rise to the top, leaving larger particles at the bottom of each layer.
  4 During deposition, larger particles usually settle to the bottom faster than smaller particles do.

55 Icebergs carry material that has been eroded from the land by a moving glacier. When this material is deposited in the ocean by a melting iceberg, it is classified as
  1 lithogenic   3 authigenic
  2 biogenic   4 turbigenic

Group D — Glacial Processes

If you choose this group, be sure to answer questions 56–60.

56 At the end of the last period of glaciation, the natural environment of New York State probably looked like the present environment in
1 Alaska
2 North Carolina
3 Texas
4 Ohio

57 Which geological resource in New York State resulted from glaciation?
1 coal and oil deposits
2 sand and gravel deposits
3 iron and zinc ores
4 garnet and quartz crystals

58 Evidence that several periods of glaciation occurred in the geologic past is provided by
1 glacial erratics in New York State
2 glacial erosion in the high regions of the Adirondack Mountains
3 layers of glacial till deposited on top of each other
4 discovery of mastodont fossils in the surface bedrock of the Adirondack Mountains

59 The cross section below represents the transport of sediments by an advancing glacier. The arrow shows the direction of movement.

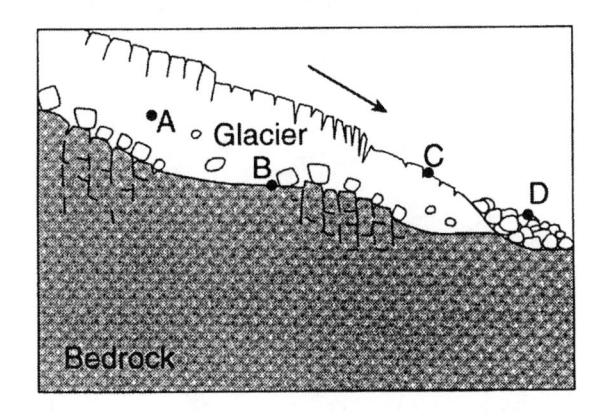

At which location are striations and glacial grooves most likely being carved?
(1) A
(2) B
(3) C
(4) D

60 On which New York State map does the arrow indicate the most likely direction of advance of the last continental ice sheet?

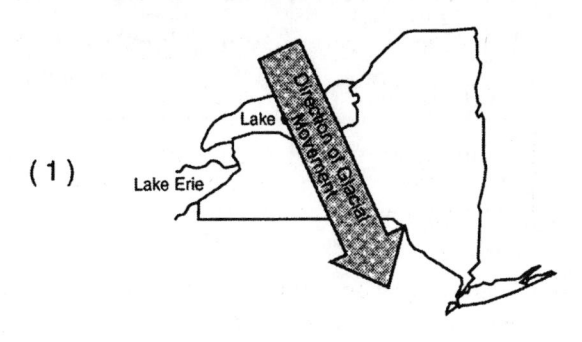

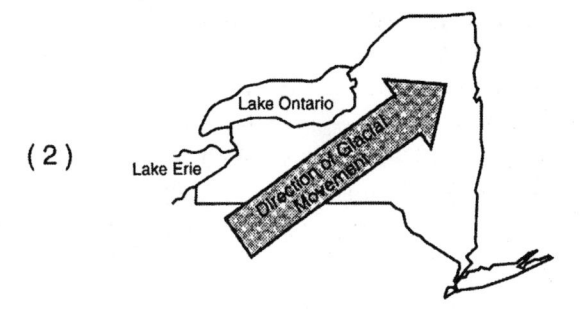

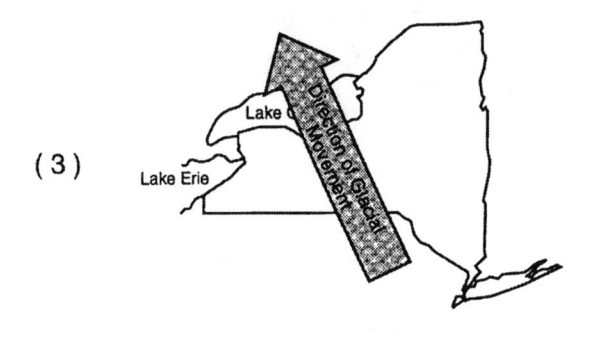

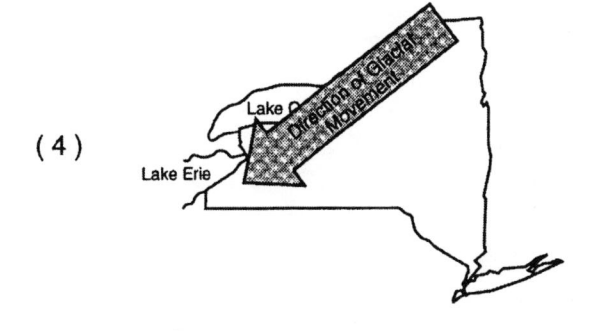

Group E — Atmospheric Energy

If you choose this group, be sure to answer questions 61–65.

61 At 1 p.m. at a location in New York State, the surface air temperature was 20°C and the dewpoint temperature was 10°C. At 3 p.m. at the same location, the altitude of the cloud base was 1.2 kilometers. Compared to the altitude of the cloud base at 1 p.m., the altitude of the cloud base at 3 p.m. was

(1) 0.5 km lower
(2) 2.0 km lower
(3) 0.5 km higher
(4) nearly the same

62 On a very hot summer afternoon, the air over Long Island is warmer than the air over the nearby ocean. As a result, the air over Long Island tends to

1 sink and cool, causing clouds to form
2 sink and warm, causing clouds to disappear
3 rise and cool, causing clouds to form
4 rise and warm, causing clouds to disappear

63 Why have weather predictions become more accurate and reliable in recent years?

1 Weather conditions now change more slowly than they did in the past.
2 More people today watch televised weather reports.
3 Scientists have developed better methods of controlling the weather.
4 Scientists have developed better technology to observe weather conditions.

Base your answers to questions 64 and 65 on the diagrams below. Beaker *A* contains 100 milliliters of boiling water. Beaker *B* contains 225 milliliters of boiling water. The hot plates are adding equal amounts of heat to each beaker each minute.

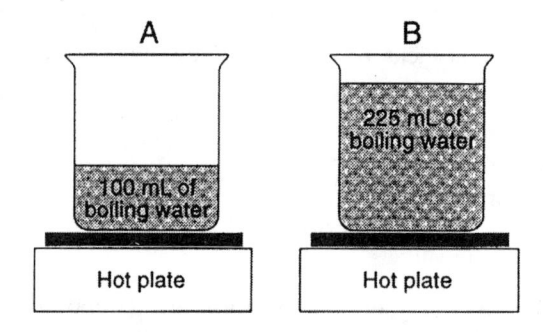

64 What is the total number of calories that must be added to completely change the 100 milliliters of boiling water (mass = 100 grams) in beaker *A* into water vapor?

(1) 54,000 cal   (3) 640 cal
(2) 8,000 cal   (4) 5.4 cal

Note that question 65 has only three choices.

65 Thermometers are placed in both beakers and allowed to adjust as the water boils. The thermometers will show that, compared to the temperature of the water in beaker *A*, the temperature of the water in beaker *B* is

1 lower
2 higher
3 the same

Group F — Astronomy

If you choose this group, be sure to answer questions 66–70.

66 Which diagram best represents the heliocentric model of a portion of the solar system? [S = Sun, E = Earth, and M = Moon. The diagrams are not drawn to scale.]

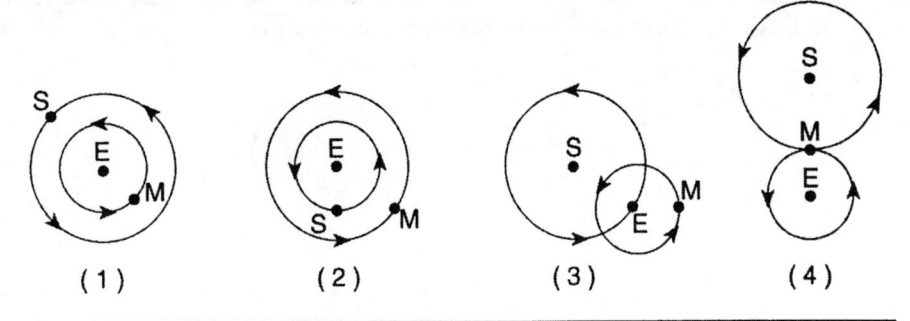

67 The planets known as "gas giants" include Jupiter, Uranus, and

1 Pluto
2 Saturn
3 Mars
4 Earth

68 A comparison of the age of Earth obtained from radioactive dating and the age of the universe based on galactic Doppler shifts suggests that

1 Earth is about the same age as the universe
2 the universe is much younger than Earth
3 the solar system and Earth formed billions of years after the universe began
4 the two dating methods contradict one another

69 The Moon has more surface craters than Earth does because the Moon has

1 no significant atmosphere
2 a surface more sensitive to impacts
3 a smaller diameter than Earth
4 a stronger gravitational force

70 With respect to one another, galaxies have been found to be

1 moving closer together
2 moving farther apart
3 moving in random directions
4 stationary

Part III

This part consists of questions 71 through 88. Be sure that you answer *all* questions in this part. Record your answers in the spaces provided on the separate answer paper. You may use pen or pencil. Some questions may require the use of the *Earth Science Reference Tables*. [25]

71 A total solar eclipse was visible to observers in the southeastern United States on February 26, 1998. The diagram below shows the Sun and Earth as they were viewed from space on that date. The same diagram appears on your answer paper.

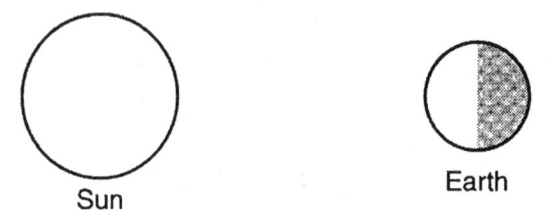

Sun   Earth

(Not drawn to scale)

On the diagram provided *on your answer paper*, draw the Moon (◯), showing its position at the time of the solar eclipse. [1]

Base your answers to questions 72 through 74 on the diagram below, which shows the Sun's apparent path as viewed by an observer in New York State on March 21.

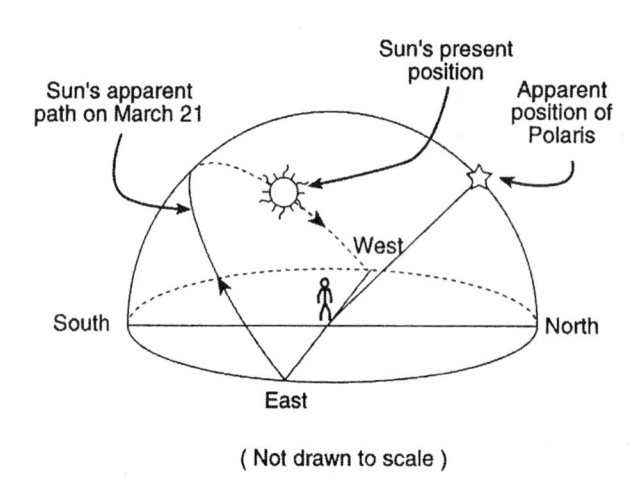

(Not drawn to scale)

72 State how the apparent position of Polaris is related to the latitude of the observer. [1]

73 At approximately what hour of the day would the Sun be at the position shown in the diagram? [1]

74 On the diagram provided *on your answer paper*, draw the Sun's apparent path as viewed by the observer on December 21. [1]

Base your answers to questions 75 through 77 on the map below. The star symbol represents a volcano located on the mid-Atlantic Ridge in Iceland. The isolines represent the thickness, in centimeters, of volcanic ash deposited from an eruption of this volcano. Points A and B represent locations in the area.

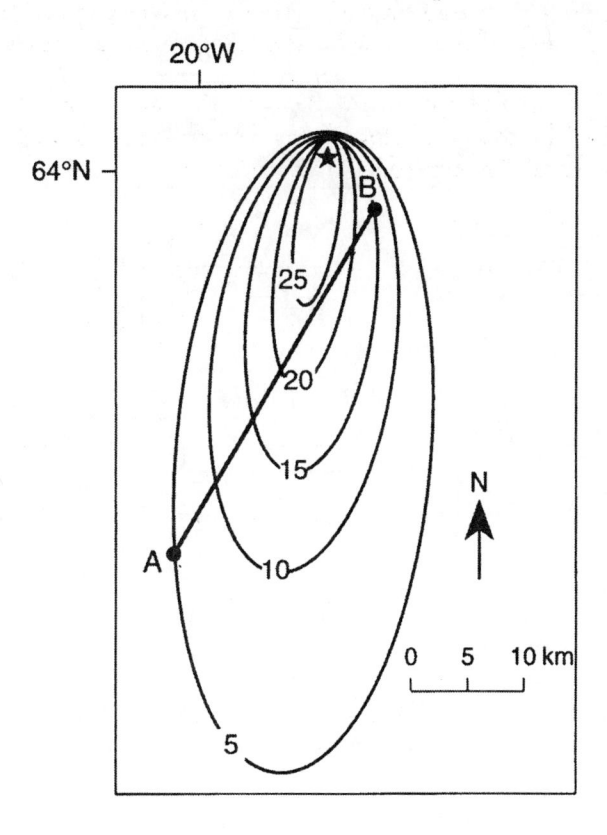

75 On the grid provided *on your answer paper*, construct a profile of the ash thickness between point A and point B, following the directions below.

   a Plot the thickness of the volcanic ash along line AB by marking with a dot each point where an isoline is crossed by line AB.   [2]

   b Connect the dots to complete the profile of the thickness of the volcanic ash.   [1]

76 State one factor that could have produced this pattern of deposition of the ash.   [1]

77 State why volcanic eruptions are likely to occur in Iceland.   [1]

Base your answers to questions 78 through 81 on the diagrams below. Columns *A* and *B* represent two widely separated outcrops of rocks. The symbols show the rock types and the locations of fossils found in the rock layers. The rock layers have not been overturned.

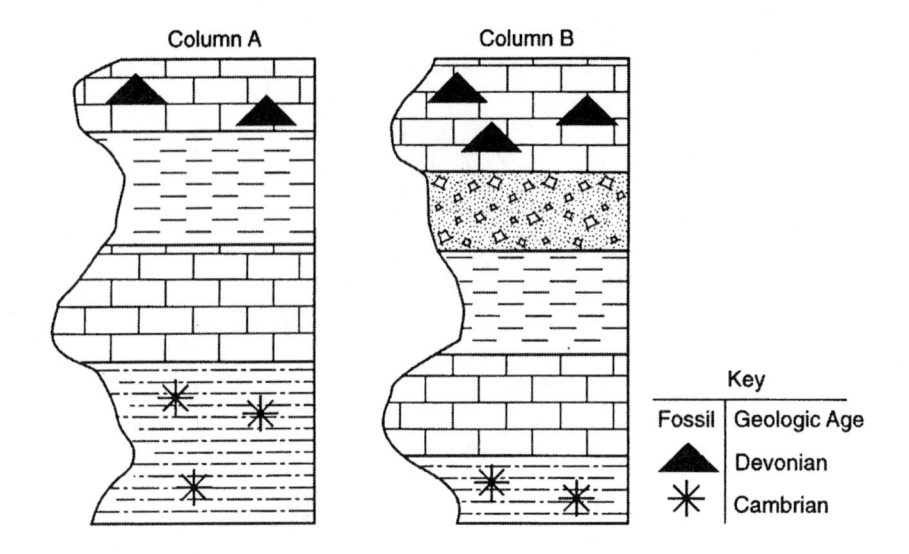

78 State one method used to correlate rock layers found in the outcrop represented by column *A* with rock layers found in the outcrop represented by column *B*. [1]

79 An unconformity (buried erosional surface) exists between two layers in the outcrop represented by column *A*. Identify the location of the unconformity by drawing a thick wavy line (∿∿∿) at the correct position on column *A* *on your answer paper*. [1]

80 *In one or more sentences*, state the evidence that limestone is the most resistant layer in these outcrops. [2]

81 State the oldest possible age, in millions of years, for the fossils in the siltstone layer. [1]

Base your answers to questions 82 and 83 on the meteorological conditions shown in the table and partial station model below, as reported by the weather bureau in the city of Oswego, New York. The diagram of the station model also appears on your answer paper.

Air temperature: 65°F
Wind direction: from the southeast
Windspeed: 20 knots
Barometeric pressure: 1017.5 mb
Dewpoint: 53°F

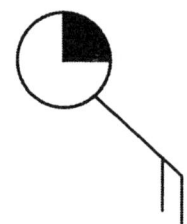

82 Using the meteorological conditions given, complete the station model provided *on your answer paper* by recording the air temperature, dewpoint, and barometric pressure in the proper format. [2]

83 State the sky conditions or amount of cloud cover over Oswego as shown by the station model. [1]

Base your answers to questions 84 through 87 on the information and data table below.

The snowline is the lowest elevation at which snow remains on the ground all year. The data table below shows the elevation of the snowline at different latitudes in the Northern Hemisphere.

Latitude (°N)	Elevation of Snowline (m)
0	5400
10	4900
25	3800
35	3100
50	1600
65	500
80	100
90	0

84 On the grid provided *on your answer paper*, plot the latitude and elevation of the snowline for the locations in the data table. Use a dot for each point and connect the dots with a line. [2]

85 Mt. Mitchell, in North Carolina, is located at 36° N and has a peak elevation of 2037 meters. Plot the latitude and elevation of Mt. Mitchell on your graph. Use a plus sign (+) to mark this point. [1]

86 Using your graph, determine, to the *nearest whole degree*, the lowest latitude at which a peak with the same elevation as Mt. Mitchell would have permanent snow. [1]

87 State the relationship between latitude and elevation of the snowline. [1]

88 The diagram below represents the elliptical orbit of a spacecraft around the Sun.

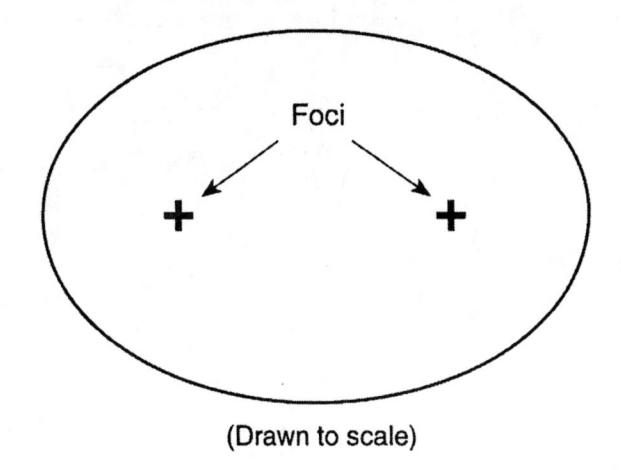

(Drawn to scale)

In the space provided *on your answer paper*, calculate the eccentricity of the spacecraft's orbit following the directions below:

a Write the equation for eccentricity. [1]

b Substitute measurements of the diagram into the equation. [1]

c Calculate the eccentricity and record your answer in decimal form. [1]

The University of the State of New York

REGENTS HIGH SCHOOL EXAMINATION

EARTH SCIENCE
PROGRAM MODIFICATION EDITION

Thursday, June 18, 1998 — 1:15 to 4:15 p.m., only

ANSWER PAPER

Part I Credits	....	....
Part II Credits	....	....
Part III Credits	....	....
Performance Test Credits	....	....
Local Project Credits	....	....
Total (Official Regents) Examination Mark	....	....
Reviewer's Initials:		_____

Student .... Sex: ☐ Male ☐ Female

Teacher .... School ....

Grade (circle one)   8   9   10   11   12

Record all of your answers on this answer paper in accordance with the instructions on the front cover of the test booklet.

Part I (40 credits)

1	1	2	3	4	15	1	2	3	4	29	1	2	3	4
2	1	2	3	4	16	1	2	3	4	30	1	2	3	4
3	1	2	3	4	17	1	2	3	4	31	1	2	3	4
4	1	2	3	4	18	1	2	3	4	32	1	2	3	4
5	1	2	3	4	19	1	2	3	4	33	1	2	3	4
6	1	2	3	4	20	1	2	3	4	34	1	2	3	4
7	1	2	3	4	21	1	2	3	4	35	1	2	3	4
8	1	2	3	4	22	1	2	3	4	36	1	2	3	4
9	1	2	3	4	23	1	2	3		37	1	2	3	4
10	1	2	3	4	24	1	2	3	4	38	1	2	3	4
11	1	2	3	4	25	1	2	3	4	39	1	2	3	4
12	1	2	3	4	26	1	2	3	4	40	1	2	3	4
13	1	2	3	4	27	1	2	3	4					
14	1	2	3	4	28	1	2	3	4					

Part II (10 credits)

Answer the questions in two of the six groups in this part. Be sure to mark the answers to the groups you choose in accordance with the instructions on the front cover of the test booklet. Leave blank the spaces for the four groups of questions you do not choose to answer.

Group A Rocks and Minerals	Group B Plate Tectonics	Group C Oceanography
41 1 2 3 4	46 1 2 3 4	51 1 2 3 4
42 1 2 3 4	47 1 2 3 4	52 1 2 3 4
43 1 2 3 4	48 1 2 3 4	53 1 2 3 4
44 1 2 3 4	49 1 2 3 4	54 1 2 3 4
45 1 2 3 4	50 1 2 3 4	55 1 2 3 4

Group D Glacial Processes	Group E Atmospheric Energy	Group F Astronomy
56 1 2 3 4	61 1 2 3 4	66 1 2 3 4
57 1 2 3 4	62 1 2 3 4	67 1 2 3 4
58 1 2 3 4	63 1 2 3 4	68 1 2 3 4
59 1 2 3 4	64 1 2 3 4	69 1 2 3 4
60 1 2 3 4	65 1 2 3	70 1 2 3 4

NOTES:

NOTES:

NOTES:

NOTES: